GUIDELINES FOR

Slope Performance Monitoring

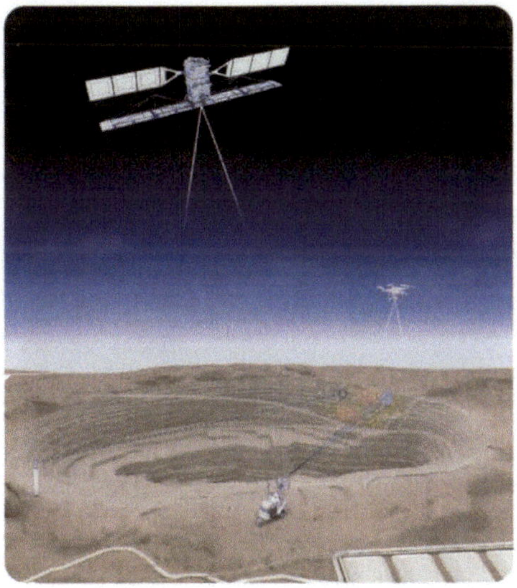

EDITORS: ROBERT SHARON AND ERIK EBERHARDT

CSIRO PUBLISHING

CRC Press
Taylor & Francis Group
Boca Raton London New York

CRC Press is an imprint of the
Taylor & Francis Group, an **Informa** business
A BALKEMA BOOK

A catalogue record for this book is available from the National Library of Australia.

ISBN: 9781486310999 (hbk)
ISBN: 9781486311002 (epdf)
ISBN: 9781486311019 (epub)

Published in print in Australia and New Zealand, and in all other formats throughout the world, by CSIRO Publishing.

CSIRO Publishing
Locked Bag 10
Clayton South VIC 3169
Australia

Telephone: +61 3 9545 8400
Email: publishing.sales@csiro.au
Website: www.publish.csiro.au

Published in print only, throughout the world (excluding Australia and New Zealand), by CRC Press/Balkema, with ISBN 978-0-367-50919-4

CRC Press/Balkema
Schipholweg 107C
2316XC Leiden
The Netherlands
Tel: +31 71 524 3080
Website: www.crcpress.com

Front cover: Conceptual image of instrumentation used for slope monitoring at an open pit mining operation. The image was generated by Motions 2 Pixels, sponsored and managed by Reutech Mining, including a model of RADARSAT-2 (with permission arranged with MDA), and images of a theodolite, UAV drone and radar acquired by Motions 2 Pixels and Reutech Mining.

Set in 10/12 Adobe Minion Pro and Optima
Technical editing by Vivian Giang
Copy editing by Adrienne de Kretser, Righting Writing
Cover design by James Kelly
Typeset by Desktop Concepts Pty Ltd, Melbourne
Index by Max McMaster
Printed in China by Leo Paper Products Ltd.

Mar20_01

Contents

Preface and acknowledgements

The Large Open Pit (LOP) project is an international research and technology transfer project focused on the stability of rock slopes in open pit mines. It is an industry-sponsored and funded project that was initiated in 2005 and managed by Dr John Read under the auspices of Australia's Commonwealth Scientific and Industrial Research Organisation (CSIRO). The project was renewed as the LOP II in 2016 under the management of Professor Marc Ruest at the University of Queensland (now with Newmont Goldcorp). At the time of writing, the LOP II is managed by Professor David Williams at the University of Queensland. The sponsors have comprised a diverse group of multinational mining companies currently including AngloGold Ashanti Ltd, Barrick Gold Corporation, BHP Billiton Innovation Pty Ltd, De Beers Group Services, Debswana Diamond Co., Newcrest Mining Ltd, Rio Tinto Group and Vale S.A.

Among the initiatives supported by the LOP sponsors is the development of a series of reference books on the state-of-practice related to the design and stability of large slopes associated with open pit mines. The first of these books, *Guidelines for Open Pit Slope Design*, was published in 2009 (Read and Stacey 2009) and presents the fundamentals of geotechnical investigation, analysis, design and monitoring of open pit slopes. As such, it represented the first comprehensive publication on this subject since *Rock Slope Engineering* by Hoek and Bray, which was published by the Institution of Mining and Metallurgy, London in 1974 and 1977, and the *Pit Slope Manual* published by the Canadian Centre for Mining and Metallurgy (CANMET) in 1977. The second book in the series, *Guidelines for Evaluating Water in Pit Slope Stability* (Beale and Read 2013), covers the key influences of groundwater on the stability of open pit slopes and includes chapters on groundwater investigation, modelling, pore pressures and inflows, dewatering and depressurisation techniques, and monitoring. The third book in the series, *Guidelines for Mine Waste and Stockpile Design* (Hawley and Cunning 2017), focuses on the investigation, design, operation and monitoring of waste rock dumps, dragline spoils, and stockpiles associated with large open pit mines. The fourth book in the series, *Guidelines for Open Pit Slope Design in Weak Rocks* (Martin and Stacey 2018) primarily addresses the challenges and design aspects associated with open pits that are developed in weak rocks (i.e. with a grade classification of R0 to ≤R2).

Guidelines for Slope Performance Monitoring is the fifth book in the series, which expands on Chapter 12 of Read and Stacey (2009), together with the performance monitoring chapters in the subsequent three LOP project books. It was recognised by the LOP Sponsors' Management Committee that details of guidance on slope performance monitoring could materially aid mining operations in managing their geotechnical risk management programs. Although most mining companies have in-house systems in place for slope monitoring, experience indicates that mining operations continue to be surprised by the occurrence of negative geotechnical events. To address this concern, the book is intended to provide guidance to site engineers on the selection and design of performance monitoring systems, with emphasis on establishing and understanding system performance and the limitations of various types of instrumentation. Available documentation is often limited in scope and lacks practical relevance to the modern mining operation. It is generally understood that the geotechnical practitioner should have formal training on performance monitoring instrumentation, data collection and compilation, data management and, most importantly, how to manage, interpret and utilise the data. Aspects of training that support system operation, data compilation, analysis and reporting results are addressed. This book compiles the current and suggested practice for open pit operators, from detailed requirements through system design and operations support. Each topic is illustrated with examples and case histories demonstrating lessons learned. An objective is to provide geotechnical practitioners with a road map that will help them to decide how to investigate and manage slope stability and the associated geotechnical risk management process in a mining operation.

The intended audience for *Guidelines for Slope Performance Monitoring* includes mining and civil engineers, geotechnical engineers, engineering geologists, hydrogeologists and geotechnical technicians involved in the investigation, design, construction, maintenance and performance monitoring of excavated rock slopes in open pit mines and civil engineering projects around the world. This book is also intended to support technical study and for undergraduate and graduate teaching.

The 10 chapters of this book draw significantly from the four previous LOP project publications for the purpose of making relevant and additive connections with those publications. The editors and several colleagues designated as 'chapter responsible persons' have generated and compiled material contributed by over 70 industry and academic practitioners who volunteered their knowledge and practical experience for this project. The responsible persons for these chapters include:

Phil de Graaf, De Beers – Anglo American, Johannesburg, South Africa
Erik Eberhardt, University of British Columbia, Vancouver, British Columbia, Canada
Naani Mphathiwa, Debswana Diamond Co., Gaborone, Botswana
Mike Ness, Freeport Mining Co., Tucson, Arizona, USA
Marc Ruest, Newmont Goldcorp, Vancouver, British Columbia, Canada
Eric Schwarz, Piteau Associates, Tucson, Arizona, USA
Robert Sharon, Sharon Geotechnical, Tucson, Arizona, USA
Peter Stacey, Stacey Mining Geotechnical Ltd, Vancouver, British Columbia, Canada

The major premise for this project involved engaging the direct support of the LOP sponsor representatives and industry-wide contributors who have shared their knowledge and experience by contributing the material used in the development of this book. Without this collective support, the goals of this project would not have been achieved. The efforts of the following people, listed in alphabetical order, are gratefully acknowledged:

Jorge Armstrong, Newmont Goldcorp, Elko, Nevada, USA
Eric Audigé, Sixense Oceania, Melbourne, Victoria, Australia
David Ball, Freeport Mining Co., Tucson, Arizona, USA
Alex Bals, Aether Mine Surveying, Johannesburg, South Africa
Isabel Barton, University of Arizona, Tucson, Arizona, USA
Geoff Beale, Piteau Associates, Shrewsbury, UK
Johan Bergé, Rio Tinto Diavik Diamond Mines, Yellowknife, NWT, Canada
Henry Chiwaye, Debswana Diamond Co., Gaborone, Botswana
Pierre Choquet, RST Instruments, Vancouver, British Columbia, Canada
Sharla Coetsee, Reutech Mining, Stellenbosch, South Africa
Niccolò Coli, IDS Corp., Florence, Italy
Davide Colombo, TRE-Altamira, Vancouver, British Columbia, Canada
Ashley Creighton, Rio Tinto, Brisbane, Australia
John Davis, Navstar Geomatics, Chepstow, UK
Jan deBeer, Reutech Mining, Stellenbosch, South Africa
Willem de Beer, ESG, Brisbane, Queensland, Australia
Adeline Delonca, University of Queensland, Brisbane, Australia
Graham Dick, BGC Engineering, Kamloops, British Columbia, Canada
Phil Dight, University of Western Australia, Perth, Western Australia, Australia
Alex Duran, PSM Consultants, Sydney, New South Wales, Australia
Chris Fagan, Sixense Oceania, Melbourne, Victoria, Australia
Giacomo Falorni, TRE-Altamira, Vancouver, British Columbia, Canada
Paolo Farina, Geoapp S.R.L., Florence, Italy
John Floyd, Blast Dynamics, Steamboat Springs, Colorado, USA
Mark Fowler, PSM Consultants, Sydney, New South Wales, Australia
Idalina Garoes, Rio Tinto, Rössing Uranium Mine, Namibia
Vivian Giang, University of Alberta, Edmonton, Alberta, Canada
Emrich Hamman, Anglo Gold Ashanti, Perth, Western Australia, Australia
Ben Haugen, RESPEC, Rapid City, South Dakota, USA
P. Mark Hawley, Piteau Associates, Vancouver, British Columbia, Canada
James Hogarth, Piteau Associates, Vancouver, British Columbia, Canada
Derek Hrubes, BGC Engineering, Montrose, Colorado, USA
Graham Hunter, 3D Laser Mapping, Nottingham, UK
Sean Jefferys, Sean Jefferys Ltd, Bristol, UK
Pat Jenks, Barrick of North America, Elko, Nevada, USA
James Jung, Rio Tinto Diavik Diamond Mines, Yellowknife, NWT, Canada
Tatyana Katsaga, Itasca Consulting, Sudbury, Ontario, Canada
Jon Leighton, 3vGeomatics, Vancouver, British Columbia, Canada
Lorenzo Leoni, IDS Corp., Florence, Italy
Loren Lorig, Itasca Consulting, Minneapolis, Minnesota, USA
Ernest Lötter, IMS, Hobart, Tasmania, Australia
Xun Luo, CSIRO, Brisbane, Queensland, Australia
Graeme Major, Major Geotech, Reno, Nevada, USA
Derek Martin, University of Alberta, Edmonton, Alberta, Canada
James Mathis, Zostrich Geotechnical, Ellensburg, Washington, USA
Jeff Mattern, Barrick of North America, Elko, Nevada, USA

M. Kim McCarter, University of Utah, Salt Lake City, Utah, USA
Moe Momayez, University of Arizona, Tucson, Arizona, USA
Mario Mons, Motion2Pixels, Somerset West, South Africa
Leon Nel, Reutech Mining, Stellenbosch, South Africa
Alex Neuwirt, Canary Systems, New London, New Hampshire, USA
Warren Newcomen, BGC Engineering, Kamloops, British Columbia, Canada
David Noon, GroundProbe, Brisbane, Queensland, Australia
Emre Onsel, Simon Fraser University, Vancouver, British Columbia, Canada
Verne Pere, Golder Associates, Perth, Western Australia, Australia
Alex Pienaar, SenseMetrics, San Diego, California, USA
Cliff Preston, IDS Corp., Tucson, Arizona, USA
Felix Ramsden, Debswana Diamond Co., Jwaneng, Botswana
Orapeleng Randall, Debswana Diamond Co., Orapa, Botswana
Rigoberto Rimmelin, BHP, Brisbane, Queensland, Australia
Nick Rose, Piteau Associates, Vancouver, British Columbia, Canada
Brad Ross, University of Arizona, Tucson, Arizona, USA
David Rutledge, Hexagon Mining, Sonora, California, USA
Peter Saunders, GroundProbe, Brisbane, Queensland, Australia
Michael Shelbourn, Barrick of North America, Tucson, Arizona, USA
Doug Stead, Simon Fraser University, Vancouver, British Columbia, Canada
Simon Steffen, Exelon Mining, Brisbane, Queensland, Australia
Ryan Turner, Barrick Golden Sunlight Mine, Whitehall, Montana, USA
P.W. van der Walt, Stellenbosch University and Reutech, Stellenbosch, South Africa
Felicia Weir, PSM Consultants, Sydney, New South Wales, Australia
Fanie Wessels, Rio Tinto Iron Ore, Perth, Western Australia, Australia
Eleonora Widzyk-Capehart, University of Chile, Santiago, Chile
Chad Williams, University of Arizona, Tucson, Arizona, USA
David Williams, University of Queensland, Brisbane, Queensland, Australia
Rocky Wu, Rocky Mining Consultants, Toronto, Ontario, Canada

Robert Sharon and Erik Eberhardt
February 2020

1

SCOPE AND INTRODUCTION

Robert Sharon, Peter Stacey

1.1 Introduction

A comprehensive and robust slope monitoring system is an essential component of the slope management program in an open pit mining operation. The purpose of the program should be both to ensure the safety of personnel and to support the recovery of the ore reserves contained in the mine plan.

These Guidelines are intended to provide a detailed description of the slope monitoring and stability management methods currently available, as well as to outline the requirements for, and design of, suitable slope monitoring programs. As such, this book builds on the general descriptions of monitoring programs contained in the companion Large Open Pit (LOP) books entitled *Guidelines for Open Pit Slope Design* (Read and Stacey 2009), *Guidelines for Evaluating Water in Pit Slope Stability* (Beale and Read 2013), *Guidelines for Mine Waste and Stockpile Design* (Hawley and Cunning 2017) and *Guidelines for Pit Slope Design in Weak Rocks* (Martin and Stacey 2018).

Focus is placed on the design of open pit slopes, although monitoring of nearby infrastructure is also addressed. While slope designs are intended to ensure stability or at least manageable deformation of the pit walls, variations in geotechnical models that are not captured in the design process can result in instability that could significantly impact the mining operation. The primary purposes of a slope monitoring program are therefore to assess the stability of the highwalls during open pit mine development and to detect the onset of any unexpected instability. This process of assessment and detection provides a basis for determining the controlling factors of the movement such that suitable management or remedial programs can be formulated.

The Large Open Pit (LOP) project, which was initiated in 2004 under the sponsorship of 12 major mining companies, had as one of its mandates the compilation of available technology related to the design and operation of slopes in open pit mines. This has resulted to date in the four companion books listed above. In its current form, the reconstituted Large Open Pit project (LOPII), which is also sponsored by major mining companies, is expanding on technology dissemination and funding specific research projects associated with slope designs and their implementation. In this context, the LOPII sponsors recognised that there is currently no single document that provides a comprehensive summary of pit slope movement monitoring systems and practice.

This book is intended to address that deficiency and to provide geotechnical engineers responsible for the management of open pit slopes with an understanding of the currently available technology and best practices for monitoring slope behaviour and assessing the resulting data. In compiling the book, the editors have received invaluable input and assistance from the sponsors, industry practitioners, and equipment and software suppliers.

The contents of these Guidelines follow the general process involved in establishing and operating a slope monitoring system for an open pit mine. Chapter 2 summarises the role of, and motivation for, slope monitoring in open pit mining operations. The following chapters address available pit slope monitoring methods and instrumentation (Chapter 3); applications and system design (Chapter 4); monitoring system operation (Chapter 5); data acquisition and management (Chapter 6); data analysis (Chapter 7); and utilisation and communication of the results (Chapter 8). Chapter 9 provides general descriptions of monitoring of other mine facilities, including tailings dams and waste dumps, as well as reconciliation of the monitoring data with the geomechanical program for the mine. Chapter 10 concludes by presenting a risk-based framework for slope monitoring, and by providing examples of innovative applications of slope monitoring together with discussion of emerging technologies, particularly in relation to automated mines in the future.

It should be recognised that the monitoring systems discussed in this book are the subject of continuing development. Consequently, practitioners should regularly review their program and update equipment, software and leading practices to provide the mine stakeholders with the optimum slope management program.

1.2 Background

The relationship of a performance monitoring system to the slope design process is illustrated in Fig. 1.1, which was originally presented in Chapter 12 of *Guidelines for Open Pit Slope Design* (Read and Stacey 2009). It is generally recognised in the mining industry that slope designs are almost invariably subject to 'uncertainties' related to the geotechnical model that supports the proposed slope configurations. This is particularly the case in complex geological environments. The uncertainties are obviously greater in the development of 'greenfield' projects, where there is no previous experience, but they can also extend to 'brownfield' mine expansions. Consequently, slope performance monitoring and stability assessment programs must form an integral part of the mine engineering and development system for an open pit operation.

Slope monitoring is an essential component in the design–implementation–verification cycle shown in Fig. 1.2. The monitoring component provides a basis for verifying critical design assumptions, giving the mine management and operators confidence that slope designs are reliable, and providing confidence that the mine plan is viable.

In broad terms, the slope management program for an open pit mine should address three requirements, namely:

- validation of the geotechnical model upon which the slope designs are based;
- assessment of slope performance in terms of design achievement; and
- monitoring of the pit slopes as they are developed to detect and quantify any movements, planned or unexpected.

Techniques for validating the geotechnical model and assessing slope performance in terms of achievement of

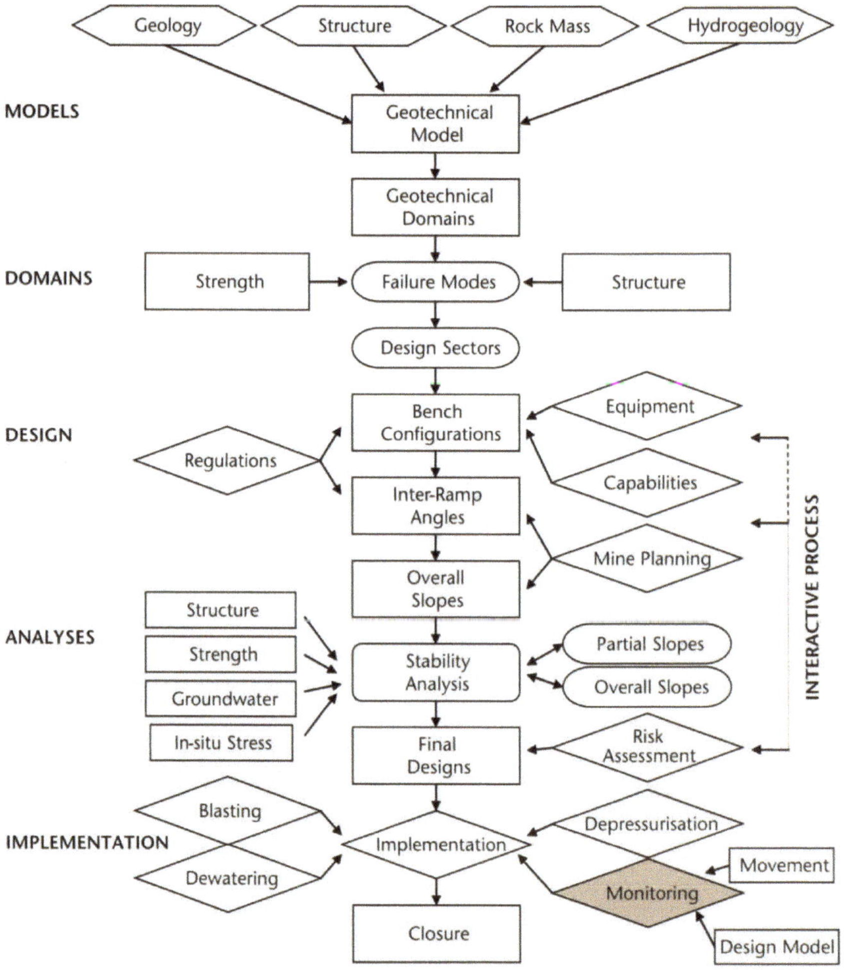

Fig. 1.1: Slope design process (source: Read and Stacey 2009).

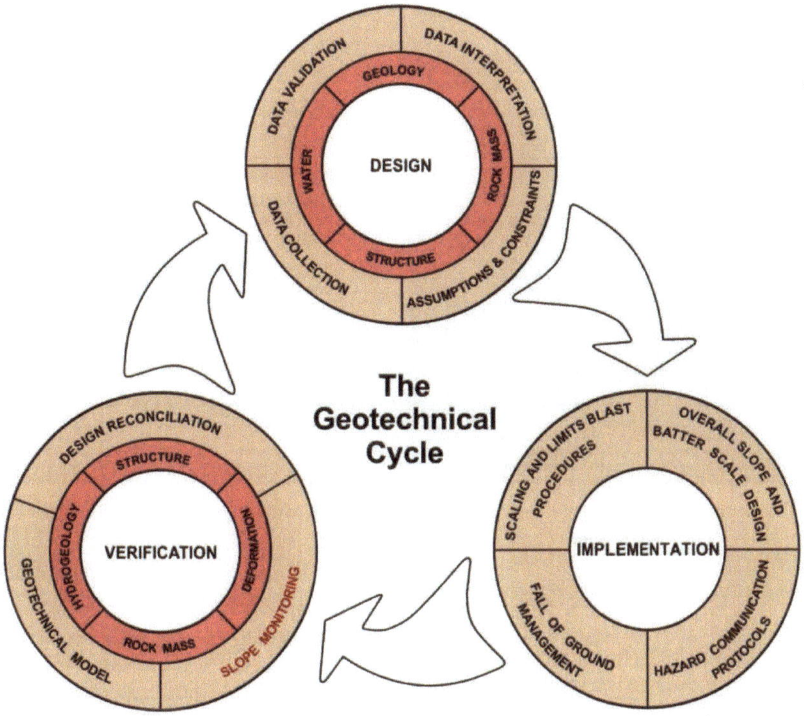

Fig. 1.2: Design–implementation–verification cycle for open pit geotechnical engineering (modified from de Graaf and Wessels 2015).

the design configurations are discussed in detail in Read and Stacey (2009). The validation of groundwater aspects is described in detail in Beale and Read (2013).

In most jurisdictions, there is a requirement in the mining regulations for a slope movement monitoring program to be in place and maintained to ensure the safety of workers. A monitoring program and additional assessment measures are also required by management to reduce the economic risk associated with design uncertainties.

The main focus of this publication pertains to the movement monitoring aspects of slope management programs, which involve surface, subsurface and satellite-based methods. The current technology, as applied to open pit slopes, has been progressively developed over the last 30 years, as discussed in Section 2.3. Typically, a combination of monitoring methods is employed for any particular slope, with variations depending on the anticipated instability modes.

In most cases, any degree of movement beyond the deformation normally associated with the stress redistribution related to the unloading resulting from excavation is considered to be 'unexpected', representing a potential risk for slope failure. As a result, the magnitude, rate (including any accelerating trend) and direction of movement must be carefully evaluated to provide a basis for ensuring the safety of the mine operators and the design of any necessary remedial measures. Slope movements can be managed under certain circumstances and may therefore be anticipated in the pit slope design, relative to the expected failure mode(s). In this case, it is critical that the monitoring program covers all potential failure modes and that movement rates and orientations are carefully monitored to confirm alignment with the design assumptions.

1.3 Avoiding unexpected events

Well-informed mine operators and senior management generally understand and accept the potential for slope instability risks associated with excavated slopes during mine development. This acceptance is normally based on the belief that the risks have been identified during the design process and that, if an instability event were to occur, the risks would be operationally and safely manageable. Slope monitoring programs must therefore be designed to provide advance warning of a potential instability in time to either mitigate the event or prepare for it. In this context, Trigger Action Response Plans (TARPs) must be developed specifically to address anticipated slope movements, through detection criteria and alarm thresholds, in order to ensure safe operational management from the time that a potential instability is identified.

Despite the implementation of slope monitoring programs and TARPs, slope instabilities continue to occur with little or no advance warning in such a way that an operation is unable to respond immediately before the

Fig. 1.3: Open pit slope failure resulting in loss of the access ramp. Small slope failures such as this 50 kt event can occur with very little warning. This slope section was mined three years before the failure event and was stable during that period (source: R. Sharon).

event. Therefore, the following discussion is intended to provide guidance on how the occurrence of such unexpected instability events may be avoided or how the residual risk may be mitigated. The slope shown in Fig. 1.3 is an example of an instability that occurred during a strong storm event where the mine operators had no advance warning. Fortunately, in this case, no one was injured and cleanup of debris on the ramp allowed the operation to resume.

Such unforeseen events can have serious implications in terms of loss of life, serious injury and/or disruption of the mining operations. It is therefore incumbent on the slope engineer to develop and maintain robust monitoring programs which can assist in the avoidance or mitigation of instability events.

Typically, the focus of a monitoring program is on active slopes where miners are exposed, problematic slopes that exhibit anomalous movement, rockfall hazard zones, slopes that contain excess groundwater pressure, and/or other features determined to be a risk to the operation. Often, less attention is given to slopes that have performed well for a long period of time, such as the slope in Fig. 1.3, and to highwalls that are located far from mining activity and are therefore considered to be a low risk to the operation. Nevertheless, unexpected slope instability events in inactive areas have contributed to serious injury, death, loss of/or damage to equipment and infrastructure, and material impact on production targets.

To address the potential for unexpected conditions, it must be accepted that disruptive and potentially dangerous slope instability could occur anywhere in the open pit. Slope monitoring programs must incorporate both investigative and predictive surveillance components that provide adequate coverage for the entire open pit. Each type of slope monitoring instrument has unique capabilities, as well as limitations such as when the data produced are susceptible to error induced by climatic changes. In order to support timely decision-making, the overall monitoring system must therefore be capable of producing data that can be validated to ensure that any detected movement is real and reliable, and to interrogate the possibility of false alarms to support timely decision-making.

Management of unexpected slope instability events may be achieved by consideration of the following aspects when reviewing the effectiveness of the monitoring program:

- recognising that uncertainties in the geotechnical design model for excavated slopes can present a level of risk that could be unacceptable if failure were to occur without warning or without action being taken to mitigate the risk of a failure;
- ensuring an adequate level of slope monitoring on all the slopes in the open pit, where the level of monitoring should be commensurate with the anticipated consequence(s) of failure;
- increasing instrumentation where needed;
- improving the performance, reliability and accuracy of instrumentation;
- confirming that the data acquisition system is capable of near real-time monitoring, thereby ensuring that the data management system can detect and flag rapidly developing changes, and can address challenges associated with the ever-increasing quantity of data to analyse;
- developing an integrated capability for interpreting context, such as adding daily rainfall records to prism time-displacement plots or adding piezometers, inclinometers, extensometers, etc. into comparative time-series plots;
- confirming that the TARP is robust (e.g. includes inspections or selective mine closure, as needed, during and following strong storm events); TARPs require periodic review and updating, at least annually, based on slope performance experience;
- ensuring the operation has sufficient staff on site to support slope monitoring functions; and
- training of all mine staff on performance monitoring and hazard awareness, as well as responses to unexpected events.

1.4 Terminology

The terminology used in these Guidelines follows current practice for slope design and performance monitoring. Particular aspects are discussed in the following sections.

1.4.1 Slope configurations

The standard terminology used to describe pit slope configurations is illustrated in Fig. 1.4.

It should be noted that there are some geographic variations in pit slope terminology. For example, bench face angle (North America) equates to batter (Australia), and bench (North America) equates to berm (Australia).

1.4.2 Slope movements

Three broad levels of movement are recognised for pit slopes, namely:

- an unloading response in the form of slope deformations or rock mass dilation associated with the mining process;
- slope movements beyond the deformation expected from unloading indicating the development of a potential instability; and
- failure, collapse or movement to the extent that the slope no longer performs as intended; the original design configuration is destroyed.

The limits between these three general categories of movement are essentially gradational, which emphasises the importance of an accurate monitoring system.

1.4.3 Technical conventions, definitions and abbreviations

A summary of technical terms, keywords and abbreviations, supplemental to those described above and that are commonly used throughout these Guidelines, is presented in a Glossary at the back of this book. Some technical terms are used interchangeably, while others are defined more narrowly for standardisation, consistency and to avoid confusion. In reference to Fig. 1.4, further details of pit slope geometries are described in Chapter 1 of *Guidelines for Open Pit Slope Design* (Read and Stacey 2009) and are not repeated in these Guidelines.

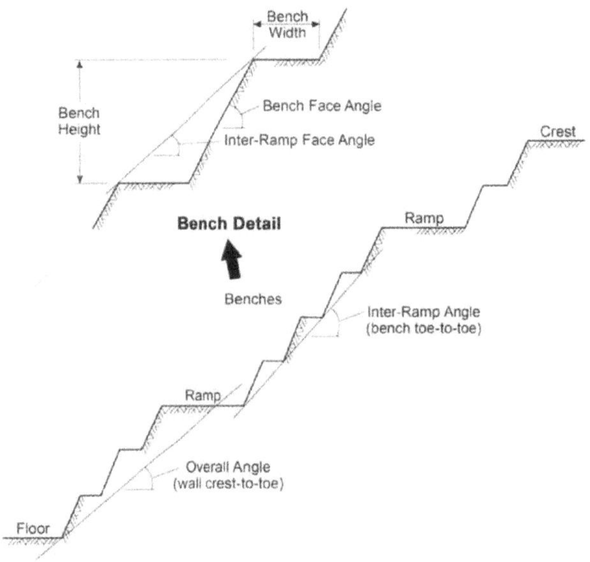

Fig. 1.4: Pit slope terminology (source: P. Stacey).

Reference is made throughout these Guidelines to the staff on site who are most responsible for the development and maintenance of the slope monitoring system. For consistency, these personnel are interchangeably referred to as 'slope engineers' or 'geotechnical engineers'. These terms imply that the slope or geotechnical engineer is qualified to perform the task of managing the monitoring system, even if the individual does not have an engineering degree. Further details regarding relevant duty of care for the geotechnical function at a mine site are summarised in Chapter 8. It should also be noted that, for the purposes of these Guidelines, the terms 'geotechnical' and 'geomechanics' are also used interchangeably, with a preference for using the term 'geotechnical'.

2

OVERVIEW OF SLOPE MONITORING

Erik Eberhardt, Mike Ness, David Noon, Eric Schwarz, Peter Stacey

2.1 Introduction

Open pit mining operations develop the largest engineered excavations of any kind, one of which is illustrated in Fig. 2.1. For pit slopes in small and large operations, there is a requirement to monitor performance during construction to ensure the safety of the workforce and that the design will be achieved with the anticipated economic recovery of the ore reserves.

Open pit slopes are often excavated in complex geological environments, with relatively limited data upon which to base the design. In the mining industry, where there has traditionally been a reasonably high economic risk tolerance, accepted levels of design reliability are relatively low compared with, for example, civil structures. This increases the requirement for a detailed and high-quality slope performance monitoring program, which extends throughout the life of the mine and into closure. In consequence, the slope monitoring program is normally an integral part of mine engineering, supporting the mine operations and the design functions.

Fig. 2.1: A typical large open pit – Chuquicamata open pit in 2018 with an approximate depth of 1050 m (source: Codelco).

The two basic requirements for a slope monitoring program, namely ensuring the safety of personnel and assessing the stability of the slopes in terms of achievement of the design, have led to the use of different, but overlapping approaches:

- *tactical monitoring*, which addresses the safety of the operating crews and equipment and therefore focuses predominantly on slopes at the bench and inter-ramp scales in the operating areas; and
- *strategic monitoring*, which relates more to the stability of the inter-ramp and overall slopes over the entire mine area on which the ore reserve recovery depends.

These approaches to monitoring involve essentially the same techniques and equipment, but emphasise the operational and design aspects of the mining program, respectively. Although not typically documented in mining regulations, the predominant requirements in regional and national mining policies implicitly include a slope monitoring program to ensure the safety of personnel and protection of the environment. In the event of a slope failure, particularly where injuries or fatalities are involved, the regulators require clear proof that the slopes have been adequately monitored.

2.2 Components of a slope monitoring program

The basic requirements of an open pit slope movement monitoring program revolve around the early detection of the onset of deformation and displacements. Where ground movement is determined to be outside the expected range inherent in the design (i.e. deformation due to unloading), the requirement increases to that of providing tracking, evaluation and management of a potential instability.

The components of such a program typically include surface and subsurface monitoring techniques. Surface techniques can be qualitative and quantitative, normally including a combination of:

- visual observations by geotechnical engineers, technicians and the mine operations supervisors, as well as all other personnel working in the pit;
- terrestrial quantitative movement monitoring using cross-crack measurements, geodetic surveying, borehole instrumentation and radar/ laser scanning methods; and
- satellite-based imaging and quantitative movement monitoring.

Subsurface monitoring techniques involve installations in drill holes located behind the slopes to detect movements and to measure pore pressures. Microseismic monitoring also falls into this category.

2.3 History of pit slope monitoring

2.3.1 Overview

Advances in the development of slope monitoring technology have essentially followed the progress of pit slope design and implementation processes, for which monitoring technology has become an important verification tool. Until the late 1960s, open pit mines were generally small by current standards with slope heights typically less than 300 m. Mining equipment was also smaller than what is currently used in large open pits. Slope designs were typically based on empirical methods and relevant development experience, since the rock mechanics engineering discipline was in its infancy. In consequence, an empirically based wall angle of 45° was a relatively common starting point for a pit design.

The steady increase in the size of open pits since the late 1960s, occasioned by increasing mining equipment size and resulting efficiencies that provided the ability to economically mine lower-grade ore, necessitated increased sophistication in the slope design process. The increasing number of slope failures was, however, causing concern in the mining industry from the standpoints of safety and economics. Accordingly, slope design research was funded by industry and governments, resulting in the publication of *Rock Slope Engineering* (Hoek and Bray 1974) and the *Pit Slope Manual* (CANMET 1977). The fundamental concepts developed in these books were the subject of additional research and improvement into the mid-1980s, by which time pit slopes exceeding 500 m in height were more common.

Between 1985 and 2000, there was little additional research into pit slope designs, even though slope heights continued to increase. During this period, basic slope monitoring methods demonstrated continued improvement in the areas of accuracy and automation. This allowed geotechnical engineers to increase their understanding of slope performance and their experience in the management of unstable slopes. Pit slopes were developed to heights of up to 1000 m, and pits as deep as 1500 m were being planned.

Several major slope failures between 1999 and 2002, some resulting in the loss of life and significant economic consequences, led to renewed interest in the slope design and performance monitoring process. This renewed interest was accompanied by an increased emphasis on safety, resulting in the inception of the Large Open Pit (LOP) project. At that time, monitoring relied predominantly on surveying of prisms distributed around the pit walls to provide information on the amount and direction of any slope movement. A further advance in the first decade of the 21st century was the introduction of slope monitoring radar, which provides rapid determination of slope movement across the entire wall sector being monitored. The historical development of these systems, as well as other methods of slope monitoring, is discussed later in this chapter. Details of the methods are presented in Chapters 3 and 4.

2.3.2 Early slope monitoring

Slope monitoring before the early 1970s was restricted to detailed visual observations, cross-crack measurements by tape or extensometer, and basic surveying methods using transits and theodolites. These methods had an inherent safety issue in that they typically required slope engineers accessing the unstable areas to complete the measurements. The resulting data had to be manually input into a computer for reduction to coordinates, which were then further analysed for movements and associated vectors. Compilation of the results could take two or more hours from the time that readings were taken. The notion of evaluating data in near real-time was never a consideration. However, the monitoring methods were effective, as demonstrated by the well-documented prediction of the actual date of the 12 Mt failure at Chuquicamata Mine in 1969, about one month in advance of the event (Kennedy and Niermeyer 1970). The monitoring system involved cross-crack measurements and transit surveys as the primary methods.

The introduction of basic wireline extensometers, in wide use by the 1970s, allowed alarming for movement through the use of limit switches attached to the wires and connected to flashing lights and/or sirens, as shown in Fig. 2.2. These extensometers were typically manufactured in the mine workshop. They used aircraft control cable to reduce the impact of loading (stretch) and temperature variations on the readings. Wireline extensometers of this type are still in common use as a 'first response' method, as discussed in Section 3.2.4. This technology presents an opportunity for mine

Fig. 2.2: Wireline extensometers with alarm limit switches at Steep Rock Iron Mine, 1974 (source: P. Stacey).

operations personnel to be directly engaged in the monitoring process.

By the 1980s, automated units that could initially retain wireline extensometer data in a datalogger were introduced, then further improved to enable wireless transmission of readings to the mine office. These units are frequently used for monitoring movement on waste dump platforms, but have also found local application on pit slopes.

Although earlier monitoring methods have largely been replaced by automated variants utilising more sophisticated technology, they arguably had an advantage in that practitioners had to analyse the data more closely through graphing trends using simple spreadsheets (see an example in Section 6.3.1). This required a fundamental understanding of the data, a more intuitive feel for error, and detailed evaluation of the raw datasets to assess their reliability. Combined with the additional effort to compile and think about the data, this brought slope engineers closer to a hands-on feel for field performance. With the automated data collection and compilation that exists today, it is easy to put too much faith in the capabilities of instrumentation, data and computer outputs, without detailed consideration of potential errors and biases, and the associated implications. It is suggested that modern slope engineers should continue to learn from traditional tools, including in-house built cross-crack gauges and extensometers, and of course employ visual observation, as supplements to merely accepting automated instrumentation and computer output without detailed examination.

2.3.3 Geodetic survey monitoring

An alternative to cross-crack measurements was the triangulation of fixed points using first-order theodolites.

Fig. 2.3: AGA Geodimeter 76 EDM unit (source: P. Stacey).

This approach was time-consuming and required very detailed surveying procedures to obtain centimetre accuracy at the ranges required for shooting across even moderately sized pits (i.e. in the order of hundreds of metres).

In the late 1960s, the introduction of the electronic distance meter (EDM) allowed the remote, accurate measurement of slope movements in the line-of-sight (LOS). Instruments, such as the laser-based AGA Geodimeter laser EDM (shown in Fig. 2.3) and the infrared Wild DI10 shooting to corner cube prisms, delivered slope distance accuracy of a few millimetres, provided atmospheric conditions were taken into account. This meant that slope movements could be rapidly determined with limited requirements for data reduction. Movement vectors could also be established to a high degree of accuracy, if the prisms were surveyed using a first-order theodolite. The combination of data from an EDM with a theodolite provided geotechnical engineers with a clear understanding of how the slope was performing and allowed for improved management of instabilities.

Soon after the introduction of slope distance measurement instruments, equipment suppliers provided the potential for mounting an EDM on a theodolite, thereby directly coordinating the distance and angular measurements. By that time, distance measurements were also being made using infrared technology and incorporating the pulse measurement principle. This was followed in 1977 by the production of instruments combining the two functions into a single unit termed a 'total station', such as the Wild TC1 (shown in Fig. 2.4). The development of the total station involved the introduction of on-board data recording and data reduction units for instruments. This configuration

Fig. 2.4: Wild TC1 total station combining a theodolite with an infrared EDM (source: P. Stacey).

remains one of the basic slope monitoring tools today, as discussed in detail in Section 3.3.

Total station units were automated by the mid-1990s to allow remote operation at predetermined time intervals. These remote operation units, which were termed robotic total stations (RTS), could be connected through a telemetry system to transmit data to the mine office, where it could be reduced by computer and plotted using software provided by the equipment manufacturer (e.g. Leica GeoMos and Trimble 4D software). These systems continue to improve in terms of more accurate equipment and software, including program systems that can analyse the data and generate alarms based on predetermined trigger levels, in effect providing near real-time monitoring capabilities. However, operating principles using total station theodolites have remained the same.

The major advantage of prism surveying for strategic monitoring (i.e. understanding the factors controlling the movement) is that it can provide three-dimensional vectors for the movements. However, it also involves points (prisms) distributed across the slopes that may or may not remain accessible for maintenance during mine development. As such, it is highly dependent upon a good coverage of prisms, which can be disrupted as movement levels increase.

2.3.4 Slope monitoring radar

The introduction of ground-based radar (GB radar) for slope monitoring in the late 1990s provided a means of establishing detailed patterns of movement on slopes without the requirement for specific monitoring points (prisms) distributed across the walls.

In 1996, the University of Queensland in Australia commenced the first research and development project to demonstrate a GB radar system for monitoring the deformation of a highwall face in a coal mine. The idea was based on the considerable success achieved by space-borne (satellite) interferometric synthetic aperture radar (InSAR) remote sensing systems that measure small ground surface movements (e.g. fault slip associated with earthquakes, surface subsidence associated with underground mining, etc.). These radars use the differential interferometric technique to measure slight movements of land masses from satellites along the LOS.

In 1999, the first ground-based slope monitoring radar was tested at an Australian coal mine (Fig. 2.5). Analysis of the data showed that an unstable section of the wall moved 2 mm over 12 days of monitoring. This result was broadly consistent with wireline extensometer readings taken during the trial period.

Fig. 2.5: First slope monitoring radar tested by the University of Queensland at an Australian coal mine in 1999 (source: GroundProbe).

Fig. 2.6: The first commercial Slope Stability Radar (SSR001) produced by GroundProbe in 2001 (source: GroundProbe).

In 2001, the company 'GroundProbe' was formed by the University of Queensland to develop a mine-operational Slope Stability Radar (SSR™) for the global mining industry. The first commercial radar, called SSR001 (Fig. 2.6), was leased to local coal and metalliferous mines. Additional units were then made available to the Australian mining industry. From September 2003, units were exported to South Africa, then to Indonesia, Zambia, the USA and Chile. Subsequently, other slope monitoring radar manufacturers, such as Reutech Mining and IDS GeoRadar, have emerged, using alternative technology but providing essentially the same general degree of capability, limitations and precision. Further information on GB radar is summarised in Section 3.4.

All slope radar systems provide LOS detection of millimetre-scale movements over small to large regions of the targeted slope. The actual minimum resolution area covered depends on the range and the type of equipment being used. The size of the resolution area, or pixel size, is typically less than 10 m^2, thus potentially providing sufficient coverage and image resolution to ensure worker safety against all sizes of slope movements, with newer system designs targeting localised rockfall events. Instrumentation developed for monitoring rockfall is described in Chapter 4.

Today, there are hundreds of ground-based slope monitoring radar systems operating in mine sites in over 26 countries. These GB radar technologies are often deployed in an integrated system involving a combination of geodetic survey methods (GNSS and RTS) and slope monitoring radars.

2.3.5 Remote sensing methods

The success of GB radar has coincided with, and inspired, parallel developments in other remote sensing technologies for open pit slope monitoring, both ground- and satellite-based.

2.3.5.1 Ground-based remote sensing methods

Ground-based laser scanning (LiDAR) and photogrammetry have been used extensively since the 2000s for establishing pit wall configurations and material volumes in open pit mines or in waste dumps. Early use saw significant development of both hardware and data-processing software for rock mass characterisation and discontinuity mapping. The techniques have also been used extensively for remote monitoring of natural slopes, including potential zones of movement, and for monitoring active landslide areas.

More recently, automated LiDAR systems with precision as high as 1–5 mm have become available. Such systems are available at prices lower than GB radar units with similar ranges, although the precision of the radar measurements is currently significantly higher. With continued technology development, it is possible that laser scanning and change detection software will become a viable competitor to slope monitoring radar. Comparative capabilities and limitations of radar, LiDAR and photogrammetry systems, as well as other major slope monitoring systems, are described in detail in Chapters 3 and 4.

2.3.5.2 Space and airborne remote sensing methods

Satellite-based global positioning systems (GPS) saw early research use in the 1990s to monitor landslide movements and behaviour. This evolved to practical use in open pit mines in the 2000s, where GPS was combined with global navigation space systems (GNSS) to provide three-dimensional coordinate measurements with accuracies of a few millimetres when operated in a differential mode with multiple receivers. As such, GNSS has found application in the monitoring of pit slopes, waste dumps and high-accuracy control of survey monitoring base stations. An example of a GPS station, utilised to monitoring movement at the crest of an open pit slope, is shown in Fig. 2.7.

For current slope monitoring purposes, utilisation of basic GNSS units require access to the monitored points, creating a potential safety issue. More sophisticated remote GNSS monitoring is available in the form of continuous operating reference systems (CORS; an example is provided in Chapter 9). The units are powered by solar panels and are capable of transmitting data to the mine office using a telemetry system; as such, it can be used for an alarm system. Research in the area of GNSS for surveillance monitoring and first-order survey control for surface mines should advance this method significantly for slope monitoring in the coming years. Further discussion of GNSS is presented in Section 3.3.4.

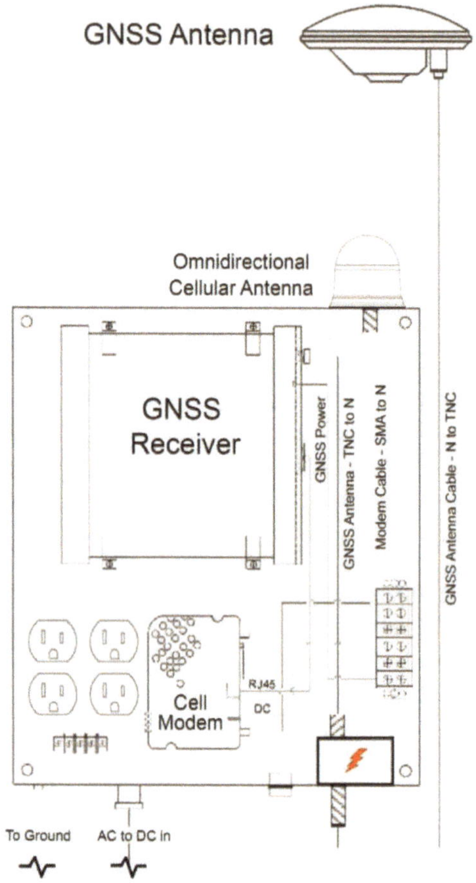

Fig. 2.7: GPS receiver located on the crest of an open pit (source: R. Turner and D. Rutledge).

Another satellite-based monitoring system that has found significant application in the mining industry is InSAR. As noted in Section 2.3.4, this methodology was initially introduced in the early 1990s as a method of detecting and monitoring ground movements over large areas, with a focus on subsidence monitoring and geological processes. It was subsequently used in the mining industry as a subsidence monitoring tool in the early 2000s, enabling remote monitoring of surface movements over wide areas to better than a centimetre precision.

Although primarily used for monitoring mine facilities, such as tailings dams and waste rock dumps, with a typical frequency of several days to monthly, satellite InSAR monitoring has successfully detected long-term creep movements in open pit slopes. While normal slope monitoring systems such as geodetic survey monitoring and GB radar are typically focused on monitoring trends over periods of one to several months, an annual InSAR monitoring frequency has indicated a constant creep of 35 mm/year on the highwall of a Chilean copper mine over a two-year period. Satellite InSAR and its application to open pit slope monitoring are discussed in Section 3.5.3 and Section 4.5.

The success of both ground- and satellite-based imaging and surveillance methods has led to the development and use of unmanned aerial vehicles (UAVs, or drones), which have rapidly gained traction as a tool for surveying open pits and mine facilities. They offer a low-cost means of measuring volumes and developing high-resolution images of pit slopes. These images can be used for remote mapping of the orientation of structures exposed in the pit walls, as well as for providing historic records of open pit development, including unstable areas.

Current experience and experiments show that centimetre-level precision can be achieved. This capability is likely to improve as the technology continues to develop. While drones can be operated by mine staff and deployed at any time during daylight hours, they are somewhat dependent upon weather conditions for accuracy, and time is required to process the acquired data. In consequence, while UAV-based mapping is likely to become a standard tool in most open pit mines, its application to real-time slope monitoring will likely require further development. Details on drones are presented in Section 3.5.2, and their potential capabilities are further examined in Chapter 10.

2.3.6 Subsurface monitoring systems

Subsurface monitoring instruments can provide invaluable information on the mode and extent of failure, indicating whether it is controlled by discrete structures or deformation of the rock mass. Early quantitative subsurface deformation monitoring was achieved with inclinometers, originally developed for use in soil slopes. Later methods have included semiquantitative probing of standpipe piezometer pipes using 'poor-boy inclinometers' and time-domain reflectometry (TDR) cables. Significant advancements in subsurface deformation monitoring techniques include quantitative measurement methods, including Shape Accel Arrays (SAA) and Smart Markers, as discussed in Sections 3.6.4 and 3.6.5, respectively.

The development of subsurface monitoring methods is described briefly in the following sections.

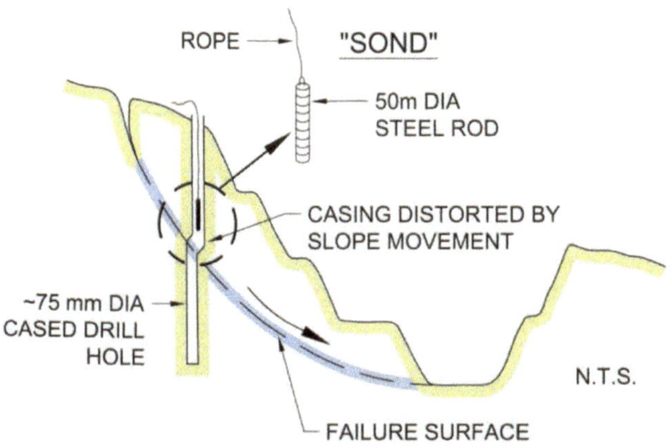

Fig. 2.8: 'Poor-boy inclinometer' (source: P. Stacey).

2.3.6.1 *Inclinometers*

In the late 1960s, the only subsurface monitoring instrument commonly available was the borehole inclinometer, which was first used commercially in the late 1950s for monitoring civil slopes and landslides (Stark and Choi 2008). This system involved a probe containing servo-accelerometers and guide wheels that was lowered down a grooved casing grouted into a drill hole collared behind the slope. By traversing the probe down the casing, displacement magnitudes and directions can be calculated via comparison with the lower end of the guide casing, which acts as a reference (datum). Relative displacement over time can then be used to determine the ground deformation over the profile of the hole, as discussed in Section 3.6.3. Localised deformation of the casing can be interpreted as the location of shearing on a planar structure.

A more rudimentary inclinometer was termed a 'poor-boy inclinometer', which took advantage of existing standpipe piezometers. A rigid steel rod pipe would be connected to the surface by a flexible line, typically thin aircraft control cable, and lowered down the standpipe to determine the depth of any borehole casing obstruction (Fig. 2.8). A similar steel rod left at the base of the standpipe and attached to a cable could be pulled up to determine the depth of the base of any shearing.

A variation of the traversing inclinometers, introduced in the 1990s, was the in-place or stationary inclinometer (IPI), where one or more probes are positioned at fixed locations in the casing. IPI probes are combined with automatic data acquisition systems (ADAS), allowing for continuous monitoring. These instruments are also described in Section 3.6.3. Enhancements to IPI units included the use of in-ground wireless units (Smart Markers), introduced in 2009, to provide a means of communicating the results of subsurface deformation measurements made by a network of IPI units.

2.3.6.2 *TDR cables*

The use of TDR measurements on coaxial cables grouted in drill holes around open pit mines is reported from the early 1980s. Stress on the cable created by movements in the surrounding rock, principally along structural features, results in deformation or breakage along the cable, with resulting signal impedance or reflections.

TDR measurements are essentially qualitative, since they are not able to provide reliable or accurate measurement of the deformation direction or magnitude. However, the method has an advantage over manually operated inclinometers in that it can be equipped with alarms that respond to changes in the impedance of the cable, indicating deformation or breakage, as described in Section 3.6.2.

Since the 1990s, it has been common practice to include a TDR cable as an integral part of a vibrating wire piezometer installation in drill holes behind pit slopes.

2.4 Current status of slope monitoring

As discussed in Section 2.3, since the 1990s the mining industry has experienced rapid advances in the technology available to support slope monitoring programs, including improved instrumentation, increased accuracy, and automated systems with data transmittal and software for the compilation and analysis of the data.

Some technological advances have been specifically developed for geotechnical monitoring purposes, such as inclinometers and TDR, but many have been transferred from other areas, including land surveying, GNSS, radar technology and satellite imagery. As noted in the introduction (Section 2.1), these advances continue at a rapid pace, primarily in the form of improvements to surveillance and communications technology and to software systems. Slope engineers are encouraged to remain current through regular reviews of the literature and discussions with equipment and software manufacturers.

Monitoring techniques typically being used at the time of writing these Guidelines are summarised in Table 2.1 with reference to the potential failure sizes and speed of development, as well as the implications to the mining operation. The various methods are described in Chapter 3.

2.5 Role in the mine

2.5.1 Overview

Typically, the slope monitoring program falls under the responsibility of the mine operation's geotechnical section, which is generally a component of the technical services or

Table 2.1: Summary of monitoring methods by potential failure size and implications

Block size (m³)	Speed of failure	Implications	Monitoring for detection	Typical remedial measures
1–10	Immediate	Safety (rockfall)	**Visual**	**Catchment**
10–1000	Very rapid to rapid	Safety	**Visual** Radar/laser scanning	**Catchment**
1000–100 000	Rapid to slow	Operational	Visual Geodetic surveying **Radar**/laser scanning Microseismic	**Manage** Modify slope (step-out)
100 000–1 000 000	Moderate to slow	Operational/financial	**Geodetic surveying** **Radar**/laser scanning InSAR TDR/inclinometer Microseismic	**Manage** Modify slope (step-out or re-cut)
>1 000 000	Slow to moderate	Operational/financial/force majeure	**Geodetic surveying** TDR/inclinometer Radar/laser scanning InSAR Microseismic	**Modify slope (re-cut)** Mine closure (>10 million m³) Manage

Speed of failure
Very rapid – Immediate to minutes
Rapid – Minutes to hours
Moderate – Days to weeks
Slow – Weeks to months
Bold – Most likely monitoring and typical remedial measures

Source: updated from Martin and Stacey (2018)

mine engineering departments. In many cases, the mine's survey section also provides direct support to the program.

In view of the geotechnical section's role in ensuring the safety of the mining operation, frequent detailed communications between the slope monitoring team and the operations department are essential. This support is often achieved through slope monitoring personnel participating in operations planning meetings, as well as the publication of hazard plans and regular reports on the condition of the slopes. These reports should be circulated to the operations and mine planning teams, to all other sections that have personnel working in the pit, and to management. Details of the reporting process are presented in Section 8.6.

Details of the slope monitoring program in the open pit should be contained in the mine operation's ground control management plan, which is discussed in detail in Section 2.6. This document should be freely available to the mine planning and operations teams, as well as to site management, who are responsible for approving the plan.

2.5.2 Observational approach to mining

The observational approach to geotechnical design, which was developed for use in soil mechanics as applied to construction (Terzaghi and Peck 1948; Peck 1969), is finding increasing application in open pit mining to address design uncertainties, or as a basis for the management of predicted marginal stability or unexpected instability in a pit slope. It involves adjustment of the slope designs, and consequently the mine plan, to the observed (measured) performance of the slopes, as discussed in Section 8.2. Implicit in observational mining is the availability of contingency plans, each of which should have its own monitoring system.

When a mining company formally adopts the observational approach, the monitoring system forms the critical control for the operation. Not only is slope monitoring the basis for ensuring the safety of the operation, but it also provides data for the design of the contingency plans that are an essential part of the adopted approach. It is therefore essential that the monitoring system is conservatively designed to address all potential conditions. As such, the system is best formulated through a detailed risk assessment.

2.5.3 Integration into operations

In the simplest sense, geotechnical or slope engineers use their knowledge and experience to correctly interpret slope monitoring results and to convey necessary information to the mine operations team to provide protection from geotechnical hazards. All of the principles and methods discussed in these Guidelines are directly applicable to mine operations, but awareness of a few extra considerations related to the mine environment enables a better, more effective, and easier to manage monitoring program. This does not imply that slope monitoring in mining is complicated or specialised. Rather, any open pit slope monitoring program will require a unique implementation that is practical and provides value for the specific operation.

The unique demands on slope monitoring systems at active mine operations are largely the result of four key conditions which define considerations for the integration of slope monitoring into mining.

1. *Environment of active destabilisation.* By virtue of the ongoing open pit mining process, terrain is being aggressively steepened and, at least theoretically, being rendered less stable; at the very least, stability is a moving target. The result is a life-of-mine working environment subject to unplanned instabilities largely, but not completely, mitigated by appropriate mine slope designs (due to inherent uncertainties).
2. *High cost of controlling exposure, demands correct interpretations, decisions, and actions.* In mining, as elsewhere, safety of personnel is the highest operating criterion, taking priority over all risk (cost)/benefit analyses. Control of personnel and/or equipment exposure to hazards can partially or completely halt mining operations. This in turn possibly affects revenue, but not in terms of large costs. Slope engineers need reliable, timely and high-quality monitoring data to make good interpretations of stability conditions and correctly assess hazards and potential impacts, and thereby contribute to good mining business decisions.
3. *Evolution of open pit mining.* Open pit mining is evolving from interim, shallow working angles for most operations with only steep ultimate slopes to steep, aggressive slope angles on both interim and ultimate cutbacks. Modern open pit designs are steeper, deeper and typically mined at a faster rate. To keep up (actually, to get ahead), slope monitoring programs require similar increases in intensity, reliability and rapid response to achieve risk mitigation objectives. At some mines (e.g. Grasberg and Cerro Verde), slope monitoring efforts are staffed full-time (24/7), with dedicated slope engineers and technicians on every mine shift. At times, more traditional mining personnel need to be convinced that conditions have changed and old paradigms are no longer adequate to achieve risk management objectives.
4. *Traditional thinking and false risk logic abound in mining.* Long-term personal mining experience without significant incidents does not apply under changing mining conditions, nor is a long mining career an adequate sample duration for encountering risk potential associated with high-consequence, low-likelihood instability events. But it is human nature to think that this is the case. Consequently, geotechnical engineers sometimes have the burden of not only providing correct and timely recommendations, but also 'selling' seemingly conservative recommendations to recalcitrant miners who believe they know better.

Certain practices discussed in these Guidelines can overcome these and other impediments to successful slope monitoring for mine operations. Slope engineers should normally have complete, unencumbered control of access to hazard areas, so that the risk of inadvertent exposure to hazards is minimised. Additional details pertinent to practical aspects of performance monitoring applicable to mining operations are presented in Chapter 5.

2.6 Ground control management plans

Ground control management plans (GCMPs) are designed to describe all of the geotechnical systems and procedures in place at a surface mine. The objective of the GCMP is to ensure systematic planning and implementation of ground control systems to safely and efficiently develop open pit mine slopes. The GCMP documents the slope risk controls and demonstrates how the operation intends to monitor the controls to manage slope risks, making certain that the controls function effectively at all times. As such, the document presents a detailed definition of the geotechnical monitoring strategy and defines the allocated resources required to pursue this strategy. Slope monitoring is a key element of the mine engineering program, and the monitoring strategy should be outlined in the GCMP. The GCMP provides stakeholders with information regarding what to monitor, how to monitor and who is responsible for key components of the monitoring system, which must involve a systematic and well-managed approach.

The GCMP for the mine site should define the ground movement monitoring strategies, including:

- mapping of geological features to confirm the geotechnical model;
- the operating monitoring methods, including which instruments are in service and records of general ground performance; and
- appropriate alternative monitoring instruments for the mine site.

For slope monitoring to be successful, the slope monitoring instrumentation plan and strategy outlined in the GCMP must be appropriate for the local ground conditions and the life-of-mine plan. This requires a detailed understanding of the geotechnical model and the associated slope designs, as well as the mine development plan and excavation practices unique to the operation.

Verification of the slope design is an iterative process in which slopes are monitored, the actual geological conditions determined, and as-built plans are generated and compared to predicted conditions and to mine plan assumptions. Slope designs are modified accordingly, completing the cycle. Effective design validation requires a systematic process that includes periodic reviews of the monitoring data and updates of the mine design during development.

A key component in the GCMP is the strategy defining ground movement monitoring. The monitoring plan should be clearly defined in terms of:

- identification of the anticipated hazards associated with the slope designs, including a list of historic instability occurrences, which forms the basis for the monitoring plan;
- monitoring instrumentation used at the mine – types, models, locations (using a map) – and required support (power, network, communications). Subsurface instrumentation should also be included, even if data are manually collected;
- the plan for future monitoring instrumentation as mine development advances;
- accountabilities and responsibilities with respect to monitoring instrumentation, including support requirements, monitoring data collection and interpretation, and communication of the results;
- the data acquisition system used to collect, validate, store and analyse monitoring data trends;
- a full explanation of the trigger action response plans (TARPs) for each slope in the mine;
- assigned responsibilities for reactions to any unexpected movements and indications of instabilities;
- descriptions of alternative monitoring instrumentation to provide data validation (e.g. the use of prism data to validate radar measurements), interpretive context and back-up support;
- a description of the steps that mine operators must take to ensure that all monitoring systems are performing as designed, including instrumentation recalibration or servicing at the manufacturer's recommended intervals (including communication systems, remote data acquisition gear and digital storage maintenance);
- a summary of the instrumentation replacement plan and budget; and
- a flowchart or process for geotechnical engineers to create and maintain an instability database (initial observations including maps, photos, results of geologic mapping, monitoring trends and back-analysis results, and final impacts to the mine).

In addition, the GCMP should summarise the slope monitoring data analysis process. This includes the frequency of analysis, any data alarming procedures and/or criteria developed from experience, and reporting of results to management and stakeholders. The interaction of the monitoring data relevant to the TARP and emergency preparedness and response plan (EPRP) should be clearly defined in the GCMP.

The monitoring plan summarised in the GCMP should also define how the site uses rates of ground movement as a tool to predict imminent wall failure. Monitoring data from slope displacement monitoring instrumentation can be used to validate and verify geotechnical design assumptions, assist in assessing levels of risk, and provide a level of comfort to operators by confirming that the ground is stable. Usually, monitoring is linked to a TARP to safely manage risk potential and identified instability hazards in the mine.

It is important that management and the operations teams are fully involved in the performance monitoring process, participate in process development, and understand the GCMP. Details regarding observational approach mining and the utilisation and reporting of results generated from the performance monitoring system are presented in Chapter 8.

3

MONITORING METHODS AND INSTRUMENTS

Alex Bals, Geoff Beale, Pierre Choquet, Sharla Coetsee, Niccolò Coli, Davide Colombo, Ashley Creighton, John Davis, Jan deBeer, Willem de Beer, Adeline Delonca, Phil Dight, Erik Eberhardt, Giacomo Falorni, John Floyd, Idalina Garoes, Emrich Hamman, Ben Haugen, P. Mark Hawley, Sean Jefferys, Jon Leighton, Lorenzo Leoni, Ernest Lötter, Xun Luo, James Mathis, Naani Mphathiwa, Leon Nel, Mike Ness, Alex Neuwirt, David Noon, Alex Pienaar, Marc Ruest, David Rutledge, Peter Saunders, Robert Sharon, Simon Steffen, P.W. van der Walt, Eleonora Widzyk-Capehart

3.1 Introduction

Methods used for open pit slope monitoring can vary significantly with respect to their versatility in:

- how the various instruments used can be deployed (on surface, in boreholes);
- what they are meant to measure (slope displacements, water pressures);
- how they measure (total vector movement, relative positions, line-of-sight); and
- how the data are recorded (manual, automated) and transmitted (manual downloads, wireless) and their frequency (periodic, real-time).

Performance monitoring extends from initial mine development (pre-stripping) through to final mine closure, serving strategic (investigative, surveillance) and tactical (predictive, safety-critical) purposes to ensure the safe and economically successful operation of the mine. For example, investigative monitoring may be carried out to provide a baseline for open pit development (to understand the slope behaviour over time and the depth of movement) and to assess typical responses to external triggers. This basic longer-term understanding of the slope behaviour can then be used to aid predictive monitoring to establish early warning alarm thresholds. Experience gained from the collective performance monitoring program is then used to establish that the open pit slope is behaving as expected, as per its design, or to provide a warning of diverging behaviour, enabling mitigation and safety measures to be taken.

Monitoring systems, whether manual or automated, can also be divided into two categories, namely systematic and campaign monitoring. Systematic monitoring is the long-term component that is deployed to ensure safe and continuous economic operation of the mine. Such systems require permanent to semi-permanent equipment installations to ensure continuous monitoring for the service life of the excavated slope or other structures under surveillance. Requirements for systematic monitoring are further described in Chapter 5.

Campaign monitoring involves relocating slope monitoring equipment as needed to focus on active mining areas, or other areas identified by the slope engineer, which require more direct attention for a specific amount of time (Harries *et al.* 2006). The equipment designated for this type of monitoring should be easily moveable. Infrastructure such as instrument shelters, power supply and communications tools should be designed so that they are quick to dismantle and easy to redeploy.

When choosing a monitoring system, the slope engineer must consider and balance the instrument specifications relative to the monitoring requirements. These are discussed in more detail in Chapters 4 and 6, and include the following.

- *Range:* the maximum distance over which the measurement can be reliably performed, with greater range usually being obtained at the expense of resolution
- *Resolution:* the smallest numerical change an instrument can measure
- *Accuracy:* the degree of correctness with respect to the true value, usually expressed as a ± number or percentage
- *Precision*: the repeatability of similar measurements with respect to a mean, usually reflected in the number of significant figures or decimal places quoted for a value
- *Conformance*: whether the instrument is properly fixed to the ground it is meant to be measuring or if the presence of the instrument affects the value being measured
- *Robustness*: the ability of an instrument to function properly under harsh conditions to ensure data accuracy and continuity are maintained
- *Reliability*: synonymous with confidence in the data; poor-quality or inaccurate data can be misleading and may be worse than having no data.

In manual and automated monitoring, standard procedures such as having a well-established and maintained survey control network need to be adhered to at all times. Measuring principles, like working from whole to part and check surveys, are also applicable. Equipment maintenance processes, such as instrument calibration, are common and necessary practices.

3.2 Movement detection

3.2.1 Overview

Movement detection using a strategic approach is typically a first step before installing a more advanced (and expensive) tactical monitoring system. Its purpose is to alert the geotechnical engineer to a slope that has started to move and may present a hazard, thus requiring the allocation of extra time and resources for safety-critical monitoring. The open pit mining industry uses several types of manual and automated movement detection systems. Although many of these can be automated, it is useful to have a good understanding of manual techniques, since they are commonly used as a practical, initial validation indicator where stability risk is considered to be low. They can also be employed in an investigative monitoring strategy as a short-term information-gathering exercise, for example, to understand the size and limits of the moving slope. The most commonly used manual monitoring methods include:

- field or visual observations and inspections;
- geotechnical instruments including tiltmeters, crack gauges and extensometers; and
- geodetic surveys including total station and precise levelling.

The following describes visual inspections, cross-crack measurements and manual geodetic precise levelling surveys. Other monitoring methods, such as total station surveys, are described in the next section that address automated displacement monitoring systems. Practical applications are presented in Chapter 9 and elsewhere in these Guidelines.

3.2.2 Visual inspections

Visual field inspections play an important role in identifying risk areas where previously inactive slopes have begun to move and, subsequently, in implementing the appropriate monitoring systems (Little 2006). Field inspections are considered to be a basic form of monitoring. They can be classified into the following five types.

1. *Incident reporting* is conducted by anyone working on site, provided they have received basic training or awareness orientation on stability risks. Reportable incidents may include the identification of potentially unstable areas by spotting cracks (Fig. 3.1) or potential rockfalls at working sites. It is good practice to conduct slope hazard awareness training at least twice a year for the workforce (both supervisors and operators).
2. *Daily inspections* of active areas are conducted by the working crews (mining, survey, geology, etc.) with observations of hazards being reported to the geotechnical engineering department. This is a routine inspection which forms part of the procedural risk analysis done as a planned job observation at the beginning of every task.
3. *Detailed inspections* are conducted by the slope engineer with the main focus being the high-risk areas. These inspections enable the slope engineer to select the correct monitoring devices for the hazardous areas (Little 2006). The example shown in Fig. 3.2 is a flash report from inspections carried out by geotechnical engineers at Debswana's Jwaneng Mine.
4. *Comprehensive monthly inspections* of all areas are conducted by the slope engineer, who uses the data to determine risk ratings for each area.
5. *Final highwall inspections* are conducted by the slope engineer and blast engineer to determine the cause of damaged final walls and to remedy the problem ahead of future blasts (Little 2006).

3.2.3 Cross-crack monitoring and tell-tales

Crack meters are a low-cost and simple means for detecting slope movements by monitoring across tension

Fig. 3.1: Examples of cracks that can be easily spotted and reported to the geotechnical department as an indicator of slope movement (source: E. Eberhardt).

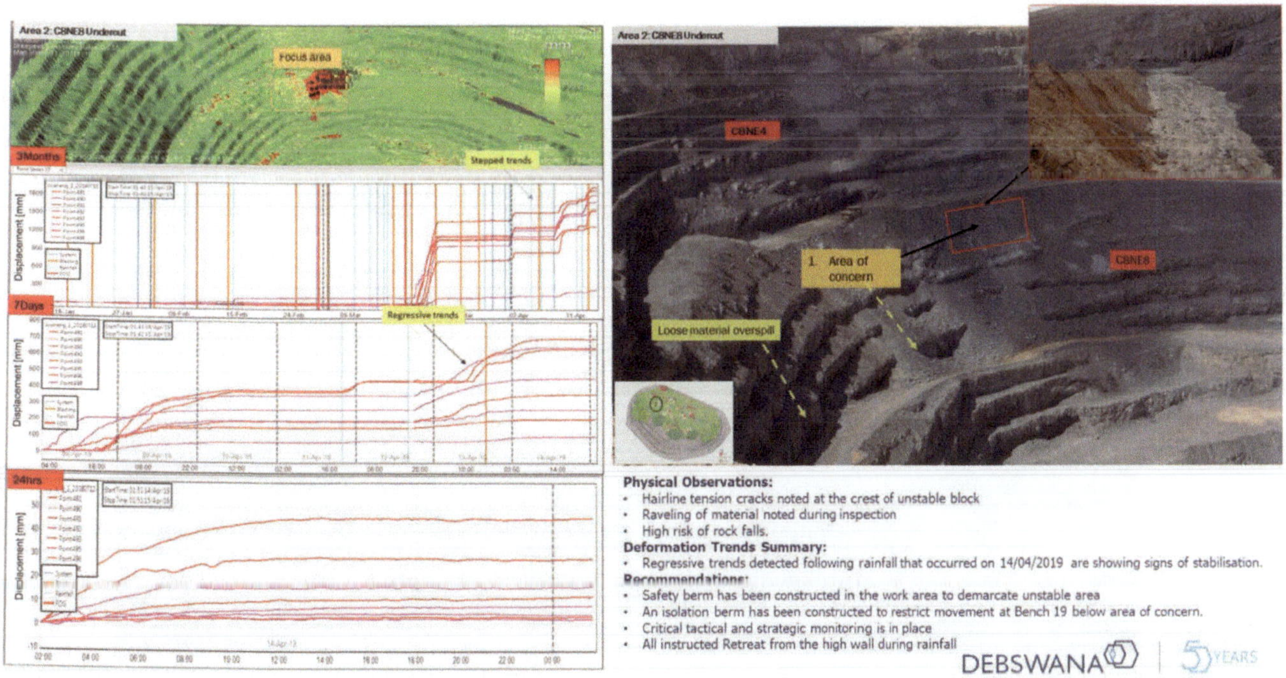

Fig. 3.2: An example of a flash report from the visual inspections at Debswana Jwaneng mine in Botswana (source: N. Mphathiwa).

Fig. 3.3: Example of a crack meter capable of automatic measurement with digital output (source: E. Eberhardt).

cracks, especially near critical crests such as on ramps or geotechnical berms. A simple construction involves anchoring two mounts on either side of a tension crack or open joint, and measuring the relative movement between the two fixed mounts. A sensor such as a potentiometer, vibrating wire or laser can be used to provide automatic digital readouts to feed remote transmission. An example is shown in Fig. 3.3. Measurement accuracy and repeatability depend on the type of sensing device and the distance between the monitoring points. Typical accuracies range from submillimetre to millimetre over distances of less than a metre when using stiff sliding rods spanning fixed monitoring points.

For understanding more complex movement in two or three dimensions, more sophisticated crack meters can be constructed using two- and three-roller potentiometers that rotate along axes that are mutually perpendicular.

Tell-tales are simple devices placed to span a crack or series of cracks. They are composed of two posts (or pickets) anchored on opposite sides of a crack and connected by an indicator (or pointer). Movement along the crack causes a relative displacement that can be measured in terms of rate and magnitude. A tell-tale can also be a paint or picket line where subsequent dislocation can be observed and measured (Hawley and Cunning 2017).

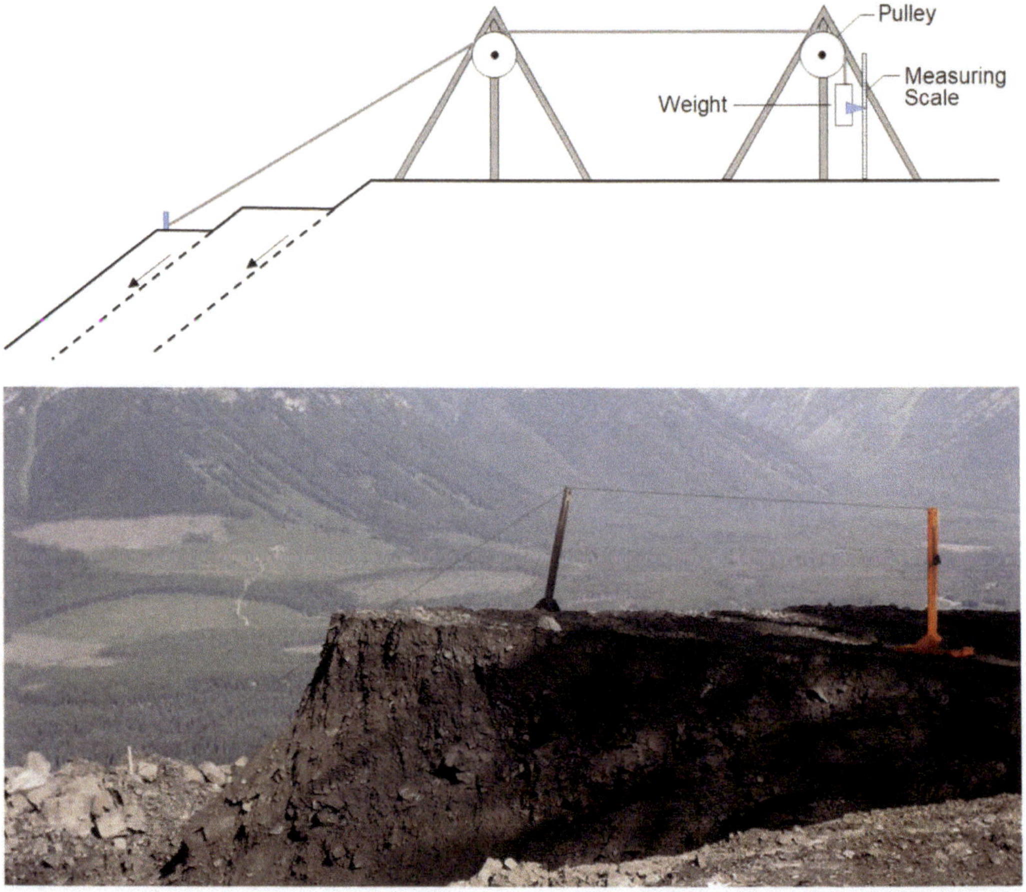

Fig. 3.4: Schematic diagram of a wireline extensometer (modified from Hogarth *et al.* 2017), and photo showing monitoring of a waste dump slope (source: E. Eberhardt).

3.2.4 Wireline extensometers

Wireline extensometers are commonly used to monitor movement behind slope crests where a high measurement resolution is not required (e.g. in monitoring waste dump movements where displacements of several metres might occur before failure). They can be constructed to provide either manual or automatic readings. A semi-rigid wire with high tensile strength is fixed in place at or near the crest of the moving slope and attached to a monitoring unit positioned on a tripod at the other end, as illustrated in Fig. 3.4. The monitoring unit is located on stable ground where little to no movement is expected. A constant tension is maintained on the wire using a counterweight such that the weight will move in response to changes in the distance between the anchor point and tripod pulley as the slope moves. A visual indicator or sensor in the housing converts the movement of the wire to displacement. The measured movement is a function of the angle of the wireline with respect to the true direction of slope movement. Advanced monitoring units can also provide temperature and rainfall measurements which, along with displacements, can be transmitted by radio to a remotely located receiver. These devices can be quickly positioned and easily moved, but care must be taken to minimise sag or thermal expansion/contraction in the wire, which can produce measurement errors. Rose and Hungr (2007) describe the use of wireline extensometer data at several mines in the USA to forecast the time of slope failure using an inverse velocity technique to interpret the data (see Section 7.5).

3.2.5 Precise levelling

Levelling is the operation required in the determination or comparison of elevations of fixed monitoring points on the ground surface (Bannister *et al.* 1998). Precise levelling, also known as the highest order of levelling, is the most accurate and is suitable for measuring vertical deformation or subsidence. It can be used to detect slope movements through conventional manual surveys or automatically with the use of digital (automatic) levels. An application of precise levelling, where a process plant is located near the crest of a large open pit, is shown in Fig. 3.5 and described in Chapter 9.

The following should be considered when applying the precise levelling method:

- a levelling network covering a defined area should be designed to traverse the area to be monitored for elevation changes, as shown in Fig. 3.5;

Fig. 3.5: A levelling network around the process plant area at Debswana Orapa mine adjacent to the open pit (source: N. Mphathiwa).

- the levelling network should have a benchmark as a starting point and transfer benchmarks (TBMs) along the network using permanent (fixed) structures to ensure consistency of the measuring points. In Fig. 3.5, BM1 is the original mine benchmark, while BM2, BM3, etc. represent the TBMs;
- the original mine benchmark should be established well outside the deformation zone. All elevation measurements will be referenced to the original benchmark. For convenience, the datum reference elevation for benchmarks should agree closely with that used for the general mine survey (Watt 1996);
- the TBMs should be connected by a series of levelling loops to improve the reliability of the precise levelling measurements. It is good practice to have the instrument setups positioned mid-way between TBMs and marked with wooden pegs to facilitate the recovery of future measurement cycles (Watt 1996); and
- to minimise the impact of temperature variations on the measurements, it is recommended that precise levelling be done during the time of day when the temperature variations are minimal. In many locations, midday is the preferred timing.

3.2.6 Laser distance measuring systems

Laser distance measuring (LDM) systems use a fixed, tripod-mounted laser to measure the distance between the laser and a remote target. The target is typically a circular, rectangular or triangular flat surface made of metal sheeting or plywood with a smooth, painted surface, and it is mounted on a stake or tripod located at or near a target of interest, such as the crest of an open pit slope or stockpile (Hawley and Cunning 2017). An LDM system can be configured to take repeated measurements at a specific frequency then transmit the data using telemetry for digital storage and processing. Line-of-sight (LOS) movement rates can be calculated based on the change in distance and time between measurements, to produce time-series plots. An LDM system produces results similar to that generated by wireline extensometers without the need for a wireline, which is an advantage for monitoring over large distances or in areas of active operations. While this system works well in cases where the movement direction is predominantly along the LOS of the laser, Hawley and Cunning (2017) report that it does not work well where there is a significant component of settlement. In these cases, depending on the size and position of the target, frequent manual retargeting of the laser may be required, which may limit practicality. An example of an LDM system is shown in Fig. 3.6.

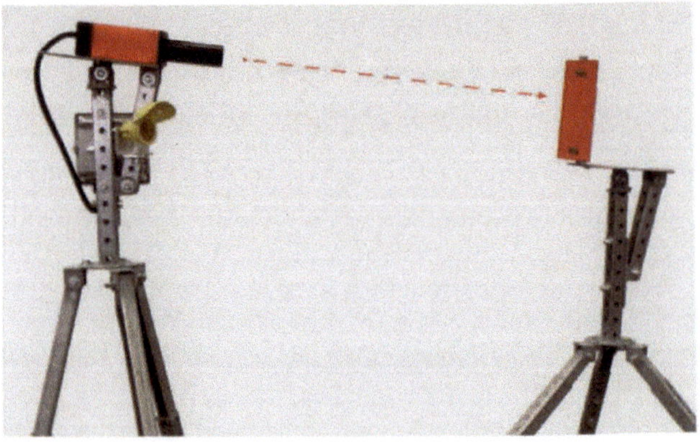

Fig. 3.6: Example of a laser distance meter system (source: RST Instruments Ltd).

3.2.7 Tiltmeters

Tiltmeters are devices used to monitor the change in inclination of a ground surface point (see Dunnicliff 1993 for a detailed description). Accordingly, they are used to monitor slope movements where the pit slope failure mode is expected to contain a rotational component. The device consists of a gravity-sensing transducer (servo-accelerometer, microelectromechanical systems (MEMS), electrolytic tilt sensor, pendulum-actuated vibrating wire, etc.) capable of measuring changes in inclination as small as 1–2 arc seconds (0.0003° to 0.0006°). This high level of resolution equates to ~0.01 mm per metre. Their range is typically up to a rotation of 15°.

Because of their high resolution and sensitivity, tiltmeters are often deployed as permanent installations to provide long-term observations. Tiltmeters may be uniaxial or biaxial, allowing measurement of tilt in two orthogonal directions across either a horizontal or vertical plane. Other advantages of tiltmeters include their light weight, simple operation and relatively low cost; additionally, tiltmeters can be read manually or automated by connecting to a datalogger. Tiltmeters may be combined with inclinometers and extensometers in an integrated pit slope monitoring system (see Section 4.2.6). An application for tiltmeters is described in Section 9.5.2.5.

3.3 Displacement monitoring: point measurements

3.3.1 Overview

Displacement monitoring provides a means for measuring the magnitude and rate of horizontal and vertical slope displacements. A well-established, low-cost, reliable means for making such measurements is through repeated geodetic surveys of a series of strategically placed monitoring points. Geodetic surveys encompass a range of spatial technologies and measurement principles that include using total stations, robotic total stations (RTS) and global navigation satellite systems (GNSS).

The primary focus herein is the application of geodetic surveys for displacement monitoring, including geodetic monitoring considerations for mine operating system design. A description of the fundamentals that govern all types of surveying or geodetic engineering is beyond the scope of these Guidelines. However, it is recognised that geodetic monitoring almost always forms the core of a performance monitoring network. In the context of slope stability monitoring, the fundamental components of a geodetic monitoring program should include:

- repeat measurements of the established monitoring points, ensuring design considerations and displacement monitoring requirements are satisfied;
- data processing relative to a choice of coordinate system and datums; and
- presentation of data in a manner that allows timely and accurate decision-making (Cawood and Stacey 2006).

Measurement accuracy and reliability are controlled by the characteristics of the reference datum, benchmarks and monitoring points. A geodetic datum is a coordinate system defined with respect to a reference surface (e.g. sea level) that serves to provide known locations to begin a survey. When designing a geodetic monitoring program, a robust survey reference network should be developed that includes checks to make sure reference benchmarks are located on stable ground. The design and layout of geodetic survey networks are described in Chapter 4, including details on how these systems are utilised for strategic and tactical monitoring. Instrument shelters and integrity checks are described in Chapter 5.

3.3.2 Total station surveys

3.3.2.1 Total stations

Total station survey instruments typically combine an electronic theodolite and electronic distance meter (EDM) into a single unit. They measure and digitally record the horizontal and vertical angles and LOS distance to a target or network of targets (e.g. prisms), from which their 3D positions are adjusted (i.e. corrected), computed and transformed to a reference coordinate system (e.g. local x, y, z coordinates or easting, northing and elevation). Through repeat surveys, for example, of prisms installed across the face of an open pit slope, changes in position of the prisms are used to calculate slope displacements and movement directions. A total station can be operated manually, with the surveyor physically moving the instrument into place to measure the locations of defined targets in the field, or automatically from a permanent location using software to drive the theodolite and carry out frequent repeat surveys of a network of prisms (e.g. robotic total stations, which are discussed in detail in Section 3.3.3). An example of a total station is shown in Fig. 3.7.

Fig. 3.7: Example of a total station housed in a protective shelter (source: Freeport McMoRan).

Typically, slope monitoring using a total station comprises three components (Afeni and Cawood 2013).

1. A network of reference beacons. These should be installed on stable ground that can be observed from the total station to ensure that the computed measurements remain in a consistent reference frame.
2. The total station (or transfer station) is also installed on stable ground at a location from which the slope surface, or target region, is visible. It is important to establish that the displacements being measured are those of the prisms and not the total station. In large operations where all slopes in the area of interest might not be visible from one instrument location, multiple total stations can be used together as a network of transfer stations. If the positions of the monitoring points are to be measured, then the transfer stations should be arranged so that they form a robust survey network that includes an optimal LOS.
3. A network of monitoring prisms installed systematically across each highwall is supplemented by added prisms to provide a higher concentration across zones at risk or areas of interest. It is preferable that the measurement direction is in the likely direction of movement so that the LOS distance readings approximate the actual slope movement.

Total stations have ranges of 1500–3500 m and long-range modes of up to 10 km. The accuracy depends on the target (e.g. reflective surface, prism) and the distance to the target. For most measurements using prisms, this is 1 mm plus 1.5 ppm; for long-range measurements (>3500 m), this is 5 mm plus 2 ppm. Total stations used for manual measurements offer features such as automatic instrument height detection using a laser under the instrument pointed vertically downwards to measure the distance to the ground surface to eliminate critical errors during instrument setup. Another source of error can be direct sunlight heating one side of the instrument relative to the other. For the highest accuracy, it is recommended that the total station be protected by an umbrella if the instrument is being set up and used manually, or by a protective shelter in the case of a permanent installation (Watt 1996). At least one stable point (but preferably three or more) is needed to orient the total station and to account for rotation due to uneven heating and cooling of the instrument or, in the case of a permanent installation, to account for the pillar it is mounted on (Fig. 3.7).

3.3.2.2 Prisms

Obtaining the maximum level of accuracy and reliability from a total station measurement depends not only on the specifications and accuracy of the total station, but also on the reflector characteristics. There are two main reflector properties that should be considered (Mao and Nindl 2009). The first is the reflector type and its general geometry, which is purely dependent on the model chosen. A 360° prism will reflect measurement signals from all directions, whereas a round prism with directional reflectors will provide higher precision but must be aligned with the instrument's LOS. The second main reflector property is the reflector constant, which is related to the prism housing. The reflector constant defines the referencing of the front surface of the prism (glass face) to the designated reflector centre, recognising that glass has a different refraction index from that of air (Mao and Nindl 2009). In principle, the reflector constant, if correctly considered, will have no influence on the measurements. However, this means that all prisms (or reflectors) must be identical, to maximise consistency by ensuring the uniformity of error propagation (Cawood and Stacey 2006).

3.3.3 Robotic total stations

An important extension of the use of total station surveys is the incorporation of RTS into the open pit mine monitoring network. The RTS is simply an automated total station that systematically performs more measurements in less time (or around the clock) than a conventional manual total station survey. The RTS incorporates automatic target recognition (ATR) to monitor a network of points by sighting prisms and tracking them as movement occurs. After the RTS has 'learned' the location of the prisms, it returns to re-determine the location of each prism during successive monitoring cycles, angular and distance measurements are made, and then the new prism locations are calculated (Eberhardt and Stead 2011). Advantages of RTS over a conventional total station survey include faster and more accurate quality checks, reduced errors, greater accuracy, increased numbers of inspection points (hundreds of targets can be included in the survey network), real-time measurement, and data accessibility. Under optimal conditions, an accuracy of 0.6 mm plus 1 ppm is achievable up to a range of 3500 m.

The improved accuracy of an RTS measurement reflects the ability to automatically lock onto a prism and continuously make real-time measurements. By using the same instrument and same procedures for each repeat survey, calibration errors and other constant systematic errors can be easily differenced out and removed from the measurements (Chrzanowski and Wilkens 2006). This leaves varying air density (due to changes in meteorological conditions) as the main factor affecting repeat RTS measurements. Rain, fog, dust, sun, heat shimmer, etc. can each contribute to atmospheric refraction.

Another consideration is that, although the RTS network is used to track the 3D movement over time of a discrete monitoring point on a slope, the three inputs to that movement (horizontal angle, vertical angle and LOS distance) have differing levels of accuracy, and this accuracy varies with distance. In trying to analyse the accuracy of their RTS monitoring system in an operating open pit mining environment, Klappstein *et al.* (2014) found that:

- horizontal movements had the greatest error with distance, likely due to errors in calibration of the horizontal angle;
- vertical movement had slightly less variance error with distance;
- the slope distance measurement was the most predictable (precise) of the three outputs; and
- the atmospheric conditions were the most significant contributor to measurement noise.

Other considerations to minimise measurement errors in an RTS network include building a shelter for the RTS instrument, equipment for measuring atmospheric conditions, fixed pillars to mount the RTS as well as any reference prisms (where the x, y and z coordinates are known), identical prisms, and a stable and reliable power source for the total station and other auxiliary equipment. Examples are presented in Chapter 5. The prism installation procedure must also be considered, including setting up different control points (reference prisms) and ensuring unobstructed LOS between the RTS, reference prisms and monitoring point prisms (e.g. bench face prisms). The latter requires proper drilling, grouting, maintenance to clean dust from the face of the prisms, and checking to ensure they remain properly coupled with the rock face they are meant to be monitoring (see the Jwaneng example in Chapter 9). The design of an RTS network is described in Chapter 4.

3.3.4 GPS and GNSS

The ability to make measurements describing the positions, velocities and accelerations of objects and natural features in three dimensions and in an absolute sense is of great value to the pit slope engineer. The global positioning system (GPS) and other global navigation satellite systems (GNSS) provide the tools to do this. A GPS receiver provides 3D positions by measuring the travel times of electromagnetic waves transmitted by orbiting satellites whose locations are known. At the heart of this process are oscillators: devices that work by converting a non-alternating current, that cannot be used to generate electromagnetic waves, into an alternating current that can. A transmitter is defined as a device (like a GPS satellite) that uses its oscillator to create an electromagnetic wave for propagation through space. A receiver is defined as a device (like a GPS receiver) that receives this wave. This section will briefly review the GPS system and explain the basics of satellite ranging.

3.3.4.1 *The GPS constellation*

The GPS satellite's purpose is to broadcast a navigation message and a navigation timing signal that enables the user receiver to estimate the distance between the two; this process is known in navigation as ranging. Each estimated range is derived from a direct measurement of the wave travel time. The GPS satellites are in a slightly elliptical Keplerian orbit, obey Newton's laws of gravity, and broadcast weak but sophisticated modulated signals to support ranging. The entire system exists within a Cartesian coordinate system, which is made up of three axes that are pair-wise perpendicular and pass through a common point called the origin. Such a system can be used to assign any point in space with coordinates, usually given in the form (X,Y,Z) where the origin is usually set to $(0,0,0)$. A Cartesian system in metres is used for GPS. Its origin is not set at a point on the surface; instead, it is set at the centre of mass of the Earth. This is known as an Earth-centred system. It can be used to identify any point in space with (X,Y,Z) coordinates. The distance between two points in 3D is represented by the Euclidean formula:

$$d = \sqrt{(X_2 - X_1)^2 + (Y_2 - Y_1)^2 + (Z_1 - Z_1)^2}$$

The Z axis of the system passes through true north; the X axis passes through the point where the prime meridian and the equator meet; and the Y axis is set at 90° to the X axis and also intersects the equator. This entire coordinate frame rotates with the Earth, making fixed points appear stationary. This is known as Earth-fixed; taken together, the entire system is termed Earth-centred, Earth-fixed (ECEF). While Cartesian coordinates are the native language of the GPS receiver, they are of little interest to users. The receiver transforms them into more familiar measures such as latitude, longitude and height. This is the conceptual framework of GPS. Satellites in an elliptical orbit with highly accurate oscillators use the electromagnetic field to transmit modulated broadcasts that are timed by the user on the ground to compute (X,Y,Z) coordinates in a Cartesian system whose origin $(0,0,0)$ is at the centre of the Earth. These are then transformed to reference systems that users can more easily access.

The following information about the GPS constellation, the GPS antenna and the GPS receiver is based upon Richard Langley's Innovation articles (Langley 1990, 1991a, 1991b, 1998). Each satellite in the constellation can be thought of as a 'navigation star', the origin of the contraction Navstar. Each is oriented so that its transmitting antenna is always pointing towards the Earth. A GPS receiver takes simultaneous measurements from its antenna to all satellites that it can see above its local horizon. These are then converted into ranges. With this information, the receiver is able to estimate its own position. The location of the satellites themselves are established by the US Air Force and its Operational Control System (OCS or OCX) and are provided to users in the navigation messages broadcast by the GPS satellites. This control segment is a global network of facilities that manages the Navstar system. It consists of four dedicated ground-based transmission stations with special antennas for communicating with the satellites and uploading data to them. These are located at Cape Canaveral, Florida; Ascension Island; Diego Garcia; and Kwajalein. There are also 16 monitoring stations that track the satellites and collect carrier data, modulated carrier data and atmospheric data. These are spread fairly evenly across the world. A Master Control Station (MCS) located at Schriever Air Force Base in Colorado collects these data and uses models to predict future orbits. These are called an ephemeris and are uploaded to each of the corresponding satellites using one of the four ground transmission facilities. The ephemeris data are then broadcast by the satellite to the user base. Each satellite also broadcasts the orbit information for all of the other satellites. This is known as the almanac. It is used by the GPS receiver to calculate where all of the satellites are in the sky at any given time. This accelerates the initial position solution by minimising the area of sky that has to be searched. This OCS is managed by the 50th Space Wing of the Air Force Space Command, located at Schriever Air Force Base in Colorado. This makes GPS instrumentation unique because it is underlain by a network of experts who constantly study system integrity.

The current Navstar constellation consists of 32 satellites (a mixture of Block II/IIA/IIR/IIR-M). The guaranteed minimum operational constellation is 24 satellites, but the typical operating number at the time of writing is around 30. These occupy six equally spaced medium Earth orbital planes (MEOs), each of which is inclined 55° to the equator. At least six satellites are visible at any place on Earth at any time; because of their high altitude, each can see ~1/3 of the Earth. Even though each satellite is moving at roughly 3000 m/s, they are high enough that they are visible for around four hours.

3.3.4.2 The GPS antenna

It is the task of the receiver antenna to receive the signals from the orbiting navigation satellites. This process begins with the oscillator that is onboard the satellite and is connected to its broadcast antenna. The oscillator's job is to convert direct current from the satellite battery to alternating current, which then moves the electrons in the metal antenna back and forth. This oscillating electric field produces an oscillating magnetic field, and so on. The result is a coupled electromagnetic wave that is self-propagating and continuously radiating outward. Each satellite carries four atomic oscillators that are used to generate these pure sine waves (also known as carrier waves, since they are used to carry data). The oscillators can also be thought of as frequency standards or clocks. Each oscillates at 10 230 000 cycles per second (i.e. 10.23 MHz). The carrier waves that they generate are multiples of this. One is formed by multiplying 10.23 MHz by 154, which is 1575.42 MHz. The other is formed by multiplying 10.23 MHz by 120, which is 1227.60 MHz. These two – 1575.42 MHz and 1227.60 MHz – are the broadcast frequencies. They are termed Link 1 and Link 2, respectively. Each is located within the L band of the electromagnetic spectrum, and each can be converted from frequency to wavelength using the equation

$$\lambda = \frac{c}{f}$$

where λ is wavelength, c is the speed of light and f is the frequency. Link 1, frequently abbreviated to L1, has a wavelength of 19.0293 cm and Link 2, or L2, has a wavelength of 24.1240 cm. L1 antennas are widespread and track only L1 carrier waves. L1/L2 antennas allow the user to track both carriers. Both L1 and L1/L2 antenna types are found in mining applications.

The signal spreads out as it leaves the satellite antenna, weakening in strength with distance. This can be illustrated by considering an imaginary transmitter in deep space emitting electromagnetic energy. In the absence of a focusing antenna, the energy will radiate equally in all directions. As the energy moves outward at the speed of light, the leading edge of its wave front will always be a sphere centred on its transmission point source. The original transmission power is continually distributed over an ever-larger spherical surface; its power density is diluted as the radius of the sphere increases. The received power of the GPS broadcast is therefore related to the surface area of a sphere, or $4\pi r^2$, where r is the radius. The most basic power transmit equation is

$$P_r = \frac{P_t}{4\pi r^2}$$

where P_r is the received signal strength, or intensity, at the receiver, and P_t is the original transmit power. To compensate for this inverse-square loss, the satellite focuses the transmit power which is around 25 W. Its antenna has what is called gain in the direction of the Earth. Antenna gain is the ratio of the full area of the sphere to just a portion of the sphere, known as the spot area. Including a gain term, the equation becomes

$$P_r = \frac{PtG_t}{4\pi r^2}$$

where P_r is the power received and G_t is the gain at the satellite antenna. The satellite gain value is ~10.5 dB. This increases the effective transmit power to around 300 W. This may sound like a lot but, at 20 200 km away, the GPS satellite is more than 36 times higher than the Hubble telescope. The received power on Earth is around only 0.00000000000000014 W or -158.5 dBW. The GPS signals are weak compared to other broadcast signals, so the GPS antenna (and its receiver) must be carefully designed.

The GPS satellites are not stationary with respect to the Earth, so the antenna must track them at a wide variety of elevations above the horizon and at almost all azimuths. The receiver has a gain pattern which is nearly omnidirectional for this purpose. The receiver antenna has a gain term that is usually around 3 dB isotropic, where isotropic refers to an imagined isotropic emitter. An L1/L2 antenna will often have two distinct metal antenna patches – one for tracking L1 and one for tracking L2 – to minimise out-of-band interference. The use of two patches within an antenna means that the L1 and L2 electrical positions are not perfectly coincident. These abstract locations are termed phase centres and are mean values that are determined in a laboratory setting. Since the signals arrive from an orbiting satellite, the phase centre of each antenna may vary slightly throughout the day (subcentimetre). It is good practice to always use the same antenna types and to orient them all in the same direction (magnetic north) to minimise this effect. Well-crafted antennas are manufactured such that the antenna's physical centre nearly coincides with its electrical phase centre.

The incoming electromagnetic waves from the satellite wash over the receiver antenna and, within it, induce replicas of the back-and-forth electric currents from the transmitter. These are then sent along to the receiver, but they are weak because of the distance travelled. The antenna usually has an amplifier to compensate for this, termed a

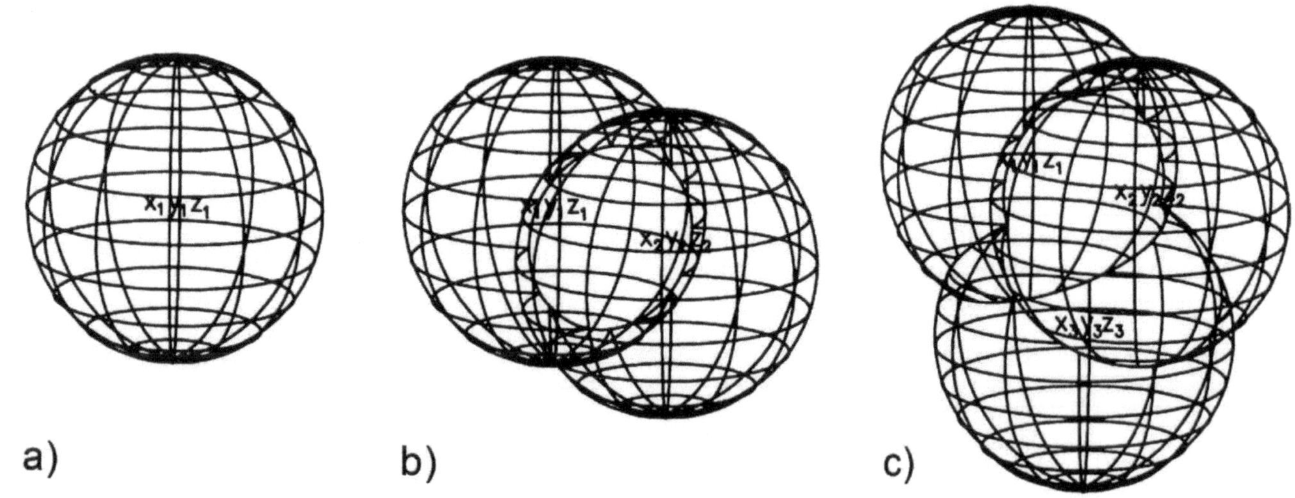

Fig. 3.8: Illustration of how a GPS receiver is located by utilising at least three GPS satellites (source: D. Rutledge).

low noise amplifier (LNA). When these electric currents are present, the antenna is known as an active antenna. The induced currents contain the various signals from the satellite, and it is from these that the receiver estimates the range to each satellite. These estimations are based upon a direct arrival to the antenna, but sometimes the path is indirect; that is, the waves reflect off local objects. These multiple path arrivals are known as multipath and contaminate the range estimates by additional distance (by increasing the travel time). Multipath can be mitigated, but it is preferable to avoid them in the first place. Most antennas use a metal plane extending out from the antenna patch that lowers the antenna's sensitivity (gain) at the lower angles that multipath often come from (e.g. ground reflections). Multipath from above the antenna are difficult to mitigate, so the antenna should always be mounted in such a way that there is nothing above the ground plane or immediately adjacent to it. The antenna converts the weak satellite broadcast into internal electrical currents, amplifies it then sends it along to the receiver for processing. It also protects itself from out-of-band noise, and it attempts to reject multipath. The signal it captures is sent to the receiver through a coaxial line. There would be no GPS positioning without the antenna's ability to do all of this.

3.3.4.3 The GPS receiver

A comprehensive review of how a GPS receiver acquires the signal from the orbiting satellites then processes these data into a solution crosses many disciplines and is outside the scope of these Guidelines. Only the essential concepts are presented here. The key measurement taken by the receiver is the travel time from the GPS satellite. Signals sent from the satellite broadcast antennas travel at the speed of light, c (i.e. 299 792 458 m/s). Travel time can be converted into a range, or distance, by simply multiplying it by this value. It takes roughly 67 ms to traverse the ~20 200 km to Earth. Imagine that this transmission travels at exactly c all the way from the satellite to the receiver (this is not true because of the slight delay when it passes through the charged ionosphere and the troposphere). The signal has to be timed on its transit from space to Earth, but doing so with the accuracy required is not straightforward because GPS transmission has to deal with two oscillators (clocks) rather than one. That they are separated by space means that they are not coordinated. The GPS receiver is passive by design; it is not allowed to communicate directly with the satellite to synchronise its clock. This is a military requirement that ensures that soldiers do not announce their presence to an enemy force and it also shields the satellites from having to manage a fleet of users. Such ideal conditions (synchronised oscillators and no speed of light delays) make it relatively easy in theory to measure the travel time from the satellite in space and thus compute the range between the two.

A geometry illustration helps to demonstrate what is then done with the range. Each satellite position (x_1,y_1,z_1) is known, but the receiver position (X,Y,Z) is not. Location information for each satellite is calculated by military ground stations, uploaded into space, then broadcast back to the receiver. This ephemeris information is used by the receiver to calculate where each satellite will be at any given moment in time. The receiver decodes the ephemeris from the navigation message. It consists of six orbit determination parameters used to position the satellite to within ~1 m. Once the receiver calculates the first range, imagine that it creates a sphere around that satellite's x_1,y_1,z_1 coordinates with a radius equal to the range r_1. This provides the very first position information. The GPS receiver must lie somewhere on the surface of this sphere, as shown in Fig. 3.8a. Computing the range r_2 to a second satellite will generate another sphere, and these two spheres will intersect to form a circle, as shown in Fig 3.8b.

The receiver must lie on this circle, known as a line of position. Measuring yet another range, r_3, creates a third sphere that intersects the others at just two points, as shown in Fig. 3.8c. Only one of these points will make sense; the other is dismissed.

This geometry-based visualisation shows that at least three satellite's ranges must be determined. This is done inside the GPS receiver with algebra rather than geometry. The simplified example from above is represented with these three Euclidean distance equations:

$$r_1 = \sqrt{(X-x_1)^2+(Y-y_1)^2+(Z-z_1)^2}$$

$$r_2 = \sqrt{(X-x_2)^2+(Y-y_2)^2+(Z-z_2)^2}$$

$$r_3 = \sqrt{(X-x_3)^2+(Y-y_3)^2+(Z-z_3)^2}$$

Here r_1, r_2 and r_3 are the three true ranges, x_1,y_1,z_1; x_2,y_2,z_2; x_3,y_3,z_3, which are the coordinates for the three satellites; and X,Y,Z are the unknown receiver coordinates. Since the satellites have broadcast their position information to the receiver, and the receiver has made an estimate of the ranges, all that is left is to calculate the receiver coordinates. The geometry exercise in Fig. 3.8 demonstrates that three ranges are sufficient to narrow this down to a single point. We have three equations and three unknowns, and this is enough information to solve for each component of position X,Y,Z.

However, the timing offset between the satellite oscillator and the receiver oscillator must be accounted for. It exists because they drift apart relative to one another and because the receiver is not allowed to periodically communicate with the satellites to synchronise. When it is reintroduced, the measurements become contaminated by it, and the ranges to each satellite will be false by some amount (they can no longer be described as true ranges). This forces a change in how they are referred to. They become known as pseudoranges. This turns r_1 into p_1. Since the satellites have highly accurate atomic clocks, this timing error is identified with the receiver and is known as the receiver clock error, dT. It is common to every satellite range measurement, which means that it can be added to the range equations as a distance by multiplying it by c, like so:

$$p_1 = \sqrt{(X-x_1)^2+(Y-y_1)^2+(Z-z_1)^2} - cdT$$

$$p_2 = \sqrt{(X-x_2)^2+(Y-y_2)^2+(Z-z_2)^2} - cdT$$

$$p_3 = \sqrt{(X-x_3)^2+(Y-y_3)^2+(Z-z_3)^2} - cdT$$

Note that there are now four unknowns (X, Y, Z and dT), but only three equations. A fourth equation must be added to solve for the receiver clock offset, which means that a fourth satellite range must be taken, as shown in Fig. 3.9. Including it gives four equations and four unknowns. These pseudorange equations can now be used to solve for the clock offset:

$$p_1 = \sqrt{(X-x_1)^2+(Y-y_1)^2+(Z-z_1)^2} - cdT$$

$$p_2 = \sqrt{(X-x_2)^2+(Y-y_2)^2+(Z-z_2)^2} - cdT$$

$$p_3 = \sqrt{(X-x_3)^2+(Y-y_3)^2+(Z-z_3)^2} - cdT$$

$$p_4 = \sqrt{(X-x_4)^2+(Y-y_4)^2+(Z-z_4)^2} - cdT$$

This algebraic discovery can be related back to the geometry illustration; the three spheres will not intersect at a common point unless the clock receiver offset is solved for. This is because the clock offset will make the ranges either too long or too short. This state needs to be solved for, which adds a fourth range to the algebra tableau. Once dT is known, the pseudorange can be corrected and the position of the receiver estimated. Another source of error that must be accounted for is called a satellite clock error. The receiver oscillator and the transmitter oscillator are not the only clocks that are out of synchronisation. Each satellite carries four atomic clocks that are referenced to what is called GPS time. It would be nice if these clocks were synchronised to GPS time, and to each other, to a couple of nanoseconds (equal to ~60 cm of range error), but they are not. They are biased by drift by as much as 1 ms, which equals almost 300 km of range error. Military ground stations determine these offsets and the drift rates, and upload them. The satellites in turn broadcast them as part of the navigation message where it is used by the GPS receiver to correct for the satellite clock errors. This correction, plus the correction for the receiver dT error, makes a solution possible.

It has been assumed up to this point that the ranges are just there. They are not: the receiver must determine them. A timing comparison needs to be made between the separated oscillators. This is accomplished by using the very signals that the orbiting satellites broadcast. Each transmits not one but two (soon three) carrier frequencies that are similar in concept to the carrier frequency broadcast by a ground-based AM or FM transmitter. That two are used instead of just one is part of the legacy of Sputnik and Explorer 1, and the early studies of the ionosphere that these launches stimulated. The ionosphere is a shell of charged particles that extends from 50 km above the Earth to over 1000 km. It is opaque to certain frequencies (less than 30 MHz), a trait that amateur radio enthusiasts use to skip broadcasts off its lower surface. The higher frequencies used by GPS are capable of passing through it, but not without being affected. The ionosphere delays the signal and can be the largest source of error in a GPS solution. With two frequencies, the delay can be directly measured and removed by combining them at the receiver (single-frequency receivers use a model broadcast in the navigation message). The carriers have been chosen

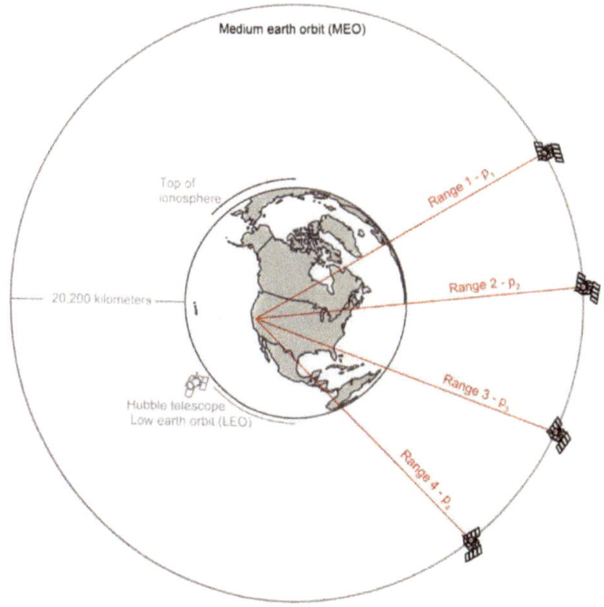

Fig. 3.9: Geometric illustration of satellite ranging (source: D. Rutledge).

so that they can effectively propagate to Earth, so that they can be used to eliminate the ionospheric error, and so that they can still carry the message bandwidth. Both frequencies lie in the L band of the electromagnetic spectrum, a region that offers the right compromise for the above competing interests. These are the L1 at 1575.42 MHz and L2 at 1227.60 MHz. They are great at propagating to the Earth, but they are not ideal for computing the range between the satellite and the receiver. They are so pure and featureless that neither provides a good timer for the travel time determination. They appear as an endless stream of identical sine waves, making it difficult to determine the total number of wavelengths between the satellite and the receiver. Their exact number is said to be ambiguous. Additional information is modulated onto them to create a 'stopwatch'. This process is conceptually similar to voice modulation in AM radio, but much less organic. Where a radio-transmitter maps the chaotic frequencies of human voice onto a single carrier using a telephone transducer and a simple mixer, the GPS satellite uses microprocessors to map complex computer binary information onto its two (soon three) carriers.

Each satellite modulates a navigation code and a navigation message onto the carriers. The navigation L1 code is known as the coarse/acquisition code (frequently abbreviated to C/A) and is a sequence of 1023 digits that repeat every millisecond. This fast repeat time allows the receiver to quickly acquire it and then to track it. Each digit is known as a chip and is ~1 microsecond in time, or 300 m long. These ride along the carrier wave down to the GPS receiver. The receiver can resolve the arrival of each chip to ~1% of its width (~3 m). The coarse/acquisition code is not carrying actual data. It is a timing signal and, while it is made to appear random, it is instead carefully generated by the satellite and is made to repeat after a certain time period. Signals like this are termed pseudorandom. The receiver contains the exact same pseudorandom code generation information as the satellite and generates the exact same signal. This internal replica is synchronised to the drifting receiver clock, while the incoming signal is synchronised to the more precise satellite atomic clock and, by extension through the ground control stations, to GPS time. A special property of the pseudorandom signal is that it is made to autocorrelate. This means that when the receiver slides its replica code back and forth in time, a strong peak will appear when all of the +1s and -1s from the two signals line up. The receiver is said to be tracking the signal once this lock occurs. The amount of time that the receiver must slide its code to achieve correlation is the travel time. Each GPS satellite broadcasts a unique pseudorandom code so that the receiver always knows which signal it has acquired. The pseudorandom modulation is so important that each satellite is named by it. Satellite 21 is pseudorandom number 21, typically shortened to PRN 21. The pseudorandom codes have one additional important property: low cross-correlation. This ensures that pseudorandom codes from different satellites do not autocorrelate on top of one another in the receiver.

The navigation message is also modulated onto L1. It contains the code timing information (timestamp) used by the receiver during autocorrelation. It is a slow data stream of 50 bits per second that also contains the six Keplerian parameters (these define the orbital ellipse, orient it to the Earth and locate the satellite in its orbit), the GPS satellite clock offsets, the satellite health information, and almanac data on the entire constellation. The navigation message is ordered into 300-bit subframes and takes 12.5 min to download in its entirety. The GPS receiver simultaneously receives both the coarse/acquisition code timing signal and the navigation message from a given satellite, and can also receive this pair simultaneously from all satellites in the entire constellation. The information is used to compute the pseudoranges for all satellites in view, then algebra is used to estimate the receiver's location. The pseudorange equation system removes the clock error and synchronises the GPS oscillator to GPS time, but there are still some errors left over. The satellite orbit is only known in real-time to ~1 m. The atomic clock has ~1 ns of error which equates to ~0.3 m of range. Reflected arrivals of the signal are not represented in the Euclidean straight-line distance equation and can contribute many metres of error. Ionospheric modelling removes most of the delay error, but there is still ~1 m left, even when this is done well. The delay in the signal caused by the troposphere also contributes error, but typically less than that caused by the ionosphere. Finally, there is natural measurement error plus any unwanted radio interference. These error sources also need to be represented in the range equation:

$$p_1 = \sqrt{(X - x_1)^2 + (Y - y_1)^2 + (Z - z_1)^2} - cdT + v$$

A pseudorange equation is reintroduced here with the term v, collecting all of these remaining error sources together. The ionosphere and multipath can vary widely and should be thought of as rogue errors. The notation above is simplified, but still provides the mining engineer with a good idea of the range equation concept. The GPS receiver uses a version of this algebra (in reality more complicated) to solve for the four states (*X*, *Y*, *Z* and *dT*) which then yields a position location to ~5 m or so. Figure 3.10 shows a typical scatter plot of the two horizontal components.

The Block IIR-M satellites, first launched in 2005, added a new civilian code (L2C) on Link 2 to support high-end civilian receivers. Receivers that can track both the L1 and L2 carriers are known as dual-frequency receivers and are capable of measuring the ionospheric delay in real-time. This provides an advantage over single-frequency receivers which rely on the broadcast ionospheric model (which reduces the ionospheric error by only ~50%). There is also an encrypted military P(Y) code that is not discussed here. The L2C navigation code and navigation message pair also provides a redundancy to the user. A third carrier, located at 1176.45 MHz, is known as Link 5 (L5). This signal was added to the Block IIF satellites that were first launched in 2010. It lies in a special area of the L band called the Aeronautical Radionavigation Services band, a region of the spectrum that is protected for safety-of-life applications such as aircraft landings. As more satellites are launched with L5, the modern GPS constellation can increasingly be viewed as a triple-frequency system.

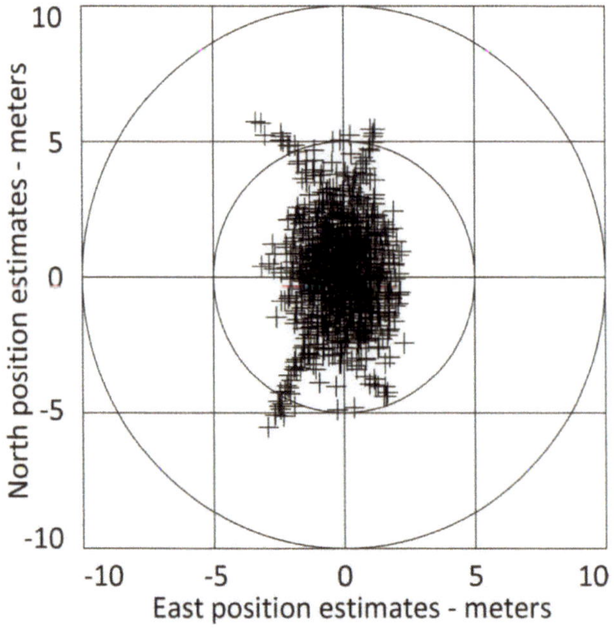

Fig. 3.10: Scatter plot of independent GPS position estimates (source: D. Rutledge).

3.3.4.4 Dependencies

The GPS system was launched a mere 21 years after Sputnik. Around 3 billion civilian users rely on the L1 coarse/acquisition signal to navigate from place to place at the 5 m level of accuracy. Such performance exceeds the original design specifications, a point of pride within the US military. Numerous other countries have launched their own systems, but the coarse/acquisition code remains the highest ranking and most trusted navigation signal. While the system was conceived from the start as both a military and a civilian system, many aspects of its design were guided primarily by military requirements. Military users use GPS almost exclusively in what can be thought of as an independent mode. This means that they use only the signals coming from the satellites. Since they directly control these through their Air Force control stations, they are deemed secure. Several dependent modes of operation are favoured by the civilian community. Accepting dependencies can greatly increase system performance, and this approach is what is most commonly used in mining environments to achieve centimetre and subcentimetre accuracies. A widely used type of dependent positioning involves the use of a reference GPS receiver located (intentionally) at known coordinates. The user is then dependent upon this station for each monitoring station position estimate.

Those interested in additional information about GPS and GNSS systems are referred to *Global Positioning System: Signals, Measurements, and Performance* by Misra and Enge (2012). Likewise, the electrical engineering aspect of a GPS receiver involves a world of mixers, correlators, intermediate frequencies, down converters and multiple other electrical engineering and signal processing topics. Those interested in this aspect are referred to *Engineering Satellite-Based Navigation and Timing: Global Navigation Satellite Systems, Signals, and Receivers* by Betz (2016).

3.4 Displacement monitoring: full-spatial measurements

3.4.1 Overview

An important limitation to geodetic monitoring is the reliance placed on point measurements that require movement and displacements between prisms, or GNSS receivers, to be interpolated. This may result in:

- the boundaries of areas with high displacement rates being poorly defined;
- smaller scale, structurally controlled movements such as wedge or planar sliding being overlooked; and/or
- the kinematics behind larger and more complex pit-scale failures being misinterpreted (Severin *et al.* 2014).

Furthermore, safety concerns also arise with the frequent need to replace damaged prisms or deploy new prisms on a moving slope. The need to manually install prisms, and therefore expose workers to hazards such as rockfall, often prohibits maintenance, resulting in the loss of spatial coverage and continuity of critical data for early warning. To address these issues, remote sensing techniques such as GB radar, laser scanners (LiDAR) and digital photogrammetry are frequently added to conventional geodetic prism monitoring programs. The advancement of GB radar since the early 2000s represents an important step-change in how open pit slopes are monitored and managed.

The main advantages of these techniques over conventional geodetic point monitoring is that they allow:

- high-resolution, full-spatial measurement of a highwall over a broad area;
- near real-time slope deformation monitoring; and
- remote measurements without the need to install prism reflectors, reducing workforce exposure to rockfall hazards.

Further, slope deformation alarms can be set when using GB radar or LiDAR within a large scan area to alert mine operators to sudden changes in pit wall behaviour. These alarms can be incorporated into the mine operation's Trigger Action Response Plan (TARP), formalising pre-determined responses to likely events based on alert levels prompted by exceeding certain thresholds (see Chapter 8).

3.4.2 Ground-based radar

GB radar is widely utilised by the open pit mining industry, and its implementation with regard to slope performance monitoring is widely regarded as leading industry practice. The utilisation of radar technology as part of an integrated slope monitoring program is well established and provides unique strategic and tactical capabilities:

- GB radar can provide high-resolution and full-area coverage of a slope with submillimetre precision at distances exceeding 1 km in combination with near real-time acquisition, with repeat measurements taking a few minutes for an entire highwall;
- radar slope stability monitoring systems require no preparation of the slope or access to the slope, since no ground reference points are needed;
- radar works equally well during the day and night, and is largely unaffected by dust;
- radar slope stability monitoring systems have been deployed in regions experiencing ambient temperatures of -30°C to more than 60°C;
- alarms of incipient slope failure can be generated automatically, and operators can control the systems remotely; and
- radar equipment and software facilitate the maintenance and management of long-term databases.

3.4.2.1 How radar works

The principal components of a radar system for slope stability monitoring are shown in Fig. 3.11. The system works by generating a sinusoidal waveform with a precise frequency that is transmitted through an antenna as a narrow beam of electromagnetic waves. This is directed at the pit slope being monitored, where a small fraction of the energy is reflected back to the radar (also referred to as

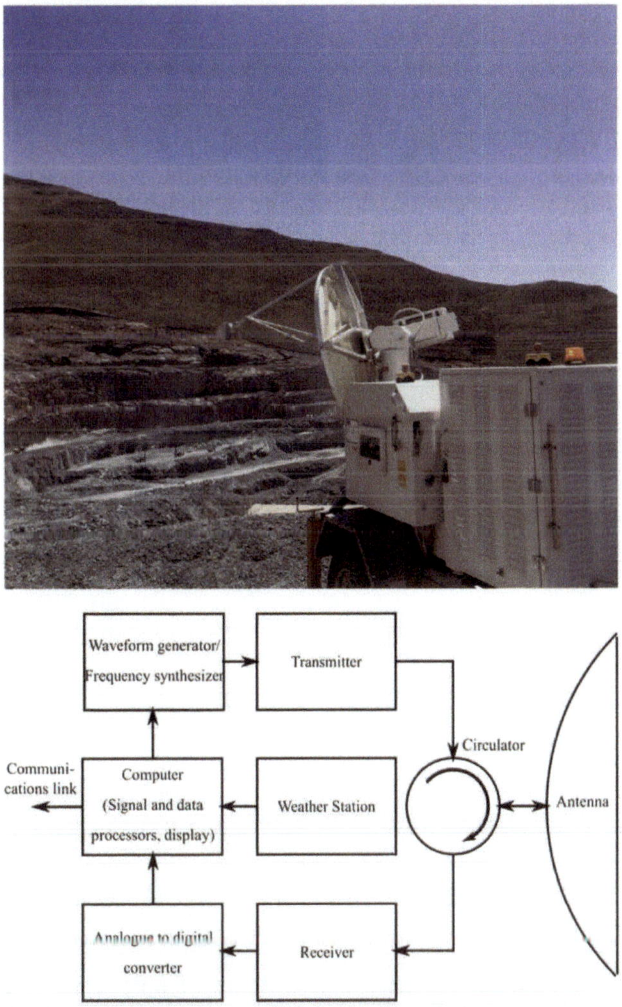

Fig. 3.11: Photo of a GB radar system, together with its basic components, including a weather station (source: Reutech).

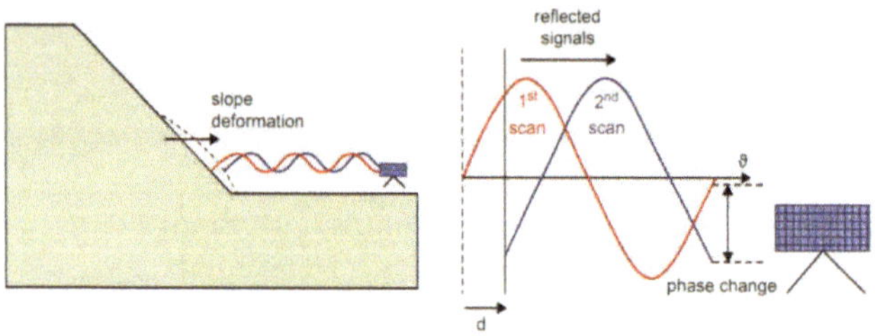

Fig. 3.12: Measurement of displacement using radar interferometry calculated by comparing changes in phase of the reflected signals between two consecutive radar measurements (source: E. Eberhardt).

backscattering). The backscattered wave, or echo, is intercepted by the antenna and converted to a signal that is then processed and analysed using interferometry to determine the differences in the phase values of the backscattered waves between consecutive measurements. In simple terms, if a reflective target is stable, then the phase registered by the radar on the reflected signal will be the same between two consecutive measurements. If a phase change is registered, then a movement of the target has occurred between the two acquisitions, the magnitude of which can be calculated based on the phase shift, as illustrated in Fig. 3.12.

Figure 3.13 provides a detailed example of how radar measurements work. Consider an experiment where a wave with a wavelength (λ) radiated by an antenna is reflected by a slope face some distance away, as illustrated in Fig. 3.13. The phase of the reflected sinusoidal wave is compared with that of the transmitted wave at the antenna. In the left example, this phase difference is 180°. In the right example, the reflecting slope face moves away from the antenna by a distance (*d*) that is a small fraction of a wavelength. This increases the path length from the antenna to the reflector and back by a distance 2*d*, resulting in a change of relative phase of the reflected wave by 360° × (2*d* / λ), or ~60° in the lower example in Fig. 3.13. If the reflecting slope face moved closer to the antenna by the same distance, the change in phase would have been +60°. Since phase angles can be measured accurately, this interferometry principle allows a radar to measure submillimetre changes in distance to a reflecting object hundreds of metres away from the radar.

The collated radar information is used to generate a display, such as a displacement map, an example of which is shown in Fig. 3.14. The information is communicated to auxiliary equipment (such as alarm beacons) and the control centre via telemetry, where it can be immediately reviewed. Data may also be viewed at the unit itself. By comparing the signals received from each radar pass, movement in the LOS is determined. Because radar is an LOS measurement, the actual direction of slope movement (e.g. provided by geodetic monitoring) should be used to align the radar unit to minimise the underestimation of real, or total, displacement. The application of GB radar for open pit slope monitoring is provided in Section 4.4.

3.4.2.2 *Measurement limitations and corrections*

The measurement capabilities of radar systems are ultimately limited by ever-present random disturbances called noise. The measurement precision of the radar depends on the strength of the echo relative to the noise floor of the radar. This is called the signal to noise ratio (SNR) of the measurement and is determined by the energy of the returned echo. In modern instruments, the phase shift introduced by the intrinsic noise is negligible and usually below the instrument's accuracy.

Another limitation is that it is not possible for the radar measurement to determine the number of wavelengths making up the total path length from the radar to the target and back. The radar system simply measures (very accurately) the phase change between consecutive measurements, but it has no means of detecting if the

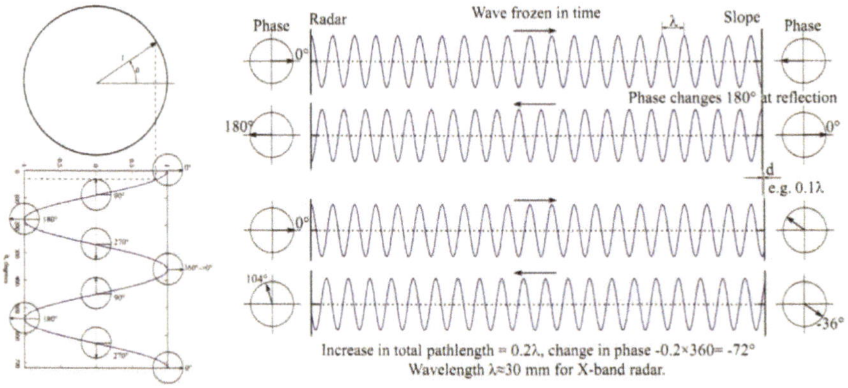

Fig. 3.13: Transmission of a sinusoidal wave towards a reflecting object and measurement of the phase of the reflected sinusoid signal. In the left illustration, this phase difference is 180° relative to the transmitted signal. When the reflecting object (e.g. a slope face) moves a small distance (*d*) away from (or towards) the radar antenna, the reflected signal received at the antenna shows a change in phase that can be used to calculate differential movement in the LOS (source: Reutech).

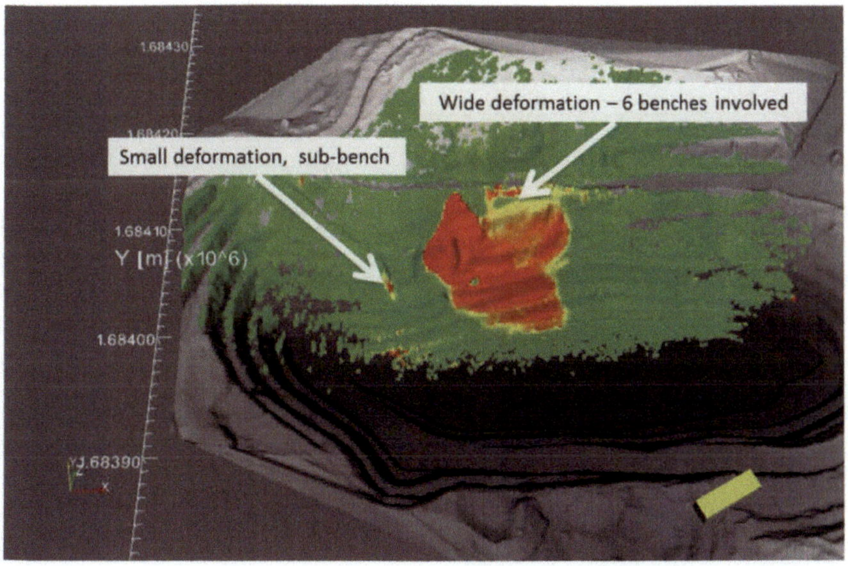

Fig. 3.14: Cumulative displacement map covering three days of radar measurements. Negative values indicate slope movement approaching the radar (source: IDS).

change is less than or greater than one wavelength. This can lead to ambiguous measurements if the movement of the targeted pit wall is too large between any two consecutive measurements (see also Section 3.5.3.5 and Section 4.5.6).

With respect to corrections, it is important to note that, even though GB radars can work with any meteorological conditions, night and day, the propagation of the electromagnetic waves through the atmosphere is affected by variations of temperature, humidity and pressure. Changes in atmospheric refractivity present slope monitoring radars with their most difficult challenge. This disturbance causes a shift in the phase of the backscattered signal and can be significant, requiring a correction of the data. A weather station, as diagrammatically shown in Fig. 3.11, can help feed atmospheric properties such as air temperature, humidity and pressure to the computer. Standard correction algorithms assume stability in the atmospheric conditions over the monitored area of interest, which are assumed as reference points for the calculation of the atmospheric correction (see also Section 4.3.5).

A linear estimation of the atmospheric contribution to measurement error is assumed as true only at close to medium ranges and, most of the time, is not effective in areas with strong atmospheric changes. More advanced algorithms are based on the Permanent Scattering technique, originally developed for satellite radar (InSAR) applications. Data processing for this technique is based on the evaluation of the atmospheric contribution by performing an automatic selection of an array of hundreds of stable pixels by exploiting the different spectral behaviour in time and space. This approach makes it possible to extend the operating range of the radar to up to 5 km, while maintaining a high level of precision and reliability in the final results.

However, the Permanent Scattering technique can inadvertently suppress real slope movement, if the automatic selection of pixels by the algorithm misinterprets real slope movement for atmospheric changes. Based on experience, up to 60% of the real wall movements can be suppressed by the automatic algorithm. The algorithm also requires enough time to determine the spectral behaviour in space and time to automatically select pixels for correction. Therefore, with this limitation in mind, the Permanent Scattering technique may be considered more suitable for long-term surveillance monitoring and less suitable for safety-critical monitoring, especially for the detection of rapid brittle slope failure.

3.4.2.3 Real and synthetic apertures

There are two main types of radars used for slope monitoring: real aperture radar (RAR) and synthetic aperture radar (SAR). These are depicted in Fig. 3.15. Both are interferometric radars (i.e. they exploit the phase difference of the backscattered signal to measure the movement of the target between two consecutive acquisitions) but differ in the way they record the radar image of the monitored scenario.

In the case of RAR, a directive antenna is rotated to scan all directions. This uses a parabolic antenna that produces a narrow wave beam that is emitted towards the reflective surface, illuminating the wall with a circular footprint that is inversely proportional to the size of the antenna (i.e. the larger the radius of the antenna, the smaller the detectable area of movement on the wall). With a wider dish, RAR achieves a better resolution. By

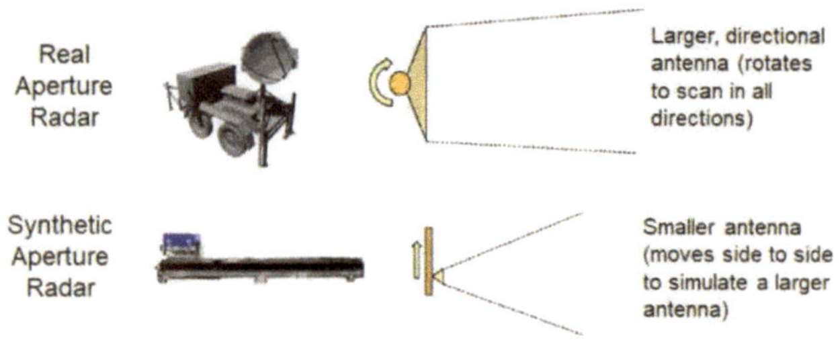

Fig. 3.15: Working principles and antenna movement for real aperture radar (RAR) *v.* synthetic aperture radar (SAR) (source: E. Eberhardt).

exploiting the physical movement of the dish antenna, RAR covers the monitored scenario by means of consecutive footprints until the entire area of interest is covered. The scan time for RAR is therefore proportional to the covered area.

In the case of SAR, a small low-gain antenna is moved along a guide or linear rail to 'simulate' a larger real aperture antenna; it takes several radar measurements at constant intervals. The individual measurements are then processed in a signal processor to produce a beam as if measured with a real aperture antenna. This allows SAR systems to use an antenna a few centimetres in size moving along, say, a 2 m linear rail to achieve the same spatial resolution of RAR with a 4 m wide dish. This results in a wide beam that can encompass a large area of interest, such as an entire highwall sector. The scan time of a SAR is not proportional to the covered area, such as for RAR, and usually does not vary appreciably with the extension of the monitored area.

Both systems have unique capabilities and limitations. RAR has superior angular discrimination performance, as the beam discrimination acts on both the transmitted and received signals. With SAR, the discrimination is only on the received signal. RAR also has excellent antenna side-lobe performance and high SNR measurements, as well as 360° azimuth coverage and ± 60° elevation coverage. In the case of a linear SAR, the slope coverage is limited to an included angle of 90° or less in azimuth and elevation. However, the azimuth coverage of a SAR can be efficiently increased up to 360° by modifying the SAR acquisition from a linear path to a circular path, but with no change in elevation coverage.

A shorter scan time is beneficial to increase the maximum detectable velocity and to reduce the noise introduced by atmospheric changes. Typical scan times for modern commercial SAR systems are on the order of 2 min for a full resolution image at 2.5 km of operating distance. RAR systems with a 3D scanning antenna can take 15–30 min to scan the area of interest line by line. However, the scan time of RAR can be reduced by using a 2D scanning antenna, which can scan a full 360° in azimuth in the order of 2 min.

3.4.2.4 Resolution and range of detectable velocities

The resolution of radar varies with the distance between the radar unit and the target; in general, it corresponds to the radar footprint for RAR dish antennas and to the radar pixel for SAR. The footprint of the RAR (perpendicular to the radar beam) and the radar pixel of the SAR (parallel to the radar beam) should be projected on the real reflective surface to obtain the real resolution in the field.

The displacement range that a radar system can measure between two consecutive scans depends on its working frequency. As previously noted, radar cannot determine the number of wavelengths that make up the path from the antenna to the reflector (target) and therefore can measure only very small changes within its range; it cannot measure absolute range. RARs working in the X band frequency (10 GHz) can measure up to 9 mm of unambiguous movement between two scans before becoming ambiguous. SARs working in the Ku-band frequency (17 GHz) can measure up to 5 mm of unambiguous movement between two scans before becoming ambiguous. To calculate the maximum detectable slope velocity before ambiguities become a problem, it is necessary to relate the unambiguous movement values to the sampling frequency (i.e. minimum scan time) of the radar system. Radars with a shorter scan time, or larger unambiguous measurements between two scans, can track higher deformation rates due to the higher sampling frequency. It should also be noted that the minimum detectable velocity depends mainly on the specific processing chain and on the effectiveness of its algorithm for dealing with atmospheric-induced disturbances.

3.4.3 Ground-based laser scanners: LiDAR

GB or terrestrial laser scanners (TLS) were first used in mining to obtain accurate 3D surface models for survey and design conformance purposes. Advancements in software and hardware have since enabled geological and geotechnical slope characterisation and, more recently, near real-time displacement and velocity measurements. The usefulness of TLS instruments for these purposes largely depends on the type of instrument and software used. Automated TLS monitoring tools, such as the one shown in Fig. 3.16, measure movement along the LOS between the instrument and the target. It is thus necessary to carefully position the instrument such that the maximum portion of slope movement can be measured.

There are two main types of TLS instruments: phase-shift (PS) and time-of-flight (TOF). There are important differences between scanner types, and the degree to which they can be used to improve mine site geotechnical

Fig. 3.16: Example of a terrestrial laser scanner (TLS) (source: E. Eberhardt).

monitoring programs depends on three main factors: scale of interest, usability and software.

3.4.3.1 Accuracy and scale of interest

TLS instruments have a wide range of accuracies. These depend primarily on the design of the instrument.

- *Phase-shift* (PS) scanners utilise changes between transmitted and received portions of a continuous laser signal. Because multiple frequencies can be used in the signal and the wavelength is very small, very fine measurements can be made with PS systems. PS scanners usually have accuracies of 2–6 mm at 100 m range. They cannot be used at longer ranges because the laser power required would be enormous and because phase ambiguities become more significant as the laser passes through more atmosphere. Most PS scanners have ranges in the order of 50–350 m. This limits their application for pit slope monitoring.
- *Time-of-flight* (TOF) scanners use a single laser pulse (i.e. one wave cycle) to measure the time required for the pulse to travel to the target and return to the detector. This method requires less power because the laser is not continuously running, enabling the use of more powerful lasers. As a result, TOF scanners have significantly longer ranges (500–2000 m) and therefore are more appropriate for pit slope monitoring applications. In terms of accuracy, most TOF scanners have accuracies of 4–15 mm at 100 m range. At the range required to monitor most open pit mine slopes (~500–1500 m) TOF scanner accuracies are usually between 20 mm and 80 mm.

For both PS and TOF scanners, accuracy decreases with increasing range between the scanner and target, as the influence of atmospheric conditions compounds and the laser signal is refracted and absorbed by objects. Scanners with higher-powered lasers and more sophisticated sources/detectors will have longer ranges. The reflected signal is also easier to detect if it has higher intensity, which depends in part on target reflectivity. White or 'shiny' (high albedo) targets will invariably be detected at longer ranges than dull-coloured targets.

Laser scan spatial resolution is a function of the instrument's angular resolution (i.e. the angular 'step' between measurements) and range to target. The angular resolution of most laser scanners is between 0.002° and 0.025°, with the finer spatial resolutions usually belonging to PS scanners. The spacing between laser scan points increases linearly with range. Thus, if spatial resolution is the size of the 'square' formed by four adjacent scan points, its area will increase as the square of the range. For example, if at 100 m range a scan's point spacing is 10 cm, then at 1000 m range the point spacing will be 100 cm. For the same ranges, the spatial resolutions will be 100 cm^2 and 10 000 cm^2, respectively.

The maximum and minimum spatial resolutions of a laser scanner are largely a function of scan speed and the applications for which it is designed. Most PS scanners are faster than TOF scanners and likewise can use a smaller angular step to achieve finer spatial resolution without having excessive scan times. This is useful for short-range, detailed measurement and analysis applications. TOF scanners, on the other hand, were designed for longer-range, less detailed applications and, in part because they are slower, can afford to have a larger angular step and slightly coarser spatial resolution.

Beyond range and accuracy, for a given area of interest, feature scale and the required scale of displacement and velocity measurements are important considerations. For example, where submillimetre joint roughness values for a highwall must be measured, a PS scanner would be more appropriate, given the fine feature scale and accuracy needed. Where near real-time displacement and velocity measurements are needed to monitor ongoing creep movement spanning multiple benches, a TOF scanner is likely the better choice, given that TOF scanners have the range required to monitor the entire slope while providing sufficient accuracy.

Thus, an appropriate TLS instrument must be chosen and its settings adjusted to meet the user's slope monitoring scale needs. Failure to do so will result in either too much or too little data, and/or erroneous measurements.

3.4.3.2 Usability and measurement limitations

In the open pit mining environment, TLS data need to be collected rapidly and reliably. Slow scanners may increase

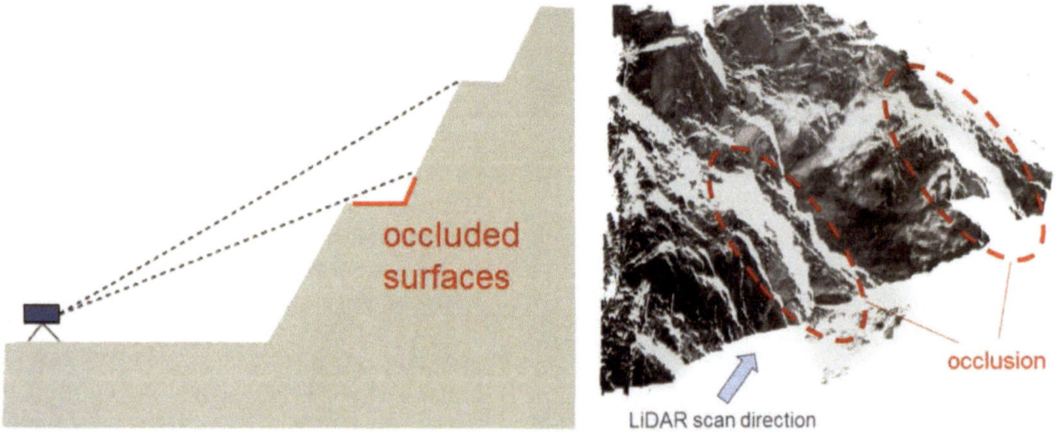

Fig. 3.17: Illustration of occlusion when the scanner's LOS is directed at an angle relative to the slope face that prevents part of the slope from being imaged (source: E. Eberhardt).

the 'field-to-finish' time and unnecessarily delay operational activities while data are being collected. Scanner downtime for repairs due to rough handling or exposure to dust and precipitation may also cause an operational disruption that could result in incomplete surveys or delays in end-of-month reconciliations. Thus, choosing the faster, more durable and reliable scanner for monitoring or other applications will increase its value to the user (more details on scan speed and data volume considerations in Section 4.6.4).

Mining environments are harsh and laser scanners are highly sophisticated survey instruments. While most scanners are designed for at least some degree of exposure and abuse, the more reliable a system is, the better. The international Ingress Protection (IP) rating of an instrument is a good indicator of reliability. The first number in an IP rating can vary between 0 and 6: it indicates protection from dust ingress, with 6 being completely impervious. The second number varies between 0 and 8: it indicates protection from water ingress, with 8 being submersible beyond a depth of 1 m. Higher IP ratings (e.g. IP65 *v.* IP54) therefore indicate instruments that are better protected from dust and water ingress. Scanners used on mine sites should be able to withstand frequent, heavy use and some rough handling. There is no specific measure of how well a scanner might hold up to rough treatment, but shock, vibration and impact testing are performed by some manufacturers. Trusted vendors may be able to provide information on these types of tests and the general durability of the instrument.

As an LOS instrument, consideration also needs to be given to occlusion. Sturzenegger and Stead (2009) describe occlusion as occurring when parts of a rock face cannot be imaged because it is obscured by protruding features (Fig. 3.17). In some cases, this might not be important as it is usually the tops of benches that are occluded. However, occlusion can be both horizontal and vertical, and therefore may require TLS surveys to be carried out from different positions for the same rock face. The multiple datasets can then be merged into a single point cloud, overlapping to fill in the occluded shadow zones for the slope.

3.4.3.3 Software

The software used to process point cloud data may be the most important part of the TLS-based slope monitoring workflow. Slow, cumbersome software or limited functionality can make data management and analysis difficult. At a minimum, point cloud processing software should enable the user to register the data in a local coordinate system, create surfaces, measure changes between surfaces and extract structural data, such as strikes and dips and geometry of structural features. Tools that accomplish these tasks are highly desirable in point cloud processing software.

Registration (i.e. location data) is important as many mines are surveyed in a local coordinate system that does not conform to standard coordinate grids. Tools that enable automated registration of one cloud (or surface) to another based on their relative geometry are essential to correct for small inaccuracies in reference point positions (e.g. vertical offsets in GPS positions).

Similarly, because slope monitoring and analysis are generally performed on surfaces (not directly on point cloud data) it is important that processing software enables the user to create accurate surfaces from points, including not just topographic surfaces but also models of steep, undulating slopes with overhangs and fissures. Unfortunately, because they have been purpose-built for point cloud data, the programs used to process TLS data may produce surfaces too complex for CAD or mine planning software to import. The point cloud processing software should likewise include robust ways of simplifying and exporting surfaces that enable cross-platform data integration.

By definition, slope deformation monitoring is the process of detecting and analysing changes in slope geometry and position. A software application that lacks the ability to compare changes in point or surface position is of limited value for slope monitoring. Ideally, the software can be used to measure the distance between surfaces or points as a proxy for deformation magnitude and/or rate response. Deformation response is essential information for predictive modelling and analysis, and is one of the primary goals in any slope monitoring program. To the same end, the ability to visualise and export this information clearly and quickly for use in reports and other software is essential.

Slope characteristics such as discontinuity orientation, spacing and persistence are important considerations in any monitoring program. Point cloud processing software should have tools that enable the manual or automatic extraction of these features (Fig. 3.18), as well as the ability to measure distances and angles. Object creation tools for modelling blocks, creating fault traces and other tasks are also essential. Built-in kinematic analysis tools are also helpful. These types of data products are key components of geomechanical slope characterisation; the more quickly and accurately they can be created, the more efficient a given slope monitoring program will be.

Lastly, laser scanners are extremely useful for volume calculations and for the creation of triangulated 3D surfaces for reconciliations. Efficient and accurate means for calculating volume changes and measuring conformance to design can be used to improve monitoring programs. At a minimum, the user should be able to calculate differential volumes and create cross-sections. These tools are particularly useful for post-failure analysis, where changes in surface geometry can be used to interpret stress-strain behaviour of the slope. Sectioning and volumetrics tools are also useful for conformance reporting and periodic reconciliations. The reconciliation process is further explored in Chapter 9. Application examples of TLS for slope monitoring are provided in Section 4.6.

3.4.4 Digital photogrammetry

3.4.4.1 Capabilities and limitations

In general, photogrammetry has limited capacity as a tool for slope monitoring. The techniques described in the previous subsections provide better estimates of location, motion rates, displacement vectors, accuracy and precision. That said, photogrammetric techniques have a place in slope performance monitoring, especially in

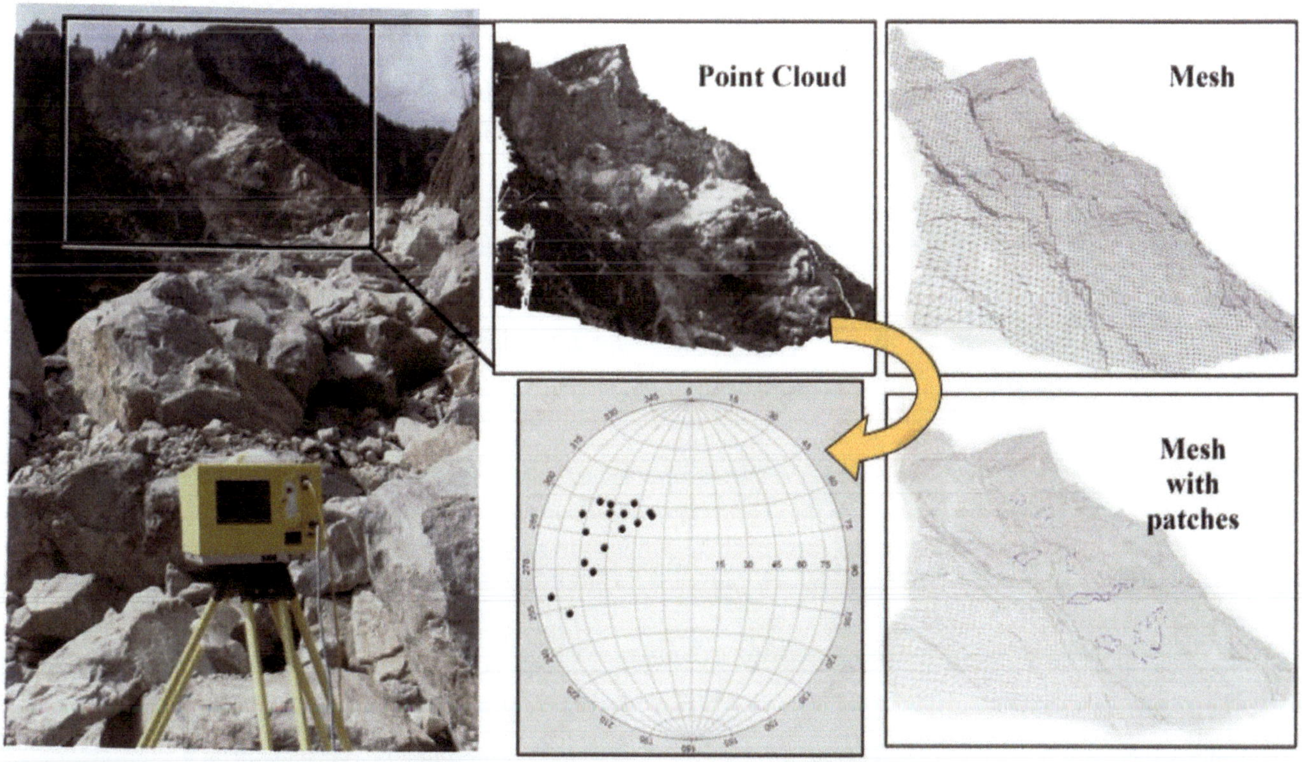

Fig. 3.18: TLS point cloud processing procedure, including the generation of a meshed surface from the point cloud and automatic recognition of persistent discontinuities (source: Strouth and Eberhardt 2006).

mapping and modelling geologic structures, assessing the mechanics of structurally controlled slope failures, and for visualisation. In many ways, photogrammetric modelling is essential to adequately characterise the latter features, both pre- and post-deformation, to better understand displacement mechanisms.

Photogrammetry's role revolves around its relatively high accuracy and its ability to accurately model geological structures and their orientations. The latter capability is due to the digital terrain models (DTM) that can be generated from photogrammetric images, intrinsically linking the DTM with the draped imagery.

3.4.4.2 *How photogrammetry works*

Rendering 3D models from photogrammetric images is based on taking images of the same area from different viewpoints. A common point on the rock face in the image will be seen by the two cameras, as illustrated by Fig. 3.19. The closer the cameras, the greater the number of matching points the user will be able to obtain. Conversely, the closer the cameras, the worse the depth accuracy to the point. A happy medium is one where sufficient matching points can be obtained to create a model and to provide adequate depth accuracy. This is crucial to any photogrammetric modelling, especially that which may be utilised for slope monitoring.

The 3D points calculated from the image matches are in relative coordinates. These are then mathematically transformed into true spatial coordinates using known point locations (in the image) or known camera locations. Multiple matching images may be combined to generate a point cloud, reducing the blind spot problem (i.e. occlusion) associated with having just two viewpoints.

To create accurate photogrammetric models, location data are necessary (i.e. registration). These can be obtained from accurately surveying the camera locations and using these known coordinates for transforming the model. Ground-based control points incorporated into the images provide another, and possibly simpler, means to ensure the accuracy of the model. Control points need not be permanent and, depending on the size of the area to be monitored, need not be on any displacing slope.

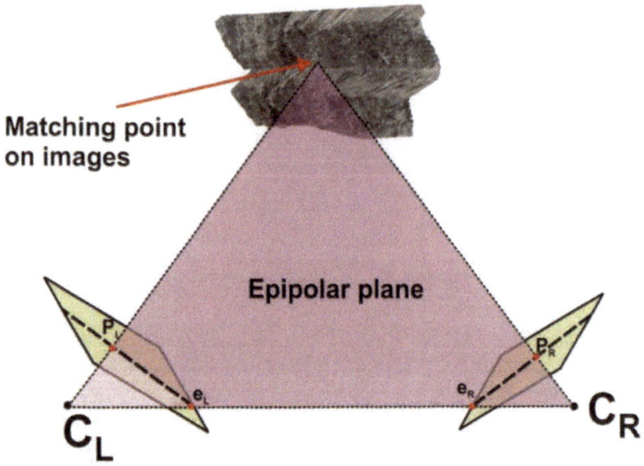

Fig. 3.19: Simplified transformation of stereo images into 3D coordinates (source: J. Mathis).

There are, of course, multiple platforms for photogrammetry. Terrestrial photogrammetry involves imaging the pit slope from the ground surface. The best imagery is obtained when a user can look down on the area to be modelled, with the slope displacement moving towards the camera position. This method will normally provide the highest accuracy in terms of ground resolution. If image fan mapping is utilised, this allows images from any single camera station to be merged in post processing; in essence, this creates a very large single image from a single point.

An aerial platform, while very flexible, is inherently unstable. Each image is essentially taken from a new camera location. As such, the user is generally limited to strip image mapping or multiple image combination techniques. Strip image mapping provides high-quality image pairs that are relatively accurate, as long as sufficient ground control points are available. However, this limits the individual model areal extent and draped image coverage.

Point clouds can be created from either terrestrial- or aerial-generated models. However, experience suggests that the merged/fanned image pairs from terrestrial photogrammetry are better matched as they are essentially 'primary' matches, especially for large areas. Resampling multiple images from overlapping image pairs, then draping this combined image over the combined point cloud, can result in a 'fuzzy' model that may or may not be useful from a structural interpretation standpoint. This is a critical consideration when evaluating structurally controlled displacement events.

3.4.4.3 *Modelling technique effects*

While the creation of the photogrammetric model and resulting DTMs may seem 'automatic', this is far from the case. Significant mathematical calculations are required to create the model, and the baseline assumptions and mathematical techniques utilised are critical. Simply bringing a bunch of images together to create a model, while potentially resulting in a visually appealing rendering, may well be far from fit-for-purpose.

While far from an exhaustive list, the following should be considered when evaluating the model generation techniques to be utilised.

- *Model error analysis.* When creating a photogrammetric model, a root mean square (RMS)

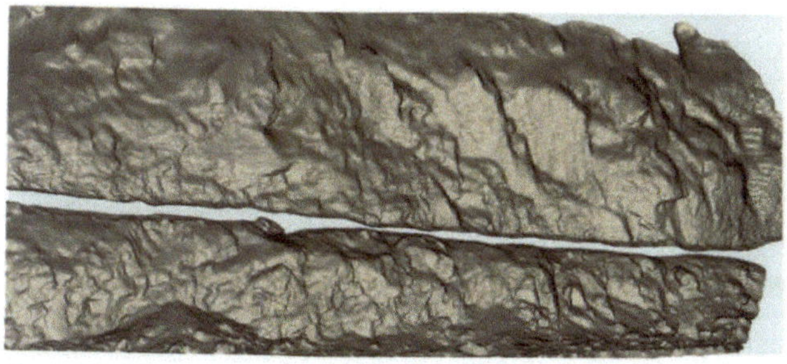

Fig. 3.20: Photogrammetric DTM without draped image showing over-smoothing (source: J. Mathis).

error will generally be reported. This is a measurement of the goodness of fit either to the raw pixels or to the ground-based coordinates (control points or camera stations). However, the method of calculation is important, as internal errors compared to global errors are different. With some software packages, while the residual error may indicate submillimetre precision, checking the model control point (calculated) could indicate an error in the order of tens of metres.

- *Smoothing/registration.* Care should be taken to make sure any mathematical treatment applied to the photogrammetric data does not result in over-smoothing. An example is shown in Fig. 3.20. This depicts a model of two bench faces on a pit wall. Note that the sharp edges one should see on the rock blocks are rounded. Any cross-cutting joint traces are absent as the algorithms have removed them in the smoothing process. Figure 3.21 shows a generally acceptable DTM/draped image model. Note that in this figure, even the low relief discontinuities are registered on the DTM surface. In some cases, the draped image can be incorrectly registered to the triangulated irregular network (TIN). This should not be the case, as the TIN should be created from the imagery. However, not all software packages are equal. Errors can vary from millimetres to, in extreme cases, over 200 m. If one is attempting to track a boulder moving down the slope on a model, it is nice to know that the visual representation of the boulder is actually that, and not an incorrectly registered DTM point.

3.4.4.4 Slope monitoring

Although other techniques such as geodetic monitoring and radar provide a more effective means to monitor near real-time displacements, photogrammetric tools can add significant value in certain situations. A common deficiency in many pit slope monitoring programs is that the slopes are generally not adequately characterised (geotechnical, structural, etc.) before slope displacements are detected. This is where photogrammetry, if properly implemented, can be a valuable tool.

Photogrammetric models can be created as the slope is excavated, providing a detailed and accurate picture of the rock mass. Geologic structures can, and should, be interpreted and traced down the slope as excavation progresses, so that surface features are connected to those found behind the slope in drill holes. In other words, a complete structural interpretation and excavated slope surface can be created before any displacement occurs. Of course, if this is coupled with a proper geotechnical analysis, issues regarding predictive behaviour, estimating the potential failure mechanism and deployment of an appropriate slope performance monitoring system can be addressed before slope displacement occurs. After all, understanding the potential failure mechanism is the most important consideration for developing the monitoring system.

That being said, photogrammetry can, and has been, used as a method of monitoring slope displacements. The difference in shape and volume between the original slope and the deformed slope can be established through differential DTM evaluation, sectional analysis, etc. This also allows local slope deformation velocities to be calculated. Structural block dislocations (slickensides on fault surfaces, translation, etc.) can be visually observed

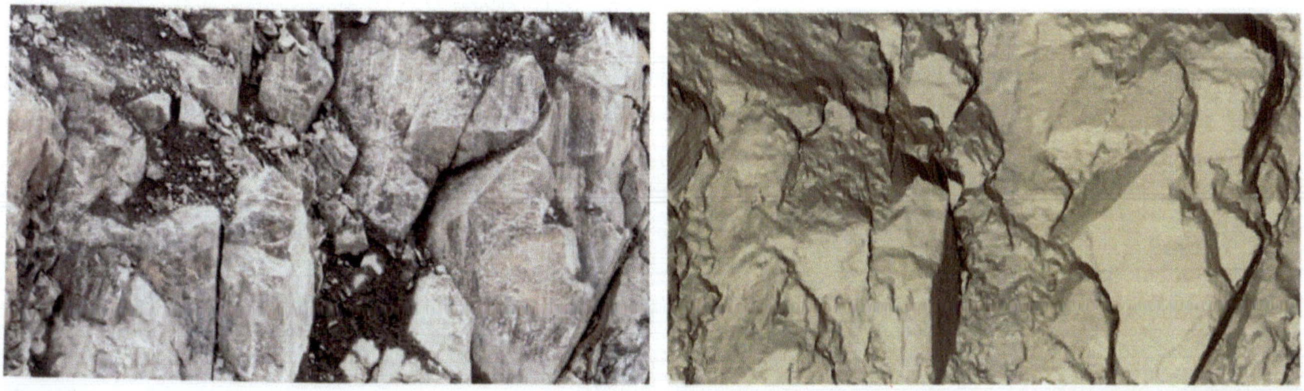

Fig. 3.21: Acceptable draped image/DTM compared with the underlying DTM (source: J. Mathis).

and measured on high-resolution photographic images far more accurately than with most monitoring tools. Repeatable model accuracy can be less than 1 cm (model-dependent), which allows relatively precise measurements. Visually distinct blocks, or artificial targets placed on the slope (balls or barrels tossed onto the slope), provide targets that are similar to survey prisms in terms of ability to calculate slope displacement vectors. Geologic structure rotation, coupled with block translation, is also interesting to observe and can assist in understanding the internal mechanics of slope deformation.

3.5 Displacement monitoring: site-wide measurements

3.5.1 Overview

The ground-based instruments previously described are especially effective at monitoring the behaviour of a pit wall slope but, being ground-based, it is recognised that they are restricted to the angle of view: from toe to crest, whether the instrument is positioned at the bottom or top of the open pit. In contrast, air- and space-borne instruments have the power to cover significantly larger areas (tens of square kilometres) and at high precision (several millimetres). These include airborne LiDAR and photogrammetry, and space-borne InSAR and imagery.

These techniques provide an advantage where full areal coverage is required to identify movements in a large open pit where ground-based monitoring may not be focused or even possible (e.g. in areas of the pit that are not actively being mined or accessible). The advantage of site-wide monitoring afforded by a look angle from above is advantageous where movements encompass an area that extends outside the open pit. This ability is important for protecting key mine infrastructure located near the pit crest, especially where subsidence caused by underground mass mining below the pit may be interacting with the open pit slopes. Examples are presented in Chapters 9 and 10.

3.5.2 Unmanned aerial vehicle drones

The availability of low-cost autonomous unmanned aerial vehicle (UAV) platforms, together with technical advances in desktop processing software, mean that UAVs are increasingly used as a specialised technique to map open pits to determine changes related to volumetrics (mined volumes) and to detect ground deformations related to slope movements. Their advantage, relative to other airborne platforms, is that they are capable of flying much lower and hence can collect imagery at much higher resolution, often under 10 cm, and even as detailed as 1 cm/pixel. In contrast, the highest spatial resolution data available from manned aircraft or satellite platforms are typically in the range of 20–50 cm/pixel.

In the case of photogrammetry, UAVs can provide comprehensive, feature-rich data using a normal RGB camera that can cover the entire open pit area (including waste dumps, tailings dams, etc.). Sectors of geotechnical concern can be included in the routine data collection, providing a source of medium accuracy (3–5 cm), comprehensive, high-definition surface data. Through combinations of purpose-suited UAVs and sensors, higher accuracies (<1 cm) and detection of changes to other indicators, such as moisture content and surface temperature, can be achieved.

3.5.2.1 Accuracy and precision

For long-term monitoring systems, absolute accuracy and survey control integrity is essential to maintain data continuity over the life of the operation. For medium, short-term and near real-time systems, repeatable precision is most important for change detection. A key strategy that best suits the use of UAVs for slope monitoring applications is to reference data acquisitions over time by identifying 'stable' areas in the dataset, independent of survey controls. The software references all data in a single cycle to these stable areas, allowing relative movement relevant to the stable areas to be detected to a high degree of confidence.

The geometric accuracy of 3D models, and the digital elevation models (DEMs) and ortho-photo mosaics constructed, can be further improved by locating control points at strategic positions using GNSS receivers (Fig. 3.22). Horizontal accuracies of 7 cm and vertical accuracies of 6 cm have been achieved using this methodology. The magnitude and direction of the displacement vectors can be derived from correlating two hill-shaded DEM layers corresponding to a visual interpretation of pit slope movement. These can include the use of visible displacement features (tension cracks), soil masses and vegetation patches on top of the slope.

3.5.2.2 Flight planning

Light-weight UAVs capable of fully autonomous flight, weighing just under 1 kg, have a cruise speed of ~40 km/h for safe and easy operation. To begin a mission, the UAV will circle a pre-set starting waypoint to gain altitude and then carry out its pre-programmed mission. The UAV will land itself with a reasonable degree of accuracy. The UAV detects how far it is from the ground when it comes in for a landing. Software is used for mission planning and image processing. Achievable accuracies vary for fixed-wing (3–5 cm) and quadcopter (<1 cm) UAVs.

Ground control points (GCPs) should be clearly marked (Fig. 3.22) and accurately surveyed for use as reference points in aerial images. These are essential for

Fig. 3.22: Use of GPS/GNSS to provide ground control points for UAV photogrammetry (source: S. Jefferys).

ensuring that the processed data (DEM and ortho-photo) are in the same site coordinate system, allowing comparison of the data over time. A pre-marked inventory of GCPs is maintained across the mining area to minimise delays before execution of a project. The GCPs should be placed outside the zone of influence of the targeted area for deformation monitoring. An example of GCP placements is shown in Fig. 3.23.

The flight path should always be planned in the office before field execution. This includes:

- ensuring that there are sufficient GCPs to cover the area selected (e.g. Fig. 3.23). Permanent GCPs will improve the repeat accuracy; and
- determining the flying height by setting the ground sampling distance (GSD). The smaller the GSD, the higher the data resolution and, in the case of monitoring, the smaller the sizes of the individual resolution cells, or pixels, that will be able to be detected (Fig. 3.24). The GSD is calculated as:

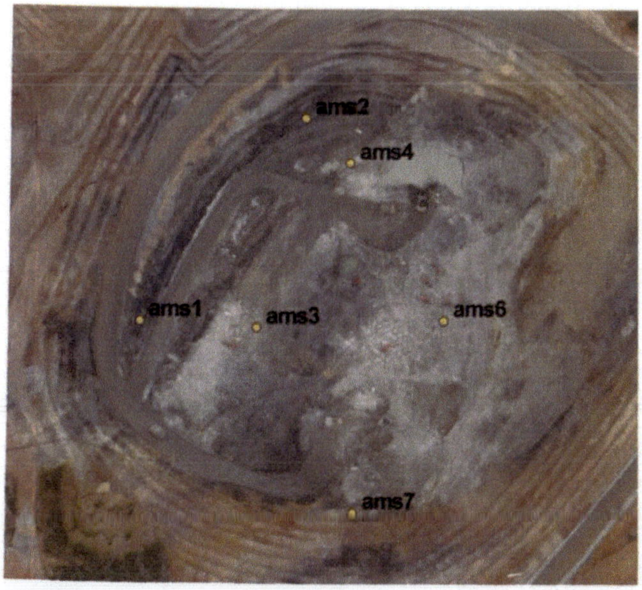

Fig. 3.23: Ground control point distribution (minimum of five GCPs per project) (source: S. Jefferys).

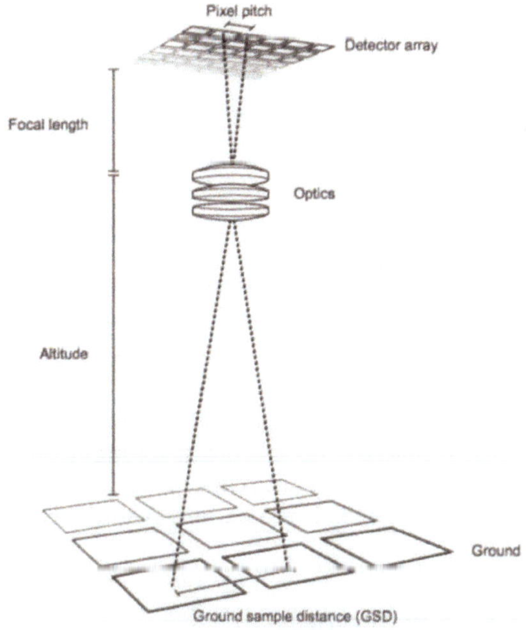

Fig. 3.24: Ground sample distance v. flying height (source: S. Jefferys).

$$\text{GSD} = \frac{(\text{pixel size} \times \text{height above ground level})}{\text{focal length}}$$

Simulating the flight ensures that the UAV is programmed correctly. It also allows for advanced planning such as options for in-flight waypoint editing and simulating the effect of high wind conditions without putting the UAV at risk. Interfaces with software, such as Google Earth, also allow the flight path to be viewed in 3D. This helps to identify any possible obstructions. Once the flight plan is approved, it can be saved and used for repeated data collection of the project area.

At the time of writing, data processing is performed manually, but developments in the software will soon support the automation of GCP identification. This will further improve the repeatable accuracy. Software options exist to automatically identify barcoded printed targets for close-range photogrammetry. However, the algorithms at the time of writing do not work for longer-range photogrammetry where larger-sized GCPs are needed.

After processing, the data are output to a raster image and DEM. The output data can then be imported into geographic information system (GIS) software for comparison and analysis. Analysis can be as simple as 'eyeballing' the images, as would be done on a field inspection where indicators are identified such as surface cracking, material moisture and water pooling. The DEM can be imported into surface modelling software where surface comparisons generate heat maps based on surface differences, as shown in Fig. 3.25.

Technical advances in UAV systems and photogrammetry software have provided an economical source of feature-rich slope monitoring data. Further development of automated GCP recognition, data processing and surface analysis will further improve data reliability. The following lessons learned from experiences to date should be noted by open pit operations planning to introduce the use of UAVs in their workflow:

- precision is directly related to local conditions: keep as many variables unchanged as possible, maintain the same GCPs, collect data at the same time of day and under similar conditions;
- GCP quality is important; further research with real-time kinematic GNSS-enabled UAVs will eventually liberate the reliance on ground control; and
- there is collective value in incorporating all sources of monitoring information (UAV, prism, LiDAR) into a single integrated platform, which greatly enhances the analytical value of UAV data.

3.5.3 Satellite radar

Space-borne interferometric synthetic aperture radar (InSAR) provides maps of surface displacement based entirely on remote-sensing radar satellites. These are typically provided as easy-to-integrate layers that complement existing monitoring methods. The unique value of InSAR is the capability to measure ground displacements of a few mm/year (precision) over large areas from satellites in orbit hundreds of kilometres above the Earth's surface. The bird's-eye view of satellites can detect previously unknown displacements over an entire mine site.

Satellite InSAR functions on the same principles as GB InSAR. The radar sensors on satellites operate in the microwave domain, with wavelengths of a few centimetres. Compared to optical imagery, the longer wavelengths of radar sensors allow the signal to penetrate clouds, rain, fog and dust, although snow cover does create a significant limitation. Images are created by illuminating an area of interest (AOI) with electromagnetic pulses and recording the echoes backscattered from natural and man-made objects to the radar antenna. This allows day or night imaging with very little interference from solar activity, cloud cover or other atmospheric conditions. In forested regions, InSAR can still be effective if a longer wavelength is used, as provided by the L band ALOS satellites for instance. Longer wavelengths are less sensitive to small displacements.

All radar satellites travel along a near-polar orbital path and acquire images, usually with a right-looking viewing geometry. The radar beam looks eastward or westward, depending on whether the satellite is travelling north (ascending orbit) or south (descending orbit). As they revisit the same areas with a fixed frequency, radar satellites provide a stable source of non-intrusive remotely sensed data.

While travelling along their polar orbit, the radar satellites observe the ground surface at an angle. All ground displacement measurements are one-dimensional, either towards or away from the satellite along this angle, known as the LOS. Note that any movement perpendicular to the LOS is not detected and cannot be measured. Consequently, LOS measurements have good sensitivity to vertical and east/west movements, but poor sensitivity for north/south movement.

Utilising data from ascending and descending orbits offers better coverage of the site, especially for mines with highly varying topography, and provides 2D displacement vectors in a plane formed by the vertical and east/west directions. This vector is a better representation of the true magnitude and direction of displacement than a single LOS measurement, as illustrated in Fig. 3.26.

3.5.3.1 *Satellite constellations*

The ERS constellation was launched in the 1990s by the European Space Agency (ESA). ESA began acquiring multi-temporal data over many areas of the Earth's

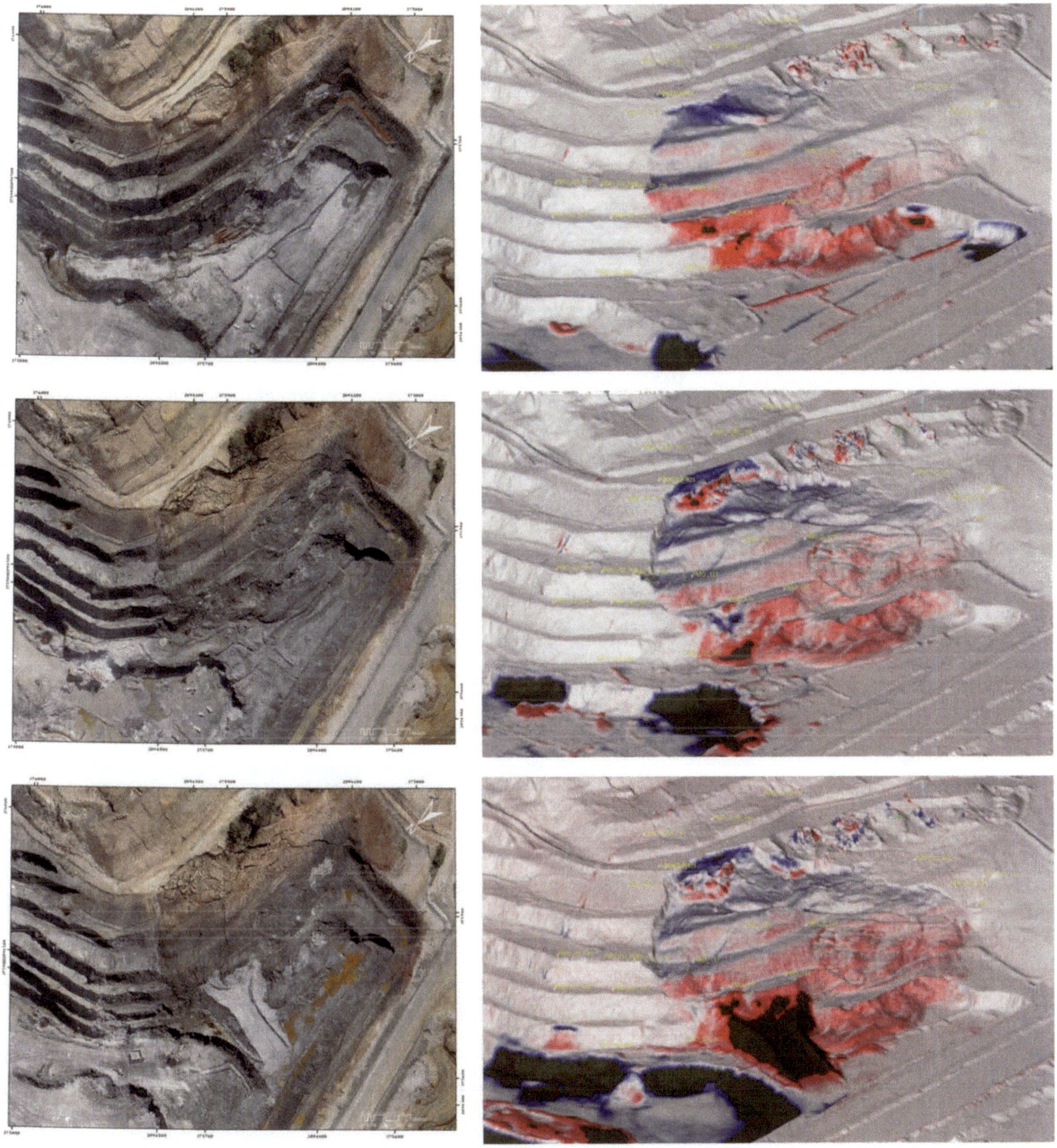

Fig. 3.25: Examples of ortho-photos (left) and DEMs (right) derived from UAV photogrammetry, comparing Day 1 (top), Day 7 (middle) and Day 18 (bottom) (source: S. Jefferys).

surface, which allowed the development of InSAR techniques based on the comparison of subsequent images. Only one to three satellite systems were available up to the early 2000s, which has increased to five at the time of writing, as shown in Fig. 3.27. In the future, additional SAR constellations will become available with further improvements in acquisition frequency. The basic technical specifications of each satellite system are summarised in Table 3.1, including revisit time, ground resolution and operating band.

Mine operators typically utilise high-resolution satellite systems that operate in the X band of wavelength-frequency, including the German TerraSAR-X and the Italian COSMO-SkyMed systems. For operations that do

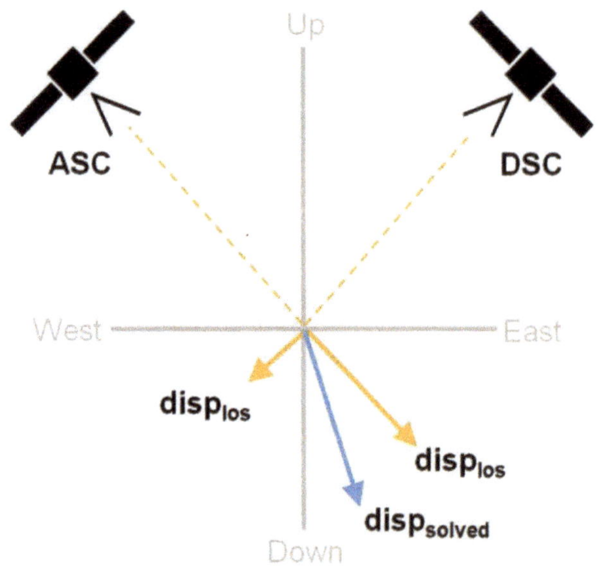

Fig. 3.26: The use of ascending and descending data to monitor ground movement not only improves coverage of the complex topography of mining operations, but also provides a movement vector that is much closer to the true magnitude and direction of displacement (source: 3vGeomatics).

not require such a high degree of displacement precision and acquisition frequency, the Canadian RADARSAT and ESA SENTINEL systems are available; these operate in C band. ALOS-2 in L band has high spatial resolution, but lower measurement precision. Its ability to penetrate vegetation, however, can be useful.

All SAR satellites have near-polar orbits and are sun-synchronous, in that the sensor passes over the same area on Earth at the same local solar time. It is possible to capture images along any point of the orbit in ascending or descending mode, as shown in Fig. 3.28. Most sensors acquire right-looking data, meaning that, for example, if the platform is traveling from south to north, the radar sensor is pointing towards the east.

3.5.3.2 *Basic technical principles*

Today's SAR satellites obtain high-resolution images even though they orbit the Earth at altitudes of 500–800 km. Their polar flightpath, combined with the rotation of the Earth, makes it possible to acquire data over the same area, with the same viewing angle and fixed number of days.

By definition, radar measures the distance between the antenna and a target on the ground surface. InSAR

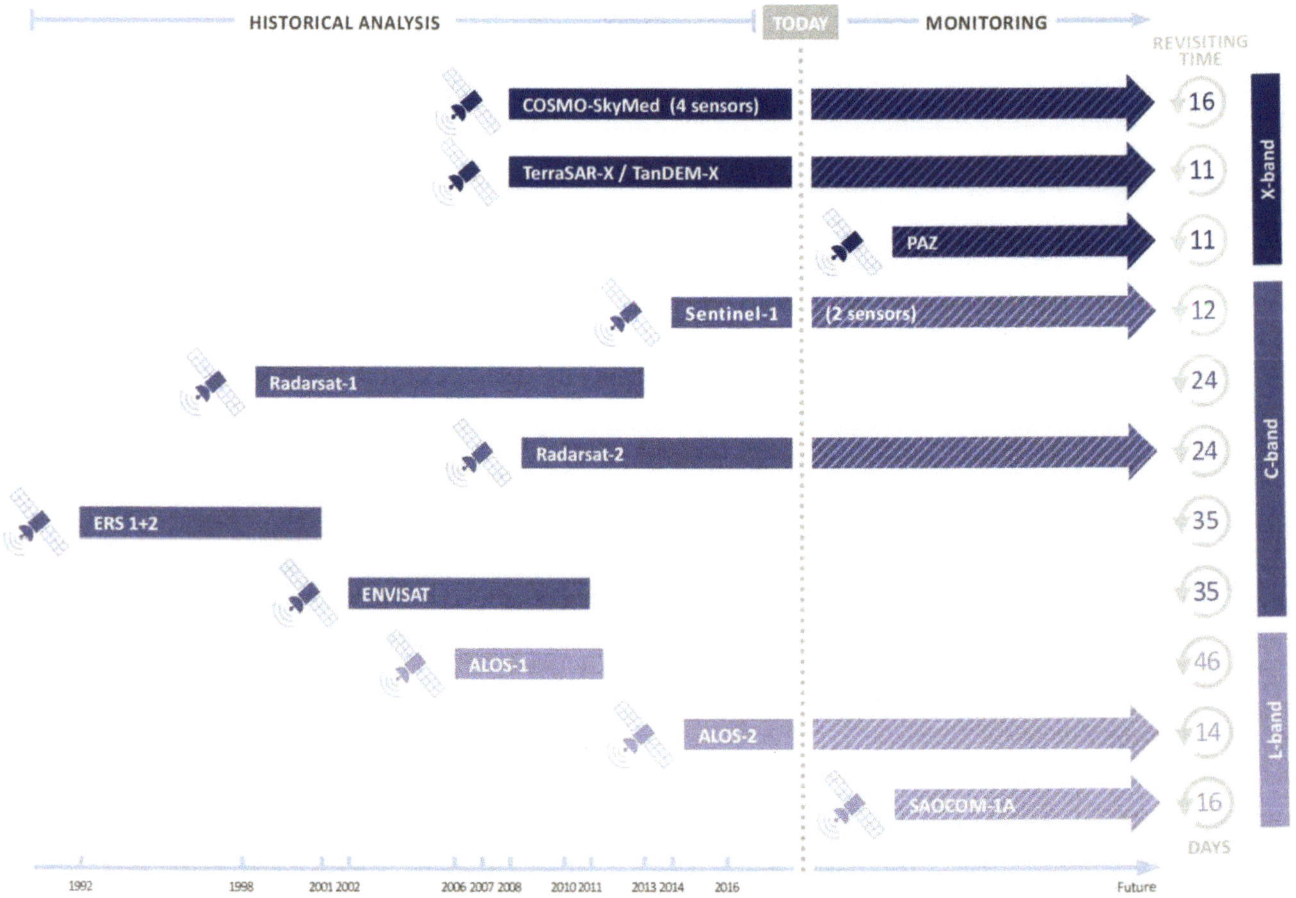

Fig. 3.27: Past, present and upcoming SAR satellites, their operating band and revisit frequencies (source: TRE-Altamira).

Table 3.1: SAR satellites and their basic specifications

Satellite	Revisit frequency (days)	Max resolution (m)*	Band (wavelength cm)	Achievable precision
ERS	35	20	C (5.6)	sub cm
ENVISAT	35	20	C (5.6)	sub cm
ALOS-1	46	10	L (23.6)	few cm
RADARSAT-1	24	8	C (5.6)	sub cm
RADARSAT-2	24	1	C (5.6)	sub cm
TerraSAR-X	11	0.6	X (3.1)	few mm
COSMO-SkyMed	8	3	X (3.1)	few mm
Sentinel-1	6	15	C (5.6)	sub cm
ALOS-2	14	3	L (23.8 cm)	few cm

Source: TRE-Altamira and 3vGeomatics.

provides a way to monitor range variations (changes in distance) using multiple SAR images of an area gathered at different times. To detect very small surface displacements, on the order of 1 cm or less, compared to the metre-scale resolution of the radar image, it is necessary to obtain accurate estimates of the phase shift between successive SAR images acquired from the same orbit (Fig. 3.29).

As SAR sensors are active systems, the radar creates images by the emission and reception of electromagnetic waves in a frequency range that does not require sunlight to illuminate the ground surface. A key feature of satellite radar is the ability to penetrate clouds and to obtain information about possible targets regardless of local weather conditions. These two features – the ability to work both day and night, and the ability to penetrate cloud cover – have made satellite radar a major Earth observation tool for the remote-sensing community.

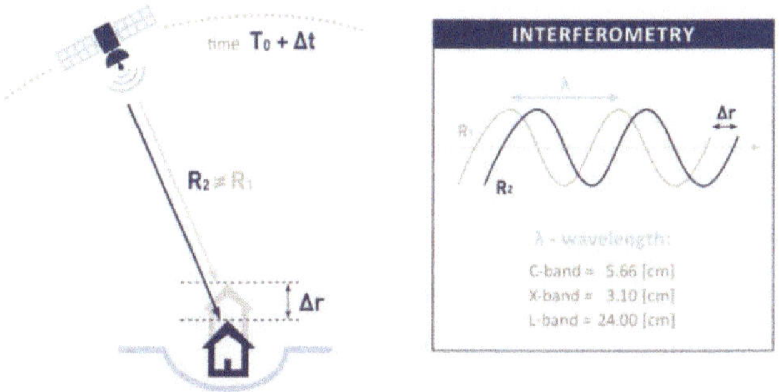

Fig. 3.29: Illustration showing the relationship between ground displacement and signal phase shift (source: TRE-Altamira).

3.5.3.3 SAR data processing

Radar sensors use a complex signal that records a combination of the amplitude and phase value of each resolution cell (pixel) in the AOI. When the satellite records a subsequent image from the same orbital geometry, the process is repeated, providing a new amplitude and phase measurement. As the image archive grows over time, the record of phase information can provide increasingly higher levels of ground displacement precision. For each pair of images, an interferogram can be created and converted to an easily readable displacement map, as shown in Fig. 3.30.

Advanced InSAR processing methods use statistical algorithms to mitigate atmospheric and orbital effects,

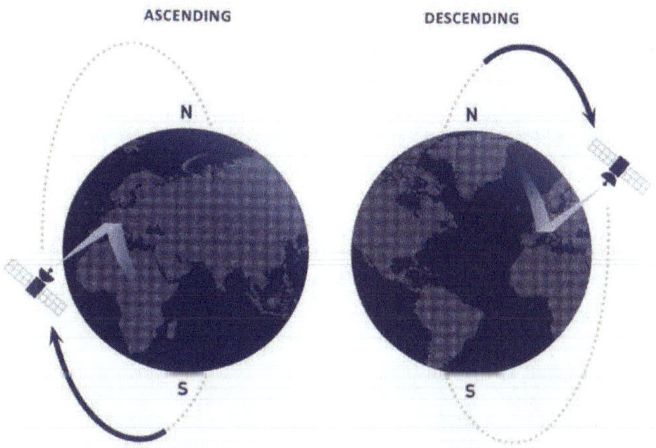

Fig. 3.28: Ascending and descending orbits (source: TRE-Altamira).

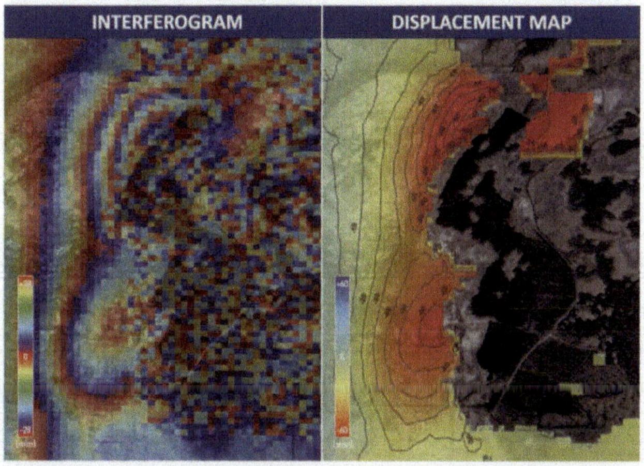

Fig. 3.30: Example of a SAR interferogram and associated displacement map (source: TRE-Altamira).

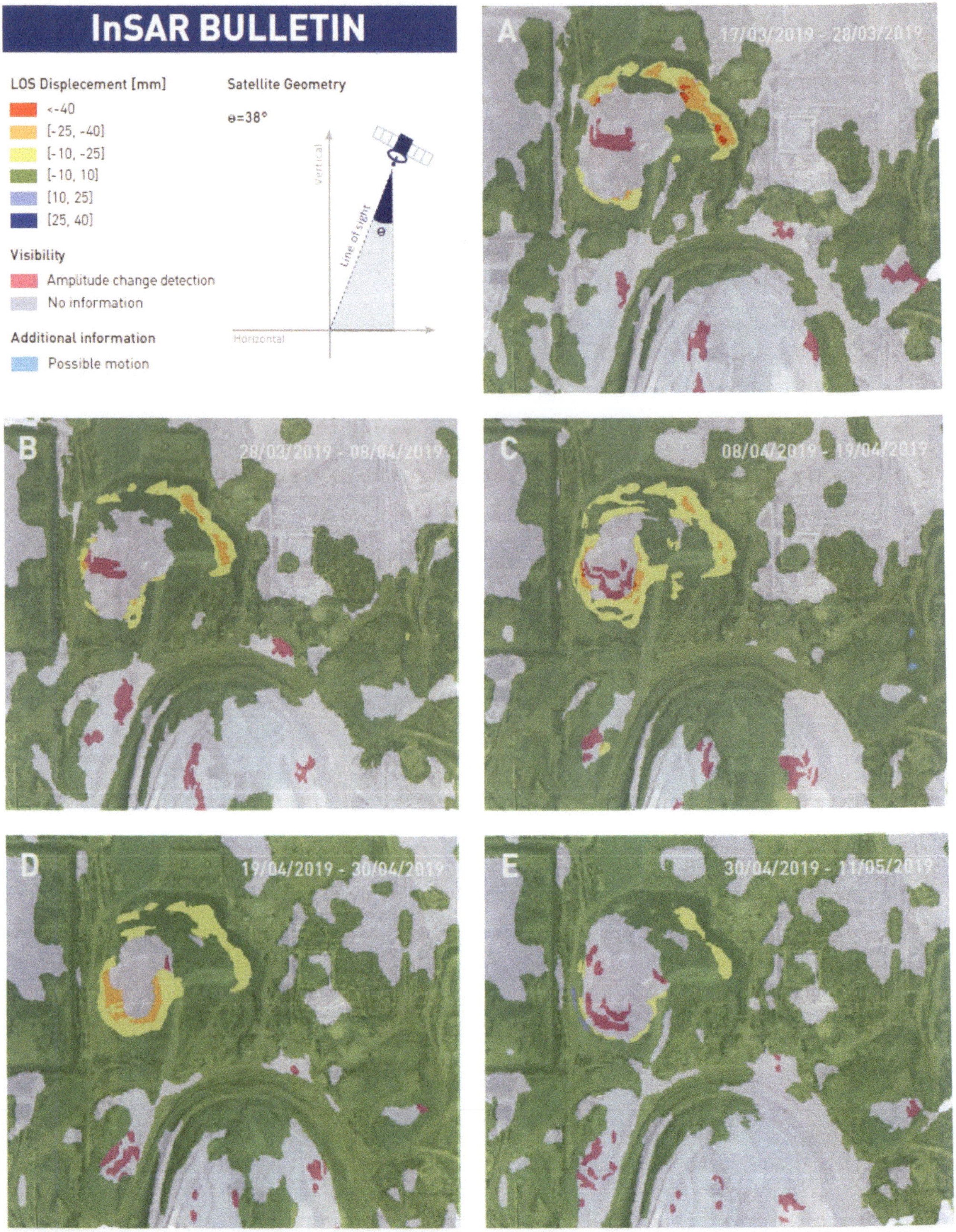

Fig. 3.31: Series of InSAR reports highlighting progressive displacement on waste rock piles at the Orapa diamond mine (source: Debswana Orapa mine and TRE-Altamira).

following which the displacement of target areas is estimated over time. Targets can be naturally occurring features, such as rock outcrops or infrastructure (see Section 3.5.3.4), that return a stable signal to the satellite and exhibit a signal quality termed coherence. Processing 15 or more successive images helps to increase the measurement precision to the millimetre range. It typically requires three to six months of data to develop a baseline for establishing precise movement trends using a satellite like the TerraSAR-X, which acquires images every 11 days.

Displacement maps can be quickly produced following the acquisition of each new image and used to identify areas of recent displacement or anomalous change. Reports can be delivered to the slope engineer in accordance with the acquisition frequency, say, every 8 or 11 days (see satellite revisiting time in Table 3.1). An example of an InSAR report for the Orapa diamond mine in Botswana is shown in Fig. 3.31 (see also Section 4.5.7).

With advanced InSAR techniques, it is possible to extract the evolution of displacement and velocities by analysing an archive of many images collected over the AOI. While the achievable precision of InSAR is limited to the centimetre range, more sophisticated processing approaches can provide information related to movements ranging from millimetres per year to metres per year.

3.5.3.4 *InSAR targets*

Any high-quality mining InSAR analysis will extract all of the available displacement information from the data. Key to this is a thorough consideration of the targets or scatterers on the ground. For any individual SAR pixel, there may be a single object, such as infrastructure or a boulder, that dominates the return signal, which consequently exhibits low noise characteristics over time. These are termed point targets (PT). Pixels with no dominant object often still provide useful information by being aggregated with nearby pixels; these are termed distributed targets (DT).

Another important factor is a target's persistence through time. Targets that remain present throughout a stack of radar images, whether they are PTs or DTs, are termed persistent and can act as anchor points underpinning the wider analysis. For mining operations with ongoing surface activity, many targets will be non-persistent, as excavation and dumping reset the targets from time to time. Non-persistent targets may easily represent 50% of all the monitorable targets at a mine (Fig. 3.32), and approaches which do not consider this information will provide much lower coverage than those that do. Target types are illustrated in Fig. 3.33.

3.5.3.5 *Monitoring rapid movement*

InSAR service providers have developed a method for tracking rapid movement at rates that exceed those measurable with SAR interferometry. This limitation was previously discussed with respect to GB radar. It is related to displacements between consecutive measurements exceeding half a wavelength, which leads to ambiguity in the phase measurements. Rapid movement of surface features can be tracked by utilising the amplitude value of the signal to evaluate the shift of recognisable features

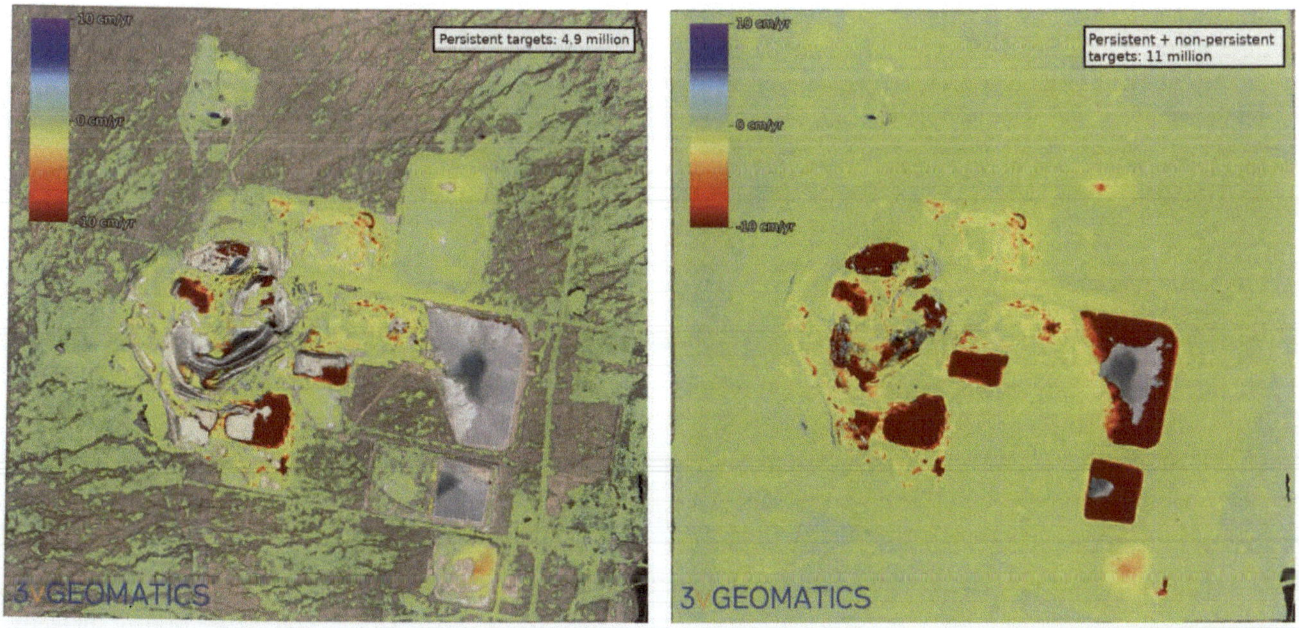

Fig. 3.32: Left: InSAR displacement results for a mine site, considering only targets which persist for every radar image in an analysis. Right: The number of measurable targets is more than doubled when non-persistent targets are also considered (source: 3vGeomatics).

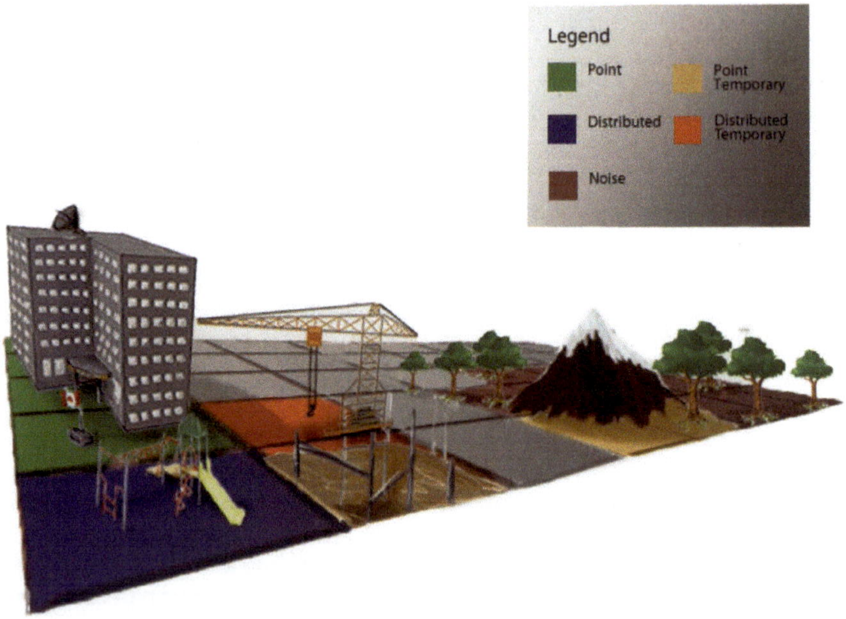

Fig. 3.33: Pixels may be dominated by a single target (point) or contain contributions from multiple objects (distributed). Targets may persist in time or be reset through ground surface activity. Any mining InSAR analysis should consider all permutations of these target types (source: 3vGeomatics).

between SAR images. This is analogous to tracking features in a series of photographs. It is illustrated in Fig. 3.34.

The method can be used for tracking movement of surface features ranging from tens of centimetres to tens of metres per year in the LOS and the azimuth (north–south) directions. The precision of this technique is related to the pixel size of the image, as it corresponds to around 1/10th of a pixel or better. Heat mapping of relative ground surface displacements, combining interferometric SAR and rapid displacement monitoring techniques, is illustrated in Fig. 3.35. Like standard InSAR displacement, rapid displacement results can be visualised as maps of surface displacement or as time-series of individual measurement targets.

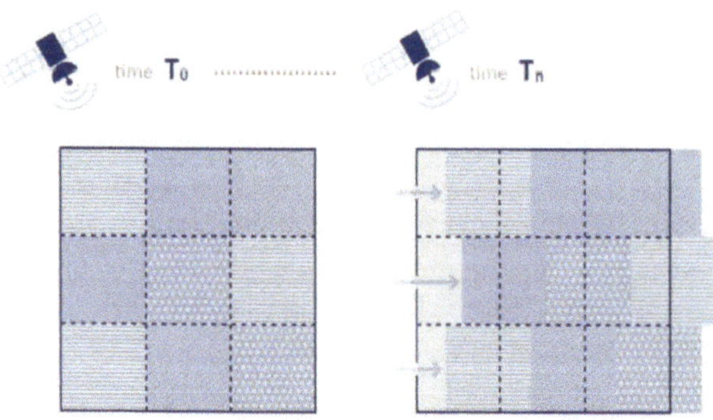

Fig. 3.34: Rapid movement can be monitored by identifying sub-pixel shifts in the amplitude information (source: TRE-Altamira).

3.6 Subsurface deformation monitoring

3.6.1 Overview

Open pit mining projects often present a significant monitoring challenge, in part because of their initial lack of definition and the sheer scale of the problem. Monitoring programs are further constrained by physical limitations and project economics. Not all parameters can be measured with equal ease and success. Subsurface monitoring requires expensive boreholes to be drilled and affords only a localised measurement point where significant effort is required to ensure that the data are reliable. Because of this, most pit slope monitoring programs focus on surface displacement measurements where accessibility allows for more thorough monitoring coverage. As was discussed in Section 3.4 with remote sensing techniques (e.g. GB radar), full coverage monitoring of pit slope surface displacements is achievable at high spatial and temporal resolutions.

However, this reliance on surface-based monitoring neglects the changes in behaviour that occur in the subsurface where the potential failure surface is developing. Generally, the kinematics and causes of failure are not well defined; instead, the surface manifestation of the instability (i.e. surface displacements) is relied upon for predictive analysis. This introduces significant subjectivity and uncertainty into the interpretation of the monitoring data, often raising the question, 'How well do we

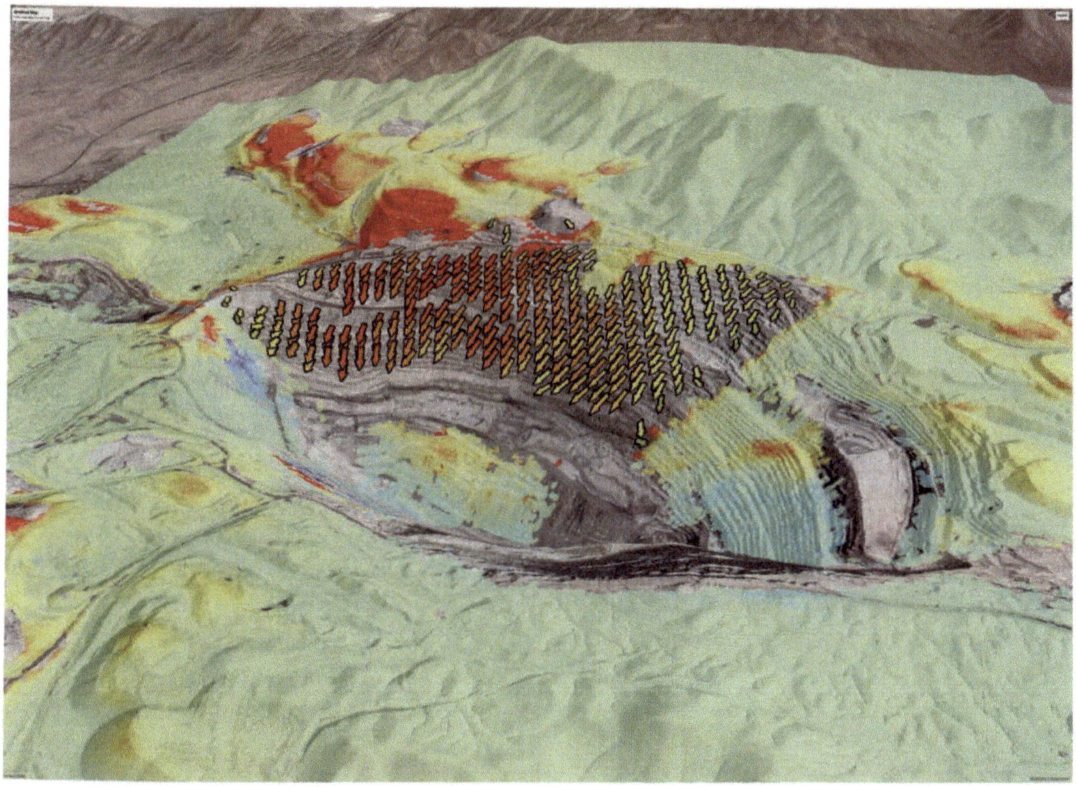

Fig. 3.35: Displacement maps obtained by combining interferometric SAR and amplitude rapid movement tracking (source: TRE-Altamira).

understand the geologic factors and mechanisms of slope movements?'

Thus, although the costs of subsurface monitoring are greater than those of surface monitoring, they can produce valuable information on the slope behaviour at depth, especially the location and nature of the sliding surface. Subsurface measurements also improve estimates of the potential volume of a pit slope failure, failure mode and rate of movement at depth. These data allow a more reliable model of the pit slope hazard to be incorporated into the engineering decision-making process.

Several different technologies are available to monitor subsurface displacements. Although they all use boreholes for installation, they vary in sensing technology and balance between simplicity, sensitivity, capabilities and reliability.

3.6.2 Time domain reflectometry

Time domain reflectometry (TDR) is increasingly being used in geotechnical applications for deformation monitoring in rock slopes. It provides a means to detect the location of shearing across a grouted borehole, which can be used to approximate the depth of sliding for a moving slope. Its primary advantages are that it is relatively inexpensive compared to other means of subsurface deformation monitoring and that it can be installed with other borehole instruments (see Section 4.7.3).

TDR technology was originally developed by the telecommunications industry as a method for detecting faults, or breaks, in electrical cables. The method involves grouting a coaxial cable in a borehole and conducting measurements where a signal (electrical pulse) is transmitted down the cable. If the cable is crimped or breaks, for example due to shearing along a sliding surface that crosses the borehole, a portion of the signal is reflected back. By measuring the time between transmission and reception of the reflection, and knowing the speed at which the signal is travelling, the distance to the reflector(s) can be calculated.

In geotechnical applications, the coaxial cable typically takes the form of a rugged coaxial cable with a diameter of 12 mm. Coaxial cables comprise an inner and outer metallic conductor separated by an insulating material (Fig. 3.36). The impedance of the cable is dependent on the distance between the inner and outer conductor; deformation of the cable will locally change the impedance, causing a reflection to be returned to the signal generator. Note that different coaxial cable types may be used, offering options in terms of cost, cable sensitivity (to deformations) and robustness in terms of how easily the cable will break. Data units returned by

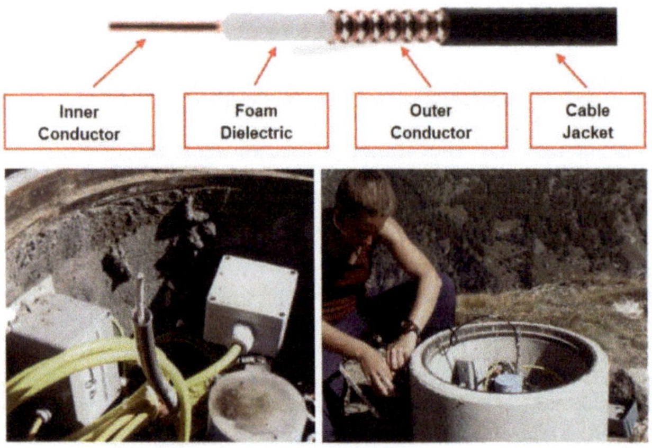

Fig. 3.36: Example of coaxial cable used for TDR installation for slope monitoring (source: E. Eberhardt).

readout devices can vary. Some equipment returns a reflection coefficient, expressed as a percentage of signal returned; others may return an impedance, where a high value indicates the end of the cable. The measurement resolution is related to the number of samples of data being collected for the waveform.

A typical installation for slope monitoring is shown in Fig. 3.37. Multiple cables may be embedded in an unstable slope to allow accurate profiling of the depth and progression of a failure mode.

It is important to keep in mind that all distances measured using TDR are 'apparent', not absolute. A key configuration parameter of the monitoring system is the velocity of propagation (Vp), which is the percentage of the speed of light at which the electrical signal travels down the cable. *In situ* modification of the cables, using a few crimps applied to the cable at measured distances during installation, can be useful to help establish more accurate distance measurements; however, this should be kept to a

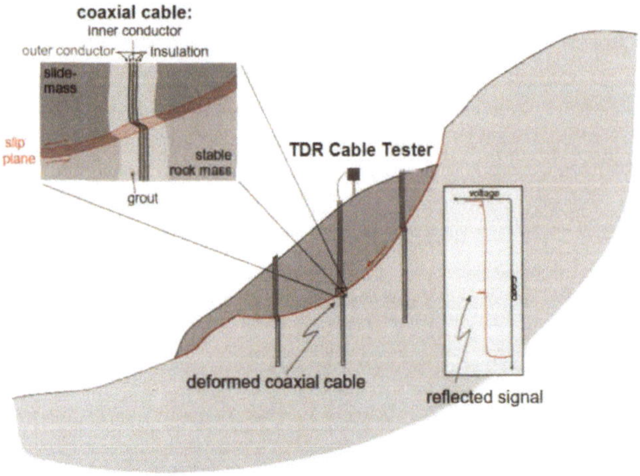

Fig. 3.37: Example use of multiple TDR cables to determine the location and progression of a slope failure surface (source: Singer and Thuro 2009).

minimum with long cables (>100 m), as crimps can cause signal loss or cable breakage during installation.

Longer cables can cause issues with measurement. As the signal propagates down the cable, it is absorbed by the dielectric; this means there is less signal to reflect back to the transmitter. In this case, some interpretation of the data may be required to determine the end of the cable or location of a reflector. Deformation monitoring cables must be properly terminated and sealed, generally with an epoxy seal kit, to provide for infinite impedance at the end of the cable. This ensures that a reflection from the end of the cable will be measured. The following list describes typical cable abnormalities/failures, together with the resultant change that would be seen in the TDR data:

- an open cable end will return a maximum reflection coefficient or infinite impedance;
- a shorted cable end will return a minimum reflection coefficient or zero impedance;
- cable abrasion causing abnormalities in the outer conductor, including holes or breaks, will typically result in a positive spike in the coefficient or impedance;
- crimping of the cable, usually the result of deformation or pre-applied during installation, will increase capacitance, causing a negative spike in the coefficient or decrease in impedance;
- corroded cable connections, often due to splicing, will typically cause a reduction in the impedance and hence a negative spike in the coefficient or decrease in impedance. If the damage is severe enough, no reflections past the location will be received, essentially ending monitoring of the cable beyond that location; and
- dielectric property changes due to water ingress will typically cause a positive spike in the coefficient or increase in impedance.

An example of how TDR data are used to locate the depth of a sliding surface, indicated by an increasing reflection at a specific depth, is shown in Fig. 3.38. It should be noted that TDR monitoring does not provide absolute movement data. The reflection coefficient or impedance is simply a relative measure of the signal reflection or cable impedance. However, continued observation of data can provide an indication of continued deformation or stabilisation.

3.6.3 Inclinometers

An inclinometer is an instrument used for measuring changes in the inclination of a borehole as a result of deep-seated slope movement (Fig. 3.39). In this sense, they serve a similar purpose as TDR, but have the added advantages of being more precise, reliable (with respect to the measurement) and able to also measure the magnitude of slip and its direction along each sliding surface.

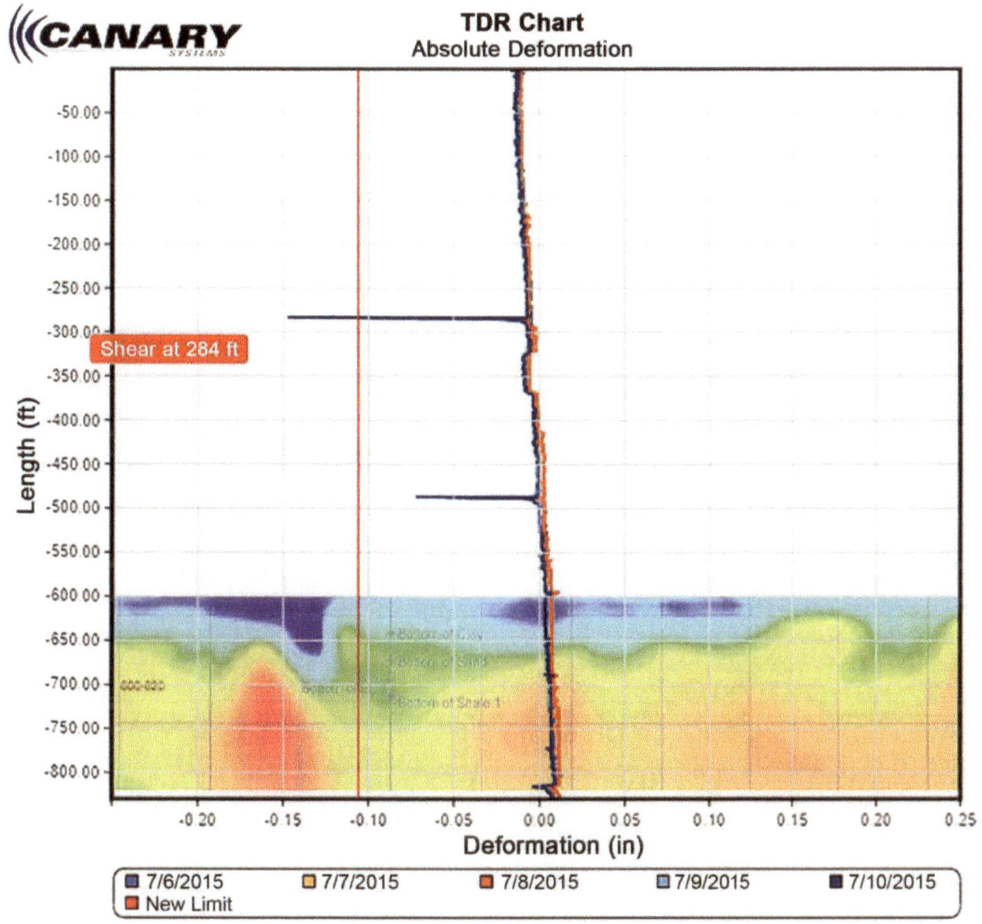

Fig. 3.38: Example of TDR data and detection of an increasing reflection interpreted as being caused by a developing sliding surface (source: N. Mphathiwa).

Installation requires drilling a borehole that is deep enough to confidently intersect the deepest anticipated sliding surface, then grouting into place a plastic casing with longitudinal grooves cut into the inside wall at 90° intervals (Fig. 3.39). One pair of inclinometer casing grooves should be oriented in the direction of the anticipated movement. It is important to select the appropriate diameter casing. Larger diameters are more suited to discrete sliding surfaces or multiple shearing zones where subsurface displacements are expected to concentrate. Moderate to small diameters can be used for short-term installations or where smaller displacements distributed along the borehole are anticipated. Correct installation of the casing is important; in deep holes, the influence of borehole deviation and twisting of the casing must be considered (Fig. 3.40). A digital spiral sensor probe can be used to check for the torsion profile of the casing (Willenberg *et al.* 2008).

Inclinometer measurements can be performed either manually by using a probe inclinometer or automated by using in-place inclinometers. Operation of a probe inclinometer (Fig. 3.41) involves lowering the probe down the inclinometer hole with the wheels on the probe running along one pair of the casing grooves. Inclination is then measured through a transducer in the probe designed to measure tilt with respect to vertical (typically an accelerometer or MEMS sensor). Tilt measurements are made at several fixed points as the probe is pulled back from the bottom of the borehole to the surface (Fig. 3.42). Comparison of repeat periodic surveys, relative to an initial zero survey after installation, provides a measurement of differential displacements at depth. Displacements are usually summed from the bottom of the borehole, assuming that it is anchored within stable rock or soil (i.e. beyond the anticipated slip surface), and this fixed point acts as a reference from which displacements are measured.

A check of the results is then made by rotating the probe 180° and repeating the survey, to average out small measurement errors related to the probe. Careful handling and storage of the inclinometer probe is required, as shock and rough handling can lead to scaling, zero offset and

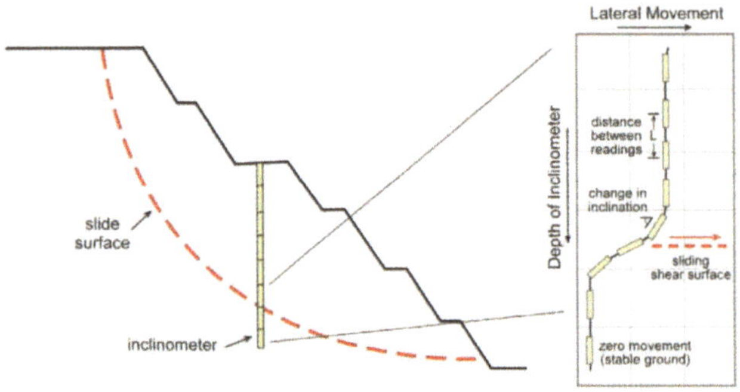

Fig. 3.39: Working principles of inclinometer monitoring to detect the location and amount of displacement developing along a sliding surface (source: E. Eberhardt).

azimuth rotation errors. These can lead to differences in the measured displacement profile between repeat surveys, that may be misinterpreted as subsurface ground movements. The error range is generally reported as 0.25 mm per reading or 6 mm accumulated over 50 readings.

A common source of poor-quality data is related to poor conformance of the casing with the surrounding rock. This is usually due to air pockets that might form around the outer casing annulus in response to a poor job of cementing the casing in place. Air pockets that occur along distinct shear surfaces can produce more gradual deformation than that which is actually occurring at a specific location.

Deformation at the distinct shear zone will cause bending of the casing, and excessive bending will prevent passage of the probe. This points to a key limitation of probe inclinometers: when deformations exceed the limits required for the inclinometer probe to pass freely, subsequent measurements below that point will no longer be possible. This might require a new inclinometer hole to be drilled to continue monitoring of the sliding surface from the same location. As previously noted, where substantial deformation is anticipated along distinct shear zones, a large diameter casing and small diameter probe should be used.

Another limitation of probe inclinometers is that they need to be manually operated, and therefore tracking displacements at depth over time will be limited to a temporal resolution corresponding to the frequency at which the measurements can be made by the geotechnical personnel. To address this limitation, in-place inclinometers can be used to monitor continuous movements at fixed points along the borehole. In-place inclinometers may consist of one or two sensors located at depths of known shear or sliding surfaces, if established beforehand using probe inclinometer surveys, or may involve several probes placed in a chain to detect the location of sliding surfaces as well as to monitor for continuous movements. The latter obviously increases the cost of an installation as it requires purchasing multiple inclinometer sensors. Figure 3.43 provides an example comparing real-time continuous data collected by an in-place inclinometer with data corresponding to the same point extracted from periodic conventional inclinometer probe surveys.

Fig. 3.40: Inclinometer casing installation, including alignment of inclinometer casing grooves with expected direction of slope movement (source: E. Eberhardt).

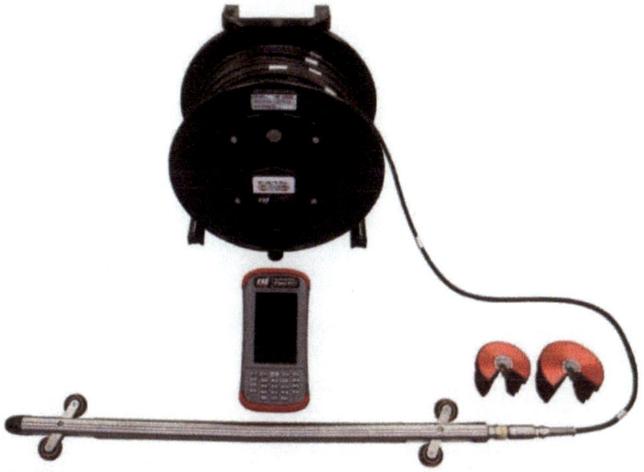

Fig. 3.41: Inclinometer probe with cabling and readout units (source: RST Instruments Ltd).

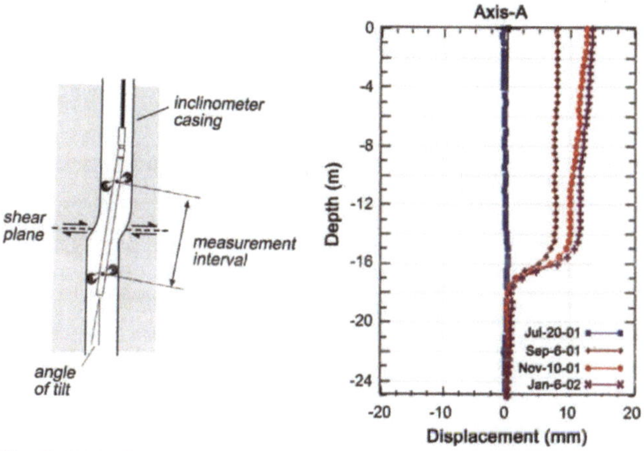

Fig. 3.42: Inclinometer measurement of change in tilt at the measurement point and example output from a probe inclinometer (source: E. Eberhardt).

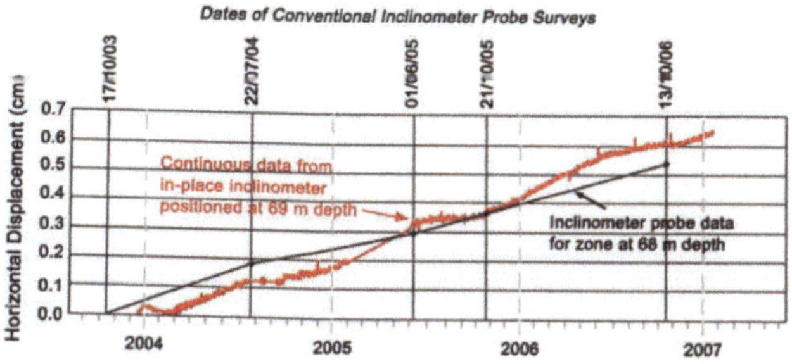

Fig. 3.43: Real-time continuous displacement data provided by an in-place inclinometer (in red) compared to periodic inclinometer probe data extracted at the same depth (in black) (modified from Willenberg *et al.* 2008).

3.6.4 Shape accelerometer arrays

Similar to in-place inclinometers, shape accelerometer arrays (SAA) involve a string of rigid segments separated by flexible joints that move in any x,y,z direction, but do

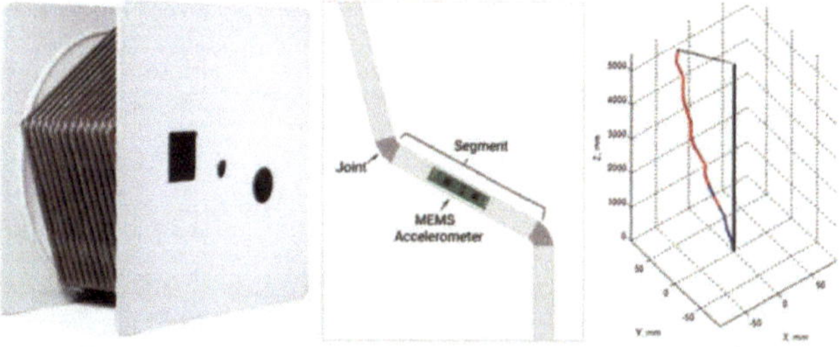

Fig. 3.44: Shape accelerometer arrays (SAA) with sample 3D deformation output (source: Measurand Inc.).

not twist. Each segment houses three MEMS tilt sensors capable of measuring tilt in two directions (Fig. 3.44). Data processing transforms the tilt of each segment to produce the shape of the array and any changes to this shape. Specifications list an accuracy of 1.5 mm over a 32 m length and precision of 0.5 mm for a 0.5 m segment.

Installation of SAA is similar to that of TDR or in-place inclinometers. The string of sensors is unrolled from a drum in lengths of up to 150 m and is installed inside a casing that is grouted into a borehole. This presents the opportunity to install SAA in inclinometer casings that are too deformed for probe inclinometers to be used. Because the sensors are already joined to one another, there is only a single wire output that needs to be managed. SAA can be used in vertical, inclined or horizontal boreholes.

Advantages of SAA include high-resolution data and high survivability under large deformations. In addition, SAA sensors can be read automatically to provide real-time measurement (typically 10- to 20-min cycle time to monitor a 50 m array) and set with a trigger alarm. The sensor casing is waterproof to 200 m of groundwater head and relatively robust. The main disadvantage of the MEMS sensors is that accuracy may vary from sensor to sensor within the array due to the differing orientation of the accelerometers. This is because MEMS sensors do not exhibit linear sensitivity across their measurement range. However, 360° calibrations are carried out on each sensor axis across the full operational temperature range and offsets are applied to correct for this. Other disadvantages are that the length cannot be easily altered and a higher level of technical expertise is required for installation and operation. Figure 3.45 provides an example of their use to track the subsurface displacements below an active haul road in response to caving-induced subsidence during concurrent operations of the open pit and a sublevel cave at the Telfer gold mine (Nicoll *et al.* 2017). The installation of Smart Markers is also indicated in Fig. 3.45; these are described in Section 3.6.5.

3.6.5 Smart Markers

Installing and maintaining traditional in-ground monitoring systems in an open pit mining environment

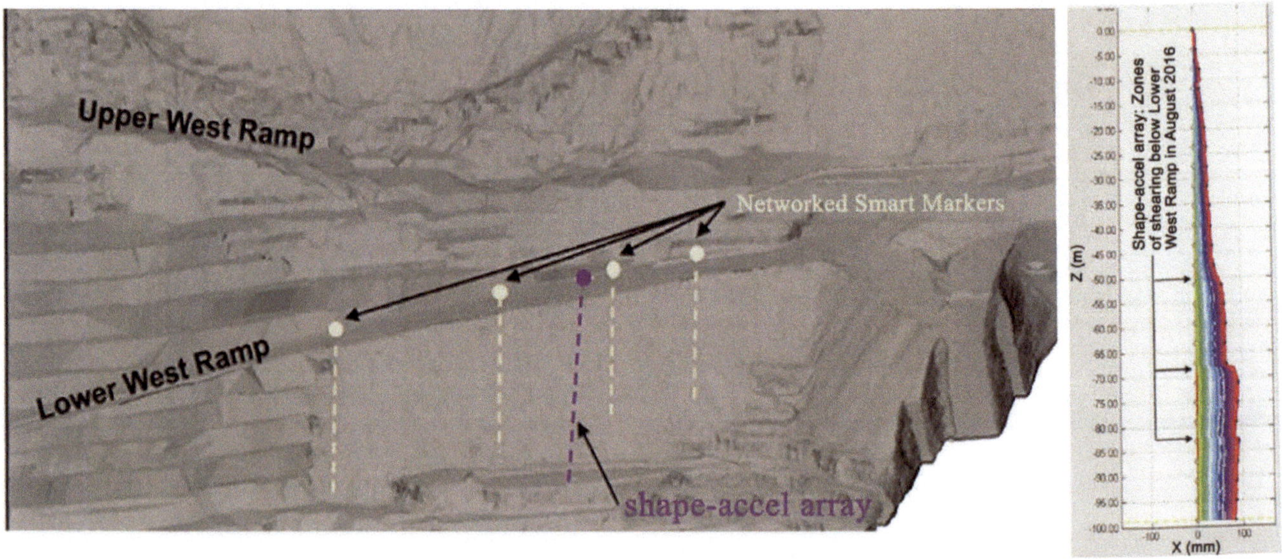

Fig. 3.45: Example use of shape accelerometer arrays (SAA) to provide high-resolution subsurface deformation data below an active haul road in response to an adjacent sublevel caving operation (source: Nicoll *et al.* 2017).

can be challenging. Conventional subsurface monitoring systems depend on a physical connection (cables and casings) between sensors that are installed in the ground with data acquisition systems (dataloggers) located on the surface. Often, even small ground movements may damage the cable, rendering the system inoperable. Reliance of monitoring systems on cables for power, data acquisition and data transfer presents particular challenges, including the following:

- the ground near the pit wall crest may experience movement due to the influence of mining activities and destabilisation of a slope that damages the cables;
- if the ground behind the pit wall crest is being monitored and the pit wall is pushed back, the cables contained in the wall will be damaged or disconnected. Finding and reconnecting the cables will be difficult; and
- tailings embankment raises are constructed to expand the tailings impoundment's capacity. Performance monitoring instrumentation is installed within the foundation and embankment. During embankment raise construction, the previous stages are covered with new material. Cables need to be protected and extended during the construction process and reconnected, which adds to the complexity of the extension.

As the disturbance zone around the open pit is experiencing ground movement, insights into subsurface rock mass behaviour and slope performance are needed. Yet, this is also the time when traditional monitoring systems may be damaged due to cable or casing damage. When this occurs, the operation relies mainly on surface monitoring systems to evaluate performance, which may limit the geotechnical engineer's ability to execute informed mitigation or remedial stabilisation measures.

To address these limitations and potential risk-related implications associated with conventional cabled subsurface monitoring systems, there are wireless in-ground monitoring platforms that do not rely on in-ground cables for data transfer. This allows a greater degree of freedom in the installation and maintenance of the monitoring instrumentation.

3.6.5.1 Functionality and measurement principles

Networked Smart Markers (NSM) make use of battery-operated radio transceivers built into a 34.5 cm long and 6.35 cm diameter plastic casing filled with epoxy for improved blasting resistance (Fig. 3.46). These have been developed to include the following characteristics (see also Whiteman 2010; Widzyk-Capehart *et al.* 2015):

- wireless transmission of data from each installed sensor to surface;
- multi-year battery-powered operation;
- several sensors in a single enclosure and multiple sensors in each borehole;
- high density of data and ability to process data in near real-time; and
- robust instruments with a rugged enclosure to survive the mining environment (including blasting).

NSMs operate as a mesh-network and at the time of writing have a typical communication range of ~5 m through rock. In a typical installation, Smart Markers are installed in drill holes at 2 m spacing intervals. A sensor

Fig. 3.46: Smart Marker in-hole devices and datalogger (source: Exelon Mining).

array for down-hole installation is shown in Fig. 3.47. They can also be installed along the ground surface, such as in the sub-base of a haul road, as indicated in Fig. 3.45. This interval allows each marker to communicate with at least two neighbouring markers. Dataloggers installed at the hole collar control communication with the markers in the hole. Any marker in the hole can be addressed with instructions through handing data-packets from one marker to the next until the instructions reach the destination marker. The destination marker will execute the instructions then send the results back through the chain to the datalogger.

NSMs are able to issue a radio broadcast to which any neighbouring markers within radio range will respond. The issuing marker will record the responding markers' addresses and the signal strength that is associated with the radio communication. The complete list of

Fig. 3.47: Smart Marker system before installation with in-hole devices spaced and attached to a grouting tube (source: Exelon Mining).

neighbouring markers and their radio signal strengths are then sent back to the datalogger. Regular querying of all markers in the monitoring holes allows analysis of changes in radio signal strengths over time. The level of radio signal strength is mainly a function of distance and alignment between markers. Changes in radio signal strength over time are indicators of ground movement between specific markers. Experiments have shown that this system has the sensitivity to detect ground movement on the order of 5 cm or more.

3.6.5.2 System performance

NSMs were initially developed for block-caving operations to track the propagation of the cave back. Wireless in-ground monitoring offered an improvement over cabled systems such as TDR, which would shear off at points well above the cave back, limiting the value of these much-needed data for safe and optimised cave management. For open pit slopes, the NSM's ability to communicate wirelessly through rock is similarly useful for monitoring slope movements within the highly sheared damaged zones that can develop within the slope and limit the service life of cabled monitoring systems. However, the sensitivity of the NSM system (<5 cm), although sufficient for its use to monitor cave backs, is insufficient for the smaller ground movements associated with developing slope failures in some rock mass environments.

To enhance the system's ability to provide early warning of pit slope instability, it was necessary to incorporate the functionalities that in-place inclinometers provide, while also considering the opportunity to measure pore pressures, which is typically done using vibrating wire piezometers (VWPs) (see Section 3.8). For tilt-sensing, a three-axis accelerometer and magnetometer were added. For pore pressure sensing, an industrial pressure sensor was added. As a result, a new generation of NSMs (called Geo4Sight) was developed to enable rock movement detection and pore pressure measurement. These devices allow the datalogger to instruct any device in the hole to take measurements and report the data back to the datalogger. Querying of the devices can be performed *ad hoc* or on a scheduled basis. The data are collected by the datalogger. If the datalogger is connected to the network, then the server software can be used to collate, analyse and utilise the data for visualisation and further remote data processing. Tilt and pore pressure sensing devices can be installed in any combination in the same hole. This achieves a high density of sensors within one hole and co-locates the measurement of deformation and pore pressure. The integration of other sensor types, including extensometers and strain gauges, is also possible.

At present, these multifunctional markers are sensitive to changes in tilt of ~0.02°, or 0.3 mm in 1 m. This sensitivity is not dependent on the orientation in space of the sensors, in contrast to the requirements of traditional in-place inclinometers. NSMs can be installed in any orientation in the ground. The accuracy of the pore pressure sensors used in NSMs is 0.125% of full scale, which is comparable to existing VWPs. The integration of data measured by a chain of sensors within a hole allows the interpretation of ground movement and pore pressure distribution over the length of a hole.

3.7 Surface water monitoring and management

3.7.1 Overview

Surface water is an important consideration for pit slope stability. There may be a need to protect the slopes from erosion damage, and it is usually important to divert water from the slopes as quickly as possible to minimise the amount of recharge that can potentially infiltrate and increase pore pressures within the slope. This may be particularly important for slopes excavated in low permeability, poorly drained, weak rock masses.

Uncontrolled surface water flow may lead to:

- erosion and back-cutting of weaker materials within the slope;
- shallow recharge, causing transient pore pressure in the disturbed zone that may create bench-scale instability which, in turn, may lead to larger pit slope movements; and
- deeper recharge into the slope, creating inter-ramp-scale instability that is often associated with a transient pore pressure increase within adversely oriented primary or major structures.

A well-managed surface water control program will reduce the potential for transport of eroded materials into the lower part of the pit, which may itself lead to damage of dewatering infrastructure or loss of mining access. The required scale of the surface water monitoring program depends on the climate, topography and site setting, dimensions of the pit, and pit slope geology. As a general rule-of-thumb, a detailed surface water management plan is required for all operations where the mean annual precipitation is greater than ~250 mm. A storm water management plan is required for virtually every major mining operation, even those in the driest of climates.

3.7.2 Meteorological monitoring

All open pit operations require some level of understanding and monitoring of climatic variables. The goals of the monitoring program may include:

- input to surface water analysis and design of ex-pit diversions;
- input to the surface water management plan for the pit;
- the design basis for in-pit sumps and pumping systems;
- an assessment of groundwater recharge;
- an assessment of infiltration to the pit slopes and the potential for transient pore pressures; and
- closure planning for the pit, including pit lake analysis.

The amount of detail required will depend on the site setting and topography, the size of the operation and the prevailing climate.

3.7.2.1 *Types of data*

The following types of data are usually important for open pit operations.

- *Mean monthly precipitation and evaporation.* For active operations, monthly data are fundamental for the prediction of external and in-pit surface water runoff flows, and for the design and monitoring of diversion systems. Prior to mine development, this information is typically used to provide input for planning, screening level assessments, and analysis of alternatives. At an early stage in the project, before establishing an on-site weather station, precipitation data can be obtained from regional stations. In some dry climates, monthly precipitation and evaporation data may be the only meteorological information that is required.
- *Daily (or hourly) precipitation.* Daily precipitation data are sometimes required to assess the peak rainfall. Runoff analysis is necessary to support the surface water management plan for the pit, and for the planning, design and monitoring of any required up-gradient surface water diversions. Daily precipitation data may also be required to assess the potential for infiltration of water and transient pore pressure in the pit slopes. There are an increasing number of operations capable of achieving a correlation between daily rainfall, piezometer response and slope deformations (Fig. 3.48). For early-stage planning where the pit is located in a new mining area, it is often necessary to use existing weather stations or published data as input to the studies.
- *Daily evaporation.* Evaporation is spatially less variable than precipitation, so it is normally possible to provide reasonable estimates using regional data, provided that the localised effects of topographic elevation, aspect and wind exposure are understood. Topographic aspect (including orientation aspect of the pit slopes) may also be important for assessing local-scale variability. The use of solar radiation, wind speed and direction, and other micro-climate data obtained from an on-site

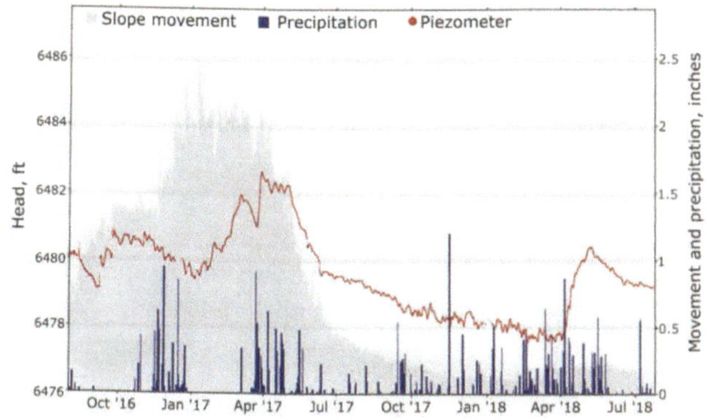

Fig. 3.48: Example of daily precipitation and piezometer response correlated with slope deformations (source: J. Weigel and G. Beale).

weather station may be important for calculating evaporation in areas where reliable pan evaporation data are not available. Daily evaporation data are also often a fundamental part of any closure studies for the open pit and other mine facilities.

- *Ambient air temperature.* Air temperature data are important in colder climates for determining how much of the precipitation may occur as snow and over what period of time the snow may accumulate. A knowledge of air temperature may also be used to assess the potential for freezing of the pit walls which may lead to (i) a reduction in the seepage rate that may contribute to higher pore pressures behind the wall, and/or (ii) the development of ice formations which may require management as part of the geotechnical program, particularly where bench face angles are steep (Fig. 3.49). Daily, monthly and annual records may be available from regional stations, before the establishment of an on-site weather station. A daily record of temperature and solar radiation may be generated using a stochastic approach. In cold climates, some operations have been able to correlate air temperature (and snowmelt potential) with deformation using radar, prisms and InSAR.
- *Wind speed and direction.* Wind data may be important for helping to monitor and control dust, particularly where the operation is located close to a population centre or in an environmentally sensitive area. Wind data may also be important for mine closure planning, to help assess limnology and turnover for the pit lake. Records may be available from regional stations before the establishment of an on-site weather station.
- *Snowfall and snowpack.* In cold climates, analysis of snowfall and snowpack data may be important to assess the potential amount of accumulating snow in the upper part of the pit, or for assessing spring thaw and runoff from the up-gradient catchment area during the seasonal melting period. The depth and nature of the snowpack can be locally variable, but indicative data are available from public records. In the USA, many open pit mines obtain Snotel data to help assess runoff potential and to correlate snowmelt with piezometer responses.

Fig. 3.49: Ice formation on the pit walls may cause a hazard, particularly where the bench face angles are steep (source: G. Beale).

3.7.2.2 Data sources

Data may be obtained from a combination of publicly available and site-specific monitoring sources. The data available through public sources will vary depending on the proximity of the project site to populated areas, the number and type of existing weather stations in a given area, the type of data and the length of data record. Available data from existing stations may range from daily precipitation or daily temperature only, to a comprehensive suite of measurements that may include continuous precipitation and temperature, barometric pressure, wind speed and direction, humidity and solar radiation. Often, data from government sources, meteorological organisations and airports can be accessed via the internet.

It is usually beneficial to establish a site-specific weather station as early as possible in the project development process, particularly if the closest climate stations are a significant distance away or at different elevations. By comparing site-specific climate data with data from nearby weather stations, it may be possible to establish empirical relationships between the site data and regional stations where the period of record is longer. This can be important for a site located at a different elevation from the nearest weather stations and/or where orographic effects may result in different weather conditions.

3.7.2.3 *Requirement for a daily record of climatic variables*

In some cases, it may be necessary to derive a daily record of climatic variables to allow:

- a reliable assessment of peak rainfall and surface water runoff from the up-gradient catchment area and/or from within the pit area; and
- an assessment of surface infiltration and recharge to the pit slopes to support pore pressure studies and piezometer monitoring results.

For the majority of sites (and particularly for early-stage studies), there is insufficient site-specific data to allow the development of a long-term dataset of daily climatic variables. Therefore, if required, it may be necessary to develop a synthetic record of precipitation, temperature and evaporation, usually over a 100-year period. When developing this record, it is important to ensure that the dataset includes peak rainfall events, wet cycles, dry cycles and other extreme conditions or trends.

3.7.2.4 *Influence of freezing conditions*

In many regions, some or all of the precipitation may occur as snow. Most model inputs assume that precipitation at mean daily temperatures above freezing occurs as rain while precipitation at mean daily temperatures below freezing is stored on the surface as snow. Snowmelt can be computed using standard procedures (e.g. the Snow Accumulation and Ablation Model, SNOW-17). Using this procedure, the melt process is divided into precipitation that occurs during non-rain periods and precipitation that occurs as rainfall. Rain-on-snow melt is computed using an energy balance approach. Non-rain melt is computed using air temperature as an index to energy exchange across the snow–air interface.

For any analysis, it is important to realise the uncertainty that may result from:

- the redistribution of snow by wind, which may have a major impact on the soil moisture balance of the natural up-gradient catchment area and the pit slope itself. The effects of wind can greatly influence the actual flux of water that penetrates below the near-surface layer and causes recharge; and
- the influence of frozen surfaces on the infiltration flux, considering that the insulating effect of snow may prevent the onset of surface freezing or may cause the materials beneath the snow to remain frozen when the surrounding surfaces have thawed.

3.7.3 In-pit surface water monitoring

3.7.3.1 *Control of runoff*

Surface water control measures may need to be considered for large open pits, including:

- diversion of larger streams that may affect the pit area or create recharge to the slope materials that are exposed in the upper pit walls;
- diversion of local up-gradient catchment runoff;
- diversion of runoff from the upper pit benches, along drainage channels constructed on catch benches or, alternatively, step-out benches that allow gravity drainage to low points on the pit crest;
- surface water collection facilities on individual catch benches or, alternatively, out-sloping of catch benches to shed localised runoff as sheet flow with low erosion potential;
- placement of surface water collection facilities below prominent zones of groundwater seepage in the slope; and
- installation of interception trenches or slurry walls to prevent shallow subsurface seepage from sources of surface water near to the pit (e.g. rivers or dams).

Details of these systems are described in Beale and Read (2013). It is usually beneficial to intercept as much surface water runoff as possible external to the pit (or on the upper benches).

3.7.3.2 *Planning the monitoring program*

The surface water management plan requires consideration of current conditions, the future mine plan, and potential future changes to the mine plan. The monitoring program will need to consider:

- rainfall, including whether specific monitoring is required for the pit area because of site-wide rainfall variations due to topography or prevailing wind direction;
- up-gradient diversions and the need for (and location of) continuous monitoring stations for flow, chemistry and sediment loading, and the need for an early-warning system for high-flow events. Monitoring of up-gradient flows and chemistry may also be important for helping to assess the magnitude of potential mining impacts;
- crest diversions and the need for temporary or continuous monitoring of flow rates, with consideration of design changes because of future pushbacks;
- collection ditches to install weir plates for flow monitoring and to allow chemistry and/or sediment sampling of the runoff waters. Consideration should be given to how conditions will change as the mining progresses;

- intermediate (higher-level) interceptor sumps and whether they would be beneficial for reducing the volume of runoff reaching the pit floor, how frequently they will need to be operated, and how they will be monitored for flow, chemistry and suspended sediment;
- pit floor sumps, including their locations, with consideration of how the locations will change during mining and how they will be monitored for flow, chemistry, suspended sediment and accumulating sediment;
- accumulation of water in the pit and whether the required peak pumping rate can be reduced by the operation of spilt floors, or mining of alternate higher dig faces seasonally or following extreme rainfall events;
- the pumping system and whether a single-stage (single-lift) pumping system is appropriate, how it will be monitored and whether telemetry controls may be required. A multi-stage system using booster pumps may be required instead, either from the outset or phased in as the pit gets deeper;
- integration of groundwater and surface water pumping, and whether it is better to integrate the pumping system for groundwater and surface water, considering that groundwater flows tend to be more sustained and less seasonal, and that groundwater pumped from wells or collected from horizontal drain systems tends to have a lower suspended solids content;
- requirement for sediment control or polishing ponds to improve water quality and reduce sediment loading before discharge;
- segregation to assess whether clean water (up-gradient runoff) and dirty water (mine contact water) need to be managed separately to minimise water quality deterioration. Sediment control ponds may be required for either clean water and/or contact water, depending on the topography and climatic conditions. Downstream monitoring and management of contact water is an increasing consideration; and
- monitoring of receiving waters and down-gradient receptors. Virtually all mine sites need to operate a rigorous down-gradient monitoring program for surface water protection, including water supplies, riparian areas and ecological systems. Establishment of the program is usually carried out as part of the baseline characterisation or early in the mine life.

Careful consideration and planning are required to determine where best to locate and manage the dewatering and monitoring infrastructure. Flow meters and control valves can often be positioned some distance from the active mining area in protected locations. Telemetry can be used where access is difficult. All monitoring installations require highly visible demarcation.

The detailed design of the monitoring program may further need to consider:

- rainfall, frequency, and the magnitude-return periods for specific storm events (which will depend on the size of the drainage area). A design based on a 24-h event is of little value for a small catchment where the time of concentration is a few hours or less;
- the catchment area, topography and nature of the geology in the up-gradient drainage area; and
- the nature of the materials that comprise the pit slope, including whether they are strong and highly permeable, or weak, deformable, of low permeability and potentially erodible.

3.7.3.3 *Monitoring system components*

The monitoring systems should involve any combination of the following:

- larger or strategic diversions, including permanent flumes or weirs, pressure transducers and dataloggers to track the stage at a flume or weir on a continuous basis, and early warning of high flows and formulation of a TARP;
- up-gradient or in-pit runoff channels, including temporary (portable) flumes and weir plates for measuring up-gradient or in-pit runoff, and synoptic (one-time) flow gauging, water quality or sediment sampling during a particular runoff event (or series of events);
- sumps, including monitoring of sump water levels using calibrated pressure transducers, monitoring of sump flows using in-line flow meters, frequent water quality monitoring for field parameters (typically pH, temperature, total dissolved solids, electrical conductivity and total suspended solids), water quality monitoring for laboratory analysis, and frequent monitoring of sediment loading and suspended solids content; and
- combined mine discharge water, which usually requires permanent in-line flow meters and includes regular or continuous sampling of water chemistry parameters, regular or continuous monitoring of sediment, and potentially upstream and downstream sediment control ponds.

The number of monitoring installations and frequency of monitoring will depend on the prevailing climatic conditions and the importance of surface water management. If surface water control structures are critical to the safety of the mining operation, these may have flumes and continuous dataloggers, which will usually be inspected daily. In arid climates, surface water flow and quality monitoring may be event-based with data collected during seasonal or instantaneous periodic runoff. It is usually beneficial to install a flow meter (cumulative and instantaneous) on all pumping systems rather than attempt to calculate flow rates based on the run times of the pumps. Where mine water discharge is sensitive to water quality, the installation of real-time water quality monitoring sensors may also be advantageous.

Surface water monitoring data will usually be presented in the form of:

- maps showing flow rate, water quality parameters and sediment loading;
- hydrographs of flow rate, water quality or sediment loading in diversions and drainage ditches;
- hydrographs of flow rate, water quality or sediment loading from sumps;
- hydrographs of total mine discharge (flow, water quality and sediment loading; potentially upstream and downstream of sediment control ponds or polishing ponds); and
- comparison of mine water discharge to receiving waters or natural catchment runoff.

3.8 Groundwater monitoring

3.8.1 Overview

The presence of pore water pressures can critically influence the stability of a pit slope as well as the stability of waste rock, tailings dams and other soil/fill slopes. Thus, pit slope designs are often based on pore pressures remaining below a certain (target) threshold. Groundwater monitoring helps to ensure the required pore pressure distribution is being achieved. This verification, including that of any pit dewatering and slope depressurisation measures being taken, requires a robust monitoring system that (i) allows responses to the pit slope dewatering and depressurisation programs to be observed, (ii) provides a comparison between predicted and actual conditions and (iii) is sufficiently flexible to respond to new hydrogeological and geotechnical data, revised geology interpretations and changing mine plans (Lomberg *et al.* 2013).

The groundwater monitoring program for an open pit typically involves the use of boreholes/wells and piezometers. Piezometers are an integral component of any pit slope monitoring system. They may take the form of:

- boreholes that are temporarily left open (with or without casing);
- standpipe piezometers or observation wells (single- or multi-level);
- VWPs placed in sand packs;
- grouted-in VWPs;
- horizontal piezometers; and/or
- multi-level measurement and sample collection systems (e.g. Westbay piezometers).

A piezometer provides a means to measure the total groundwater head at any given location. The total groundwater head is the height of the water column above a reference or datum surface. It consists of two components: (i) the elevation head, which is the height of the point above the datum level, and (ii) the pressure head (or pore pressure), which is the pressure at the point divided by the specific weight of the water column above that point. A piezometric surface can be created by contouring the total head values. The relationship between total head and pore pressure is shown in Fig. 3.50.

A transducer will make a direct measurement of pressure (though it may need adjustment for changes in atmospheric pressure). If the measurement is made by measuring the depth to water level below the top of the standpipe, and the density changes, the water level will

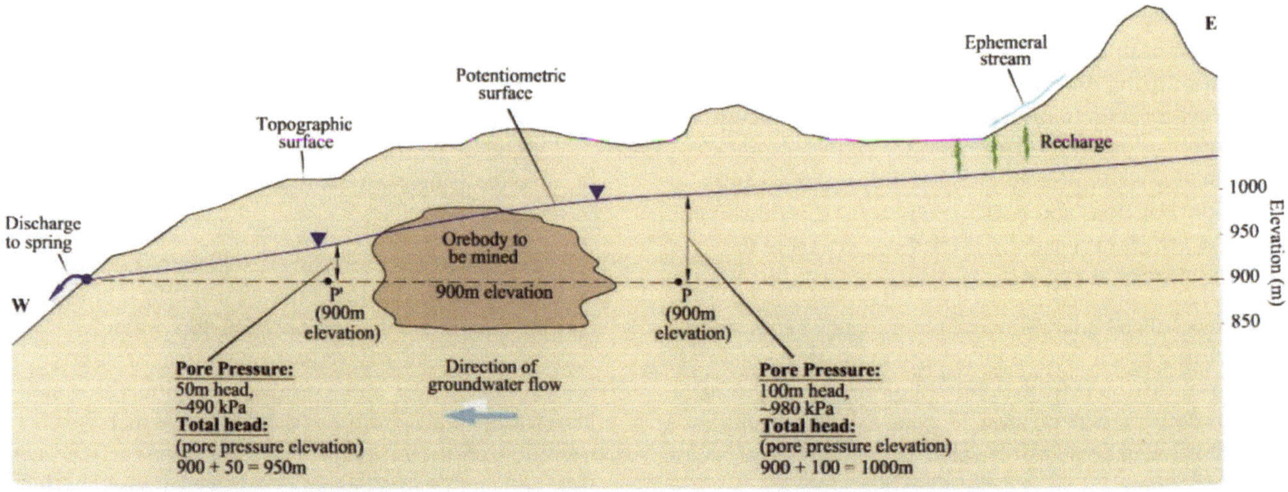

Fig. 3.50: Relationship between total groundwater head and pore pressure (source: Beale and Read 2013).

change even though the pressure does not. Care needs to be exercised when comparing head values derived from a combination of measurement methods.

The monitoring plan must consider that pit slopes above the water table may develop pore pressures as a result of recharge derived from precipitation or leaking facilities.

3.8.2 Open observation holes

The term 'open observation hole' is used to describe an open borehole which is unlined, except perhaps for the upper few metres, and is used for measuring groundwater levels. Open observation holes may have a small diameter monitoring pipe, also referred to as a standpipe (Fig. 3.51a), installed to provide stability against hole collapse, but they have no impermeable seal. Measurement is typically carried out using an electrical water-level sensing probe lowered into the observation hole. When the probe comes into contact with the water, an electrical circuit is completed and the device gives an audible and/or visual signal.

The water level recorded in an open observation hole represents an average of the pressure over the entire saturated interval. This is useful for broad-scale characterisation in the early stages of project development, where there has been little stress on the groundwater system and where significant vertical groundwater head gradients may have not yet developed. However, where vertical gradients, perched water levels or artesian conditions are present, open holes or lined holes with long open or perforated intervals can provide a cross-connection between formations or parts of formations where the heads vary with depth. Around large open pits, the vertical hydraulic gradients are often much steeper than the horizontal hydraulic gradients and, by providing a connection for water to flow vertically through a borehole or standpipe, such holes may be altering the very conditions that they were installed to measure.

3.8.3 Standpipe piezometers

The traditional piezometer is an open standpipe piezometer, which allows the water pressure above a sealed interval targeting the depth of interest to be measured as a head of water in a pipe. Historically, 'Casagrande' piezometer tips were installed, but the piezometer may consist of any filter tip or perforated interval of pipe of a desired length attached to the bottom of impermeable casing that rises to the surface.

Standpipe piezometers typically require a sand pack (or gravel pack) to be emplaced in the annulus adjacent to the perforated interval (or filter tip), followed by a bentonite seal in the annulus 3–5 m above the sand pack, and further grouting of the annulus above the bentonite seal to surface to prevent any flow of water into the filter from other horizons. It is usually necessary to install the sand pack, bentonite seal and grout using a tremie pipe that is progressively raised in the hole, as the materials are

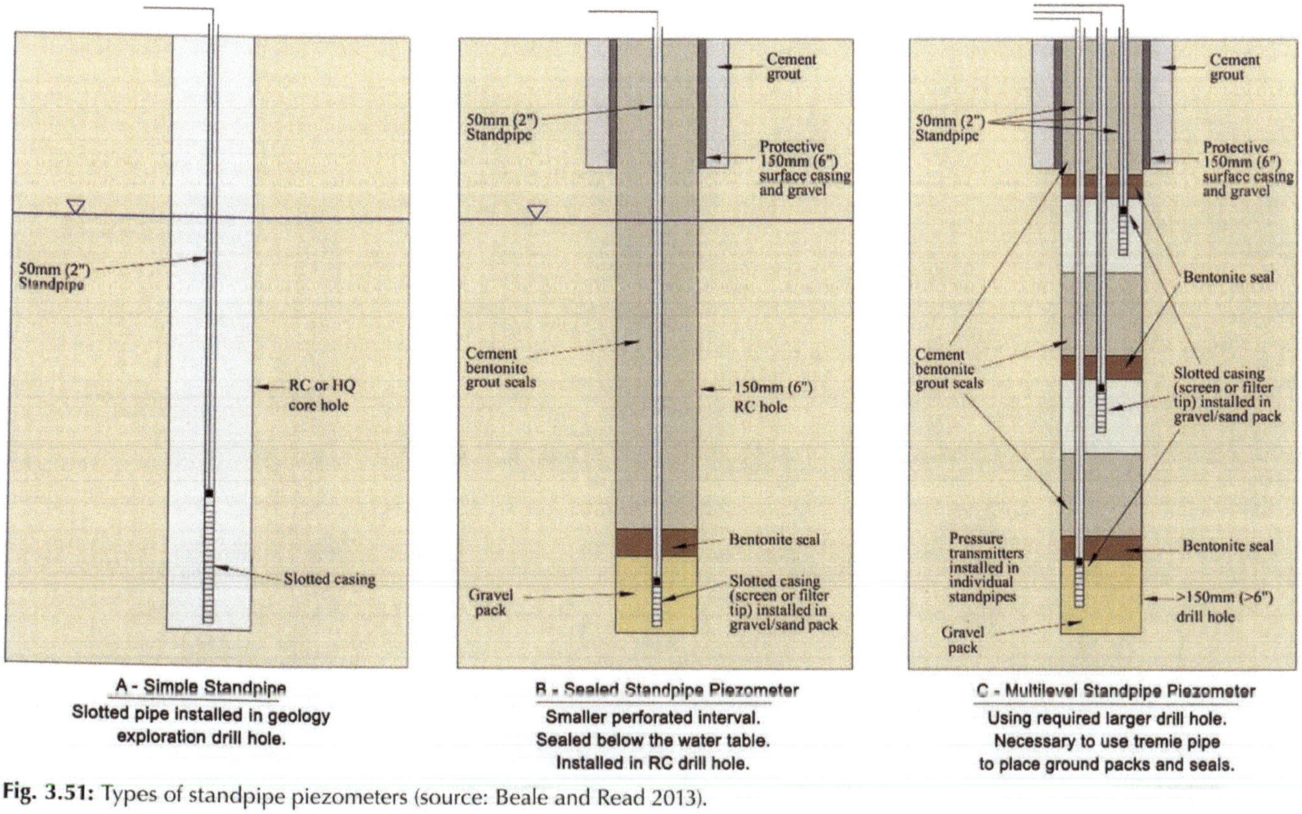

Fig. 3.51: Types of standpipe piezometers (source: Beale and Read 2013).

emplaced in the annulus and are eventually removed from the hole once the entire piezometer completion has been constructed. An electrical sensing probe can then be used to measure the water level corresponding to the groundwater pressure for the monitored interval.

Figure 3.51b shows a typical single-completion standpipe piezometer. The level to which the water rises in the pipe is indicative of the average pressure over the perforated interval. There are numerous standard hydrogeological texts that describe piezometer construction in detail, including Freeze and Cherry (1979), Kruseman and de Ridder (1994) and Domenico and Schwartz (1998). Standpipe piezometers can also be combined with inclinometer installations.

3.8.4 Multi-level standpipe piezometers

It is common practice to install multiple standpipe piezometers in a single borehole to allow the water pressure to be measured over several depth intervals from the same location. Where multiple standpipes are installed, the grout placed above the lower piezometer is brought up to just below the base of the selected interval for the second piezometer, and a bentonite layer is placed above the grout before the installation of the second pipe. Multi-level standpipe piezometers may include two, three or four installations at various depths in the hole.

Figure 3.51c shows a typical multi-completion standpipe piezometer. In cases where the hole condition is unstable or the hole diameter is small, it may be difficult to place the seals correctly in the annulus, meaning that the associated monitoring data are questionable. When considering a multiple completion, it is usually better to drill as large a diameter hole as possible (or to drill the upper part of the hole at a larger diameter).

To provide a more robust completion, it may be preferable to install the piezometers in individual holes drilled side-by-side off the same platform, rather than use a single hole. A compromise is often to place the upper and lower completion in one hole and the intermediate completions in a second (adjacent) hole.

3.8.5 Vibrating wire piezometers

VWPs provide a reliable and well-established method of measuring water pressure, in particular for characterising vertical hydraulic gradients around active pit slopes. There are several types of VWP transducers, and they may be either vented or unvented to the atmosphere. Unvented pressure transducers may be affected by barometric pressure changes. The use of unvented transducers during investigations that require a high level of sensitivity may mean that an additional barometric pressure transducer needs to be set up at a base station so that the piezometer readings can be corrected for changes in barometric pressure. This is also the case for standpipe piezometers, if the aquifer has a barometric efficiency. An explanation of how VWPs work is given in Beale and Read (2013).

Multiple transducers may be set in alternating sand/bentonite/sand packs that are placed using a tremie pipe, in

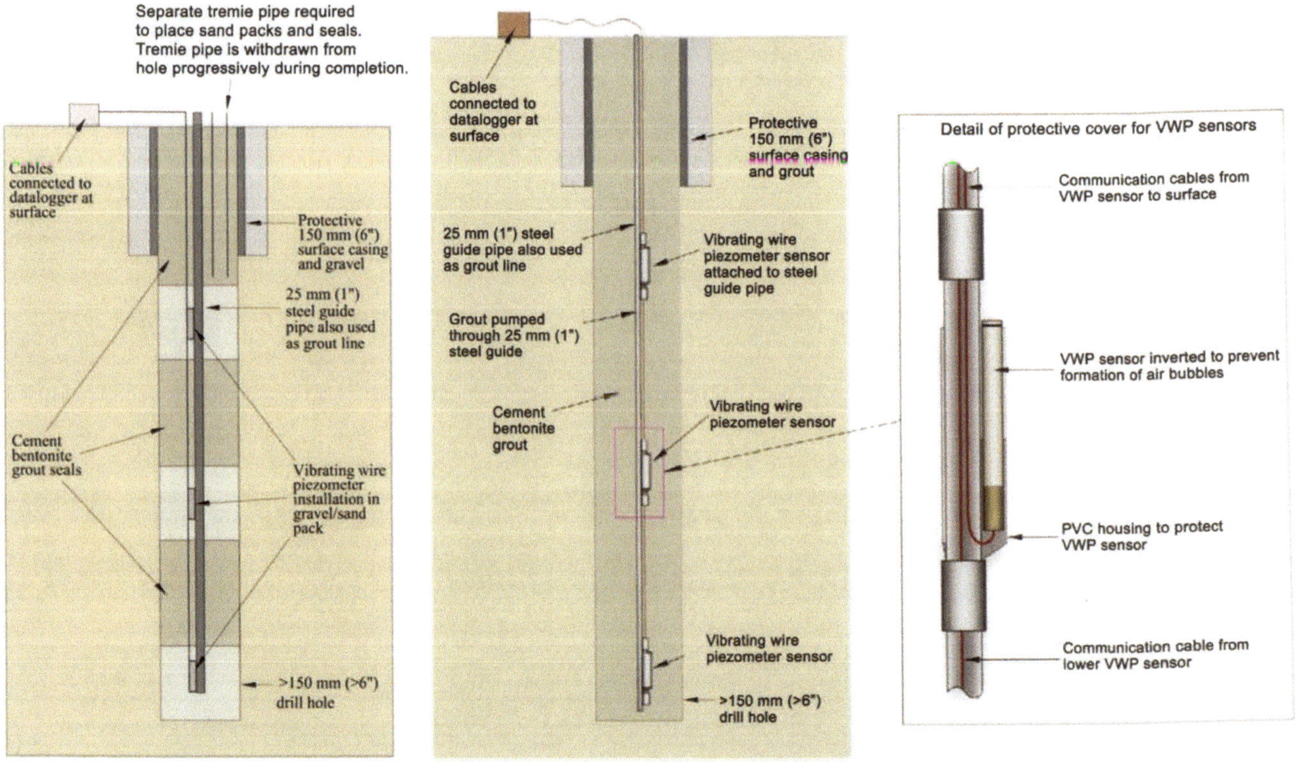

Fig. 3.52: Examples of vibrating wire piezometer (VWP) installations (modified from Beale and Read 2013).

a similar manner to multi-level standpipes (Fig. 3.52). Multi-level VWPs set in sand packs are commonly used in the construction and civil engineering industry, where installation depths are typically much less than that used for mining. Tremie pipes are commonly required for the greater depths required for open pit slope monitoring. Difficulties in placing the seals may be reduced compared with the standpipe installations because there are fewer pipes (and therefore more room) in the hole. However, poor and unreliable seals are frequently encountered, particularly in deeper holes or where the borehole wall is unstable.

3.8.5.1 Grouted-in, multi-level vibrating wire piezometers

In the late 1990s, the mining industry pioneered the application of grouted-in VWPs to provide multi-level pore pressure monitoring to help understand the complex heterogeneities and compartmentalisation common to large rock slopes. The method allows a large number of points to be instrumented in a single borehole to determine the independent pore pressure and hydraulic response at various depths, and to understand the vertical pressure gradients. The increasing use of grouted-in VWPs has greatly increased the general understanding of pit slope hydrogeology. Installations often make use of existing mineral exploration or geotechnical drill holes, lowering the cost and allowing numerous installations to be made rapidly at any given mine site.

A string of VWP sensors is grouted directly into the hole without the use of a filter pack. This provides a rapid way to install multiple piezometers in a borehole drilled at any angle or orientation, to install piezometers in holes with inclinometers or other geotechnical instruments, and to monitor pressures at discrete points. Key technical references that describe the theory and practice of installing grouted-in multi-level VWPs include McKenna (1995) and Beale and Read (2013).

Figure 3.52 shows an example of multiple grouted-in VWP sensors. Typically, two to five sensors are placed in the hole, but in some cases it is possible to install six or more sensors in HQ or RC size holes; some operators have installed as many as 10. However, where field experience at a given site is limited, it is usually sensible to start with fewer instruments in the hole (say, two to three) and slowly increase the number once the site-specific conditions are better understood and when site-specific installation procedures are developed by the mine operator.

There are several installation methods currently practised, depending on ground conditions, hole diameter, depth and angle. Several are described in Beale and Read (2013). Experience has shown that the most reliable seals are usually obtained when the hole is grouted from the bottom up using a tremie pipe. The ease of installation has increased as the diameter of the instruments has become smaller and the use of a single cable for multiple sensors has become more commonplace.

A potential concern in any multi-level completion is the quality of the grout seal between the sensors. A poor-quality grout seal may be exacerbated if there is formation damage surrounding the well bore due to drilling or where all the voids surrounding the drill hole are not fully penetrated by the grout, potentially because of entrained air. As for all piezometer installations, it is necessary to consider the actual field conditions when interpreting the results. Experience has shown that, with a properly monitored grout placement, vertical cross-connection is rarely a concern in small diameter holes because of the small length of the flow path between the side of the drill hole and the piezometer sensor relative to the length of the vertical flow path through the grouted drill hole.

3.8.5.2 Comparison of standpipe and vibrating wire piezometers installations

A standpipe piezometer allows the water level to be physically measured down the hole *below* a known reference point on the ground surface or collar casing. It also allows hydraulic testing and water chemistry sampling to be carried out. The continued performance and reliability of the piezometer relies on the integrity of the borehole being maintained. In comparison, a VWP measures the pressure of water above the installed sensor. The accuracy of the measurement relies on knowledge of the exact position of the sensor and the performance of the instrument. If the correct calibration or installation procedure is not followed, or the instrument fails once installed, there is no way of recovering it and measuring the water pressure.

Grouted-in VWPs provide a lower-cost method of measuring water levels at multiple vertically discrete points, using a smaller diameter hole than would typically be required for standpipe installations. If access to the wellhead is lost, readings from a vibrating wire sensor can be taken remotely by extending the cables or installing telemetry.

The measurement recorded by a VWP sensor is the discrete pressure at the level of the instrument. For fractured rock environments, where there is a large difference in permeability between the fracture zones and the surrounding rock mass, the lag time for piezometer response may, in theory, be large if the piezometer is not installed adjacent to a permeable fracture. However, the general experience to date when using grouted-in VWPs is that rapid responses to pumping are commonly observed even when the sensor is not positioned near a large fracture zone.

The use of grouted-in VWPs may be beneficial in low-permeability settings. Because there is no standpipe, and therefore no water in the completed hole, there are no well bore storage effects. Therefore, the response time of a VWP in a low-permeability setting is more rapid than a conventional standpipe.

3.8.5.3 Combination piezometer installations

It is possible to install standpipe piezometer completions with multi-level grouted-in VWPs installed at various levels in the annulus between the standpipe and the borehole wall. This approach may allow hydraulic testing or water sampling of the deeper interval with pressure measurements at multiple higher levels in the borehole.

The hole is completed by placing sand packs around the perforated interval of the standpipe and grouting the annulus of the hole above the perforated interval so that the VWPs become grouted in. It is often advisable to employ centralisers on the casing when using this method, provided there are no unstable zones within the hole where the centralisers may become hung up. Grouting of the annulus is best carried out using a separate tremie pipe.

3.8.6 Horizontal piezometers

During active mining operations, piezometers can be installed sub-horizontally from the pit slope, often as part of a horizontal drain drilling program. Similar to vertical piezometers, the target zone for pressure monitoring in a horizontal piezometer needs to be isolated and sealed. There are two practical ways to achieve this. The first is to use grouted-in VWPs, which allows multiple sensors to be installed within the hole. The second is to use a packer system and to grout the annulus inward from the packer. Examples of these installations are shown in Fig. 3.53.

Horizontal piezometers can be drilled at a slight downward or upward angle. However, down-drilled holes are typically easier because the grout can be pumped into the hole using a sacrificial tremie pipe and the hole is grouted from the bottom up, similar to a vertical piezometer. For up-drilled holes, the VWP sensors are attached to the sacrificial tremie pipe in a similar way, but the pipe needs to be pushed into the hole and the method of grouting needs to be modified. A second, short grout pipe is usually placed from the collar, and the collar is sealed. The grout is then pumped through the short pipe, and the hole is filled from the collar upward. The displaced air and water flow out of the hole through the deeper

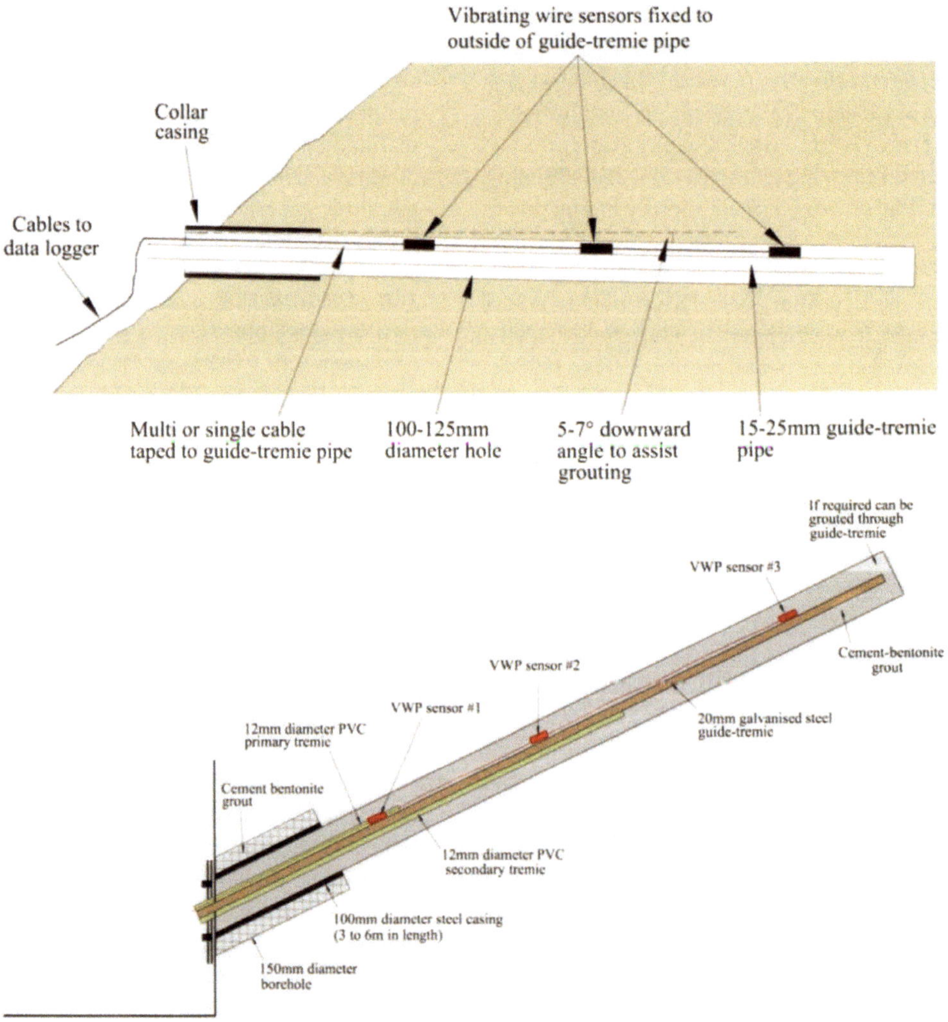

Fig. 3.53: Example of horizontal piezometer installations: down-drilled (upper) and up-drilled (lower) (modified from Beale and Read 2013).

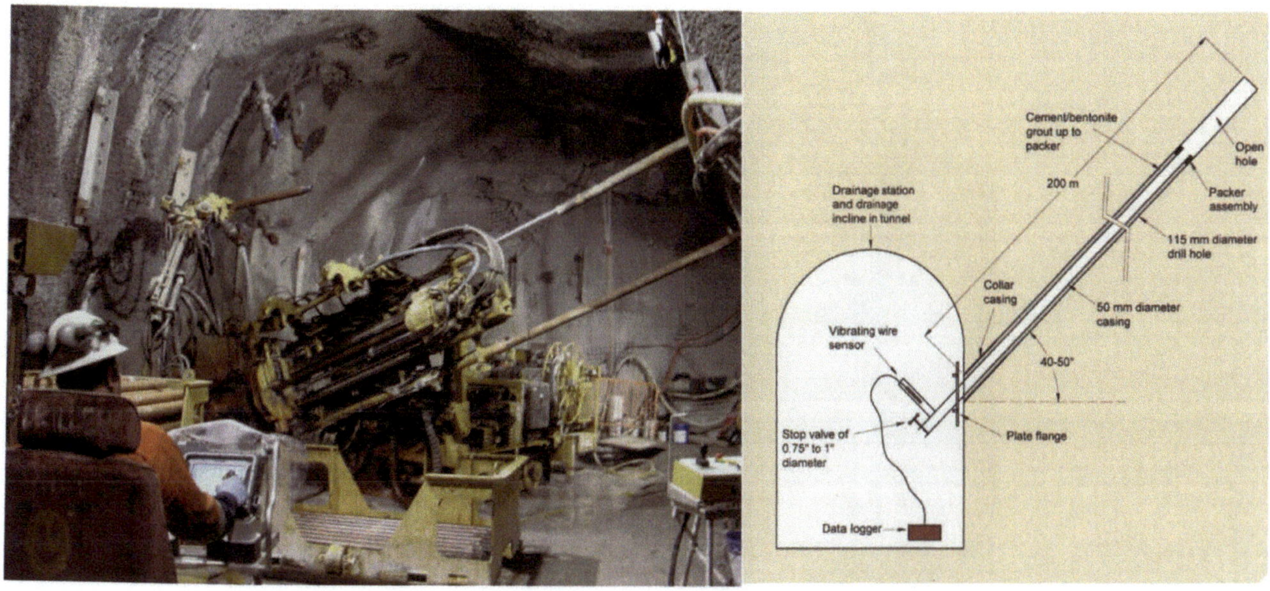

Fig. 3.54: Example of up-drilling (from a drainage tunnel) and the use of a packer to provide a discrete measuring interval (source: G. Beale).

tremie pipe, and the grouting is complete when grout returns are observed from the deeper pipe.

An issue with horizontal piezometers is that the weight of the grout does not exert any significant pressure on the borehole wall, so the grout seal may be less reliable. It is therefore better if the grout pump is used to provide positive pressure within the hole (say, 10 bar or more) to improve the quality of the grout seal. Pressuring the hole will help ensure good grout emplacement within the damaged and/or local fracture zones immediately surrounding the hole.

The designs shown in Fig. 3.53 can also be applied to piezometers installed in underground drainage tunnels excavated below the pit slopes (Fig. 3.54). Depending on the capability of the underground drilling setup, such piezometers can be installed at any angle to best target specific zones. Most underground drill holes are drilled at a relatively narrow diameter compared with surface-drilled holes. Although the piezometer string is best placed in holes of HQ size or larger, it is perfectly feasible to install multiple piezometer sensors in NQ size holes drilled from underground. For any VWP installation, because knowledge of the exact sensor location is essential for determining the groundwater head, all holes should be surveyed for deviation.

3.8.7 Westbay piezometers

Another option for multi-level groundwater monitoring is the Westbay system. The system was originally developed in the 1980s, before the widespread use of VWPs. It is a multi-level groundwater monitoring system that can be assembled in a wide variety of configurations, providing the ability to monitor virtually any borehole feature in an almost unlimited number of vertically discrete zones in a single drill hole. The Westbay system allows the user to measure formation pressures, obtain water chemistry samples and perform hydraulic conductivity testing from any of these zones (Fig. 3.55).

Some Westbay systems have remained in operation for over 25 years. Installations have included monitoring 50 or more discrete zones. For geotechnical applications, the system has been used to depths of over 450 m. Although the use of grouted-in VWPs has provided a more rapid and lower-cost method of multi-level data collection, a Westbay system can be considered where dynamic injection or cross-hole testing is required or where the collection of multi-level discrete groundwater samples, as well as pore pressure, is important.

3.9 Geophysical monitoring

3.9.1 Blast performance monitoring

Most open pit monitoring methods focus on the measurement of slope displacements brought about by mining-induced stress changes. Blast energy measurements seek to quantify the stresses being introduced into the pit slopes which have the potential to produce a reduction of shear strength that may impact local slope stability. Blast monitoring data are used in two ways.

1. In the analysis of complex mine slope displacement monitoring data, the slope engineer wants to understand acceleration events in terms of known specific destabilising factors, including excavation undercutting, pore pressure changes and blast energy impacts to stability. Key inputs to this analysis include blast energy and blast vibration data located in time and space.

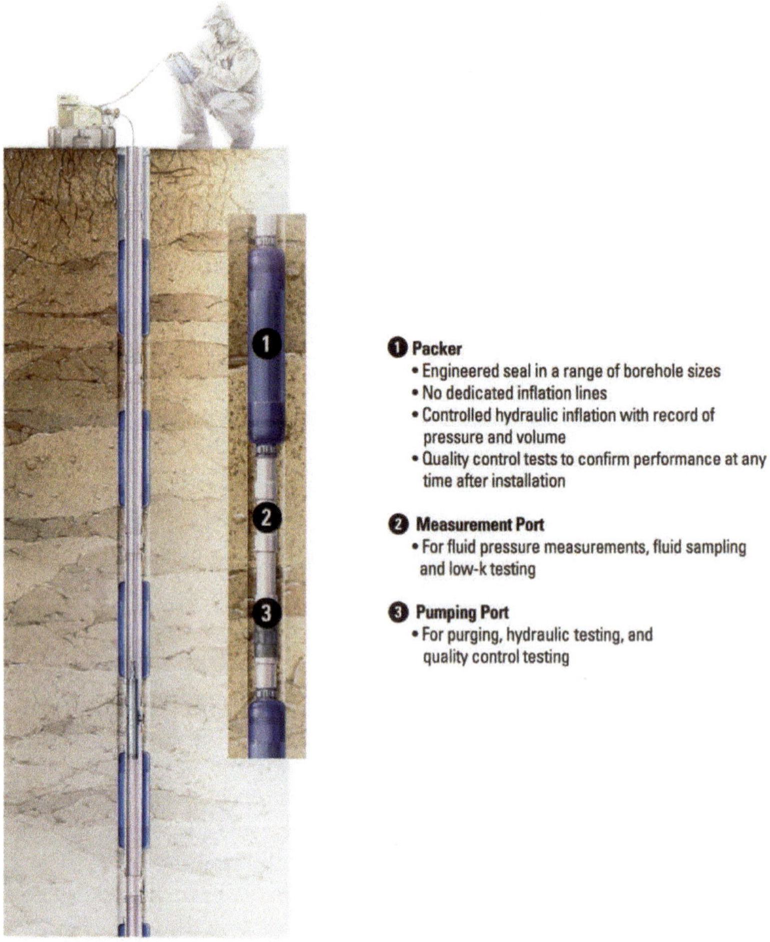

Fig. 3.55: Example of Westbay multi-level piezometer system and capabilities (source: Beale and Read 2013).

2. As a member of the team implementing mine designs, the slope engineer wants to understand and quantify the negative impacts of the blasting process on short- and long-term slope stability. Blast monitoring data are used to drive controlled blasting processes that achieve necessary fragmentation while minimising damage to slopes.

In these applications, blast vibration monitoring is not typical stability monitoring: it is used to interpret other stability monitoring data and as a process control to minimise bench damage and multi-bench slope instability. An explanation of the objectives and methods of blast performance monitoring and analysis is given in Chapter 11 of Read and Stacey (2009). This section will provide a brief summary of that work as it pertains to open pit slope monitoring, and describe advances and trends in the field.

3.9.1.1 Observations

In most open pit mines, various forms of controlled blasting are used to minimise damage to rock slopes, to achieve fragmentation objectives and to control costs. The performance of each controlled blast is monitored and analysed to ensure that blast designs continue to be refined to fit actual conditions. This performance should be evaluated in terms of damage in the near- and far-fields, as well as the vibration levels produced.

In the near-field, one of the best methods to study overbreak is with the use of standard-speed video cameras that focus on specific regions of the blast (e.g. the bench between the shot and the highwall bench face to observe displacement and overbreak) (Fig. 3.56). It is becoming commonplace to use high-resolution video cameras mounted on small UAV drones to obtain the images needed to effectively observe these blast regions. Typically, the drones hover in place over the wall side of the pattern (for a good view of the catch bench area) at an altitude that will minimise potential fly rock encounters. Drones are not cheap, but it might be reasonable to risk losing one or two a year in exchange for high-quality video that can clearly show impacts to the catch bench. It is possible to extract and register video frames that show overbreak fractures so that they can be correlated with geological mapping of known structures. After the shot is completed, the drone can perform a closer fly around to inspect the shot performance overall or specific damage areas. Videos

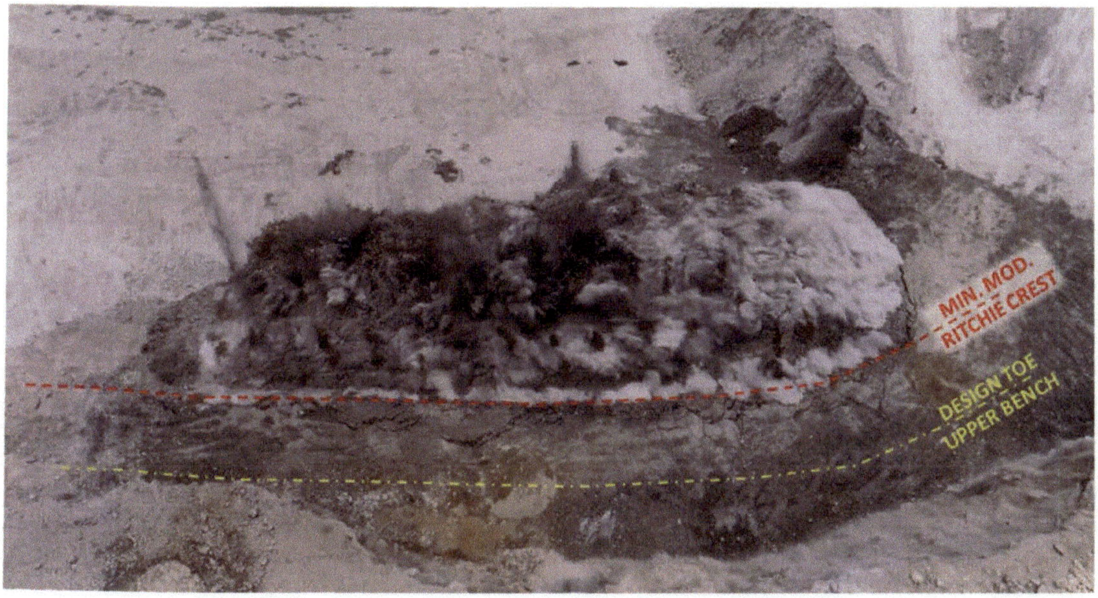

Fig. 3.56: Typical video analysis information from a blast monitoring drone (source: M. Ness).

are saved with all other blast and blast monitoring data for the shot.

At Freeport mines, the UAVs used for blast monitoring are mid-range commercially available models with built-in cameras and semi-automated flight systems. In US mines, all UAV pilots are trained and licensed to fly them as required by the FAA. More sophisticated UAVs perform other functions at most mines but, as mentioned, occasional encounters with fly rock make the mid-price range UAVs the best choice.

Motion analysis targets (cones or empty primer boxes painted orange) are placed on the desired crest and towards the wall at known increments (typically ~3 m). These targets are not as critical for near-vertical UAV videos, but are still quite important for observing and estimating overbreak using ground-based video cameras. A detonation indicator (shock tube or detonating cord) is placed in the blast hole immediately adjacent to the target set, to define when that hole fired. If ground cameras on tripods are used, it is not unusual for the video operator to set the camera up, turn it on during the final clearing of the blast and retreat to a safe location, allowing close framing without zooming for better images.

For closer inspection, borehole cameras can be used to determine the extent of subsurface damage behind the crest. Typically, a series of unloaded blast holes are drilled behind the designed crest, as shown in Fig. 3.57a. Images are taken before and after the blast to identify subsurface damage, as shown in Fig. 3.57b.

3.9.1.2 *Vibration monitoring*

Blast vibrations should be monitored in the near- and far-field, as shown in Fig. 3.58. A reference seismograph should be placed a set distance from the last hole to fire in each production and controlled blast (~50 m to avoid over-ranging in geophone-based instruments). Far-field units are placed at distances between 50 m and 500 m from the blast at points of concern. Placing the far-field unit next to a slope monitoring prism can help establish the link between vibration levels and slope response. The locations of the seismographs should be surveyed to determine the slope distance to the nearest blast hole or the centroid of the blast.

The following information should be recorded to assist with future analysis:

- date of shot, time of blast, pattern ID, location and geotechnical zone;
- blast type (production or controlled) and maximum instantaneous charge;
- seismograph ID, slope distance from nearest blast hole and scaled distance;
- PPV (mm/s): radial, vertical, transverse, maximum, vector sum;
- dominant frequency (Hz): radial, vertical, transverse;
- peak displacement (m): radial, vertical, transverse;
- peak acceleration (g): radial, vertical, transverse;
- slope response;
- blast hole diameter (mm), burden (m), spacing (m), stemming (m) and subdrill (m);
- relative confinement (free faces), number of holes per shot and total charge weight; and
- delay interval between holes, delay interval between rows and blast duration.

This information will be used to perform linear regression analysis of the vibration data to help establish the relationship between blast design and slope response. Blast vibration control is an important part of optimising

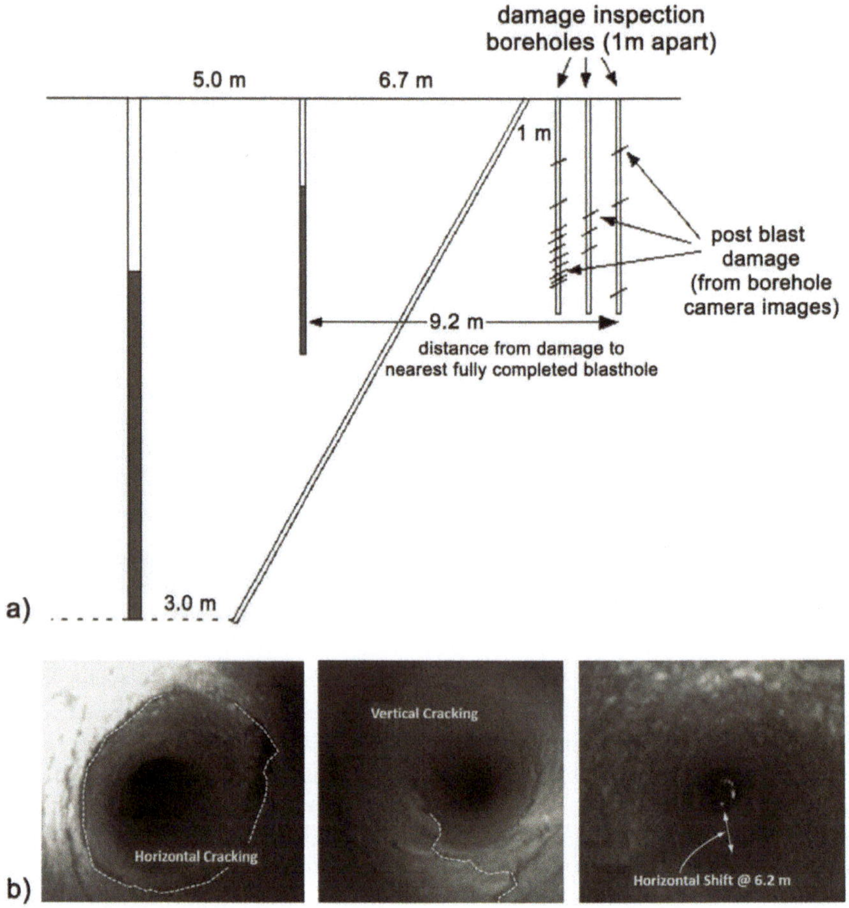

Fig. 3.57: a) Schematic drawing showing suggested layout for bench crest damage inspection boreholes. b) Subsurface damage inspection using a borehole camera (source: J. Floyd).

blast designs for slope protection. The dominant frequency content of the vibrations resulting from production and wall control blasting should not match the resonant frequency of the pit slope. If possible, delay sequences that minimise the production of unwanted frequencies should be used. This generally requires specific electronic delay times (as opposed to generic pyrotechnic delay times) and a good understanding of the vibration characteristics of the site. This understanding is gained by conducting single charge tests and measuring the frequency content and attenuation rate of the blast vibrations. The test procedure is as follows.

1. Detonate a single blast hole with similar charge weight and confinement conditions as for typical blast holes.
2. Monitor the vibration levels produced up the slope at locations of concern (tension failures, etc.) with a line of seismographs placed between 50 m and 500 m from the charge. The minimum charge weight is related to the distance away, as indicated in Table 3.2.
3. Record the charge weight and distance from the hole(s) to the monitoring locations.

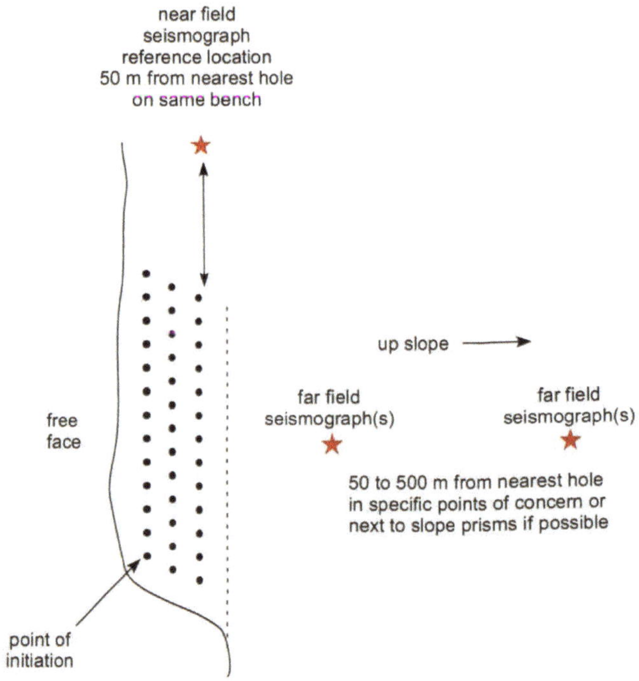

Fig. 3.58: Near- and far-field seismograph locations (source: J. Floyd).

Table 3.2: Single-hole charge recommendations based on a scaled distance of 35 m

Minimum charge weight for single-hole tests	
Distance away (m)	**Minimum charge (kg)**
50	10
100	40
150	90
200	159
250	249
300	359
400	638
500	996
600	1435
700	1953

Source: J. Floyd.

4. If the recorded levels are below 8 mm/s, consider shooting a larger charge to improve the quality of the vibration signature.
5. If two or more holes are needed to achieve the proper amplitude, then they should be detonated instantaneously (preferably with detonating cord or electronic delays).

A typical vibration characterisation test setup is shown in Fig. 3.59. It is suggested that at least two monitors be used to evaluate the slope's blast vibration attenuation characteristics. These units should be placed more or less in a straight line from the charge and located in the near- and far-field. The information is then processed to reveal the most appropriate delay intervals for frequency control.

Once it is safe to re-enter the blast site, the overbreak should be reviewed and recorded on the timing diagram of the blast report. This will allow the timing configurations to be critically evaluated. After excavation is completed,

Fig. 3.59: Vibration monitoring array (source: J. Floyd).

the face should be inspected and analysed for excessive overbreak. Examples of different degrees of visible damage are illustrated in Read and Stacey (2009). The damage should be classified into the following categories to help guide design refinement.

- *No visible damage*: joints tight, teeth marks in face, no loose material present, half barrels visible when presplitting, and a well-defined toe and crest.
- *Slight damage*: joints opened up, crest loss <1 m, few half barrels visible when presplitting, and excavation possible for 1 m beyond designed batter location.
- *Moderate damage*: blocks dislodged, crest loss of 1–3 m, and excavation possible for 1–3 m beyond designed limit.
- *Severe damage*: face shattered, blocks dislodged and rotated, and excavation possible for more than 2 m from designed limit.

It is recommended that a detailed record be made of the post-excavation condition of the benches. Several forms for this type of analysis are provided in Read and Stacey (2009). It has become common practice to use laser scanners with colour imaging to provide detailed scans of the final bench faces, which can be compared directly against the plan configuration to determine compliance to design, overbreak, and to perform some geologic mapping and geomechanical characterisation from the office. Field verification is obviously quite important, but the 3D face models can be very useful where it is not possible to physically access the areas below the highwalls. Also, the 3D face models are composited over time into an actual mine-shape model, which can be useful as baseline 'before' shapes to compare with scans of instabilities that arise later.

3.9.1.3 Pressure pulse

Open pit mine blasting breaks the rock in two ways: through shock wave-induced vibration (which is readily measured) and as a gas or fluid pressure pulse (not easily monitored). Visual observation of blasting events will occasionally reveal 'puffs' of dust showing up at short distances from the shot, often out of newly formed offset cracks. Beale and Read (2013) give examples of groundwater pressure pulses being measured in VWPs (Fig. 3.60), suggesting a potential opportunity to measure pressure pulses. While these pulses represent energy not spent breaking rock in the blast pattern, the concern for slope stability is the potential cumulative damage to slopes, especially if natural fractures in the blast pattern are saturated with groundwater.

Groundwater pressure pulses using VWPs can be measured by setting existing vibrating wire digital acquisition samplers located at the piezo-borehole collar to sample at the maximum rate before a blast is scheduled to take place. With accurate time measurement, any blast-induced pressure spike will become readily apparent. Experiments suggest that sample rates that are an order of

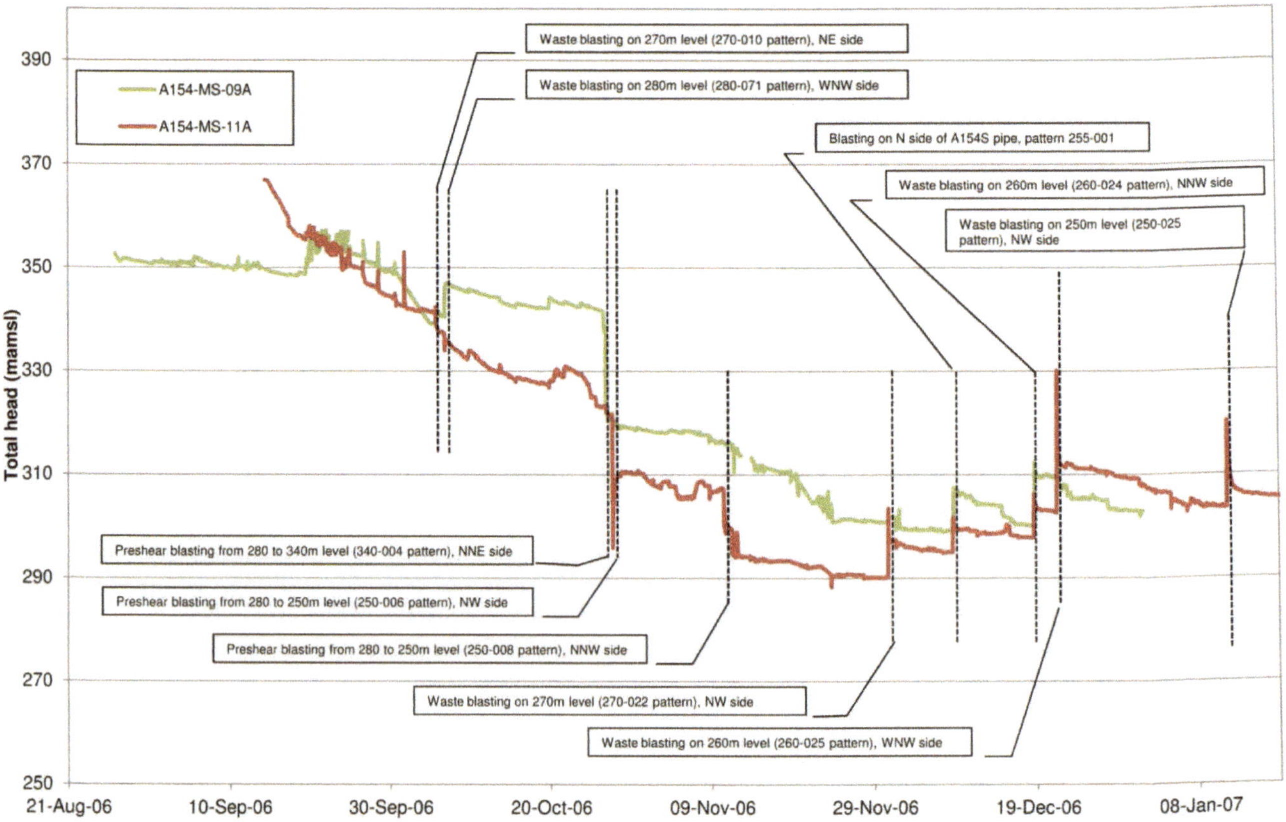

Fig. 3.60: Example of the different types of pressure pulse responses observed in clean, jointed rock due to blasting at the Diavik open pit mine (modified from Beale and Read 2013).

magnitude higher than the blast vibration frequencies (i.e. 100–1000 Hz) are needed to capture the pulse events, but more work is needed to develop this procedure. If there are regularly occurring pressure pulses that could destabilise a pit slope, it would be best to work on blasting parameters to improve relief to control or eliminate the effect.

3.9.2 Microseismic monitoring

Microseismic monitoring is recognised as an important tool for mapping rock-fracture development in rock masses and has been applied to the assessment of natural rock slopes and open pit slope stability (Eberhardt *et al.* 2004; Lynch and Malovichko 2006; Wesseloo *et al.* 2009; Luo *et al.* 2018). This monitoring technique uses distributed geophones and/or accelerometers to capture seismic signals initiated by a slope deformation fracturing event, slip on geological structures, or a sudden inelastic change in strain within the rock mass in response to mining processes. Inversion of the seismic signals allows the source location and magnitude of the event to be evaluated and the fracture mechanism to be interpreted.

Microseismic events manifest through the rock mass weakening process induced via relaxation and changes in stress in response to mining and can be measured in real-time in 4D (i.e. 3D location and time). Clusters of recorded seismic events are used to assess the association with disturbed geological structures and the shape of a fracture front, which assists in the identification and evaluation of areas of potential risk. A sudden change in seismicity and/or energy level can be used to infer and quantify seismic hazards at a particular location. Microseismicity, however, cannot pick up aseismic behaviour (e.g. slip along clay-filled fractures). It relies on rock fracture associated with shear displacement, fracture opening and fracture closure, or any combination of these mechanisms.

Microseismic monitoring has been applied widely in underground mines for risk management and production control. Routine microseismic monitoring of open pit slopes has been implemented at mines in Australia, Chile, Namibia and South Africa since 2002, with the earliest monitoring trials dating back to 1996 at the KCGM Super Pit in Western Australia. Several papers have been published on the systems installed and the events recorded at various mines, including Navachab (Lynch *et al.* 2005), PPRust (Little 2006), Sunrise Dam (Lynch and Malovichko 2006), Chuquicamata (Trifu *et al.* 2008), and Century (Salvoni 2017).

3.9.2.1 Working principles

The mining process acts to change the stress conditions present in the pit slope as the open pit deepens and its size

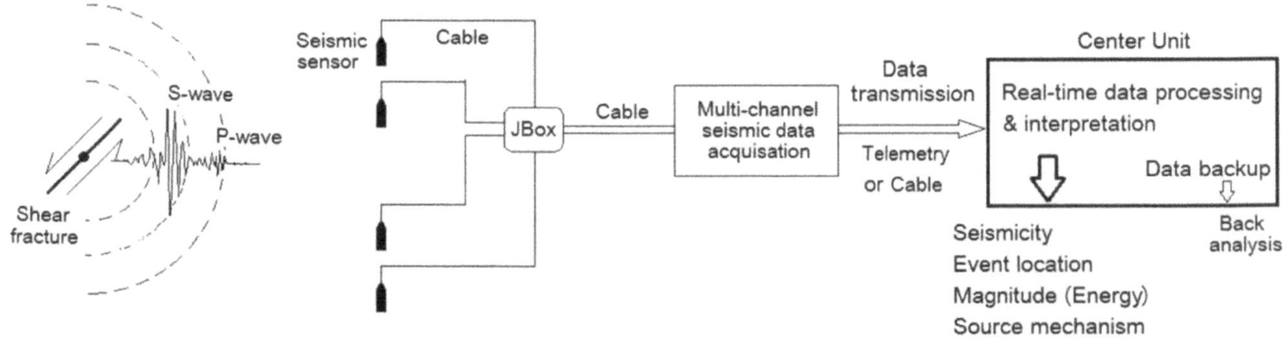

Fig. 3.61: Schematic of microseismic monitoring and the key elements of a network (source: IMS).

increases. These stress changes can cause new fractures to form or existing fractures to be disturbed and extended. When this occurs, it generates seismic energy that propagates as seismic waves through the surrounding rock mass. The two main phases in a seismic wave propagating through a rock mass are the P-waves (primary or pressure waves) and S-waves (secondary or shear waves). A fracturing event is also called a seismic event. The larger the event, the more energy is emitted that generates seismic waves with higher amplitudes, which will travel further through the rock mass (Fig. 3.61).

The seismic waves from the event can be captured using an array of triaxial or uniaxial geophones or accelerometers that are installed in the rock mass. The seismic signals captured at the sensor stations are digitised by seismic recorders (in the JBox shown in Fig. 3.61) and transmitted (either wirelessly or through cables) to a seismic acquisition system in which the seismic signals are processed using special data processing algorithms. A key step in the process is the determination of the arrival times of the different phases of a wave at a sensor. After the data have been processed, the parameters of the seismic event, such as origin time, spatial location, magnitude and potentially the failure mechanism, are obtained.

3.9.2.2 Usage capabilities for open pits

If consideration is being given to the implementation of microseismic monitoring as a component of an open pit monitoring strategy, it is important to have a clear understanding of the objectives and limitations of this technology. In the context of open pit mining, microseismic monitoring has been applied to achieve the following objectives, or a combination of objectives.

- *Mapping particular areas of instability.* Any rock-fracturing event, movement of a geological structure and sudden change of strain in the rock mass behind a pit wall can reduce the strength of the rock formation. These events can produce microseismic signals. From the alignment and clusters of the located seismic events, the weakening or weakened areas, shape and extent of movement of the surface, and distribution of hidden geological structures can be reliably mapped to assess areas of instability. The magnitudes and failure mechanics obtained from seismic signals can provide good indications of relative stress changes and principal stress directions. The latter is highly dependent on good coverage by the triaxial sensors.
- *Estimation of timing of pit wall collapse.* Microseismic monitoring records the number of events within a certain time interval and the strength of the energy release. The stronger the seismicity, the more rock-weakening events are occurring at a particular location. An increasing event frequency provides evidence for the formation of extensive rock fractures under the applied stresses. This information has been used in some underground mines as precursors for warning and assessing relative hazard levels for evacuation and re-entry protocols. However, in open pits the stress drop is very low and the number of events is also low, so its use for evacuation/re-entry is limited. Studies have also shown that the pit slope may only express some form of surface displacement a significant time after the microseismic events have been recorded.
- *Knowledge gained before visual observation.* Conventional pit slope monitoring systems such as GB radars, laser scanners, RTS and photogrammetry are routinely used to monitor pit wall surface movement. From this movement, one may infer subsurface deformation. However, such surface-sensing methods are unable to provide information on rock conditions behind the pit wall before any surface movement has been detected. Microseismic monitoring can overcome this limitation as it detects seismic events associated with rock weakening processes, which manifest as surface displacement. Pit slope engineers can use microseismic results to augment surface monitoring systems for identifying potential instability and warning of impending surface movement.

3.9.2.3 Challenges and limitations

It must be noted that unlike other geotechnical monitoring methods, which directly measure fundamental parameters such as displacement, microseismic monitoring observes seismic waveforms that, of themselves, do not directly provide readily understood information to a non-specialist. The waveforms must be interpreted using various signal-processing algorithms to generate results which are meaningful for engineering applications. These include event locations, seismic energy release, measures of rock mass deformation at the sources, and failure mechanisms. The reliability and value of microseismic monitoring is dependent on accurate knowledge of the locations and orientations of sensors. Results are then further dependent on several key factors.

- *Support from a specialist.* The design, installation, commissioning and maintenance of a microseismic monitoring system, as well as the interpretation of the data, are highly specialised activities. Microseismic data acquisition, event discrimination, arrival-time picking, event location and magnitude calculation are based on pre-set parameters and inversion algorithms. As every mine has a unique geological setting and development plan, the seismicity associated with mining-generated rock fracturing is likely to be different from mine to mine. Inclusion of a specialist at the beginning of a microseismic project is essential for setting up the instrument parameters based on their knowledge and experience of the mining environment and microseismic characteristics. Automatic data processing using traditional arrival time picking is now standard, but quality assurance/quality control (QA/QC) by an expert is still required. In some cases, manual data processing is required, which can be carried out by trained operators or a specialised external service company. Machine-learning algorithms for automatic microseismic data processing and interpretation are being refined by seismologists around the world. The development of fully automated microseismic monitoring methods is possible and is an area of continuing improvement for existing applications.
- *Data acquisition methodology. There are two modes to acquire and store data:*
  i) in continuous mode (this is standard in crustal seismology), the entire digitised ground motion from each sensor is continually streamed to data acquisition servers; and
  ii) in triggered mode, discrete blocks of data (called triggers or buffers) are transmitted. Some systems can operate in either mode (or hybrid modes), depending on the available communications infrastructure.

In the case of continuous acquisition, the continuous ground motion data are usually stored in some form of ring buffer which is continually overwritten. Pre-processing operations such as triggering and association into collections of waveforms associated with discrete events are performed on the data in the ring buffer before processing on separate servers or computers linking to the data. It is customary to store processed waveform data and the resultant event catalogues and metadata indefinitely. Backing up and storage of the ring buffer is a matter of data management practices on site; however, it usually takes a few days or weeks before a particular section is overwritten, during which the data can be re-examined.

Massive rock slope failure processes are dominantly driven by tensile fractures (Eberhardt *et al.* 2004), which might generate weaker seismic energy than shear events that occur deeper within the pit slope. Near an excavated slope, the rock mass can be highly fractured because of blasting, destressing (relaxation) and weathering, which can significantly attenuate the seismic signals that travel through the slope (Fig. 3.62). In trigger mode, small seismic events may not be strong enough to be detected by enough geophones (generally four) to trigger the system to record the event. A system in continuous mode records seismic signals from each sensor regardless of how many sensors detect an event. Many small events can occur near the slope before ground movement is observed. These events could provide crucial information for open pit stability assessment (see applications in Section 4.8). After fine-tuning of the processing algorithms, these data can be processed automatically with minimal human intervention.

3.9.2.4 Triggering

Triggering refers to the automatic detection of the onset of ground motion which may be associated with a seismic event. A trigger threshold or criterion can be pre-set or dynamic. Figure 3.63 shows an example of amplitude triggering. The dynamic threshold is usually set by specifying a ratio of the short-term average (STA) voltage level and the long-term average (LTA) voltage level. The averages are determined over different numbers of samples of digitised voltage. Care needs to be taken that the STA is not determined over so few samples that high-frequency events associated with mining activity or electrical noise continuously trigger a sensor channel.

A trace is a segment of ground motion data that starts with a certain amount of pre-trigger data and extends for a set number of post-trigger samples, either fixed or determined based on defined criteria. For example, a trace may be terminated after the amplitude envelope declines below the trigger threshold. A key difference between continuous and triggered systems is that, in the former, triggering takes place in the ring buffer on a data acquisition server and the original ground motion data are left undisturbed. In a triggered system, triggering occurs in the seismic recorders and only the traces are transmitted. Everything else is ignored.

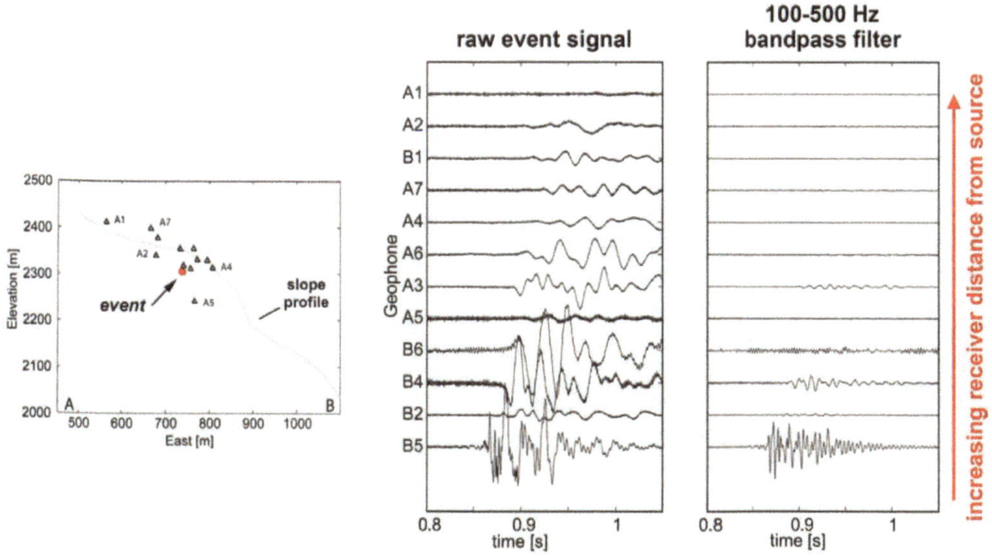

Fig. 3.62: Slope profile showing the location of a recorded microseismic event relative to the locations of the geophone in the monitoring array (projected to a 2D section) and the corresponding raw and filtered signals sorted according to the source–receiver distance (sensor A1 being the farthest and B5 the closest). These show the influence of attenuation and significant drop-off in the strength of the signal recorded, caused by the presence of large tension cracks within the slope (modified from Eberhardt *et al.* 2004).

3.9.2.5 Association

To determine the location and time of a seismic event with a high degree of confidence, the arrival times of seismic waves at the known locations of at least four appropriately placed triaxial seismic sensors must be known. Association is the act of assigning the cause of independent triggering of several sensor channels to the same event. There are at least two ways to associate triggers:

- a snapshot of all traces in the acquisition group or system can be grabbed. The snapshot (called a trigger) starts before the first trigger and continues for a pre-set number of samples; or
- only traces from sensors which triggered within a pre-set time tolerance from each other are saved. The tolerance is usually taken to be the time required for a P-wave to propagate over the largest characteristic length of the volume of interest.

If a pre-set number of sensors (the association level) meet the association criteria, they are said to associate. The association level generally, in day-to-day operation, should not be fewer than four triaxial sensors, to allow a reliable estimate of the location and time of a seismic event.

A microseismic system used for open pit monitoring normally consists of a sensor network where several uniaxial and/or triaxial sensors are installed in the ground with a volumetric spacing of ~50–200 m to obtain the 4D pattern of seismicity. Again, for confident location, the seismic signals must be recorded by four or more triaxial sensors over time. For confident failure mechanism determination, seven or more triaxial sensors, with all

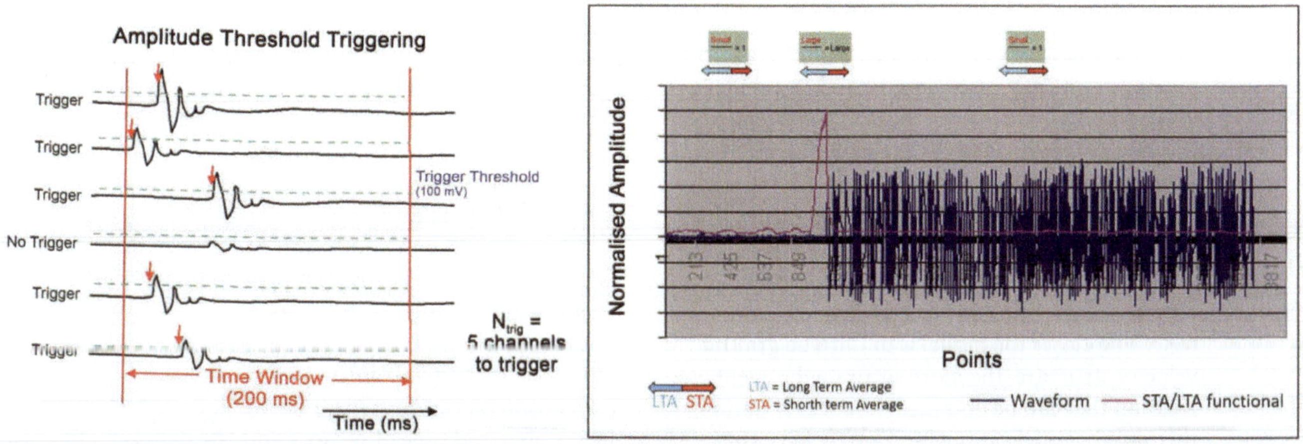

Fig. 3.63: Different types of trigger logic: amplitude threshold triggering (left) and STA/LTA triggering (right) (source: ESG Solutions).

components functioning and with known orientations surrounding the area of interest, are required.

One common system is to operate in trigger mode based on an STA and LTA algorithm. When at least four stations experience an STA/LTA ratio greater than the pre-set threshold, the stations trigger, the associator coalesces the individual seismograms and the event is recorded. This approach was developed for underground mines with poor communications infrastructure, where there has been a focus on strong events leading to hazard identification and re-entry protocols (Mendecki 1997). Another common system operates by default in continuous mode. This system was developed in a market where historically mines had fibre-optic-based underground communications infrastructure which afforded the bandwidth necessary for the streaming of continuous data.

For many mine seismic monitoring systems, event location is only attempted for triggered events presenting abrupt P- and/or S-wave arrivals. Many events, such as tensile events or strong events with weak leading signals at noise level, are not currently locatable because their arrival times cannot be reliably determined. The located events, though they reveal certain information on ground instability, may not present the whole picture about the ground condition because of current limitations in data processing. Key considerations are as follows:

1. *Location accuracy.* The accuracy of the event location is largely influenced by several main factors, including:
   i) the number of sensors that recorded an event and the sensors' spatial distribution geometry. In a mining environment, the sensor deployment geometry may be highly compromised because of access restrictions. The number of sensors may be constrained by budget limitations;
   ii) the accuracy of arrival-time picking. To achieve an accurate event location, the P- and S-wave arrivals must be reliably determined. For events with weak leading signals, arrival times are difficult to determine automatically or manually. The arrival-time picking errors may cause large biases in event location; and
   iii) the velocity model, where the event location requires a 3D seismic velocity model from which seismic ray paths are calculated for the event location. This model is normally set up based on a geological model and geotechnical parameters obtained from borehole logging and laboratory tests. The closer the model is to the true ground velocity structure, the more accurate the event location that can be obtained. In a real situation, this 'good' model cannot be constructed using limited parameters. A homogeneous or layered model is often used for determining event location. Ray paths tend to bend around the curvature of the pit shell and other excavations. In a homogeneous model, the seismic rays from the source to the sensors are assumed to be straight lines. This model simplification often yields locations either in the air or much deeper than where they occurred in reality.
2. *Prediction of the timing of an instability event.* Using microseismicity to predict the timing of the occurrence of a pit wall instability event is a challenge. The prediction is mainly based on an assumption that anomalous seismicity may be a precursor of a slope failure. It has been found that some instability events did have anomalous seismicity before the events (e.g. Senfaute *et al.* 2009; Sobolev 2011). However, many similar seismic anomalies were not followed by instability events. It is recommended that microseismic results should be integrated with other geotechnical monitoring observations for more trustworthy prediction.
3. *Poor signal-to-noise ratio.* In a mining environment, microseismic signals are always contaminated by noise signals generated by drilling, heavy truck trafficking, digging, pumping and other mechanical vibrations within or near the monitoring areas. Production blasts and earthquakes also produce strong unwanted seismic signals on microseismic records that are designed to only record changes to the rock mass behind the pit slope face. If the noise signals have similar amplitudes and frequencies to the first seismic arrivals, it can make reliable determination of the arrival times unlikely. Many microseismic events have not been spatially located because of this reason. Poor signal-to-noise ratio can also cause false triggers when noise is introduced by mining activities, such as haul trucks and excavators that generate a seismic event. Eliminating false triggers is challenging. Mining activities should generate different seismic characteristics from rock fracture events, and their signals should be identified as 'background noise' and eliminated automatically from trigger calculations. Further research is needed for efficient algorithms to more effectively avoid false triggers. At the time of writing, there is no efficient method to address this challenge. However, there are attempts to overcome the issue by using a passive system that regularly triggers so that the noise can be filtered from the data record, as well as using template-matching and machine-learning techniques.

3.9.2.6 Network design

The considerations for sensor network design are mainly the number and type of sensors and the deployment geometry that achieves the best monitoring sensitivity and most accurate event location. Factors of installation efficiency, instrument maintenance and available budget are also considered. Details are as follows.

- *Number of sensors.* At least four triaxial sensors are required to identify the location of a microseismic event. The more sensors that are used, the better constrained and thus the more reliable the identified event locations will be.
- *Type of sensors.* Typical mining-induced microseismic events can have moment magnitudes from -2.0 to 4.0. Figure 3.64 provides a conceptual image of the different bandwidths and corresponding event sizes related to the appropriate sensor. To accommodate the different frequency ranges, amplitudes and event durations, hybrid solutions consisting of sensors with different natural frequencies and different amplitude responses may be necessary. Generally, high-frequency sensors are grouped in dense arrays around the area of interest, while low-frequency sensors are distributed further away. Either accelerometers or geophones may be used. Geophones are generally cheaper than accelerometers. In mining applications, triaxial and uniaxial sensors are used (Fig. 3.65). Higher-frequency geophones (generally 15 Hz and higher) are usually omnidirectional (they may be installed in any orientation), but lower-frequency sensors (4.5 Hz and lower, as in Fig. 3.64) must be installed vertically or horizontally within the tolerances recommended by the manufacturer. The requirements for lower-frequency sensors apply to uniaxial and triaxial geophones. Piezo-electric accelerometers can be installed in any orientation.

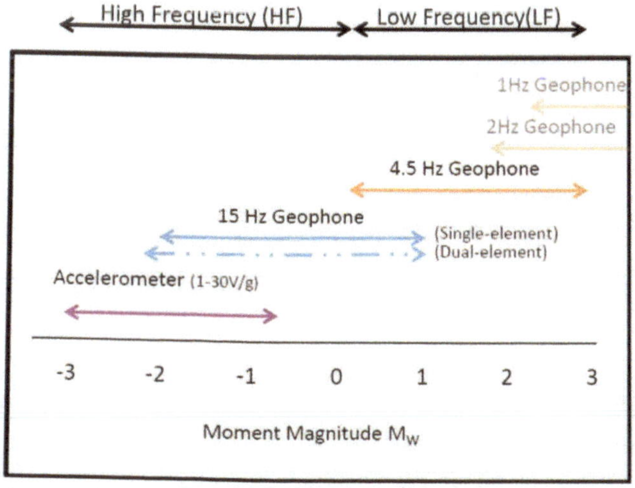

Fig. 3.64: Simplified relationship between sensor types, natural frequencies, dynamic ranges and seismic event moment magnitudes. The schematic shows the recommended magnitude range for accurate source parameter calculation from different sensor types, spanning lower-frequency strong ground motion to higher-frequency microseismicity (source: ESG Solutions).

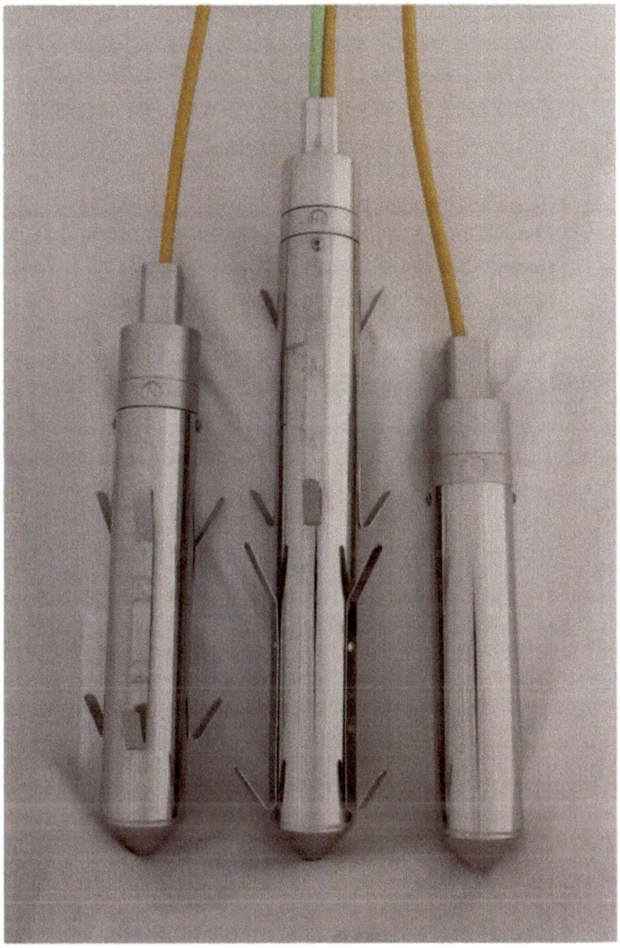

Fig. 3.65: Typical sensors used in mine seismic networks (source: IMS).

- *Sensor deployment geometry.* Ideally, sensors should be deployed in three dimensions around the rock mass in which microseismic events are to be detected. It is desirable that the sensors are arranged to form triangular patterns with sensor interval distances large enough to resolve the arrival-time differences to identify an event location. Regular, planar and excessively linear arrangements should be avoided. Due to access restrictions and progress of mining, the sensors have to be deployed in locations that protect the sensor (more particularly, protect the hole, as the sensor is far cheaper than the hole) and that form an optimal geometry for event detection and location. As most sensors are connected by cable to the surface, consideration may need to be given to cable shearing if near-surface displacements are likely to be excessive (i.e. >50 mm) adjacent to the hole.
- *Multiple arrays.* A microseismic monitoring system can be composed of several small aperture sensor arrays deployed at targeted areas of interest to capture creep-type microseismic events near the slope. Each

array may contain three or four closely spaced sensors to achieve reliable event locations (Wesseloo and Sweby 2008). These arrays form the monitoring network. Using multiple arrays allows strong and weak seismic events to be captured, recognising that high-frequency signals from small events attenuate rapidly with an increase in travel path.

- *Installation of sensors.* The sensors are normally installed in boreholes drilled from bench tops and/or crests, down into the rock mass that is to be monitored, to reduce the interference of mechanical vibrations on the ground surface. The sensors are either cemented in place (i.e. the sensor is not retrievable) or anchored to the wall of the borehole (i.e. the sensor is retrievable).
- *Data acquisition.* Seismic signals captured by the sensors are first transmitted to a data acquisition unit through cables. If the network consists of several units, each unit transmits the acquired signals to the network central site through cables or wirelessly for data integration and processing (refer to Fig. 3.61). Time synchronisation among the units is achieved by GNSS timing.
- *Calibration shots.* After the sensors are installed and connected to the data acquisition units, several calibration shots, typically 0.8–1.5 kg, should be fired at different locations within or near the microseismic network. The seismic waveforms from the shots are used for evaluating the working condition of the sensor–data acquisition system, determining the orientations of the sensor components inside the boreholes (for multiple component sensors), and calibration of the seismic velocity model for event location. For the calibration of the 3D velocity model, it is important that some ray paths between the blasts and sensors are in a single geological unit. Consideration should be given to periodically updating the calibration.

4

MONITORING REQUIREMENTS, LIMITATIONS AND APPLICATIONS

Chris Audigé, Alex Bals, Geoff Beale, Sharla Coetsee, Niccolò Coli, Davide Colombo, Ashley Creighton, Jan deBeer, Phil de Graaf, Phil Dight, Erik Eberhardt, Chris Fagan, Giacomo Falorni, Emrich Hamman, Ben Haugen, P. Mark Hawley, Graham Hunter, Jon Leighton, Lorenzo Leoni, Naani Mphathiwa, Leon Nel, David Noon, Verne Pere, Alex Pienaar, Cliff Preston, Orapeleng Randall, Nick Rose, David Rutledge, Peter Saunders, Eric Schwarz, Robert Sharon, Peter Stacey, Fanie Wessels

4.1 Introduction

This chapter describes how to integrate the instrumentation presented in Chapter 3 into a reliable and practical performance monitoring system. Slope monitoring is not merely the selection and installation of instruments, but a comprehensive step-by-step engineering process beginning with a defined objective and ending with data interpretation, results reporting, and taking appropriate actions. The goal of a comprehensive performance monitoring program is to achieve mine stability verification as a significant component of the operation's geotechnical risk management process, as illustrated in Fig. 4.1. Planning the monitoring program should be logical and comprehensive, as the measurement problem may require several different instrument types that collect information across a range of varying scales, from a single tension crack to the entire pit slope and adjacent infrastructure. Furthermore, because of physical limitations and economic constraints, not all parameters can be measured with equal ease and success.

4.2 System design requirements and considerations

4.2.1 Overview

The design of a monitoring system for an open pit mine, including survey control, communications and data management, should always be rationalised with the mine design and site-specific risk elements. System capabilities wherever possible should be in real-time, where the mine design and development performance can be continuously verified and where infrastructure and ground support can be monitored for any developing risks.

Mining operations are becoming increasingly automated, and access for installation and maintenance of instrumentation is becoming a limiting factor. In locations where access to the instrumentation may be limited for

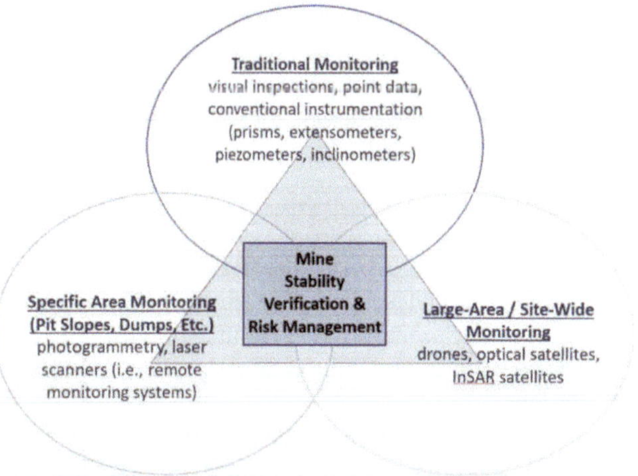

Fig. 4.1: Summary of methods used to monitor slope stability in mining operations and how monitoring methods interact in terms of scale and data validation, and for effective risk management (source: E. Schwarz).

system maintenance, monitoring and data communication systems require an increasing level of automation and performance reliability. Future operations will therefore require improvements in remote visualisation, automation using drones and/or instrumentation that is mounted on mining equipment that routinely accesses active working areas otherwise not accessible to personnel. A forward view of the requirements of performance monitoring instrumentation is addressed in Chapter 10.

4.2.2 System design considerations

When designing a monitoring system, the slope engineer must consider and balance the different objective-related requirements the system must meet. This recognises that there is a variety of instrument types that have similar purposes (e.g. measuring slope movements) but differ in exactly what they measure (i.e. tilt, relative displacement, absolute displacement, etc.) and how the instruments can be deployed (i.e. on surface, from boreholes, etc.). To manage this problem, Dunnicliff (1993) suggests several important questions that should first be asked.

- What are the objectives of the monitoring system?
- What parameters need to be measured and how will these aid mine safety, design performance and mitigation measures? How might these parameters vary spatially and temporally?
- What are the risks due to variable or poor ground conditions?
- What are the expected magnitudes of the measured parameter (e.g. expected displacements, pore pressure increase, etc.)?
- What are the optimal locations for instrument installation?

Adequate planning is required before proceeding, and these plans should be logical and comprehensive. Dunnicliff (1993) is an excellent source on geotechnical monitoring and system design, for which the following steps are recommended.

1. *Define the project conditions.* These might include mine layout, geology and major structures, geotechnical domains and rock mass properties, groundwater conditions, status and location of nearby mine infrastructure (e.g. in-pit crusher) or other facilities, environmental considerations, and mine excavation plans and blast designs.
2. *Predict mechanisms that control slope behaviour.* This may involve one or more working hypotheses based on a comprehensive knowledge of the pit slope geology and conditions.
3. *Define the questions that need to be answered.* Every instrument and monitoring system on a project should be selected and placed to assist in answering a specific question; if there is no question, there should be no instrument.
4. *Define the purpose of the monitoring.* The questions that instruments and monitoring systems can answer are so vital that we should not risk discrediting their value by using them improperly or unnecessarily (Peck 1984).
5. *Select the parameters to be monitored.* Parameters can include displacement, tilt, pore water pressure and microseismicity.
6. *Predict magnitudes of change.* Predictions are necessary to select the required instrument ranges and required instrument sensitivities or accuracies.
7. *Devise remedial action.* It is absolutely necessary to decide in advance a positive means for reacting to the results of the monitoring systems (e.g. Trigger Action Response Plan, TARP).
8. *Assign tasks for the design, construction and operation phases.* Instrumentation specialists may be mine staff or outside consultants/contractors with special expertise. When assigning tasks for monitoring, the party with the greatest vested interest in the data should be given direct line responsibility for producing them accurately.
9. *Select instruments.* When selecting instruments, the overriding requirement is reliability! Inherent in reliability is simplicity. Instruments should be reliable, rugged and capable of functioning for long periods of time without repair or replacement. They should be capable of responding rapidly and precisely to changes so a true picture of events can be maintained at all times.
10. *Select instrument locations.* The selection of instrument locations should reflect predicted behaviour and should be compatible with the method of analysis that will later be used when interpreting the data.
11. *Account for factors that may influence measured data.* Details of each instrument installation should be recorded, because local or unusual conditions often influence measured variables.
12. *Establish procedures for ensuring reading correctness.* When reading an instrument, one should be able to answer: Is the instrument functioning correctly? The answer can be provided by visual observations, duplication of instruments, checking for data consistency, or through the use of instruments that internally check their own correct functioning.
13. *List the specific purpose of each instrument.* It is useful to question whether all planned instruments are justified. If no viable purpose can be found for a planned instrument, it should be removed from the monitoring system.
14. *Prepare the budget.* A budget should be prepared for all instrumentation-related tasks to ensure sufficient funds are available. A frequent error in budgeting is to

underestimate the data collection and processing costs and time commitments.

15. *Prepare instrument procurement specifications.* There are several competing instrumentation manufacturers, each offering products designed for similar geotechnical purposes. Be aware of each instrument's capabilities, limitations and compatibilities with monitoring systems already in use at the mine. Instruments should be purchased from established manufacturers.
16. *Plan for instrument installation, regular calibration and maintenance.* Installation procedures should be planned well in advance of the scheduled installation. The step-by-step procedures should include a detailed list of the required materials and tools, and installation record sheets for documenting factors that may influence measured data.
17. *Plan data collection, processing, presentation, interpretation, reporting and implementation.* The effort required for these tasks should not be underestimated. Many mines and consulting firms have files filled with large quantities of partially processed and undigested data because they did not ensure that sufficient time/funds were available for these tasks.

4.2.3 Utilising a risk-based approach to rationalise monitoring system design

Risk-based approaches to slope monitoring system design and to performance assessment require a thorough understanding of local ground behaviour, including history, mechanism, frequency, location, size and runout distance, and experiences with these over time. Various performance monitoring methods and an array of instrumentation must be used to gather reliable information and apply results into meaningful risk or decision analyses. Risk-based monitoring enables the geotechnical engineer to calibrate monitoring zones for reliable alarming, create failure mode models and develop robust safety procedures suited to specific monitoring conditions. Improvements in instrumentation and automation of monitoring systems that make it possible to monitor and report information almost anywhere in real-time have added considerable value to a mine's capability to conduct risk-based monitoring. However, having that capacity also makes it easier to become complacent about systems that may not always perform as expected, as well as for the mine operator to become over-confident that all is well.

Local experience in ground movement behaviour should always be considered of high value in the design of a slope performance monitoring system. Experience should drive the selection and deployment of instrumentation, development of a performance monitoring program and development of an early warning system.

The importance of validating deformation behaviour and calibrating line-of-sight (LOS) movement from radar and laser scanning systems to total vector displacement values available from geodetic survey systems (e.g. total station, robotic total stations (RTS) and global navigation satellite systems (GNSS)) should never be discounted or underestimated. Not having a clear appreciation of the limitations of LOS movement monitoring, no matter how precise the measurements, can be a fatal flaw in the monitoring system. Utilisation of ground-based (GB) and satellite-based radar can help improve the interpretation of real movement, where the LOS vectors to the targeted area of interest are significantly different. Redundancy in the monitoring program, using different types of instrumentation, is required to ensure the ability to verify changing stability conditions.

Calibrating radar sensitivity for a large excavated slope can be far more difficult than intuition might suggest. Zonation of a slope into regions of similar behaviour and geometry for purposes of LOS calibration is recommended. For example, from a given GB radar position, the direction of movement near the crest of the monitored slope might be well aligned to the LOS. However, movement at the toe of a developing instability on the same slope may be perpendicular to the LOS. If slope failure initiates at the toe of the unstable section, the radar may not detect motion until it is too late for the mine operator to react to an emergency situation. This risk is relatively high for slopes excavated in hard rock masses where acceleration to failure can occur rapidly.

Slope designs in mature operations often employ an empirical approach that may supersede a conventional deterministic approach supported by design criteria specifying a minimum safety factor. Such designs may be justified from site-specific experience and where performance monitoring instrumentation is well-suited to the task, including utilisation of redundant systems for data validation and result reliability. Another important consideration is the appropriate skill-sets of the geotechnical personnel who are responsible for the maintenance of site-specific monitoring systems and for performing the standard procedures required to ensure operational safety. Depending on risk potential and the operation's tolerance to managing a slope instability event, instrumentation that measures only displacement along the LOS may be insufficient to adequately support a risk-based slope design. The operation must address the adequacy of its performance monitoring program through its risk assessment process.

4.2.4 Strategic *v.* tactical monitoring

Mining companies have an obligation to support leading mining practices so as to ensure operational safety. Slope failures are and will continue to be a reality of open pit mining. Changes in stresses (associated with deformation responses) and geometry (associated with mine development) can potentially lead to a catastrophic event, even in mines with the most seemingly conservative of slope designs. Leading mining practice calls for the timely detection of potential instability. Effectively managing instability makes strategic and tactical slope monitoring an integral part of any mine's slope monitoring program. Depending on their technical specifications, some monitoring systems may be more suitable for short-range monitoring (tactical) than for long-range and broad-area monitoring (strategic) and vice versa. Some monitoring systems are specifically designed to perform strategic or tactical monitoring.

Spatial coverage is only one element of effective strategic monitoring. The ability to detect unknown hazards begins with the capability to resolve very slow-moving areas while the widest possible portion of the slope is being monitored. The capability to determine slow movement at very detailed spatial resolutions and at a high level of precision is a valuable tool that provides geotechnical engineers and planners extra lead time to evaluate slope conditions and develop solutions before displacement begins to interfere with mining operations.

Strategic slope monitoring also deals with performance monitoring system design for the life-of-mine and the long-term stability of the overall open pit. It utilises primary monitoring methods for determining (Mossop 2014):

- overall background stability compared with areas of increased deformation;
- the effect of local geology and structure on stability;
- the effect of groundwater on stability;
- the effect of blasting on stability;
- calibration of the slope design; and
- calibration of other slope monitoring methods or systems.

GB radar systems are often deployed for strategic monitoring. However, they may be used to monitor only a specific section of the slope that is of interest because of known instabilities or the proximity of key elements at risk (e.g. areas close to active development). In these cases, mining operations use a different approach to performance monitoring, the one commonly referred to as tactical or safety-critical monitoring. This is characterised by:

- a focus on a specific critical area over a limited time-span;
- the capability of the system to rapidly measure changes in deformation rates;
- sub-bench area resolution for the detection of bench- or batter-scale failures; and
- rapid data availability and alarming capabilities.

Tactical slope monitoring deals with the management of exposure to potentially damaging slope failure events in active mining areas and aims to improve the understanding of known and suspect areas of instability. It makes use of data precision to determine the deformation response to mining activities and suitable safety-critical monitoring alarm settings (Mossop 2014). Tactical monitoring systems are best implemented after a period of investigative monitoring. Monitoring information must be assessed in the context of the physical setting and the conclusions of the investigation phase.

4.2.5 Selection of monitoring methods

The selection of monitoring methods, or individual types of instrumentation to be utilised, depends on several factors, including geometry of the ground under investigation, the purpose for monitoring (strategic *v.* tactical), block size, failure mechanism and anticipated speed of failure. It is important that the slope engineer fully understands the capabilities and limits of each instrumentation type.

All operations require daily visual inspections by qualified mine operators and selected responsible persons to ensure that areas where the workforce will operate are safe. The slope engineer must share a level of responsibility for safety-related observations on a daily basis and will utilise a variety of instrumentation to aid in monitoring deformation, groundwater, blast vibration and other parameters that may contribute to the development of an unstable condition. A primary objective of instrumentation is to allow the slope engineer to detect anomalous movement that may not be detected by visual inspection alone, and do so in time to alert personnel to a potentially developing hazard. Another objective of instrumentation is to achieve a level of data validation, in order to avoid actions based on poor or erroneous data that could be attributed to instrumentation out of calibration or changes in atmospheric conditions.

Simple, cost-effective devices available to the mine operator and slope engineer include crack meters, wireline extensometers and tiltmeters, which at times can be improvised by a resourceful individual (see Section 3.2). To obtain subsurface information, boreholes can be installed with inclinometers, time-domain reflectometer (TDR) cables and piezometers (see Sections 3.6 and 3.7). Such devices are often deployed in specific locations where the block size of the area being monitored is relatively small (typically in the order of 100 000 m^3 or smaller).

For larger excavated slopes, more sophisticated (and therefore more expensive) instrumentation is required to provide adequate protection for the operation. Conventional instrumentation for this function includes geodetic monitoring using total stations and reflective prisms, and/or GNSS, as well as GB radar, InSAR and LiDAR (see Sections 3.3, 3.4 and 3.5). These sophisticated systems are designed to complement other instrumentation, the combination of which can be used to optimise the operation's monitoring program in terms of utilisation and cost. Local experience is then applied to further optimise the collective system.

The selection of the monitoring system components requires the ability to cross-validate data, as can be done using prisms and radar, which is the typical minimum requirement for large open pit operations. Also, as indicated above, the slope engineer must fully understand the capabilities and, maybe even more importantly, the limitations of each monitoring method.

4.2.6 Integrated slope monitoring system

4.2.6.1 Requirements

Safety and productivity in modern-day open pits require the geotechnical engineer to oversee an integrated slope monitoring program, of which the most important aspects are the performance monitoring of rock slopes and the interpretation and reporting of measured data. Slope failures are typically preceded by several measurable indicators, which include displacements, tension cracks, rock noise and changes in groundwater levels. An integrated slope monitoring program should be capable of measuring these indicators, and its implementation should aim to achieve the following goals (Kayesa 2006).

- *Verify mine design* to serve as a technical motivation for maintaining, steepening or reducing slope angles with respect to the economical and safety benefits.
- *Provide technical assurance* to instil confidence in production and management personnel about the status of the operation.
- *Serve as a warning system* to alert personnel in the event that an area of the pit becomes unstable.
- *Serve as a risk management tool* that assists the operation to ensure timely decision-making in the interest of all stakeholders involved.

The slope monitoring program must have a certain level of redundancy in both type and number of measurements, primarily for purposes of data assurance, validation and reliability. Vaziri *et al.* (2010) provide a detailed classification of available monitoring systems for open pit mines, as illustrated in Fig. 4.2. Seamlessly

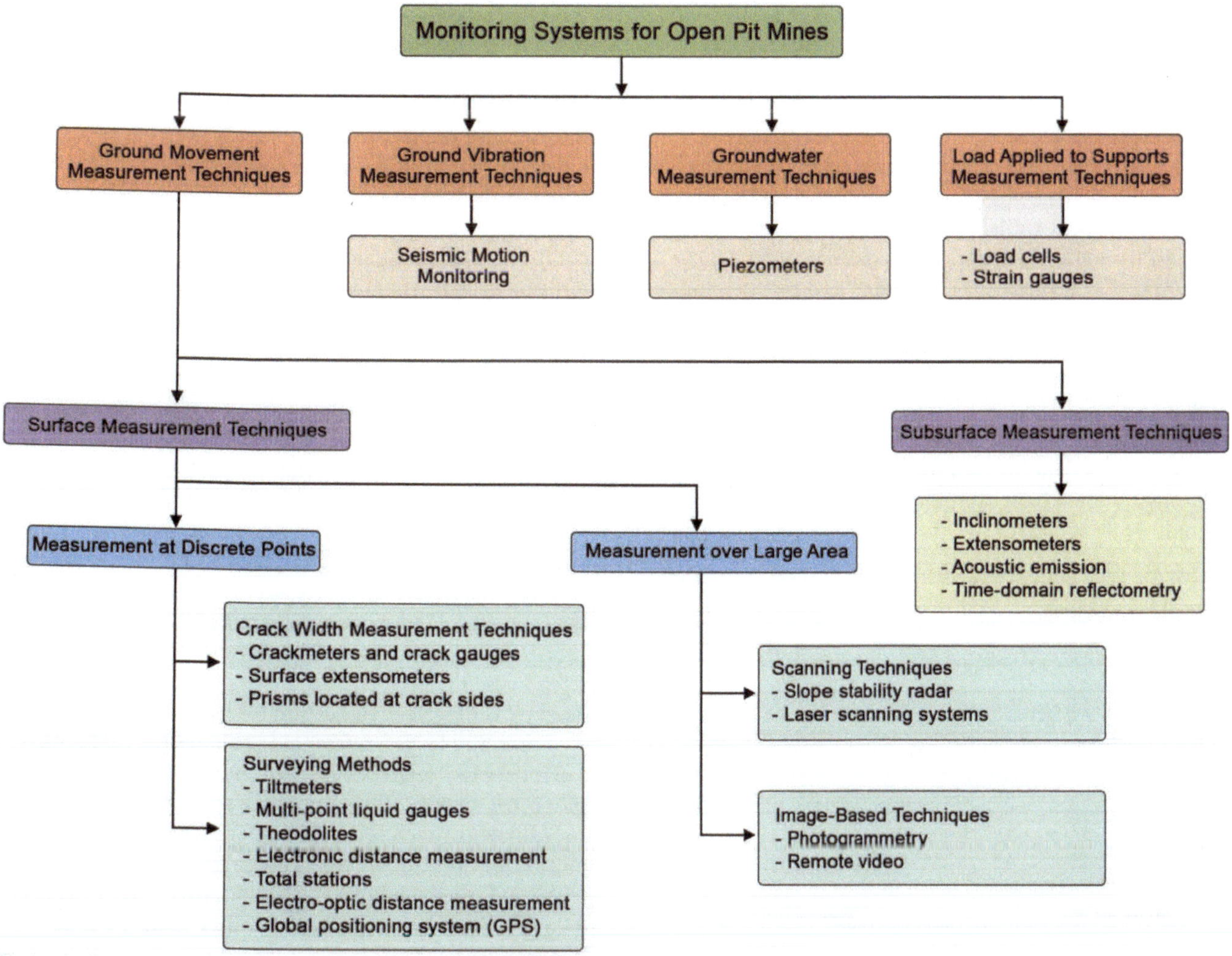

Fig. 4.2: Classification of available stability monitoring systems (source: Vaziri *et al.* 2010).

integrating data from a variety of sensors provides geotechnical personnel with the ability to quickly review and identify changing slope conditions, enabling them to respond rapidly to any anticipated change in slope performance that could impact worker safety and/or the production and development process.

It is now common practice to design and install fully automated monitoring systems. This is advantageous to the slope engineer who can access information in real-time and on demand, and can focus attention on data retrieval, interpretation and decision-making while minimising the time necessary to collect and compile the data. Improvements in data interpretation are facilitated by automating the process of gathering data from various monitoring methods, such as generating time-series plots of piezometric, prism and radar data, for data validation and for interpretation of causal factors. Details on data communications and management are addressed in Chapter 6, and the integration of data sources for interpretation and analysis is addressed in Chapter 7.

4.2.6.2 The case for an integrated slope monitoring system

An assessment of the value of various methods for monitoring slopes is reported in Bye (2009). Results using a fault tree analysis are summarised in Table 4.1, showing the success rate of different monitoring methods.

These results suggest that an effective slope monitoring program benefits from the integration of several monitoring tools available to the slope engineer. For example, seamlessly integrating radar data with established slope monitoring systems and methodologies, such as prism monitoring, helps mine personnel to identify changing slope conditions reliably by performing a side-by-side comparison of the radar with data from other sensors.

An integrated slope monitoring program allows for the integration of data from a variety of sensors and techniques interchangeably and on demand. The program should be designed for the life-of-mine, but should also be able to provide a detailed locality-specific understanding of any potential instability event. An example of a generic integrated open pit slope monitoring program from Mossop (2014) is illustrated in Fig. 4.3. The generic slope monitoring program can be viewed as being divided into six phases, namely:

- *hazard identification*: localised input and first pass monitoring indicators;
- *strategic monitoring*: stability trend analysis, hazard identification and calibration of slope design;
- *tactical monitoring*: safety-critical monitoring and alarming;
- *first response plans*: operational and technical accountabilities;
- *emergency response plans*: mine evacuation planning and accident mitigation; and
- *contingency plans*: mine recovery plan or predetermined failure scenario.

An example of a detailed integrated open pit slope monitoring program is illustrated in Fig. 4.4.

Table 4.1: Fault tree analysis results on the success rate of various monitoring methods

Method	Success rate (%)
Visual monitoring only	32
Prism/crackmeters only	45
Visual and prism/crackmeters	63
Visual and prism/crackmeters and laser	86
Radar only	93
Visual and prism/crackmeters and radar	97
Visual and prism/crackmeters, laser and radar	99

Source: Bye (2009)

4.2.7 Geo-referencing

4.2.7.1 Importance

Analysis of data acquired from slope monitoring systems often relies on datasets from multiple sensors that take into account the physical relationship between those datasets. Therefore, the correct geo-referencing of sensors and data derived from various types of instrumentation is vital. Incorrectly geo-referenced data could lead to confusion and a false sense of data validation, especially in terms of spatial relationships for data derived from different systems integrated for interpretive context.

Geo-referencing refers to associating a sensor or data with a location in physical space. It is important that every dataset integrated into a slope monitoring system is geo-referenced so that, when they are displayed in combination, the physical relationship between datasets is consistent and correct. This applies to background datasets such as computer-aided design (CAD) drawings, digital terrain models (DTM) and aerial photography; to sensors such as prisms, GNSS receivers, crack-gauges and inclinometers; and to systems such as GB radars, satellite InSAR and laser scanners. The importance of geo-referencing the performance monitoring network for an open pit is described in Section 4.3.2.

For example, geo-referencing enables the mapping of radar data to a reference grid for integrating radar data into other systems, thus giving the slope engineer the coordinates of each pixel. Geo-referenced data can be plotted on the same image as other important data such as faults and important mine infrastructure. Geo-referenced data give the slope engineer the ability to drape radar images over a DTM.

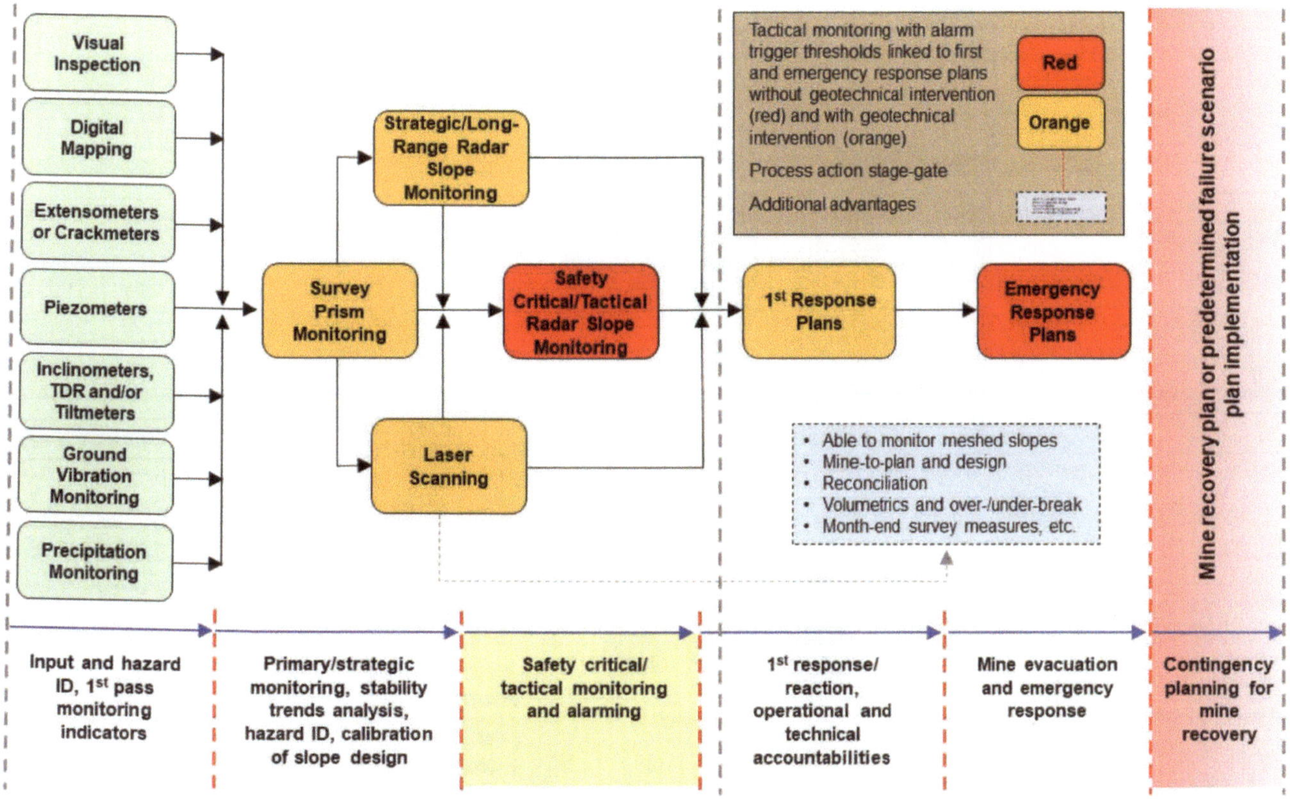

Fig. 4.3: Generic integrated slope stability program (source: Mossop 2014).

To carry this example further, the first step to geo-referencing GB radar data is to find a common reference point and azimuth. This is commonly the centre of the radar instrument, or it may have a dedicated offset survey point. On a synthetic-aperture radar (SAR) system, the azimuth is obtained by measuring two points parallel to the direction of travel. Once a location and azimuth are measured at a common reference point, the data can be remapped into the local coordinate system and assigned an elevation value for a 3D display.

4.2.7.2 Geo-referenced data and DTMs

Background datasets, such as aerial photographs, CAD drawings and DTMs generally exist in an already geo-referenced format, as geo-referencing is one of the steps taken to create the dataset. In most cases, sensors that are measured by surveying methods, such as prisms and levelling points, will also be geo-referenced by the very nature in which the data are captured. However, sometimes an offset may need to be applied to convert the coordinates from a local to global system (e.g. Universal Transverse Mercator, UTM) and vice versa. It makes no difference whether a local or global coordinate system is used in the slope monitoring system; the important factor is that all datasets are in the same coordinate system.

Sensors that are not recorded by surveying methods, such as those installed within instrumented boreholes, usually would have been surveyed for the installation record. The surface coordinates are then used to geo-reference the data in the slope monitoring system. Where depth is significant to the reading, such as for a piezometer, accelerometer or in-place inclinometer, the installation depth below the ground surface should be recorded.

GB radars can be difficult to geo-reference when integrating data as described above, for a few reasons:

- some radar systems have their own monitoring software that does not require data to be geo-referenced (e.g. radar data are overlaid on a photograph that is captured by the radar);
- radars are often moved to different areas of the pit, meaning the location cannot be considered fixed; and
- a small survey error introduced when geo-referencing a radar can have a significant impact on the displayed location of the data, because errors are proportional to the distance from the radar to the monitored surface.

Where geo-referencing is challenging, service providers and operators have included procedures to overcome these three issues; however, they require the mobilisation of a team of surveyors, and the outcome may not be accurate

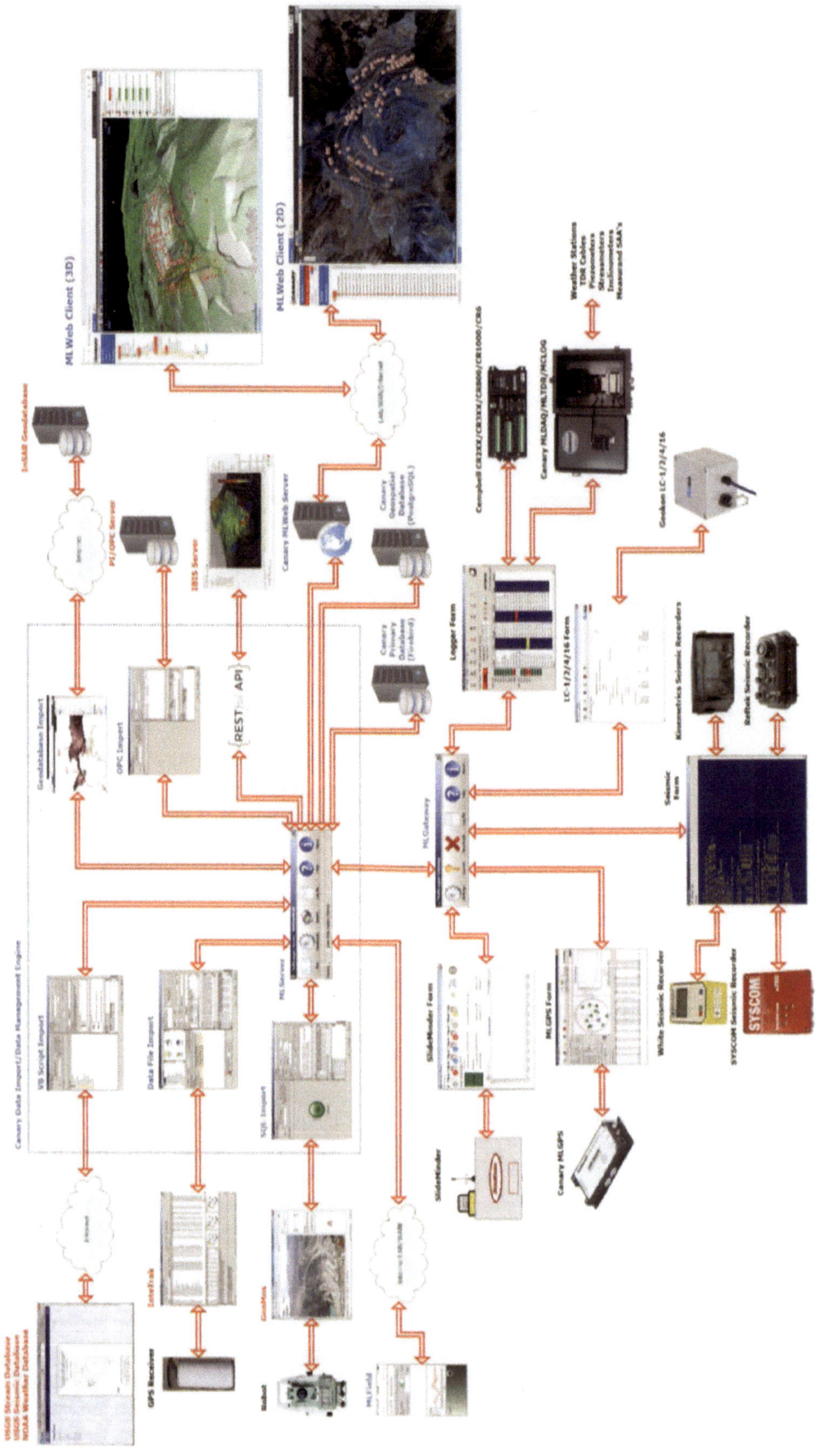

Fig. 4.4: Detailed integrated slope stability program (source: A. Neuwirt).

enough. The result can be that a frequently relocated radar is often not geo-referenced at the time of relocation or is not geo-referenced at all, meaning the data may not be useful in an integrated sense.

Once a 2D radar dataset is geo-referenced, it can be mapped onto a 3D DTM by importing the elevation data for the x- and y- data from the radar dataset. This gives the slope engineer a powerful visualisation tool for analysis and communication. New software developments directed at automating the geo-referencing of radar data include using iterative point cloud matching techniques to find a best fit for the radar point cloud within the 3D pit model. Figure 4.5 shows a selection of four iterations over a matching process, noting that the entire process from start to finish consists of many hundreds of iterations.

The quality of the fit between the radar data and the 3D DTM is checked over time to determine if the shape of the pit has changed due to ongoing mining activity. When this happens, an automatic alarm is generated and sent to the slope engineer, indicating that an updated 3D DTM should be loaded into the slope monitoring system.

4.2.8 Potential failure modes

A mining operation should design its performance monitoring system, including the types of instrumentation selected and deployed, in consideration of the potential slope failure modes. Consideration must also be given to site-specific geological and hydrogeological considerations, including structure, rock or soil mass conditions, lithology and alteration. The details of potential failure modes are addressed in key reference sources like Wyllie (2017). In

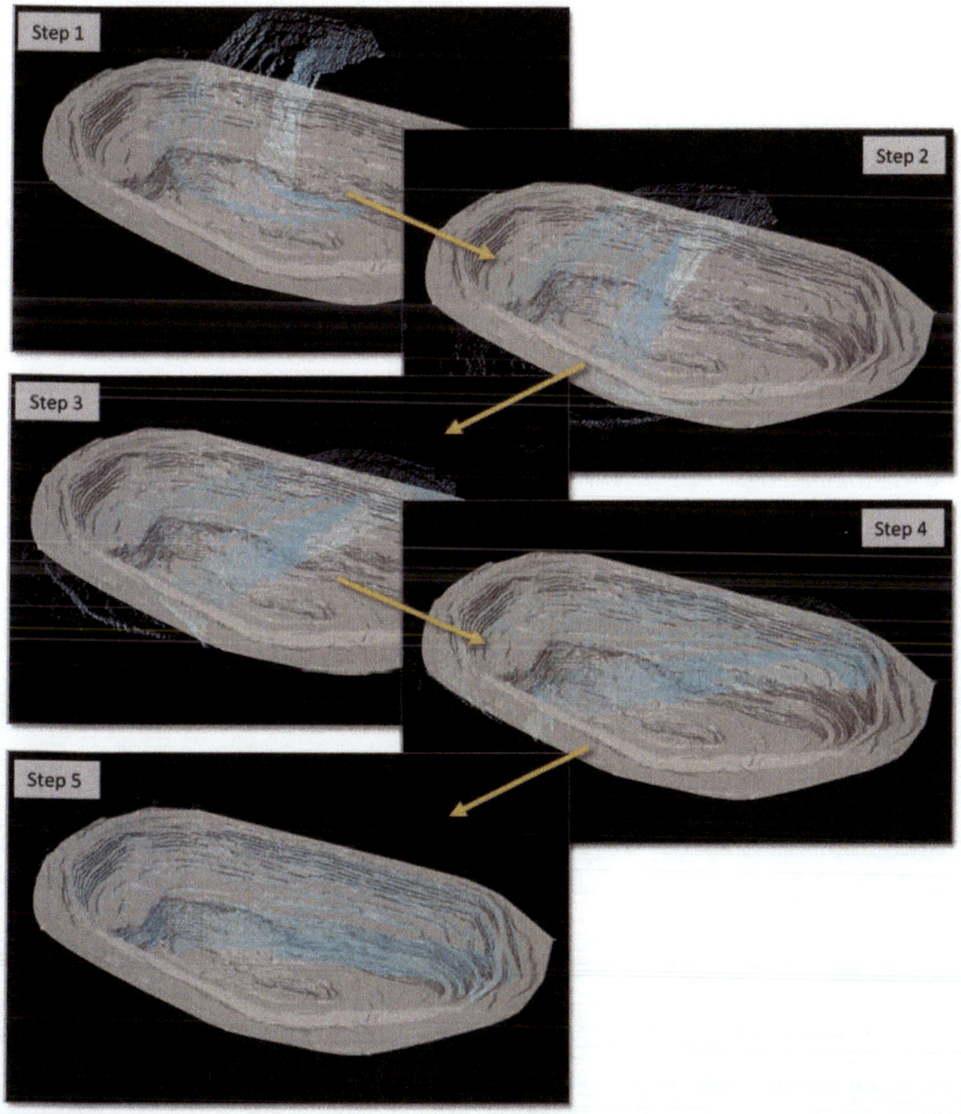

Fig. 4.5: Iterative point cloud to surface matching process for automatic geo-referencing of radar data and laser scanners (source: Sixsense – Soldata).

brief, the definition of a potential failure mode is possible through analysis of the geological structures and/or rock mass conditions and representative geotechnical parameters. The slope displacement patterns also need to be understood, in order to define a proper monitoring scheme for a particular slope.

Following the identification of the possible types of failure on an excavated slope, the displacement pattern is studied to propose an appropriate performance monitoring system. The effectiveness of the monitoring system for detecting the instability in time to implement a mitigation strategy depends on the scale of the instability, and the time between the onset of instability and the occurrence and magnitude of the event. As described previously, displacement monitoring should involve at least two different types of instruments to ensure system reliability and data validation.

In operations where slope stability and rockfall hazard potential attributed to operational controls are important, blast vibration monitoring should be a component of the slope monitoring system. This situation would typically apply in operations that excavate strong rock to create steep inter-ramp slopes. In operations that expose weak rock, including residual soils, groundwater monitoring may be an essential requirement of the monitoring system.

4.2.9 Monitoring for rockfall and bench-scale instability

Rockfall potential is the most significant geotechnical hazard to the workforce in the open pit mining environment. Accordingly, bench face scaling and cleanup operations can present the most significant hazard exposure to mine personnel, mobile equipment and other assets. Subsequent to blasting, and while a shovel or loader is working at a bench face, stress change and relaxation is occurring in the highwall; it may take time for the soil or rock mass to come to equilibrium. Where the soil or rock mass is weak or where persistent intersecting structures are present, the bench face could unravel or fail, particularly during the initial 24 h after the face is excavated. Failure of the bench face could impact one or more benches above the freshly excavated ground, thus inducing a rockfall condition that may impact more than a single bench. A small-scale rockfall event can occur quickly and without warning. Since rockfall and bench-scale instability can occur suddenly, using instrumentation to detect small-scale movement in time for mine operators to avoid the incipient hazard can be challenging.

The operation manages safety-related risk with procedures directed at limiting mine personnel access to potentially hazardous areas. To various degrees, equipment used to doze and mine blasted material protects the operators performing the excavation and cleanup tasks. However, there are limits to the degree of protection provided by mine equipment and these need to be accounted for in the mine development procedures, often referred to as standard operating procedures (SOP). In areas subject to rockfall hazards with potential for a bench face failure condition, an SOP may include the utilisation of spotters, lights for excavation at night, exclusion zones, daylight mining only, increasing frequency of inspections that may include drones or specialised surveys, and performance monitoring instrumentation for safety-critical monitoring. Techniques that include photogrammetry, laser scanners and radar have been developed for the detection of small-scale movement in real-time, that can be effective (Fig. 4.6). However, these must be utilised with extreme caution as they are sensitive to changes in atmospheric conditions, dust, rain and snowfall. Care must also be taken to deploy these instruments in strategic locations where they can be effective. Using instrumentation for the detection of rockfall and small-

Fig. 4.6: Mobile work area monitoring equipment, using a focused radar system to monitor for rockfall above workers in a high-risk area (source: Noon *et al.* 2011).

scale instability can present a high risk in that it could create a false sense of security – mine operators may trust the instrumentation to warn them in time to clear the area before the rockfall or failure event occurs.

Even with leading practice SOPs at mature operations and where operators have considerable relevant experience, the risks can be significant and low-probability events can and will occur. For example, at a North American operation, an 18 m high bench face failed 12 h after excavation by a rope shovel. The failure debris partially buried a dozer that was moving through the runout zone at a distance of 15 m from the bench toe (Fig. 4.7). Fortunately, the dozer operator was safely extracted from his equipment without injury or risk to the shovel operator who came to the rescue. The area had been inspected and cleared by operations and mine engineering personnel for cleanup in accordance with the SOP. The ground-fall event that involved the dozer was unusual in that similar ground is routinely cleaned up at this operation without incident, and the dozer was in motion and was considered to be a safe distance away from the bench face. Even though the incident occurred at night, the area was well-lit and both dozer and shovel operators were on alert for falling rock. The incident occurred without warning from the point of view of the equipment operators. Outcomes of the incident investigation included: 1) establishing the limits of proximity to the bench face where mining equipment may operate safely without the need for exclusion requirements; and 2) evaluating instrumentation that might be used in the future to assist the mine operators to detect ground movement in time to react safely at all times.

Based on the lessons learned from this event, the following improvements were made to the SOP in relation to scaling and cleanup operations in order to avoid similar incidents involving personnel and mobile equipment:

- recognition of potentially hazardous ground that requires a specific SOP for mine development;
- increased utilisation of drones to survey as-built benches and bench faces, and to detect incipient cracks and other obvious and exceptional deformations;
- utilisation of light plants where needed to improve visibility at night;
- utilisation of spotters in areas determined to be potentially hazardous;
- recognition of the spotters' attention span to ensure they are effectively utilised;
- recognition of the importance of the site's geotechnical engineer for inspections and sign-off before scaling and cleanup is approved; and
- experimentation with instrumentation that can potentially be used to detect small-scale incipient deformation and rockfall potential (e.g. photogrammetry, laser scanners and radar).

Fig. 4.7: Partially buried dozer caught in rockfall failure debris (source: J. Mattern and Nevada Gold Mines).

In summary, it is important to recognise instrumentation limitations for the purpose of detecting sudden small-scale but potentially dangerous rockfall and bench-scale slope failure events. GB radar and LiDAR can detect where rockfall has occurred, which can be useful in mapping where more rockfall is likely to occur. Larger rockfalls are often preceded by smaller rockfalls. However, an effective ability to predict where a rock is likely to fall is limited by the ability of equipment to detect the event in time to alert the mine operator. The problem is that these events can occur without warning and at such speed that the operator does not have time to react and take action. Thus, in some cases, the only remedy available to the mine operator may be utilising spotters in combination with frequent inspections by qualified personnel.

4.2.10 Summary of capabilities and limitations of major slope monitoring systems

It is extremely important to understand the specific capabilities and strengths, as well as limitations and weaknesses of performance monitoring instrumentation. Up to this point in these Guidelines, an attempt has been made to describe instruments commonly used to monitor excavated slopes, including their capabilities and limitations. The remainder of Chapter 4 and subsequent chapters is directed at application; operational considerations; data communications, management and analysis; and data utilisation and results reporting. Before moving on to applications, the capabilities and limitations of the major slope monitoring methods are summarised in Table 4.2.

4.3 Application: geodetic monitoring

4.3.1 Overview

Due to its comparatively low cost, instrument utility, reliability and ease of data collection, geodetic monitoring

Table 4.2: Summary of comparative capabilities and limitations of major slope movement monitoring methods

Monitoring method	Capabilities	Limitations
Geodetic monitoring using total stations and prisms	• Measures and tracks total vector movement. • Strategic or investigative and performance monitoring. • Validates measurements from other systems. • Interprets failure modes. • Monitors moderate to large block sizes. • Monitors speed of failure from slow to rapid. • Addresses operational and financial implications (e.g. slope design).	• Angular error in repeatability proportional to distance from target – creates short-term noise. • Safety-critical monitoring. • Monitoring limited to points where prisms are installed – must physically access area of interest. • Measurement frequency limited to cycle time of the total station. • Measurability deteriorates over time with pitting, dust and loss of prisms. • Changing atmospheric conditions.
Geodetic monitoring using GPS and GNSS	• Measures and tracks total vector movement. • Monitors moderate to large block sizes. • Monitors speed of failure from slow to rapid. • Long-term trend analysis.	• Requires access to each receiver for installation and maintenance. • Unit cost. • Monitoring limited to points where instruments are installed. • Data noise for short-term prediction (<4 h).
GB radar systems	• Measurement precision is at a very high level. • Monitoring frequency can be in near real-time. • Monitors small to large block sizes. • Can see through dust and moisture. • Monitors speed of failure from slow to rapid. • Differentiates stable *v.* limits of moving ground over large areas.	• Measures movement only along the LOS. • Unit cost. • Typically requires external support and associated costs. • Unable to see through mesh installed on the slope. • Increased rate of movement can lead to phase ambiguity that could falsely indicate an apparent reduction in velocity. • Complexity of vendor-derived algorithms that could limit system performance. • Sensitive to changing atmospheric conditions.
Satellite radar systems	• Measurement precision is at a very high level. • Monitors small to large block sizes. • Can see through dust and moisture. • Monitors at the scale of the mine property. • Monitors speed of failure from slow to rapid.	• Measures movement only along the LOS. • Imaging frequency is limited by satellite return period. • Typically requires external support and associated cost. • Geometry of excavated slopes can create gaps to visibility (i.e. blind spots). • Increased rate of movement can lead to phase ambiguity that could indicate an apparent reduction in velocity. • Correct interpretations usually require site familiarity.
Laser scanners (LiDAR)	• Measurement precision is at a very high level. • Monitoring frequency can be in near real-time. • Monitors small to large block sizes. • Monitors speed of failure from slow to rapid. • Can see through mesh installed on the slope. • Maps and measures geological structure. • Maps and measures bench geometry (batter-berm).	• Measures movement only along the LOS. • Dust and moisture inhibit system visibility. • Instrument damage can be caused by strong signal return from nearby reflectors. • Measurement/survey error proportional to distance from the target. • Survey (DTM) repeatability for displacement monitoring can be challenging. • Sensitive to changing atmospheric conditions.

remains the most commonly used slope monitoring method in open pit mining. The geodetic monitoring systems in most large open pits require hundreds of prisms to be installed and continuously monitored with subcentimetre accuracy at the 95% confidence level (Chrzanowski and Wilkins 2006). This is now routinely undertaken using RTS with automatic target recognition (ATR), supported by software that controls fully automated data collection and data processing. More recent efforts combine these systems with GPS/GNSS. Achieving the desired accuracy for a geodetic monitoring system requires a well-designed reference network, stable locations for monitoring stations and reference points, clear LOS, consideration of the distance between the total station and targeted prisms, and correction for atmospheric effects.

Newcomen *et al.* (2003) note that with the continuous deepening of an open pit, as well as the requirement for stable ground for good survey control, the challenge to keep shot distances within the range (1000 m) necessary to achieve the desired high level of accuracy is not easily resolved. The sight lengths to the prisms can be shortened by moving the RTS down into the pit (Fig. 4.8), but this has the consequence of limiting flexibility in choosing reasonably stable reference/control targets (beacons) with sufficiently short sight lengths (Wilkins *et al.* 2003). The advantage gained by reducing object point sight lengths is lost by the inability to accurately locate the RTS within the work area due to the increased reference target sight lengths. A solution to this problem is to reduce the dependence on stable reference targets by monitoring and

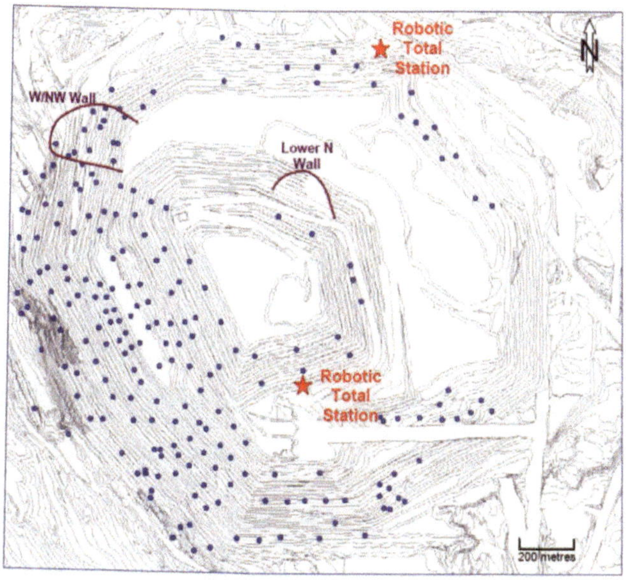

Fig. 4.8: Example of a geodetic survey network showing the location of the RTS shelters relative to the hundreds of prisms for the Valley Pit at the Highland Valley Copper Mine (source: Newcomen *et al.* 2003).

updating the position of the RTS using GNSS. The following subsections summarise important considerations in designing a survey reference network.

4.3.2 Survey reference network and system layout

The integrity of a geo-referenced deformation monitoring system depends on the accuracy of the locations of the survey stations, which form part of the survey reference network. It is therefore important to have a reliable survey reference network as the basis of any geo-referenced deformation monitoring system.

The importance of having a robust control network design is emphasised by Thompson (2005), who observed that a poorly designed network will result in orientation and resection errors outside the limit of tolerance, which will affect the accuracy of the survey. In the case of prism monitoring systems, all movements are referenced to the survey reference network, which is a set of survey beacons that are interlinked by measurements that form a network of coordinated points from which surveys can be taken (Thomas 2011). A typical survey reference network comprises the following beacon types, as illustrated in Fig. 4.9.

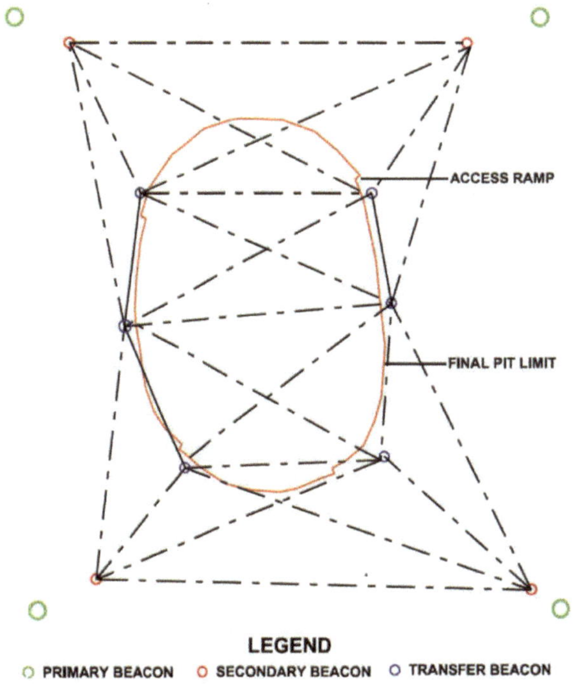

Fig. 4.9: A survey network design which shows a typical network design, including the different classes of beacons (modified after Thomas 2011).

- *Primary beacons* may be positioned between 100 m and 3000 m from the pit crest, depending on local conditions (Cawood and Stacey 2006). Primary beacons are considered the most stable points in the network, hence the requirement to have them well outside the zone of expected deformation where lateral displacement and subsidence (or elastic rebound) will more likely occur (Watt 1996). These beacons are normally used as fixed points for survey measurements and adjustments for the life-of-mine and may remain in service for reclamation and closure monitoring.
- *Secondary beacons* are typically positioned 50–200 m from the pit crest and are used as orientation reference stations. LOS visibility between the secondary and transfer beacons is required, hence the need to consider mine infrastructure such as conveyor belts, power lines, plant buildings and waste rock dumps when positioning these beacons.
- *Transfer beacons* are positioned at the pit crest in stable locations to allow maximum view into the pit. Some of these transfer beacons will be used to mount the survey equipment, such as total stations, which will be used for prism monitoring.

When designing a beacon to be used as a monitoring station, the stability of the structure is important (Henning 2011). Foundation conditions must be considered when designing the monument. Other factors to consider when designing and constructing the beacons (Henning 2011; Thomas 2011; Mphathiwa 2012) include:

- increased beacon height, at least 1.5–2.5 m above the ground surface, to overcome local obstructions;
- accessibility to power and communication infrastructure;

- positioning the monitoring beacon where stability could be compromised, such as too close to the pit crest;
- designing the beacon so that the central pillar is independent of the housing and platform structure where the beacon is used as a monitoring station; and
- using a material with reduced thermal expansion properties (e.g. plastic) for the beacon casing to lessen the effect of heat from direct sunlight exposure on the beacon.

To establish a robust survey reference network, it is recommended that surveyors consider the four-part process described below:

1. survey reference network design and construction;
2. survey beacon design and construction;
3. instrument shelter and construction; and
4. survey network measurements and integrity check.

These processes are described below. Further practical aspects related to their operation are presented in Chapter 5.

4.3.3 Survey reference network design and construction

Fundamental survey principles must be followed when designing a control network of beacons for a geo-referenced deformation monitoring system. This is critically important because no matter how sophisticated the monitoring system, basic survey methods such as resection and traversing still have to be applied during the measuring process (Cawood and Stacey 2006). As a guide to designing a control network, Bannister *et al.* (1998) recommended the following process.

- Perform a thorough reconnaissance of the area of interest using maps, aerial photographs and development plans. The primary requirement is to establish a clear LOS between the reference stations, especially between the secondary and transfer beacons. This can be achieved by avoiding areas with views obstructed by buildings and areas targeted for infrastructure development for the life-of-mine.
- Provisional positions of the beacons should be marked on the plan that is approved by key stakeholders in the design and development process. This will constitute the provisional design of the survey reference network.
- The provisional design of the network should allow for survey applications such as resection and traversing to be done with minimum geometric constraints. This means that the geometry of the beacons should allow for long distance and unobstructed sights, and avoid acute angles that would otherwise affect the accuracy of the survey measurements.
- It is critical to test the integrity of the network design before it is actually constructed. It is recommended that the least-squares method be used for the pre-analysis of the accuracy of the design. The least-squares method of estimation or adjustment of measurements is most commonly used in surveying and geodesy (Davis *et al.* 1981).
- When designing the survey reference network, it is advised to have redundancy in terms of the numbers of beacons in each class (primary, secondary and transfer). This gives the surveyor the flexibility to use other beacons if a problem develops with one of the beacons. A problem might arise from new construction blocking the LOS to a beacon or cracks emerging on the ground surface near the beacon, making the reference station suspect in terms of stability.
- The pre-analysis is done by supplying a list of input station names along with their approximate coordinates scaled from a map or photograph, indicating the positions of the planned survey beacons. Proposed measurements are then entered into a least-squares adjustment program using from–to station names rather than actual field surveying data. As with regular data, there is a need to indicate standard error values for these proposed measurements so that the proper weight can be applied (Curry and Sawyer 1991).
- The least-squares pre-analysis will take the proposed survey and generate computed accuracies (error ellipses) for all survey beacons on the network design, as shown in Fig. 4.10.
- Once the positional uncertainties have been resolved, the provisional design can be adjusted by moving 'problem' beacons until an optimal design is achieved. This is the design that will be constructed.
- Achieving an optimal network design in a mine environment is challenging because of the infrastructure that typically exists near the open pit. Structures such as waste rock dumps, process plant buildings and conveyor belts can make it difficult to achieve a robust geometry of beacons with clear LOS visibility. In the example shown in Fig. 4.10, the least-squares pre-analysis method becomes critical as it will test the integrity of all simulations done in trying to design a survey network in a constrained space.
- It is good practice to confirm the final beacon positions with a site assessment by a geotechnical engineer and a mine surveyor before construction begins. This process ensures that the proposed beacon positions are on stable ground and avoid active or potential movement features and gullies, where wash-outs might occur, which might not have been identified on the maps during reconnaissance. The mine surveyor must ensure that there is no obstruction along the LOS by newly

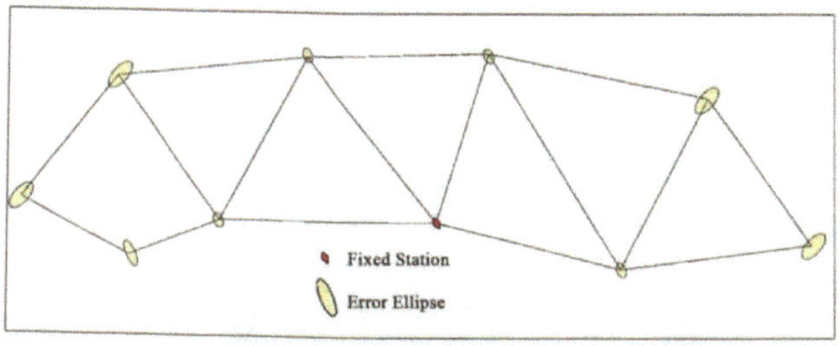

Fig. 4.10: Example of pre-analysis results for designing a survey network. The shape and size of the error ellipses indicate the positional uncertainty as the surveyor moves further away from the fixed stations (source: N. Mphathiwa).

built or planned structures, which have yet to be updated on the survey plans. This methodology is recommended by Thomas (2011).

4.3.4 Survey beacon design and construction

After completing the design of the survey control network, the focus shifts to the structural design of the beacons. There are five fundamental requirements when designing survey beacons (Mphathiwa and Cawood 2014):

1. a competent structural engineering design;
2. a foundation design suitable for the ground conditions on which the beacon will be constructed;
3. practicality and cost-effectiveness;
4. the mine surveyor must ensure the structure is constructed as designed; and
5. those constructing the beacons have the required competencies to achieve the design specifications.

It is leading practice that the beacon design is done by a qualified structural engineer in accordance with site specific requirements (Mphathiwa and Cawood 2014). For example, beacons installed at the Debswana mine operations, where thick unconsolidated cover exists, were designed to require piling through up to 17 m of loose Kalahari sand to anchor the monuments into the deeper bedrock, as shown in Fig. 4.11.

Beacon construction should be done by a qualified contractor approved by the structural engineer. The mine surveyor supervises the actual construction on site to ensure compliance to the beacon design. The beacon design is also validated for compliance with design and construction specifications as stipulated in the guidelines for real-time GNSS networks (Henning 2011).

4.3.5 Atmospheric impacts on slope monitoring system design

4.3.5.1 Meteorological sensors

Various methods have been developed for applying atmospheric corrections to distance measurements using

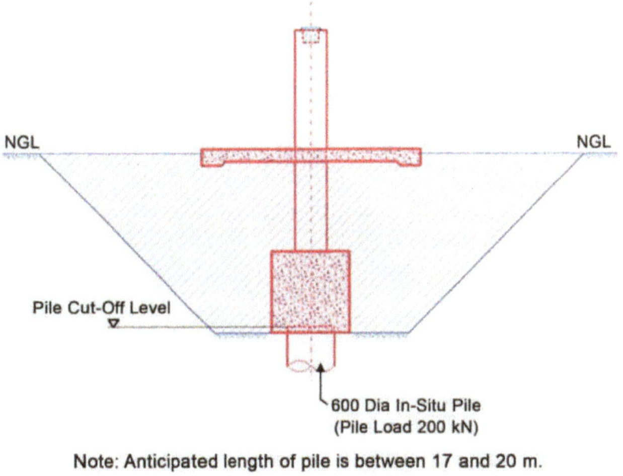

Fig. 4.11: A beacon design showing piling extending down through unconsolidated material to be anchored into the bedrock (source: Debswana Diamond Co.).

total stations. A process applied at one of Debswana's operations is used in the following discussion, to present a standard methodology for applying atmospheric correction for RTS slope monitoring. The process shows that if the meteorological sensors can be placed very close to where the actual measurements are made, a significant improvement in prism monitoring data accuracy can be achieved.

Thomas (2011) emphasises the importance of measuring the ambient temperature, atmospheric pressure and relative humidity when using total stations for measuring distances. Thomas further stated that if atmospheric conditions are not compensated for, the accuracy of the measured distances can be adversely affected because atmospheric conditions change over time. Locations of the temperature, pressure and humidity sensors have a significant impact on the final adjusted slope distances. These parameters must be recorded constantly because of varying atmospheric conditions, especially with cloud cover.

Afeni and Cawood (2013) report that the influence of atmospheric conditions on survey accuracy is a significant challenge for prism monitoring. Research on this subject has identified several accuracy effects caused by atmospheric conditions that justify further study. Afeni and Cawood (2013) report that because the accuracy of prism monitoring is affected by rainfall, dust and mist, a long-term solution is required to ensure that the RTS monitoring equipment is protected (e.g. using a sealed shelter with glass enclosures). It is important to select a suitable type of glass which will not affect the confidence levels of the measurements taken through the medium (see also Section 5.2).

Afeni (2011) carried out tests to determine the impact of taking RTS measurements through glass in an open pit

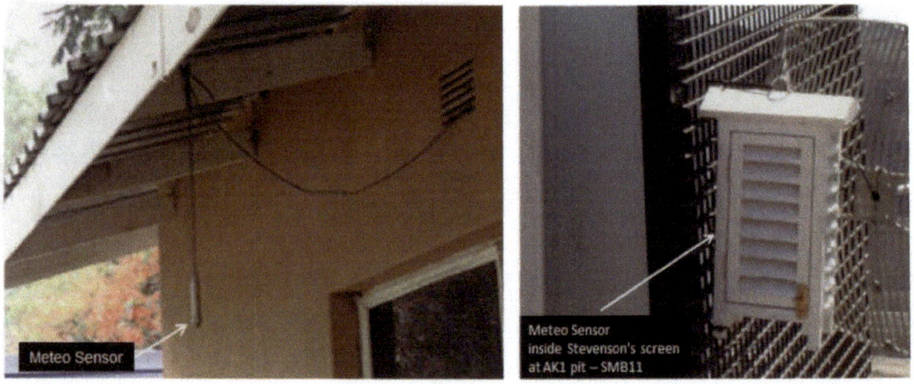

Fig. 4.12: Different positions used by Debswana to protect its meteorological sensors used for atmospheric corrections, including (left) under an eave of an enclosure housing a transfer beacon station (not visible in this view) and (right) using a Stevenson Screen (source: O. Randall and Debswana Orapa mine).

environment (i.e. with measurements also subjected to atmospheric effects). His research established that the properties of the glass must be considered when taking total station distance measurements. It was noted that glass had very little or no impact on the vertical component of the measurements; however, it had a significant impact on the horizontal distance measurements, which increased as a function of the glass thickness. This research shows that the glass effect remains the same, even after the atmospheric corrections have been applied to the measurements.

Another key result from Afeni's (2011) research was that the atmospheric corrections provided in the total station operating manuals accurately compensated for the variations in the atmospheric parameters. However, this required the meteorological sensor to be correctly positioned near the transfer beacon at the open pit, shown in Fig. 4.12 (the transfer beacon is out of view). This led to further improvements at the Debswana operation including protection of all meteorological sensors with Stevenson Screens. A Stevenson Screen, shown in Fig. 4.12, is a small wooden housing with perforated sides which mitigates the influence of precipitation and direct heat from the sun, and allows the meteorological sensors to measure temperature, pressure and humidity. The use of Stevenson Screens further improved the measurement accuracy achieved with the meteorological sensors.

4.3.5.2 How temperature and pressure impact electronic distance measurements

Changes in atmospheric conditions in an open pit environment can be natural, but can also be induced by mining activities. One side of the pit can be hot and dry while the other side can be cooler and humid, or the pit can be dusty due to wind and mining equipment. Extreme weather conditions can also adversely impact survey accuracy. Under these conditions, errors will be introduced when using optical survey instruments to measure prisms, because the carrier signal from the instrument will be propagating through different atmospheric conditions to and from the reflector. The overall distance measured may therefore be incorrect due to uneven atmospheric density distribution, as the carrier signal moves across the pit.

Ambient temperature and atmospheric pressure affect the air density, affecting the velocity of the measuring signal, which results in inaccurate distances being measured. The meteorological sensor is able to measure the atmospheric parameters at the reference beacon, but it is unable to model the atmospheric conditions across the open pit. This is one of the major challenges for geodetic monitoring in an open pit environment. The varying atmospheric conditions in the open pit affect the electronic distance measurement (EDM) signal which travels from the instrument to the reflector. Under normal circumstances, the signal would follow a straight line to the reflector, but under varying atmospheric conditions, the signal will take a more complex curved path due to refraction, as illustrated in Fig. 4.13.

Refraction is the bending of an electro-magnetic beam or change in direction of a carrier signal, where survey accuracy is affected. Varying ambient temperature and atmospheric pressure therefore affect the EDM through refraction. Although this impact cannot be totally eliminated, there are methods that can be used to minimise error. Temperature, pressure and humidity readings are recorded at the measuring instrument. The atmospheric correction parameters are then computed and applied to the measured distances.

The coordinates computed using corrected distances will usually show a relatively low displacement value whereas, when no corrections are made, the displacement is higher where the error can be more than 20 mm. Figure 4.14 illustrates this difference, where a regular pattern of minimal error is evident when atmospheric corrections are

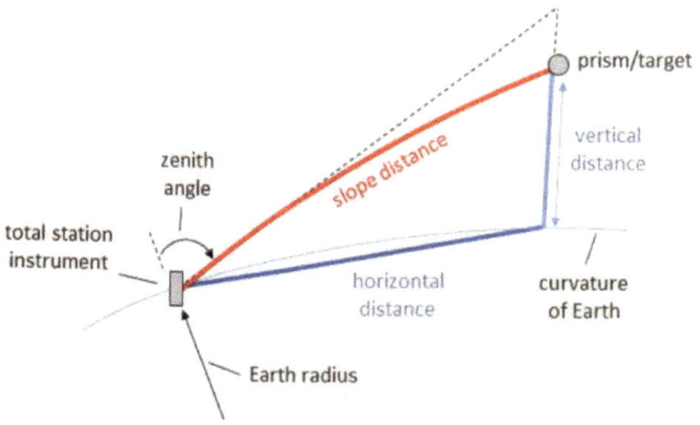

Fig. 4.13: Illustration of the combined effect of refraction and curvature (source: E. Eberhardt).

applied to the monitoring data. When the meteorological sensor is not functional, no corrections are made, resulting in a higher amplitude plot.

4.3.5.3 How atmospheric corrections are applied to electronic distance measurements

The challenge of prism monitoring in an open pit is that meteorological conditions are measured only at the reference beacon, which is the most common practice as errors of a few parts per million (ppm) will be present. To achieve an accuracy better than 1 ppm, it is important to measure meteorological conditions at shorter intervals (Bertacchini *et al.* 2011). It is also important to calculate and apply the correct atmospheric conditions (Crook 1995). The ppm application to correct for distances is outlined below (Bertacchini *et al.* 2011):

$$D_C = \left[\left(\frac{ppm}{1 \times 10^6}\right)D_m\right] + D_m$$

where D_C = the field corrected distance;

ppm = parts per million; and

D_m = the measured distance with zero ppm set in the total station.

The refraction correction is then calculated as follows:

$$d = \left(\frac{nR}{nL}\right) \times d_m$$

where d = the corrected distance;

nL = the ambient refraction index;

nR = the reference refractive index (instrument); and

d_m = the measured distance.

The accuracy of the ppm correction is directly affected by the validity of the atmospheric measurements. The most accurate measurements are obtained using meteorological sensors at the area of interest; this is the normal practice in a prism monitoring setup. The meteorological sensors must be placed away from the monitoring shelters to mitigate the effects of shadow and radiation. The sensors must be set to take atmospheric readings at intervals of 5 min or less. These measurements must be recorded at this interval to ensure that atmospheric corrections are applied to the survey measurements. In an automated monitoring setup, the last reading from the meteorological sensor is used by the monitoring software to compute atmospheric corrections and apply them to the measured distances; this process is done within the software.

When the atmospheric corrections are not applied, the displacement values shown in longitudinal and vertical plots will be high in amplitude. This effect can be misinterpreted by the slope engineer as real movement. Cawood and Stacey (2006) emphasise this point by stating, '*atmospheric corrections on total station slope distances are problematic when measuring over long distances across voids where the temperature variations along the line of measurement can vary significantly. These variations cause exaggerated fluctuation graphs and errors in elevation, and they can make interpretation of continuous monitoring difficult*'.

4.3.5.4 Alternative approach to atmospheric corrections

Applying atmospheric corrections to distance measurements can minimise the impact caused by varying atmospheric conditions across an open pit. As open pits become wider and deeper, the measurement error of prisms at longer distances increases. The most important limiting factor with an automated monitoring system is that atmospheric parameters can only be measured at the

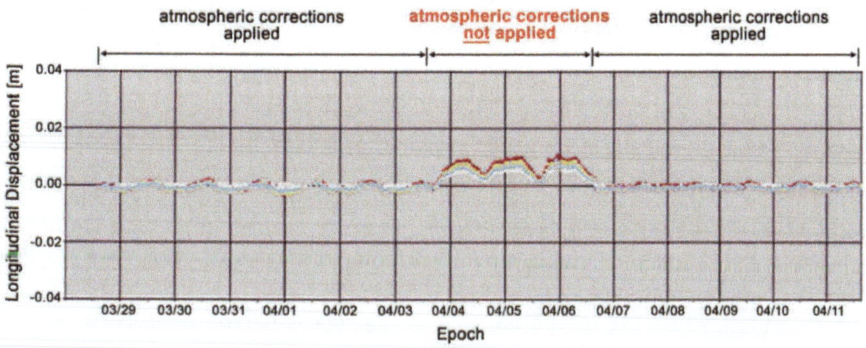

Fig. 4.14: Example of a time-series displacement plot at one of the Debswana mines (source: O. Randall).

reference beacons, not at the prisms, for practical reasons. Another limiting factor is that the temperature, pressure and humidity are continually changing, and the true values may not be known for correcting the slope distances along the LOS.

To address these issues, an alternative approach to correcting the slope distances from total station measurements was developed at the Debswana-Orapa pit expansion for when an open pit becomes deeper and wider. GNSS can provide high accuracy controls on total station measurements in changing environments (Brown *et al.* 2007). The distance between control points can be accurately computed through GNSS real-time monitoring. If there is a fixed network of beacons with a suitable geometry, then post-processed coordinates of these control points can be computed every 1, 6 or 24 h.

Once the coordinates of the control points are known, the distance between any two reference beacons within the control network can be computed. An alternative approach, which can be used for the correction of the total station distances, is to measure control beacons where the distance is accurately known so that a reference scale factor can be directly computed and used to correct the distances measured to the prisms. Figure 4.15 illustrates how this method is applied.

Monitoring software can be configured for periodic measurement of reference distances and to apply a reference scale factor to subsequent measurements of prisms. Figure 4.15 shows a network of control beacons (TS1 and BCN1–BCN5) which are continuously monitored with GNSS. Coordinates for each beacon are obtained in real-time in an automated process which requires appropriate software settings. For example, if the base or known distance between the transfer beacon (TS1) and primary beacon (BCN1) is equal to 100 m, and the new distance measured using GNSS coordinates is equal to 100.003 m, then the reference distance scale factor is computed using the formula:

$$S_R = \frac{d}{d'}$$

where S_R = the reference scale factor;

d = the originally known distance; and

d' = the new distance computed from GNSS coordinates.

Thus, the reference scale factor for this example would be:

$$S_R = \frac{100}{100.003} = 0.99997001$$

The reference scale factor is then applied to distances measured to the prisms (P1–P4) in the pit. This approach is anticipated to only work accurately when a limited number of prisms are measured, where the computed

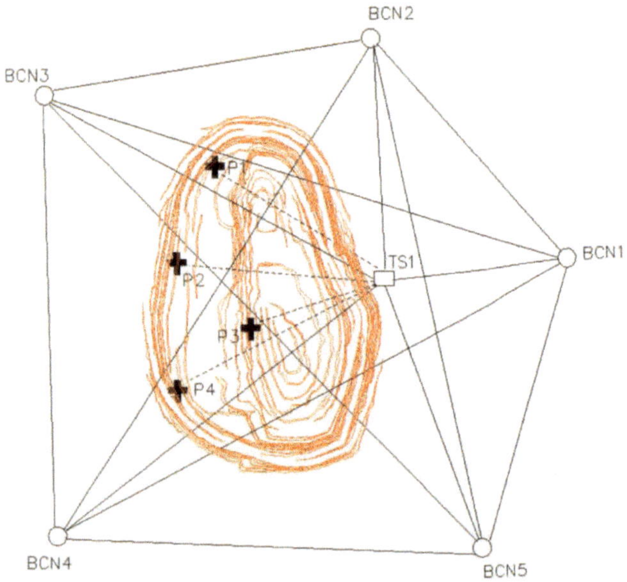

Fig. 4.15: Combined GNSS and EDM total stations for prism monitoring (source: O. Randall).

reference scale factor should be used for a limited period depending on the conditions in and around an open pit. For total station distances, the reference scale factor will require change or updating as atmospheric conditions may change rapidly. For better accuracy, the reference scale factor should be used for a maximum time of 5 min and should be continuously updated. When an open pit has at least 15 prisms, the monitoring software can be configured to have different measuring groups from the same control beacon.

Brown *et al.* (2007) report that the use of GNSS receivers to provide a stable reference frame for total stations sited in any environment is a viable option and one that has many advantages. They conclude that combining GNSS satellite receivers with total stations is an effective method for monitoring prisms.

4.3.5.5 Summary

The use of RTS is one of the best methods available for slope monitoring. However, the accuracy of the prism measurements can be highly influenced by the ever-changing atmospheric conditions in the open pit. This effect is minimised by applying atmospheric corrections to survey measurements. The corrections are based on parameters of ambient temperature, atmospheric pressure and relative humidity readings made at the reference beacon. The reference scale factor correction method is preferred in large open pits, since the baseline distance does not have to be computed where the measurements take place. Reference beacons are located at some distance beyond the pit crest, where LOS visibility is good, the ground is stable and atmospheric conditions are likely to have minimal impact on the reference scale factor when computing the baseline distances.

Meteorological stations should be located in the proximity of both total station and reference beacons for an ideal continuous monitoring system. The underlying limitation of this approach is that atmospheric conditions at the total station and monitoring prisms are not necessarily the same as those in the void of the open pit.

4.3.6 Analysis and interpretation of geodetic monitoring data

Slope engineers use time-series plots that display the recorded slope displacement over time to identify movement trends. For monitoring prisms, the cumulative displacement from the origin is calculated from recorded coordinates. Displacement can be calculated as the total displacement in: i) all three directional components (i.e. 3D displacement) – northing, easting and elevation; ii) the horizontal component only (i.e. 2D displacement) – northing and easting; or iii) only in one dimension, such as the LOS or elevation.

Figure 4.16 shows common techniques used for visualising geodetic monitoring data, including incremental and cumulative displacement plots, daily rate plots and wander plots (movement direction in plan view or section). Although displacement plots are useful for identifying movement trends, velocity values are often used as alarm triggers or to compare the movement rates between periods of displacement changes. Velocity can be expressed from any one of the displacement values described above (i.e. 3D, 2D or velocity along the LOS). Traditionally, the slope engineer uses displacement plots to calculate velocity where the amount of cumulative displacement is read off the graph and divided by the time period over which the displacement occurred (trend line method). With the increased use of software to manage prism monitoring programs and to display results, it is important for the slope engineer to understand how the velocity is calculated by the software.

Some of the available software packages calculate 3D and 2D velocity by first determining the velocity using directional components. The resultant 3D or 2D velocity is then calculated by using the velocities for each component of the vector calculation. This calculation method can be demonstrated using a hypothetical case where the horizontal displacement of a prism is shown in Fig. 4.17. Between Day 3 and Day 10 in this example, the prism moves in a non-linear manner relative to the origin. The total 2D displacement is calculated for Day 3 to Day 10 by using the formula shown below. The results are summarised in Fig. 4.17.

$$\text{2D Velocity} = \sqrt{(VelocityNorthing)^2 + (VelocityEasting)^2}$$

Using the traditional trend line method to calculate velocity will, in this example, result in a 0 mm/day velocity. However, if the calculated northing and easting displacements are plotted and the associated velocities determined, as shown in Fig. 4.18, the results produce a northing velocity of -0.63 mm/day, an easting velocity of 0.85 mm/day and a resultant 2D velocity of 1.05 mm/day.

In this example where the displacement trend is non-linear, the difference in calculated velocity is

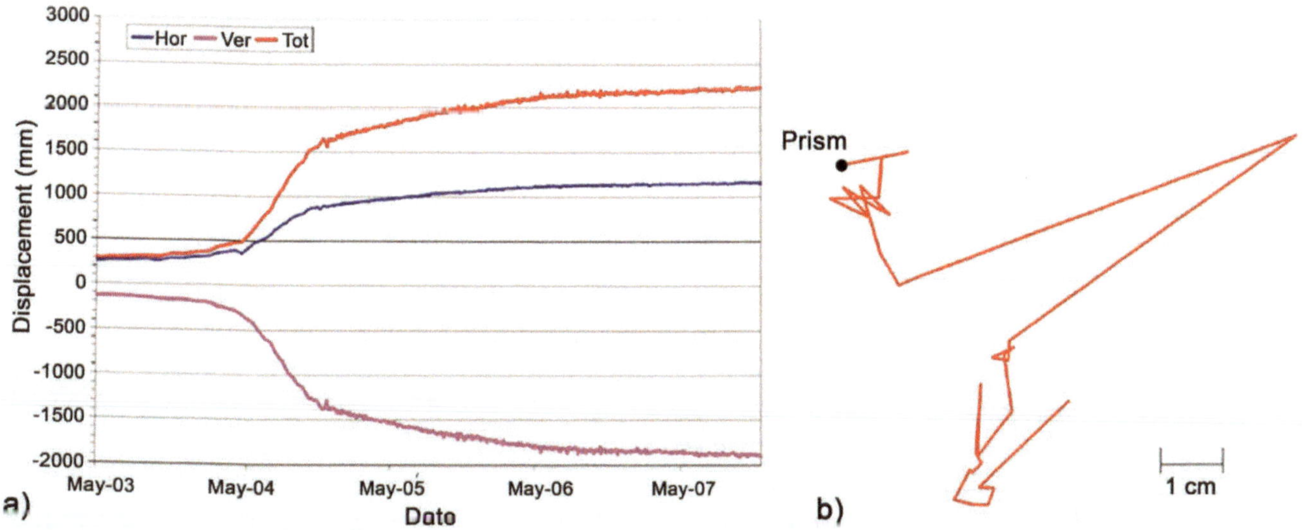

Fig. 4.16: Examples of typical methods of presenting geodetic monitoring data, including (a) cumulative displacement rate plot and (b) wander (x–y) plot (source: Piteau Associates).

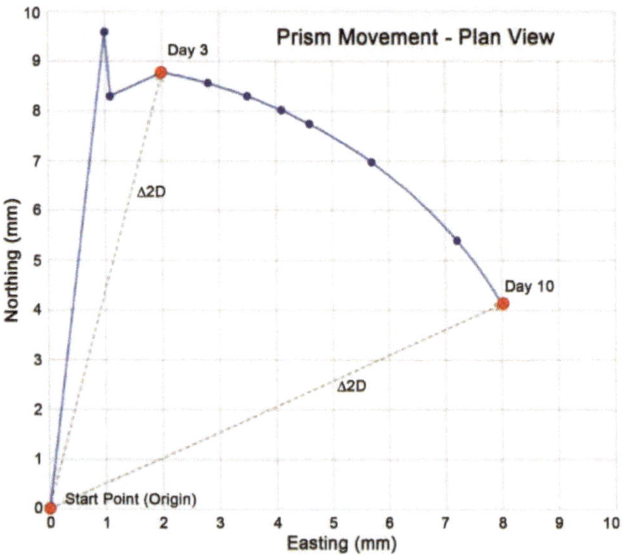

Date	ΔN (mm)	Δ E (mm)	Δ 2D (mm)
Day 3	8.78	2	9
Day 4	8.56	2.8	9
Day 5	8.30	3.5	9
Day 6	8.02	4.1	9
Day 7	7.74	4.6	9
Day 8	6.97	5.7	9
Day 9	5.40	7.2	9
Day 10	4.13	8	9

Fig. 4.17: Plan view of prism displacement for the data and calculated 2D displacements (source: F. Wessels).

significant. Where the displacement is linear (i.e. moving away from the RTS), the differences will be smaller. It is important that the slope engineer understands how the software calculates velocities, as it can assist with alarm setting and may explain 'false' alarms, as indicated from the following case study.

The case study involves an iron ore mine where a geotechnical engineer received a prism alarm triggered by a velocity threshold being exceeded. When the displacement data were examined, no obvious change in the monitoring trend was evident. However, the velocity plot for the same prism and time period displayed an increase in velocity for 16 consecutive readings, with a reported velocity of 5.7 mm/day at the time the alarm was triggered. Figure 4.19 shows the 3D displacement and velocity for the prism. If the velocity is calculated from the displacement graph (trend line method), the velocity is ~1.0 mm/day. Upon further examination of the data trends, it became evident that an apparent decrease in the elevation, with the associated increased elevation velocity, resulted in the high calculated velocity value (Fig. 4.20).

This case study illustrates how the different techniques available to calculate velocity can lead to a misinterpretation of movement trends. In this case, the relationship between the constant trend of the 3D displacement data and apparent velocity increase was not reconciled. It is therefore important that the slope engineer understands how velocity is calculated by the software and the potential impact it can have on the interpretation of movement trends and alarm thresholds.

A comprehensive review of the trend analysis to interpret changes in slope deformation patterns is summarised in Section 7.5. The discussion includes the importance of long-term surveillance slope monitoring, interpretation of various movement patterns and estimating the time-to-failure by assessing acceleration patterns and using the inverse velocity method. Caution regarding the use of inverse velocity for slope failure prediction and other cautions regarding alarm settings using deformation and velocity data are addressed in Section 8.4.

The triggering of faulty alarms or uncertainty over misleading geodetic instrument readings are frequent challenges for the slope engineer. Instrument and system reliability are therefore also of paramount importance. One key limitation after a moving slope is detected is that

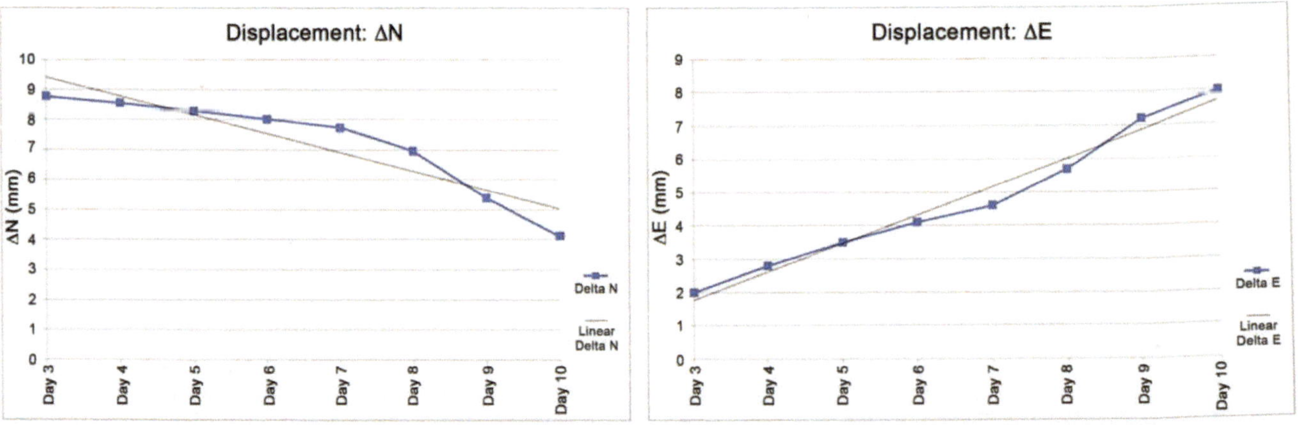

Fig. 4.18: Time-series plots showing the northing and easting displacements, and trend lines used to determine velocity (source: F. Wessels).

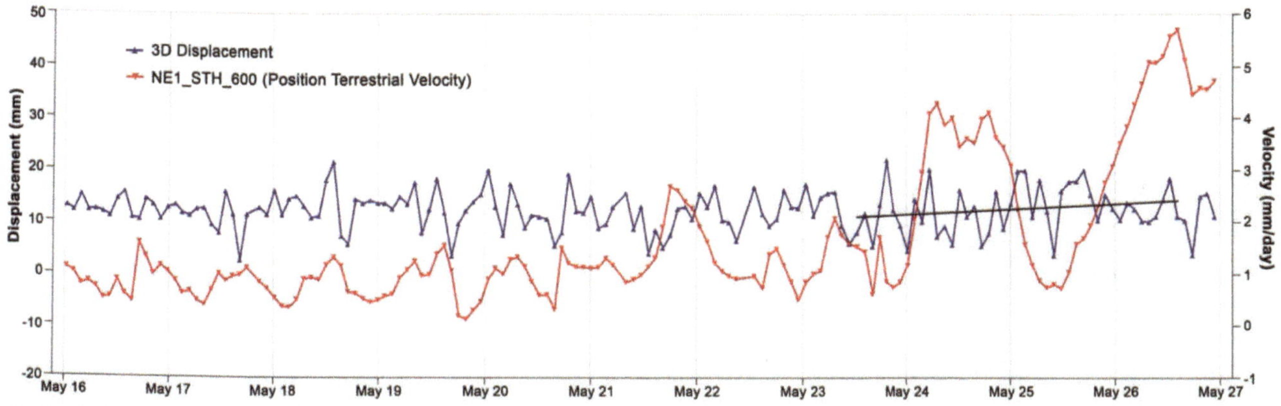

Fig. 4.19: Plot showing 3D displacement (blue) and velocity (red) for a prism. The trend line represents the velocity based on incremental displacement. Note the general increase in the velocity trend from ~24 May, especially between 25 and 26 May (source: F. Wessels).

subsequent prism installation is limited to those parts of the slope that can be safely accessed. Prisms located on benches with limited accessibility are often not maintained and have a limited operational life. Movements of the pit wall between the working prisms must be extrapolated where deformation along geologic structures that may influence stability might be missed or misinterpreted. Thus, a well-thought-out prism network is fundamental to a slope monitoring program that can further be used to calibrate and validate other monitoring campaigns.

4.4 Application: ground-based radar

4.4.1 Overview

Seamlessly integrating radar data with other established slope monitoring systems and methodologies, such as visual imaging and geodetic monitoring, plays a pivotal role in assisting the slope engineer with understanding the overall slope behaviour and assessing data reliability.

Monitoring data side-by-side or overlaid by a visual image of the slope can be used to interpret real slope movement and false movement effects. Fig. 4.21 shows a high-resolution photograph image overlaying a 3D DTM with deformation data, including limits of a slope failure. Visual images can be taken as frequently as every hour, or even every scan, to depict areas on the slope that have vegetation, snow or loose materials; cables and pipes; mining equipment; geologic structures; and water on the rock face, which can be used to facilitate the interpretation of slope conditions.

The functionality of radar monitoring directly interfacing with data from an RTS database is another example of monitoring system integration. It allows the slope engineer to identify anomalous changing conditions by doing a side-by-side comparison of the radar and total station data. This capability can be used to identify the occurrence of erroneous information when data from one

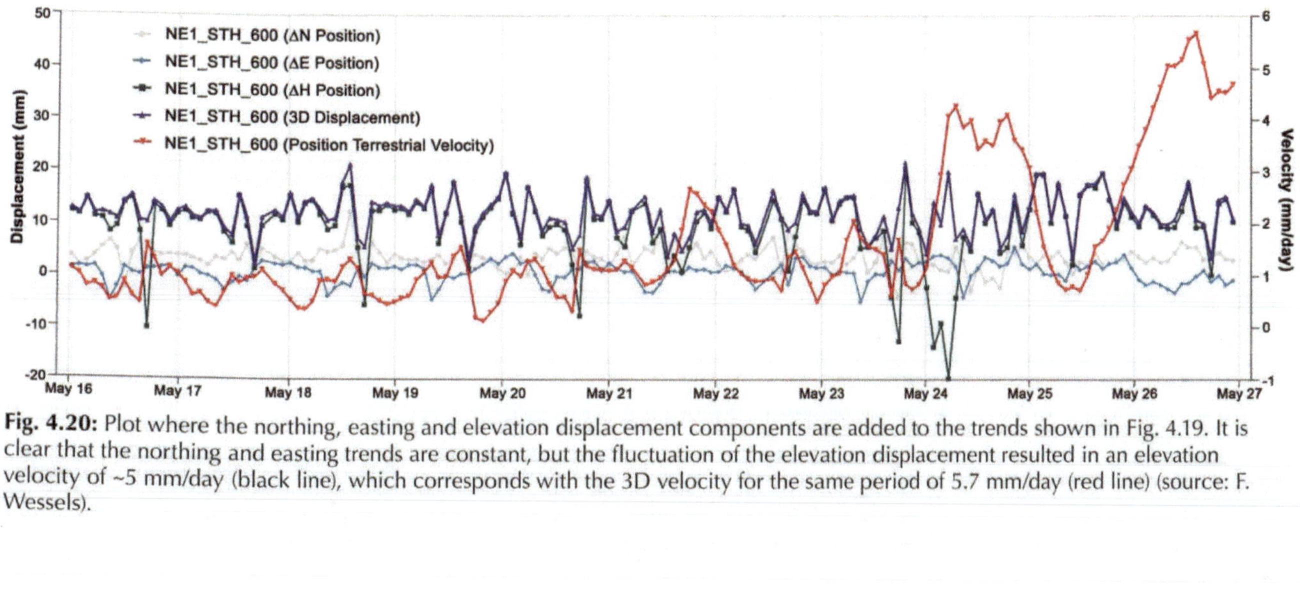

Fig. 4.20: Plot where the northing, easting and elevation displacement components are added to the trends shown in Fig. 4.19. It is clear that the northing and easting trends are constant, but the fluctuation of the elevation displacement resulted in an elevation velocity of ~5 mm/day (black line), which corresponds with the 3D velocity for the same period of 5.7 mm/day (red line) (source: F. Wessels).

Fig. 4.21: High-resolution photograph and deformation image draped over a 3D DTM (source: GroundProbe).

instrument deviate from data generated by another instrument. An example is shown in Fig. 4.22. The deviation shown between the radar data and that measured using an RTS can then be interrogated by the slope engineer to determine if changing movement is actually occurring or if there is a problem with one of the instruments, which can be remediated in time to avoid unwanted consequences or to circumvent a false alarm. The importance of confirming data reliability is further discussed in Section 7.4.5.

4.4.2 Interpretation of ground-based radar data

4.4.2.1 Effect of local geology and structure on stability

Determining the effect of the local geology and structure on the long-term stability of open pit slopes is a key element of strategic monitoring. Most radar software applications allow the slope engineer to import information pertaining

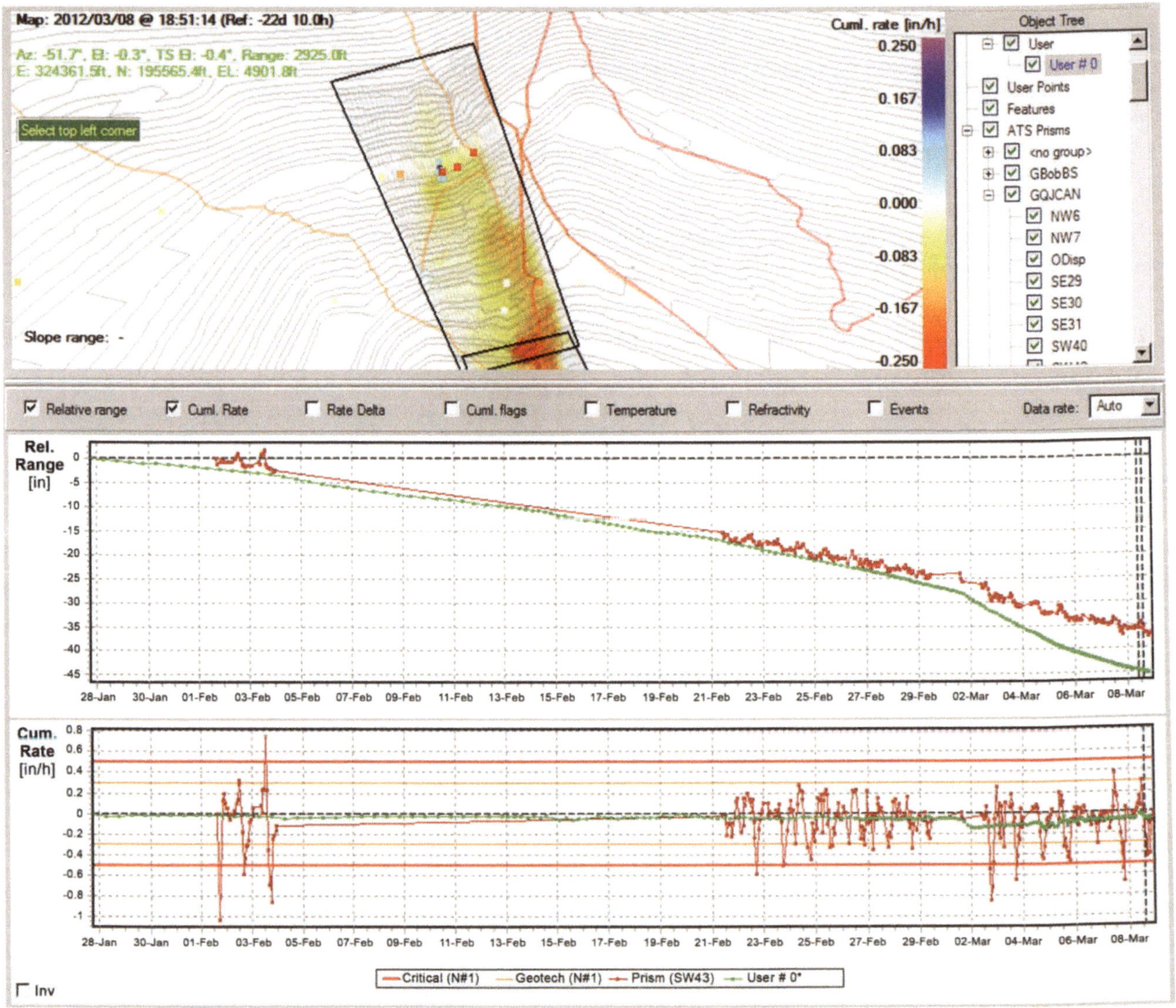

Fig. 4.22: Validation example of slope monitoring sensors made by comparing slope displacements and velocities measured using an RTS and prisms (red line) and GB radar (green line) (source: Reutech).

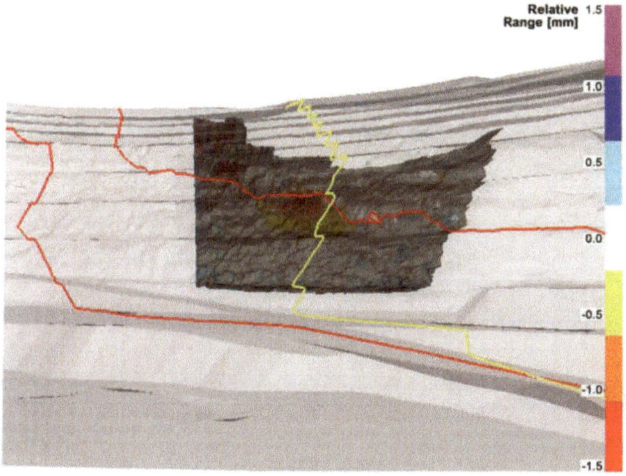

Fig. 4.23: Example of overlaying major geologic features (fault traces that may be potential stability controls) onto a synthetic radar image of the target highwall (source: Reutech).

to the local geology and structures within the slope face. Figure 4.23 shows a radar's synthetic map with an overlay of fault traces across the highwall. Movement that results from the daylighting of the structures on the highwall can be accurately located and monitored in real-time.

4.4.2.2 Effect of blasting on stability

As a result of routine blasting activity, vibration energy impacting the highwall may cause permanent deformation or temporary accelerated movement. A section of the open pit may be closed until the acceleration has slowed to an acceptable level. The time-series plot of Fig. 4.24 shows the acceleration of a slope and the effect on its stability of three separate blasting events. In this case, the slope eventually failed and the information was used to predict other potential instabilities that could occur under similar operating conditions.

Gauging the effect of blasting on overall slope stability forms an integral part of long-term strategic monitoring. It is important for the slope engineer to analyse and understand the data, as the consequence of blasting can have a major impact on the overall safety and productivity of the operation.

4.4.2.3 Slope deformation response to mining activities

Real-time monitoring is advantageous where mining activity is taking place adjacent to sensitive slopes or infrastructure. Radar is a powerful tool that can provide data that allows the slope deformation response to mining

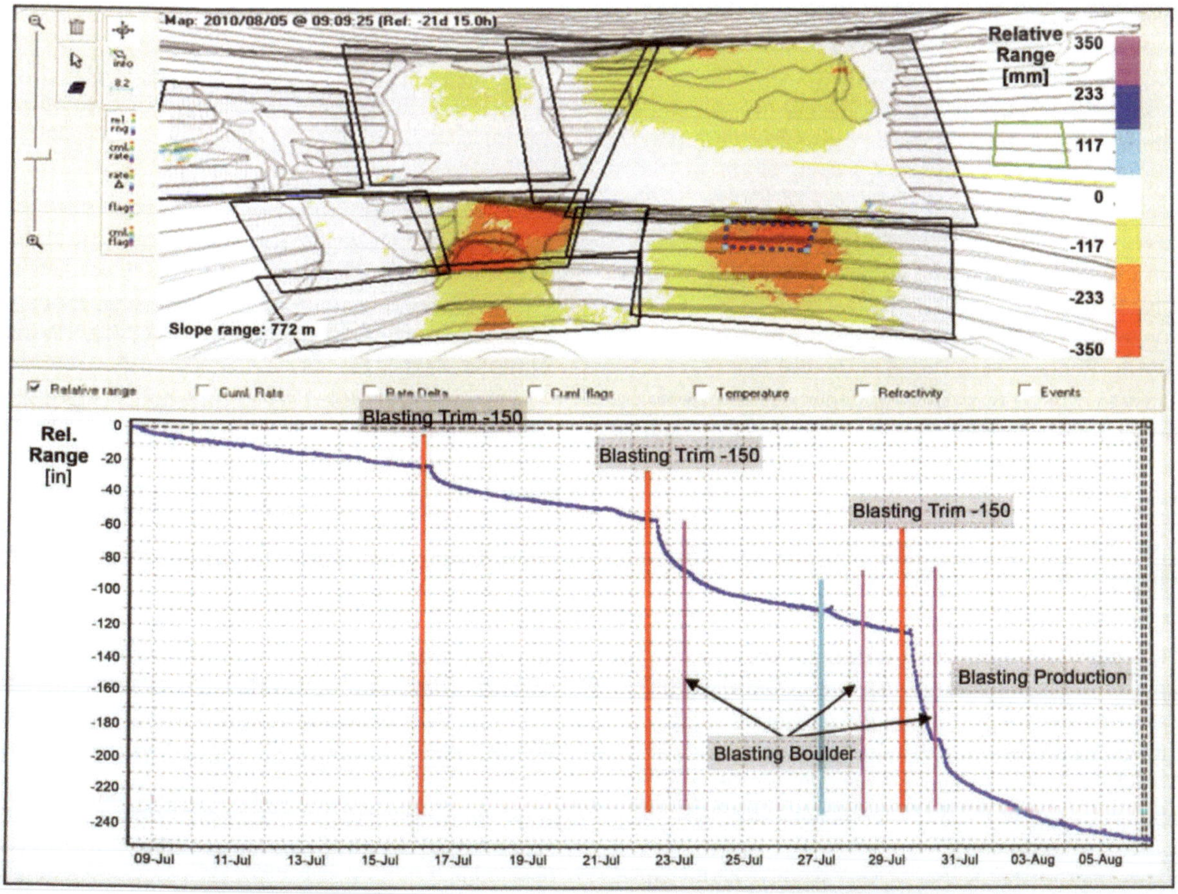

Fig. 4.24: The effect of blasting on slope stability as detected in radar monitoring data (source: Reutech).

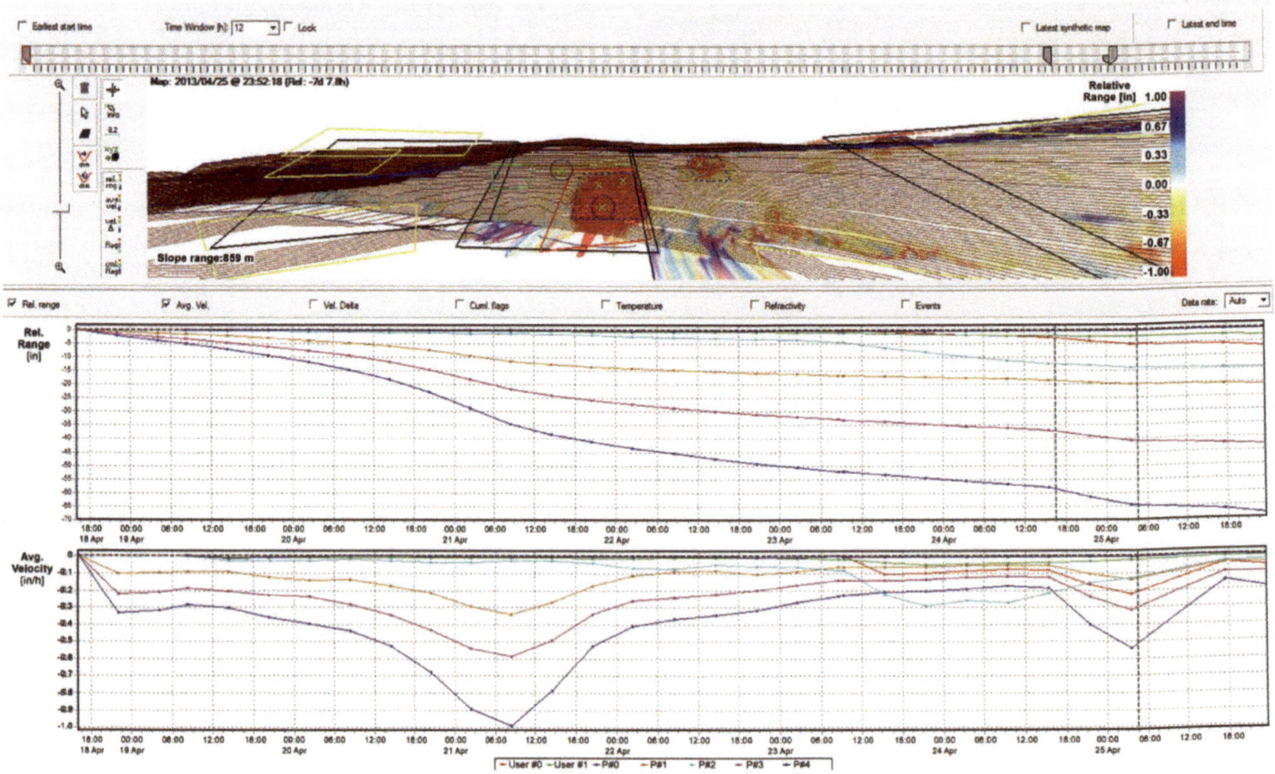

Fig. 4.25: Real-time stability response to mining activity as observed through radar monitoring (source: Reutech).

activities to be tracked continuously in real-time. This ability is demonstrated in Fig. 4.25 where site personnel determined that a slope failure was contained and that it had stabilised at approximately the angle of repose. Utilisation of monitoring results and engineering analysis led to a decision to reduce the volume of potential slide material and thus reduce the potential runout distance of the instability. An estimated 1200 t were removed from the toe by an electric shovel, thus mobilising the instability. Radar time-series data showed that the response to mining was fairly immediate as was the limiting potential energy available to drive the instability. The movement rate subsequently reduced to its original velocity within 12 h.

4.4.3 Safety-critical monitoring alarm settings

4.4.3.1 Overview

Surveillance monitoring using GB radar provides advanced warning by means of an automated alarming mechanism. This mechanism will typically signal an alarm when a predetermined alarm threshold is reached. At first glance this seems simple enough, but Sjöberg (1996a) cautions that the difficulty is not one of measurement technology, but rather of determining what the alarming thresholds should be. Zavodni and Broadbent (1978) reinforce this point in reporting that the magnitude and velocity of an unstable failure varies widely from mine to mine. It logically follows that alarm thresholds used at one mine cannot simply be transferred and used at another mine (Kostak and Rybar 1993). The effectiveness of monitoring systems and monitoring programs in general hinge on the appropriateness of mine-specific alarm thresholds that are selected by mine personnel. Mine-specific alarm thresholds must in turn trigger appropriate actions defined by the operation's TARP. TARPs are addressed in Section 8.5.

Displacement-based alarm thresholds can be used at some operations but, in the case of a constant displacement rate, they may eventually be exceeded even if slope failure is not imminent. Displacement rate (velocity) provides an improved indicator of imminent failure at many operations, as it accounts for smaller variations of potential indications of incipient failure (Sjöberg 1996a). It is therefore appropriate that at some operations, stackable alarm thresholds are based on a hierarchy of displacement, velocity, velocity ratio, inverse velocity and other radar measurements including coherence. Figure 4.26 illustrates how stackable alarms can be aligned with a site TARP.

Displacement and velocity alarms are generally the most commonly applied and easily understood alarm types for slope monitoring (an example using acceleration is presented in Section 9.5.3). Once an increased level of risk is identified, the other stackable alarms are applied.

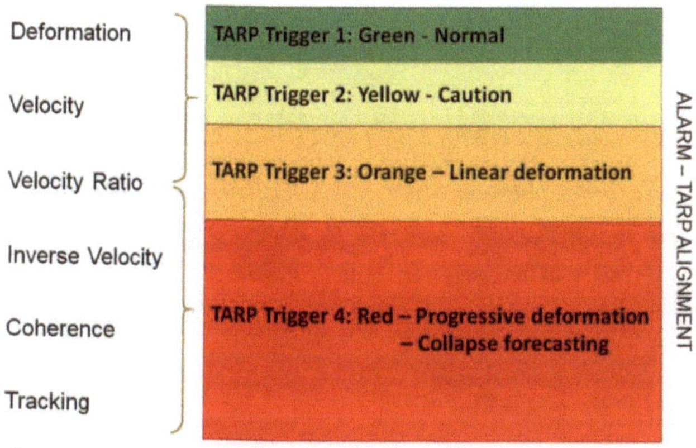

Fig. 4.26: Hierarchy of stackable alarms aligned to a TARP (source: GroundProbe).

Thresholds for early-stage alarms, when set with a broad area alarming strategy, are determined from back-analysis. A broad area alarm strategy is commonly applied where no prior significant deformation has been observed. When applying a targeted alarming approach, areas which have been seen to be deforming at rates higher than the background are isolated and alarms set, masking out areas which lie outside the area of concern. When a visible linear deformation trend is identified, the velocity is recorded and a new velocity (or velocity ratio) alarm can be set with the aim of it triggering when the velocity increases to a rate higher than that observed. As the slope begins to yield and progressive deformation is observed (acceleration), the associated risk increases and so does the assigned TARP trigger level. At this stage, forecasting methods can be applied and inverse velocity alarms are set (see cautions regarding alarm setting in Section 8.4). As the onset of failure occurs and new fractures begin to form, late-stage alarms such as coherence and tracking alarms can be set to alert the slope engineer that failure is imminent.

There are various means of calculating a threshold velocity from the sampled displacement measurements. Two general approaches are (i) calculating an empirical curve that best fits the slope movement profile and (ii) calculating the average velocity based on displacement over a certain time window.

The first approach entails methods such as using a multiple-order polynomial that best represents a known displacement curve for the specific area of concern. The polynomial can then be differentiated to yield a velocity curve for alarming purposes. Fitting a polynomial to the scaled slope movement profile for all measured points is a computationally intensive exercise and, in some cases, can lead to errors. The calculation of an average velocity (v) based on the total displacement (d) over a fixed time period (t) is a much simpler way to achieve a similar result, where $v = d/t$.

Using the average velocity approach, the user typically needs to select three parameters: the minimum size of the area expected to fail, the average velocity (or equivalent displacement), and the time period. The user can then consider all points being monitored and check whether the displacement for each possible area, of the size expected to fail, exceeds the average velocity limit.

Figure 4.27 shows the average velocity calculation graphically. Simply put, the parameters define a box. If the displacement for the current time period exceeds d (i.e. the gradient exceeds v), then it will exit the bottom of the box (red area) and an alarm will be triggered. Otherwise, if it exits the side (green area), no alarm will be triggered. In the right-hand figure, the box is shown at various time intervals; it is clear that the alarm will be triggered in period T_4.

The minimum-size area is based on knowledge of local slope behaviour and radar resolution of the target region, given the distance to the slope. To properly design the remaining alarm parameters, a slope failure profile relevant to the area of concern (a benchmark profile) is required. In other words, the slope engineer is required to have some idea of the magnitude and time-scale of the actual displacement expected for a predicted failure condition. Common practice is to use measurements derived from previous slope failures at the mine. Alternatively, a theoretical model could be considered, or even measurements from another mine with near-identical

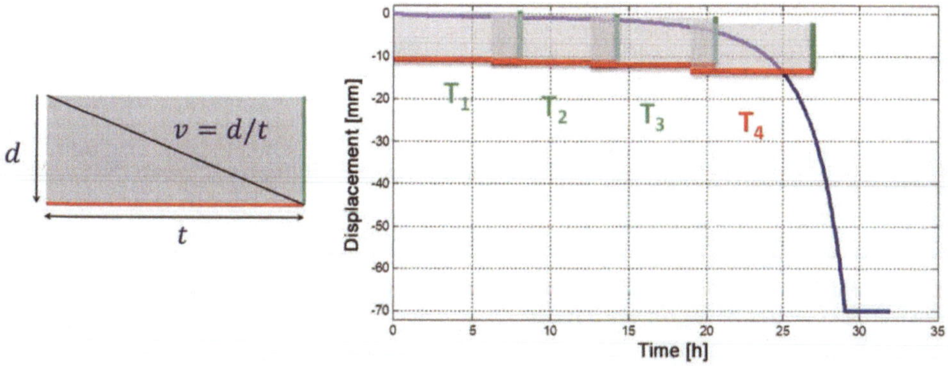

Fig. 4.27: Calculating average velocity alarms (source: Pienaar and Joubert 2011).

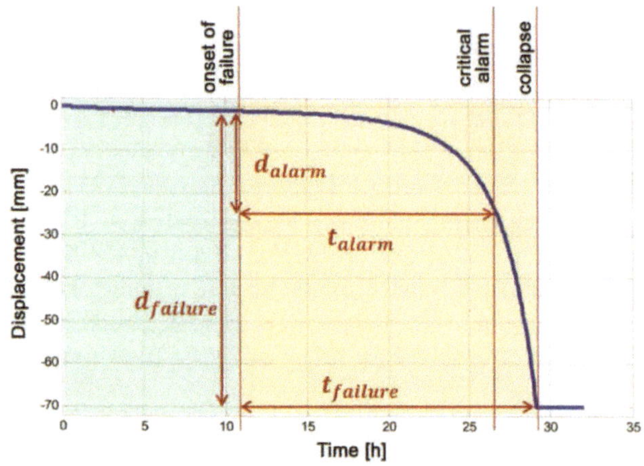

Fig. 4.28: Generic failure movement profile showing alarm parameters (source: Pienaar and Joubert 2011).

geotechnical conditions, albeit with necessary caution. The remaining decision is the desired warning time before failure. As depicted in Fig. 4.28, this will give the alarm time (t_{alarm}) and alarm displacement (d_{alarm}) which define the alarm velocity (v_{alarm}). Setting alarms and thresholds is further examined in Section 8.4.

4.4.3.2 Pit geometry and radar measurements: the scaling factor

Reliable advance warning mechanisms provided by performance monitoring instrumentation give an indication of an imminent failure condition and provide adequate lead-time to evacuate problem areas. Implementation of radar as an advanced warning tool is widely regarded as leading practice. For a radar system to provide effective early warning detection, it needs to operate in real-time and should be linked continuously to the mine's emergency communication system. It also needs to be capable of reliably detecting slope movement over the distance required by the operation. It is important to note that, however precise and reliable the radar surveillance equipment may be, its effectiveness is dependent on its correct deployment in terms of pit size and shape, as well as the angle of incidence (slope movement in the LOS of the radar) relative to the actual direction of slope movement.

An example of the potential limitation of LOS monitoring is shown in Fig. 4.29. For this example, it is assumed that the actual slope displacement is perpendicular to the surface of the DTM. The radar is deployed at two different locations, and the magnitude of the *cos*(θ) scaling factor illustrated in Fig. 4.30, or the vector component of the total movement vector, is indicated by the blue arrow (refer also to related discussion in Section 8.4). The radar position in the upper image of Fig. 4.29 provides better coverage of the pit, with most benches having an LOS component of total vector movement above 70%. In the lower image, a large portion

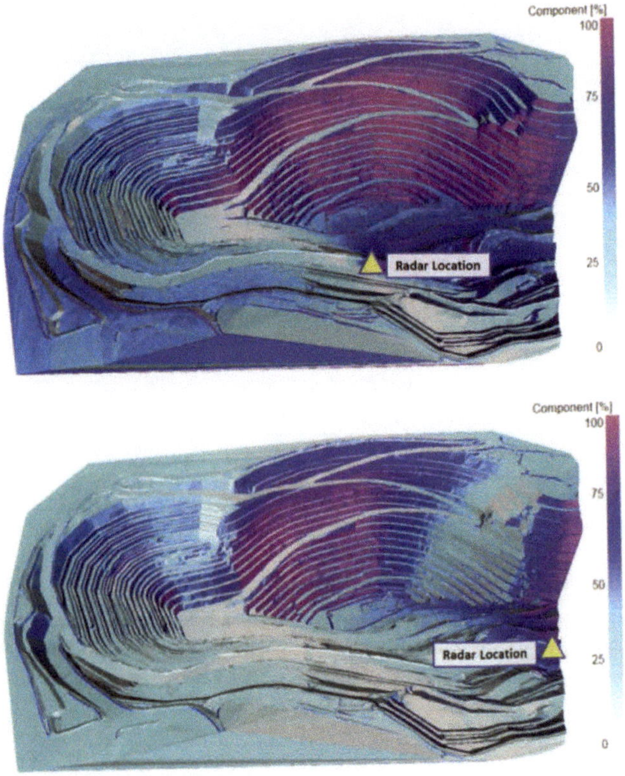

Fig. 4.29: Comparison of geometric effects for two different radar locations (source: Reutech).

of the actual movement vector of the wall on the right is very oblique to the radar, so only a small percentage of slope movement relative to the total vector will be seen from that position.

It is important that the slope engineer understands the relationship between actual slope movement and the LOS component measured by radar. Monitoring slope movement with prisms or GNSS is essential for validating and quantifying this relationship. Several practitioners who have examined this relationship report that where radar alignment produces data representative of at least 70% of the total vector movement determined from prism

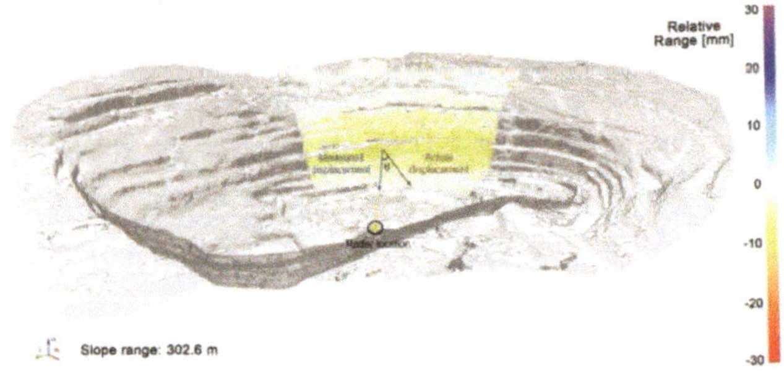

Fig. 4.30: Radar monitoring example showing the LOS movement for a given radar alignment superimposed on a mine DTM and the angle between the measured and actual displacement (source: Reutech and Storm Mountain Diamonds' Kao Mine).

surveys, this presents adequate alignment of the radar relative to the target slope under surveillance. These practitioners also discovered that many operators do not have their radars properly aligned to capture most of the actual slope movement and therefore are at risk of not detecting material changes in slope movement that might otherwise trigger an alarm. Another risk is underestimating trigger events for system alarming, a subject that is summarised in further detail below. For example, suppose that the radar unit is monitoring a slope that is capturing only 30% of the total vector movement (a common occurrence), and alarms on the system have been configured for a threshold movement rate of 3 mm/day. Given this example, the target region of the slope would actually be moving at a rate of 10 mm/day before the radar system would generate an alarm.

It is often the case that actual slope displacement is not perpendicular to the face of the slope but, without any other information, this assumption provides a reasonable starting point. If there is some idea of the slope movement direction beforehand, the radar can be deployed in a more optimal location. Total vector assessment of movement, utilising total station surveys using reflective prisms, is recommended to assist in radar positioning. If there are no prisms in the targeted area of interest, a method of dealing with a largely unknown direction of movement is described in the following section.

4.4.3.3 Calculating and applying the scaling factor to slope movement

Figures 4.30 and 4.31 illustrate an example calculation of the total vector movement and corresponding scaling factors to apply to radar data when the radar is not aligned with the actual slope movement direction. Figure 4.30 shows the difference between the measured slope displacement for a given radar alignment and the actual slope displacement direction measured independently using a total station survey of prisms. This is reported as an angle (θ). In this example, the angle is ~45°, which equates to the LOS vector component (i.e. $cos(\theta) = 0.7$), capturing only 70% of the actual movement.

Figure 4.31 shows the measured displacement *v.* time plot derived from the radar data. This corresponds with the $cos(\theta)$ term of 0.7. Also shown is the corrected displacement for $\theta = 0°$ (i.e. $cos(\theta) = 1$), which is simply the measured displacement divided by the same factor (i.e. 0.7) to estimate the actual displacement. This final adjusted benchmark profile is the one to use when designing an alarm threshold. For comparison, the displacement *v.* time plot for $\theta = 60°$ (i.e. $cos(\theta) = 0.5$) is also provided to demonstrate the sensitivity of this parameter.

In the absence of total vector survey results (from prisms and/or GNSS), the actual direction of displacement for a new failure can only be estimated beforehand. A simple way to allow this is to reduce the expected movement to what could occur in a non-ideal scenario. A reasonable, practical limit based on general experience for θ is 60°. Thus, if the radar's LOS is anywhere within ± 60° of the actual direction of movement, the slope stability alarm will be triggered as designed (or earlier). Depending on the geometry, other values of θ could be used for this adjustment. However, using a small-scale factor, or larger deviation from the actual movement direction (θ<50°), will result in reduced threshold movement rates, which may be impractical due to a likely high number of false alarms. The slope engineer must consider the trade-off between the acceptable number of false alarms and the obliqueness of the viewing geometry.

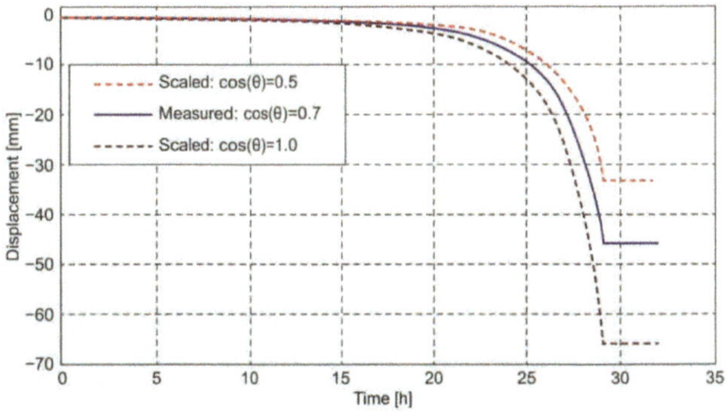

Fig. 4.31: Application of scaling factors to the measured slope movement profile (for this example, $cos(\theta) = 0.7$) to correct for the radar misalignment relative to the actual slope movement direction. The corrected total vector displacement is represented by $cos(\theta) = 1$ (source: Pienaar and Joubert 2011).

4.4.4 Scan coverage application example for surveillance monitoring

The size of the open pit plays an important role in the design and selection of a radar system utilised for large-area scan coverage. For large open pits (>2 km across), the ideal radar solution must be able to achieve a high spatial resolution, even at a long range, for the detection of bench-scale failures. It is also a requirement to achieve the widest coverage in the shortest possible scan time to minimise potential radar ambiguity. The atmospheric correction algorithm employed by the radar processing engine must guarantee reliable data, even at great distances. These considerations play an important role in the selection of the radar system required by the slope engineer. Details regarding the capabilities of real aperture radar (RAR) and SAR systems for surveillance and tactical monitoring are summarised in Section 3.4.2.

Some operations can benefit from semi-permanent installations around the crest of the pit, located away from working/blasting areas and housed in small, protective structures, as described in Section 5.2. Where possible and practical, enclosing radars to further protect

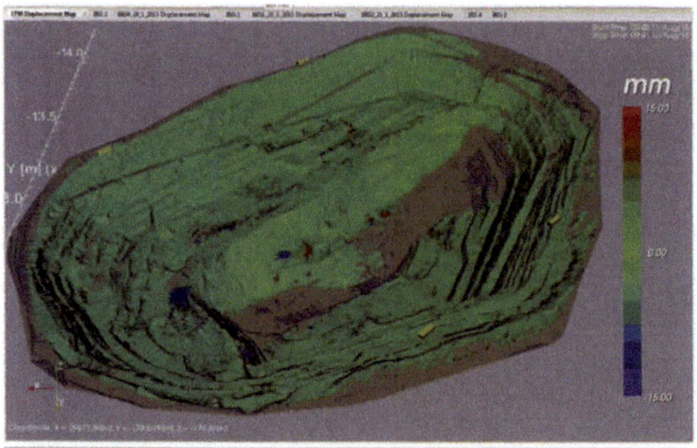

Fig. 4.32: Radar monitoring coverage of a large open pit is achieved with four semi-permanent SAR radar units located on the opposite highwalls, indicated by four yellow rectangles. Average scan range in this example is ~2 km (source: IDS).

them from the external environment can result in a significant reduction of maintenance costs. Advanced software packages in certain conditions can allow multiple radar units to be combined to effectively monitor the entire pit area.

An example of 360° full-pit monitoring coverage of an open pit is shown in Fig. 4.32. In this example, four semi-permanent SAR radar units are installed on opposite highwalls of the pit with a 3 min scan interval. A refreshed image of the entire pit is possible, while retaining sub-bench resolution for safety-critical monitoring purposes. Full-pit monitoring has the ability to spatially and temporally integrate multiple radar systems into a single 3D visual display, allowing all radar information, safety-critical monitoring and hazard alarms to be shown and managed in a single environment. The radar processing software is capable of processing rapid and slow deformation movement rates.

For small open pit operations (see also Section 9.2), flexibility and coverage can play fundamental roles in the selection of the proper radar solution. When scanning at close range from the slope (<1.5 km), the technical requirement of high maximum achievable resolution is more relaxed, since at these scanning distances the detection of sub-bench movement is easier to achieve. The radar field of view must be wider than that required for longer distances to cover the same area, due to the closer proximity to the highwall. For areas near blasting activity, system mobility plays an important role. A mobile unit that can be quickly deployed and removed to avoid damage from blasting may be preferred.

Scan time over a specific target region of the slope is an important factor in system selection. To extend the maximum measurable velocity as much as possible to avoid phase ambiguity (see Sections 3.4.2.2 and 4.5.6), the scan time must be kept to a minimum. It is recommended

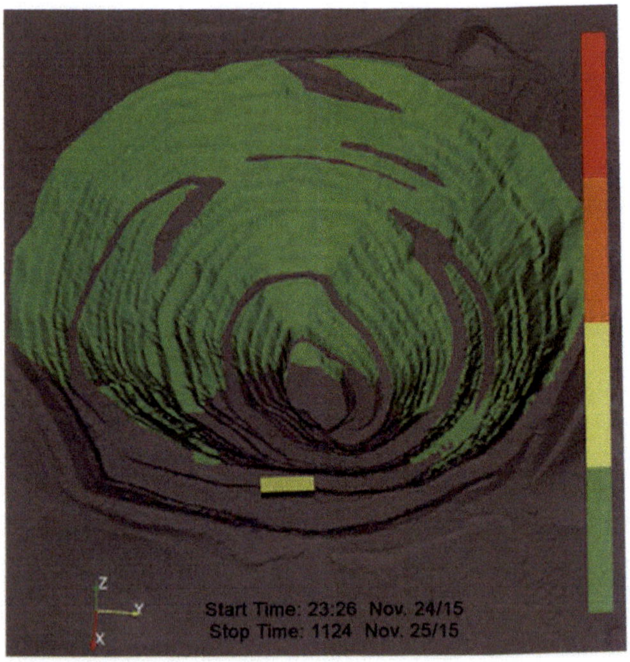

Fig. 4.33: Radar coverage over a large area of a small open pit using a single well-placed mobile SAR unit. The average scan range in this example is 800 m. The yellow rectangle indicates the location of the radar station (source: IDS).

to optimise radar systems to maximise coverage with the most rapid possible scan time.

Figure 4.33 shows an example of almost 270° coverage of a small open pit monitored with a rotating mobile SAR radar unit, with a scan time over the entire area of ~3 min. The average scanning distance is 800 m in this example. As demonstrated here, the LOS limitation can be significant for a single radar unit monitoring a small pit, as the slopes to the immediate left and right of the radar represent blind spots.

Due to the LOS limitation, for each sector of the open pit, it is important to evaluate the positions of the radar and the target region relative to the expected direction of movement. It is necessary to consider the expected movement direction and potential slope instability modes when defining the velocity thresholds that would trigger an alarm.

4.5 Application: satellite InSAR

4.5.1 Overview

Conventional methods of displacement monitoring using GB instrumentation can be labour-intensive, time-consuming and costly when detailed spatial coverage over large areas is required. Ground surface surveys may be challenging to implement in areas that are difficult to access (e.g. mountainous, forested, glaciated) or lie within restricted or unsafe areas. Sensors and other instruments may be susceptible to theft or vandalism. In such regions, satellite radar data and InSAR techniques can provide a safe and economical means for providing surveillance of critical areas. InSAR is best adopted as a complementary means to GB instrumentation; it is commonly used for GB monitoring results validation, and integration with other results. Its benefits and limitations are summarised in Section 4.5.2.

It is important to understand that radar satellites do not produce data in real-time. Acquisition frequencies available at the time of writing vary from four to 24 days or more, depending on which satellite is used, as discussed in Section 3.5.3. For users who prefer or require an even higher acquisition frequency, overlapping ascending and descending results can be combined.

InSAR is not a good choice for measuring the change in volume resulting from excavation. When the ground surface changes for any reason, measurements are not possible for that specific area and for the period of activity. Measurements return when activity ceases, and displacement due to settling can once again be monitored. The large change in the height of the ground surface due to excavation activity is not captured.

4.5.2 Capabilities and limitations

There are a variety of applications and market sectors that can benefit from ground-surface deformation monitoring with satellite radar. Results with InSAR have been used for fault characterisation and calibration of geomechanical models in the petroleum and mining sectors; monitoring landslides, volcanoes and active fault lines; monitoring activity in areas prone to sinkholes and excessive erosion; monitoring subsidence associated with tunnelling, underground mining and groundwater and petroleum extraction; and stability monitoring of large infrastructure such as dams, tailings embankments and individual buildings. Results are typically provided in commonly used spatial and temporal formats that can be viewed or interpreted in many software packages, such as ArcMAP, QGIS, Google Earth and Global Mapper. Some features and benefits of InSAR include:

- safe uninterrupted continuous surveillance;
- site-wide displacement monitoring;
- low annual cost per measurement point;
- subcentimetre precision is often achievable;
- safety of life applications;
- integration into existing geotechnical monitoring systems;
- economical monitoring of closed sites; and
- a decision-making tool for locating GB instrumentation.

Limitations of InSAR include:

- not real-time (Section 4.5.1);
- not suitable for measuring volumetric changes (Section 4.5.1);
- not recommended as a replacement for GB instrumentation and inspections;
- limited for detecting horizontal displacement in the north–south direction (Section 3.5.3);
- blind spots can occur due to terrain (e.g. geometric distortion; Section 4.5.3);
- limited in vegetated areas, unless lower-precision/longer-wavelength data are used (i.e. coherence; Section 4.5.4);
- limited on terrain with lingering ice and/or snow (i.e. coherence; Section 4.5.4);
- varying sensitivity to displacement, depending on terrain (Section 4.5.5); and
- rapid displacement can go undetected if movement exceeds range thresholds (i.e. phase ambiguity; Section 4.5.6).

4.5.3 Geometric distortion

SAR satellites operate as side-looking sensors. This is because microwaves emitted by their sensors are reflected from the surface of the Earth and are received on a first-come, first-served basis. A SAR pointing straight down would therefore receive many microwaves at similar times, resulting in very little separation between them in image space.

On flat ground, a side-looking sensor will receive reflected microwaves in the correct order and the distances in ground range measured will be scaled correctly in image space. For areas with more varying topography, such as an open pit, the situation is more complex. A slope facing away from the satellite, but not steep enough to be in shadow, makes an ideal target for InSAR monitoring (Fig. 4.34a). Conversely, a sufficiently steep slope will return signals from the top first and bottom last; thus, in image space, the slope will be represented as if it were laid on its side. This is called layover (Fig. 4.34b). Other areas are simply obscured from view, and this is called shadow (Fig. 4.34b). Areas of layover and shadow vary as a function of the angle between the satellite's nadir (the direction pointing directly down and vertical below the satellite) and its look angle. This angle is typically not fixed; for example, for TerraSAR-X it varies between 20° and 50°. Monitoring a mine site with significant topography from ascending and descending passes vastly improves the InSAR coverage, compared with only a single pass (see Section 3.5.3).

4.5.4 Coherence

InSAR is inherently a spatial technique. Lone isolated pixels are not useful (unless specialised corner reflectors are

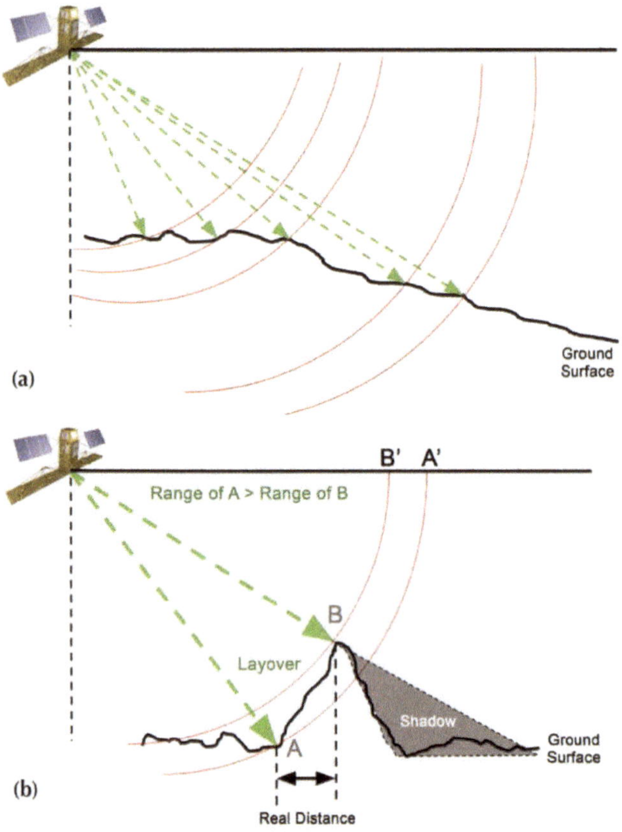

Fig. 4.34: (a) The slope is facing away from the SAR but is not steep enough to be in shadow. This represents an ideal scenario for monitoring downslope displacement. (b) The steep slope facing the satellite returns signals A and B in reversed order (layover) because of their relative distances, and the far side is shadowed by the slope itself (source: 3vGeomatics).

deployed) because measurements must be integrated spatially for them to be meaningful. An interferogram is formed by finding the difference of spatially continuous phase information from two images taken on different dates (see Section 3.5.3). For each interferogram, coherence is a per pixel estimate of how phase values vary with neighbouring pixels. Regions with unchanging or smoothly varying phases are assigned high coherence, and vice versa for low coherence. Coherence is used throughout InSAR processing for weighting measurement quality and for masking poor results and noise. Coherence is an important concept for mining InSAR: understanding it can clear up many areas of confusion for those new to the technique. Low coherence areas do not yield displacement results. Common causes of low coherence are summarised as follows.

- *Surface disturbance*: excavation and dumping between the image dates randomises the phase returns in the areas concerned, and good coherence is lost. Once the activity ceases, higher coherence returns.
- *Vegetation*: leaves and swaying trees do not reflect radar data consistently. Longer-wavelength SARs are more resistant to this, but are also less sensitive to displacement.
- *Open water or waterlogged ground*: moisture scatters the signal unpredictably from image to image.
- *Lingering snow and/or ice*: snow or ice can destroy or bias the measurements.
- *Very fast displacement*: this is analogous to displacement contours densifying to the point where they are superimposed (see Section 4.5.6). Amplitude-based techniques may be applicable here (see Sections 3.5.3.5 and 4.5.7).

The above factors are indistinguishable from one another without further information. Important context from local knowledge, radar amplitude information or optical imagery may be used to interpret the cause of low coherence.

4.5.5 Displacement sensitivity

InSAR is a 1D measurement along the LOS of the sensor. If the total vector direction of displacement happens to coincide with this, then 100% of the displacement can be recovered. However, these two vectors will typically differ, and InSAR will underestimate the true magnitude of displacement. Figure 4.35 provides two simplified scenarios of vertical and downslope displacements. Despite the implied underestimation, these represent typical scenarios for InSAR monitoring. If displacement was to occur perpendicular to the LOS (shown by the red dashed line in each subfigure), then no movement can be captured. This limitation is reduced by monitoring using overlapping ascending and descending data (see Section 3.5.3).

4.5.6 Phase ambiguity

InSAR phase measurements are expressed as angles and bounded between 0° and 360°. For TerraSAR-X (TSX), this is equivalent to a displacement of between 0 cm and 1.5 cm. When areas displace discretely, by more than 360° without a surrounding and increasing displacement gradient, ambiguities in measurements can occur. This is analogous to a sleeper awakening to an analogue clock reading 3:00 and not knowing whether 3 h or 15 h have elapsed since 12:00. This is sometimes termed phase aliasing. More commonly, however, displacement does occur spatially over smooth gradients, and the maximum displacement can be recovered by integrating measurements successively. Figure 4.36 shows how discrete displacements can be underestimated compared to more typical gradual displacement. Areas with very fast displacement (>30 cm in 11 days for TSX) can be measured using amplitude-based techniques (see Sections 3.5.3.5 and 4.5.7).

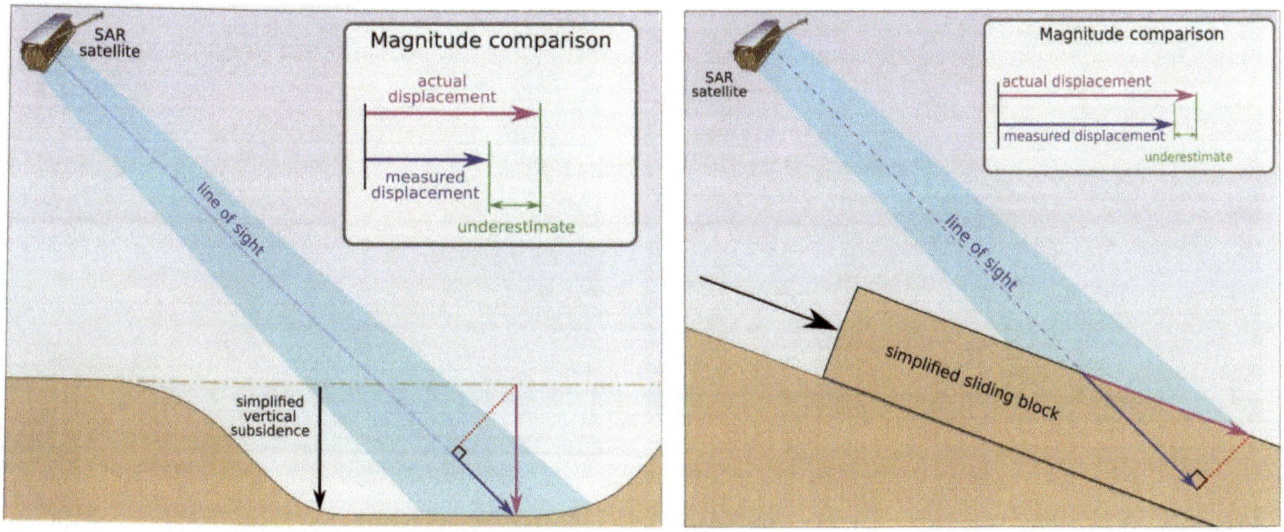

Fig. 4.35: Two monitoring scenarios: (left) vertical and (right) downslope displacement. In either case, InSAR will only capture the component of the displacement coincident with its LOS (source: 3vGeomatics).

4.5.7 Case study: Grasberg mine, Indonesia

The Grasberg mine in Papua, Indonesia, offers unique challenges for displacement monitoring. The dense equatorial rainforest rises to 4200 m at the mine site. The extreme terrain, the turbulent tropical atmosphere (with annual precipitation greater than 5 m/yr) and the threat of natural geohazards are all issues that must be managed for InSAR to be viable. South of the mine, the approach roads are at risk of landslides from densely vegetated slopes.

Further from the mine, the tailings area extends 80 km to the Arafura Sea and has its own set of monitoring challenges. These unique challenges demand non-standard, area-specific approaches. This case study focuses on the mine site itself, which includes an underground block caving operation and cave subsidence zone. The mine site is monitored using dual-look TSX data. Post-imaging reports from both look directions are generated on 11-day intervals (as discussed in Section 3.5.3.1). Figure 4.37 shows the ascending and descending pass results, which have been combined to improve coverage and, where possible, solve for the direction of displacement. Many slopes are steep enough to be in shadow from one pass and laid-over in the other (Section 4.5.3); nevertheless, coverage for important areas is good at this site. Some gaps in the coverage result from the densely vegetated terrain, which scatter the short-wavelength signals used for TSX data.

InSAR cannot measure the very high movement encountered above the block caving area, due to phase ambiguity (Section 4.5.6). Instead, radar amplitude information can be used to track the movement of groups

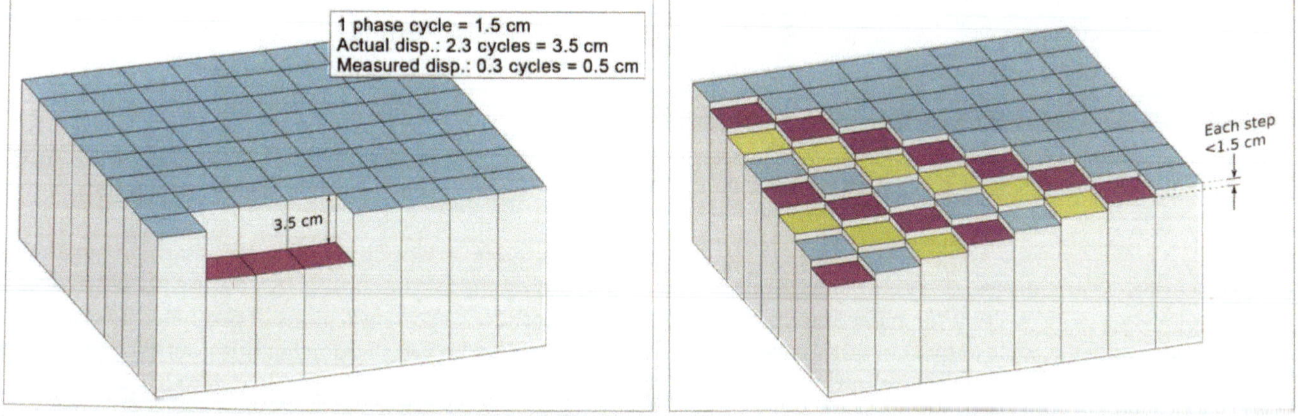

Fig. 4.36: A discrete area displaces 3.5 cm or 2.3 phase cycles for TSX (left). As only the non-integer portion of the phase is recoverable, 0.3 cycles are measured, equivalent to only 0.5 cm, which is an underestimate of 3 cm. More typically, the gradient of displacement is gradual (right), and higher displacement areas are resolvable by integrating measurements spatially (source: 3vGeomatics).

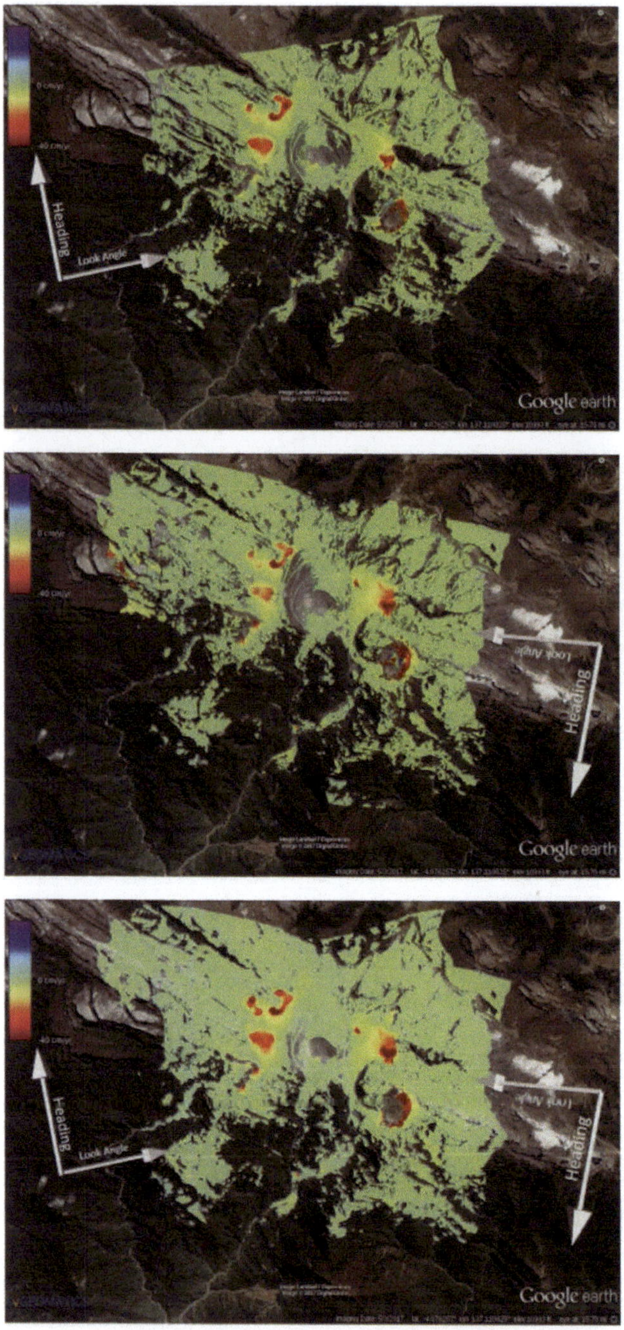

Fig. 4.37: The Grasberg mine site is monitored using ascending and descending TSX data (source: 3vGeomatics and Google Earth).

of pixels in successive images, also known as speckle tracking (Section 3.5.3.5). From the radar amplitude results, displacement can be derived in a 2D image space and projected onto the slope, as shown in Fig. 4.38. The result is less precise (around 30 cm uncertainty for TSX), but capable of mapping very large gradual displacements over wide areas.

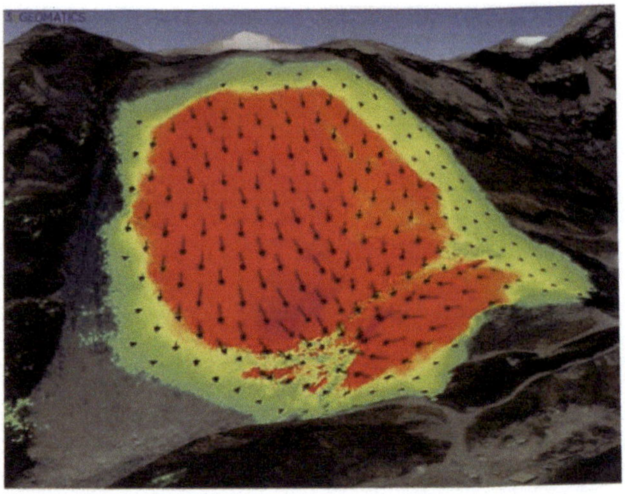

Fig. 4.38: The area above the block cave moves at rates that are beyond the capability of InSAR (phase ambiguity). Instead, using speckle tracking, groups of pixels in radar amplitude space can be tracked in successive images, providing a coarser measurement but without the velocity constraints of InSAR (source: 3vGeomatics).

4.5.8 Case study: mine in South America

InSAR has been utilised for site-wide displacement monitoring at an undisclosed mining operation in South America. The area of interest, shown in Fig. 4.39, has been monitored by the TSX constellation. Prior to an ongoing monitoring campaign, site owners may opt to analyse historical displacement using free or low-cost, low-resolution SAR data, to provide a baseline study or a proof-of-concept. Historical data at this site were available from the European ENVISAT and ERS satellites, the Japanese ALOS-1 satellite and the Canadian Radarsat-2 satellite. Since 2014, Sentinel-1 has been providing freely available and up-to-date low-resolution (20 m) data for many areas in the world and is now a common choice for low-cost baseline studies.

Following historical analysis, ongoing monitoring is typically conducted using higher-resolution data (3 m), providing areas of displacement as small as 15 × 15 m, which would be missed with lower-resolution sensors. For TSX, a small number of footprints were available covering the site, with the choice ultimately dictated by the level of geometric distortion (Section 4.5.3), which varies as a function of the angle between nadir and the look angle. Arrows in Fig. 4.39 indicate the satellite's flight heading and look angle for the chosen TSX footprint. White areas on the ground represent regions of geometric distortion which cannot be measured using this configuration. These areas are fairly limited in this case due to the modest slopes at the site and the surrounding area.

Reporting results of ongoing monitoring included post-imaging reports and periodic comprehensive

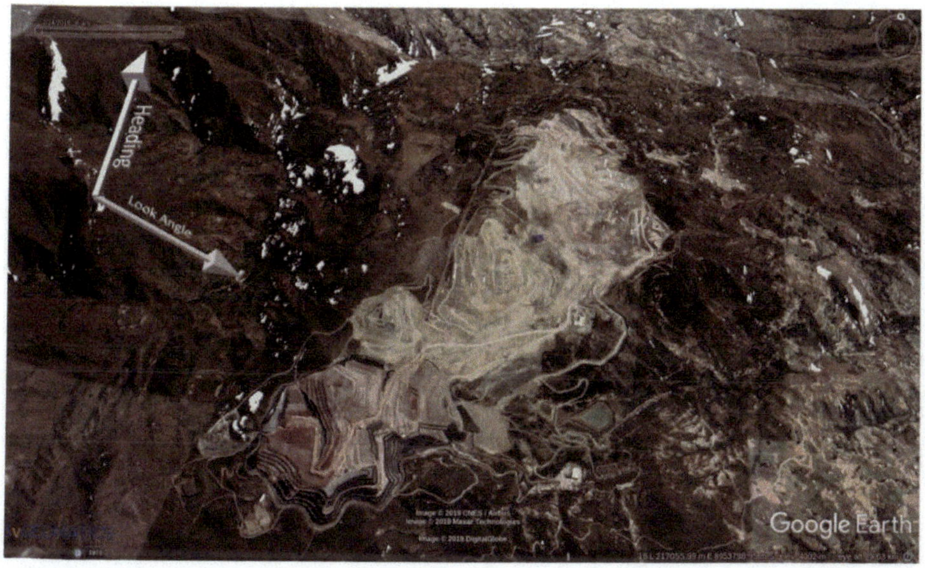

Fig. 4.39: The area of interest for a mine site in South America, as captured by TSX images. Arrows illustrate the satellite's heading and look angle. White areas on the ground represent geometric distortion, which cannot be monitored from this configuration (source: 3vGeomatics and Google Earth).

reports. Post-imaging reports are produced following each satellite pass and are designed to represent new and ongoing risk observed since the last satellite pass, which for TSX is a period of 11 days. Displacement was represented as contours, beginning at 0–0.5 cm (a range to reflect the changing noise floor across the image) and incrementing in steps of 1.5 cm. The contour interval is a function of available precision, dictated by the TSX half-wavelength, which is 1.5 cm. This information was represented in various formats, to provide easy consumption by stakeholders. Figure 4.40 shows an example post-imaging report in Google Earth. The post-imaging reports can also include maps with layers that can be turned on and off. Vector formats were also made available for geographic information system (GIS) users and in integration software, such as Canary Systems and Navstar GeoExplorer.

4.6 Application: laser scanners

4.6.1 Overview

In an open pit mining environment, terrestrial laser scanners (TLS) (described in Section 3.4.3) need to be able to collect data rapidly and reliably. Slow scanners may increase the 'field-to-finish' time and unnecessarily delay operations activities while data are being collected. Scanner downtime for repairs due to rough handling or exposure to dust and precipitation may also cause data disruptions that could result in incomplete surveys or delays in end-of-month reconciliations. Thus, choosing a faster, more durable and reliable scanner for monitoring or other applications will increase its value to the user.

The following sections describe features of laser scanners that should be considered when selecting appropriate instrumentation that will satisfy operational requirements.

4.6.2 Spatial resolution

Laser scan spatial resolution is a function of the instrument's angular resolution (i.e. the angular 'step' between measurements) and range to target. The angular resolution of most laser scanners is between 0.002° and 0.025°, with the finer resolutions usually belonging to phase-shift (PS) scanners compared to time-of-flight (TOF) scanners. PS and TOF scanners are described in Section 3.4.3.

The spacing between laser scan points increases linearly with range. Thus, if spatial resolution is the size of the 'square' formed by four adjacent scan points, its area will increase as the square of its range. For example, if the resolution for a scan results in a 10 cm point spacing for 100 m range, then at the 1000 m range the point spacing will be 100 cm.

The maximum and minimum spatial resolutions of a laser scanner are largely a function of scan speed and the applications for which it is designed. Most PS scanners are faster than TOF scanners, and likewise can use a smaller angular step to achieve finer spatial resolution without having excessive scan times. This is useful for the short-range, detailed measurement and analysis applications that PS scanners are designed for. TOF scanners, on the other

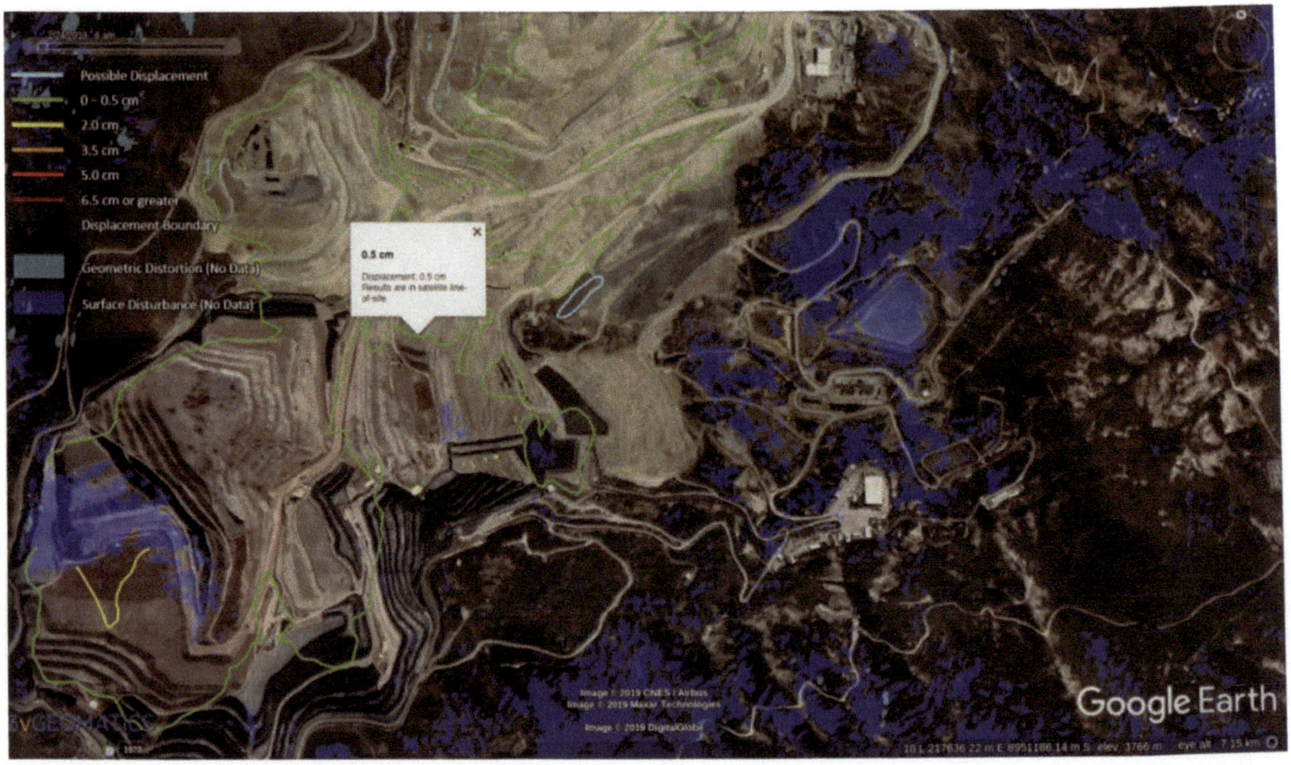

Fig. 4.40: Example of a post-imaging report for the operation shown in Google Earth. Colour-coded contours represent displacement since the previous satellite pass, which for TSX is an 11-day period (source: 3vGeomatics and Google Earth).

hand, were designed for longer-range, less detailed applications and, in part because they are slower, can afford to have a larger angular step and slightly coarser spatial resolution.

4.6.3 Reliability

Mining environments are harsh and laser scanners are highly sophisticated survey instruments. While most scanners are designed for at least some degree of exposure and abuse, the more reliable a system is, the better. The international Ingress Protection (IP) rating of an instrument is a good indicator of reliability. The first number in an IP rating can vary between 0 and 6 and indicates protection from dust ingress, with 6 being completely impervious. The second number varies between 0 and 8 and indicates protection from water ingress, with 8 being submersible beyond a depth of 1 m. Higher IP ratings (e.g. IP65 *v.* IP54) therefore indicate instruments that are better protected from dust and water ingress.

Scanners used on mine sites should be able to withstand frequent, heavy use and some rough handling. There is no specific measure of how well a scanner might hold up to rough treatment, but shock, vibration and impact testing are performed by some manufacturers. Trusted vendors may be able to provide information on these types of tests and the general durability of their instrument.

4.6.4 Speed

Typical TOF laser scanners can collect between 10 000 and 100 000 points per second. Some PS scanners can collect as many as 1 000 000 points per second or more. Because they can collect data at such a high rate, the spatial resolution of PS scanners may result in data overload. In some cases, very high data collection rates and spatial resolutions may be appropriate. However, for most open pit monitoring applications, TOF data collection rates and the resulting spatial resolutions are sufficient and may still result in more data than are needed. As the development of TOF scanners continues, improved data scan speeds will reduce field-to-finish times and increase user efficiency.

4.6.5 Performance monitoring and mapping applications

Because the point cloud data collected by laser scanners are highly precise 'snapshots' of 3D geometry, there are several other ways in which slope monitoring programs can benefit from laser scanning. The following subsections include examples of ways laser scanning can add value to slope monitoring programs.

4.6.5.1 Change detection

Pre- and post-failure analysis of surface changes can provide predictive and analytical measurements for slope monitoring. Accurate, multi-temporal point cloud

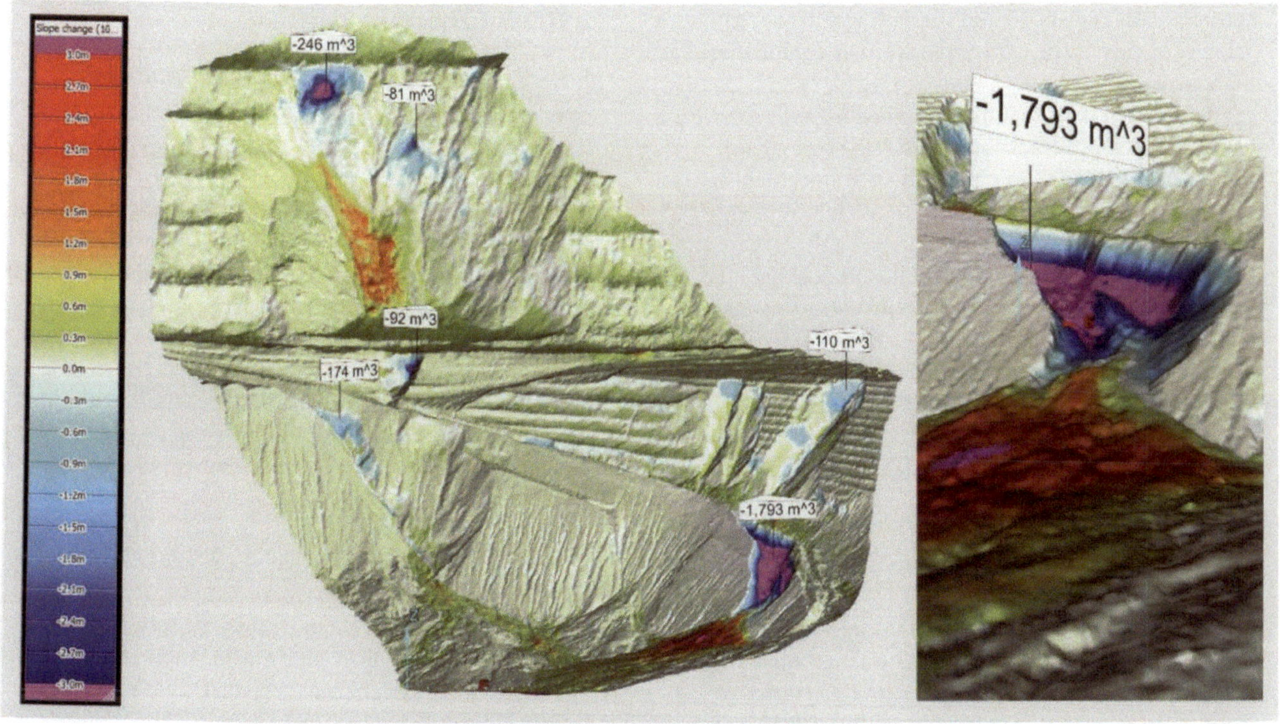

Fig. 4.41: Two views of the same change detection results for a portion of a large open pit slope where the surface-normal deformation magnitude is heat-mapped on a 3D surface created from a point cloud. Individual slope failure volumes for the larger events are shown. The image on right is a close-up view of the largest failure area visible in the lower-right corner of the image on the left (source: B. Haugen).

datasets enable the measurement of surface change through time. Figure 4.41 shows two views of change detection performed on a large open pit slope. Post-analysis in this example included estimates of surface deformation magnitude (i.e. distance between pre- and post-failure surfaces) and volume measurements for the larger failures. In this example, the user's goals were to estimate probable future event size and characterise hazards on the slope.

4.6.5.2 *Blast design and analysis*

Accurate surface models are essential for the assessment of perimeter blast designs coupled with post-blast analysis. Pre-blast surfaces generated from laser scanning point clouds can enable the accurate assessment of pattern and loading designs. Post-blast scanning enables the assessment of pre-split performance as well as blasting statistics calculations, such as cast volumes and fragmentation. These capabilities can improve blasting and excavation optimisation for a given slope design and help slope engineers monitor changes in the slope caused by blasting activities. Figure 4.42 shows an example image from a post-blast analysis report. In this example, the slope engineer's goal was to evaluate the conformity of the as-built performance of the excavated bench with the bench and blast design using as-shot loading and hole geometry data with high-resolution pre- and post-blast surfaces created with point cloud data.

4.6.5.3 *Automated and real-time monitoring*

Recent developments in TLS hardware and software have made it possible to monitor slope movement in real-time. A system controller, usually a rugged personal computer (PC) or laptop, collects point clouds at a time interval

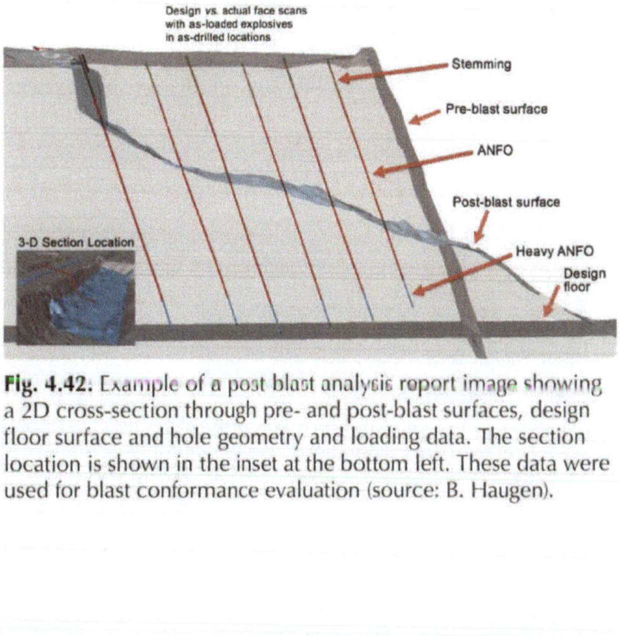

Fig. 4.42: Example of a post blast analysis report image showing a 2D cross-section through pre- and post-blast surfaces, design floor surface and hole geometry and loading data. The section location is shown in the inset at the bottom left. These data were used for blast conformance evaluation (source: B. Haugen).

specified by the user. The point clouds are then automatically registered and filtered, and displacement or velocity trends in the LOS of the scanner are measured at high spatial resolution.

Automated laser scanning monitoring tools can be used to detect and track rapid, discrete events (e.g. rockfall) as well as slower, more ductile movement trends. Because it relies on 3D geometry to track movement, TLS monitoring can be used to measure even the largest and most rapid of slope movements. Like other slope monitoring systems, critical response alarms can be configured to warn of impending and existing catastrophic failures. Moreover, because point clouds are being continuously collected, additional work, such as geomechanical characterisation and geologic mapping, can be performed on the output dataset.

4.6.5.4 Rockfall location and size analysis

Rockfall events tend to occur rapidly and with little warning. They therefore present a significant hazard. Each laser scan produces a 3D point cloud where rockfall events are easily captured as changes in the 3D geometry of successive point cloud images. Figure 4.43 shows an example of several rockfall events that occurred on an excavated slope at the bulging toe of a large rotational-translational pit slope failure. In this case, the slope engineer's goal was to track the frequency, volume and location of these types of rockfall events, and ultimately use the data to support a risk-based approach to slope safety management.

4.6.5.5 Slow-slip and continuous failures

Slow-moving soil and rock failures are common in large open pits. Rock masses in ore-rich zones are often highly fractured and/or altered, resulting in ductile slope failures that can be very large and continue to deform in creep motion. Depending on the magnitude of movement and range to target, the progression of ductile slope failures can be measured using automated or manual laser scanning-based monitoring systems. Figure 4.44 shows an example of large slow-slip failures progressing over the course of a week in two separate slope sectors. The slope engineer's goal was to track this movement, including setting alarms to warn of unexpected accelerations or other changes in these zones.

4.6.5.6 Critical alarming

Timely warning of an incipient slope failure is critical to the success of a slope performance monitoring program. Alarms and notifications must provide actionable intelligence, including event time, location and magnitude. The monitoring system must produce as few false positives as possible, while maximising the number of detectable events. Like radar systems, laser scanning-based monitoring systems now include the ability to alarm on displacement, velocity and inverse velocity. They also include the ability to filter the data to eliminate false positives and improve detection rates by reducing

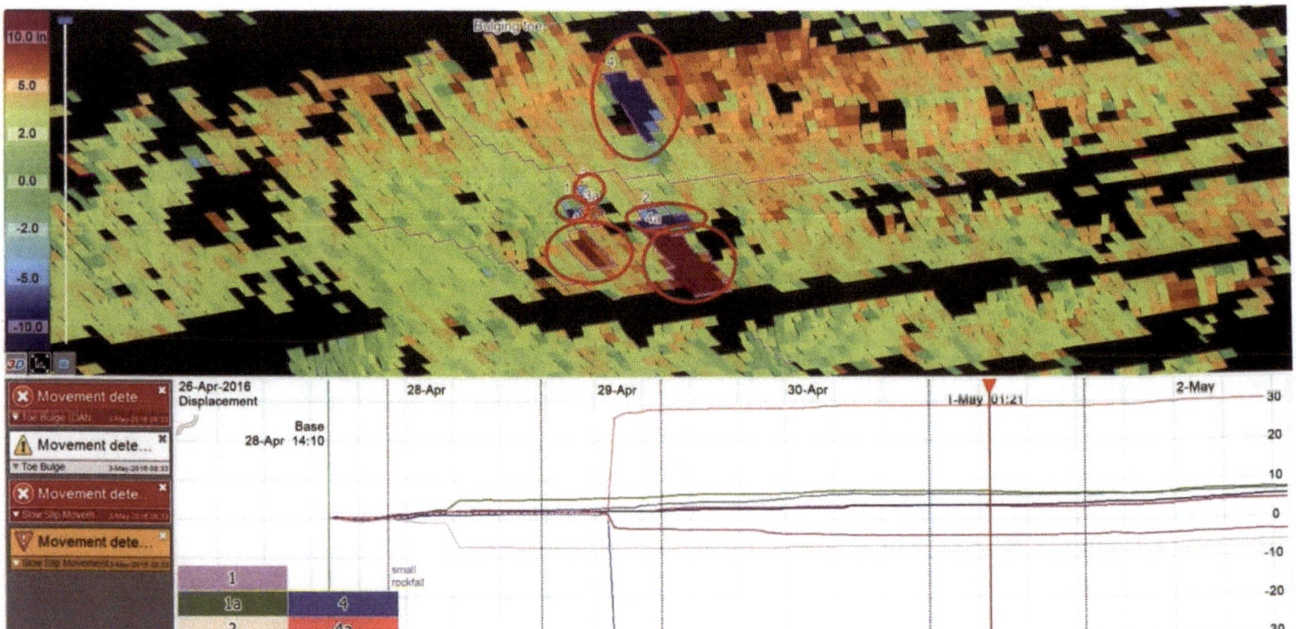

Fig. 4.43: TLS monitoring software output showing LOS displacement heat map (top), graph (bottom) and alarms (bottom left). Warm colours in the heat map indicate movement towards the monitoring position; cool colours indicate movements are moving away. In this case, the slope engineer was tracking discrete rockfall events (red circles) occurring at the toe of a large slide. Magnitude, frequency and volume information from the monitoring program were then fed into a risk-based slope management model (source: B. Haugen).

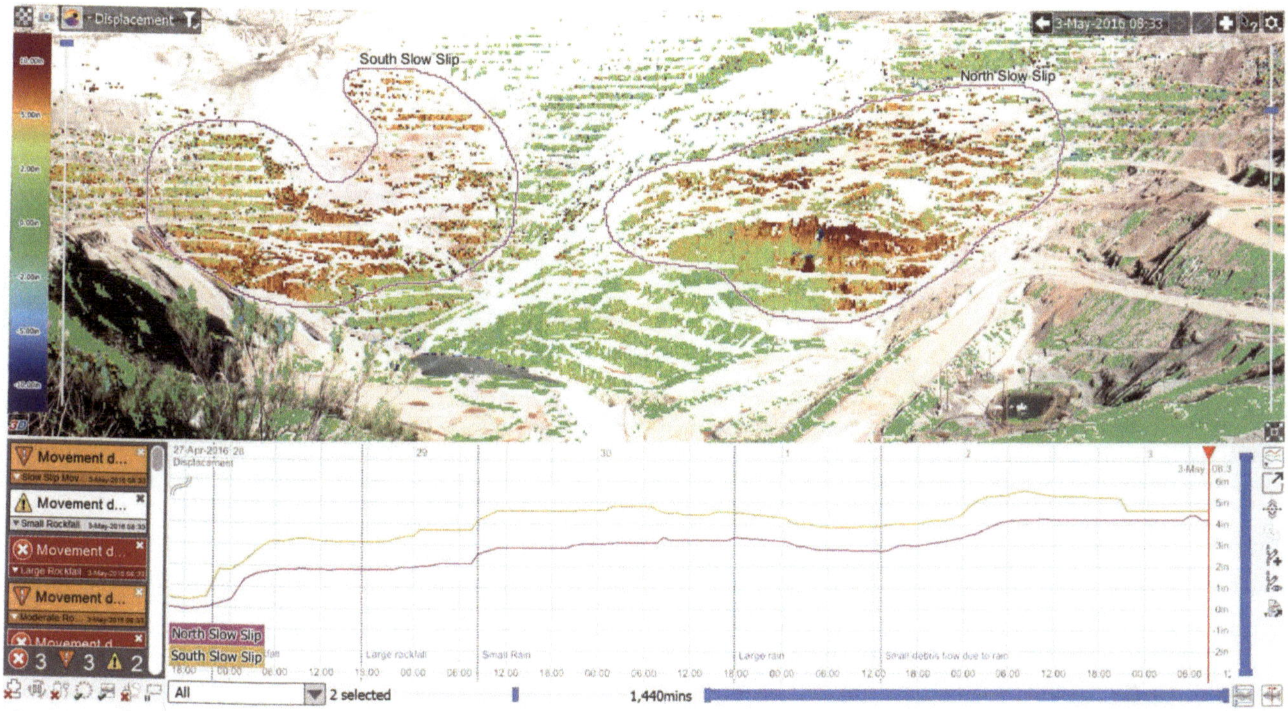

Fig. 4.44: Screenshot of near real-time TLS monitoring software showing the tracking of two large slow-slip failure zones outlined in magenta: one on the left (south) and one on the right (north) side of this large open pit. Several critical movement alarms were triggered (bottom left) (source: B. Haugen).

measurement noise. Figure 4.45 shows an example of reducing false positives and achieving a higher signal-to-noise ratio by applying data filters. In this case, the slope engineer's goal was to perform a back-analysis from previously collected data to optimise system settings. The optimised settings would then become the standard for future monitoring efforts.

4.6.5.7 Supporting risk-based monitoring

Risk-based approaches to slope monitoring require detailed event information, including frequency, location, size and runout distance. Laser scanners can be used to gather this information and use it for meaningful risk or decision analysis. From a strategic perspective, risk-based monitoring may enable the user to calibrate monitoring zones, create failure mode models and develop robust safety procedures under specific monitoring conditions. As an example, at the Kanmantoo and Savage River mines in Australia, the geotechnical teams have developed mandatory, staged safety exclusion zones beneath highwalls based in part on the data gathered from laser scanning-based monitoring tools (Fig. 4.46). Their

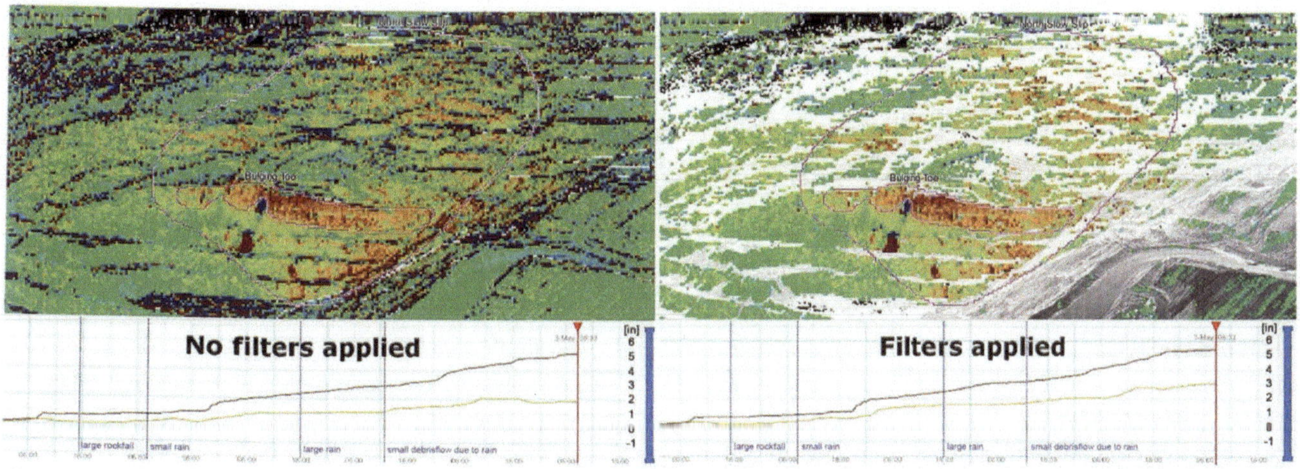

Fig. 4.45: Comparative images showing how the application of filters can improve signal-to-noise ratios and eliminate false positives from laser scan monitoring. These filters are used to improve detection rates for critical slope movement alarms and can be used to more accurately delineate failure boundaries (source: B. Haugen).

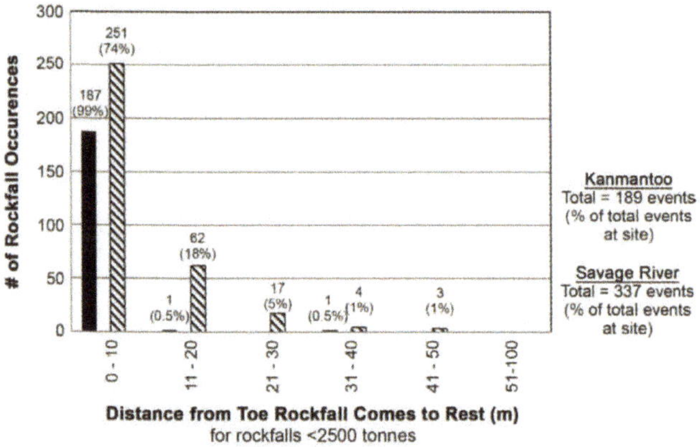

Fig. 4.46: Example of slope failure event data gathered in part using laser scanning data. Hutchison *et al.* (2015) used these data to develop mandatory exclusion zones near highwalls, by applying a risk-based approach to their slope monitoring program. Their approach has facilitated the use of vertical bench faces and steep inter-ramp slopes without compromising safety (source: Kanmantoo and Savage River mines).

approach has enabled the safe operation, maintenance and monitoring of pits with vertical to near-vertical bench faces and steep inter-ramp slopes (Hutchison *et al.* 2015).

4.6.5.8 Design conformance

Comparing as-built constructed slopes to their design basis can be used to evaluate long-term slope performance and is a key indicator of progressive risk. Figure 4.47 shows an example of design conformance analysis on a pit slope. In this case, the slope engineer's goal was to determine variance of the as-built slope with respect to design, including measurement of over- and under-excavated volumes. The slope engineer also needed to ensure that toe and crest positions and bench geometry were sufficient to provide adequate catchment for rockfall.

4.6.5.9 3D structural analysis

Structure mapping and analysis is essential to any robust geotechnical monitoring program. Figure 4.48 shows an example of a 3D structural analysis from a vertical bench face. In this case, rockfall hazards were the primary concern, where block sizes, joint set orientations and joint roughness/waviness were needed to evaluate *in situ* stability.

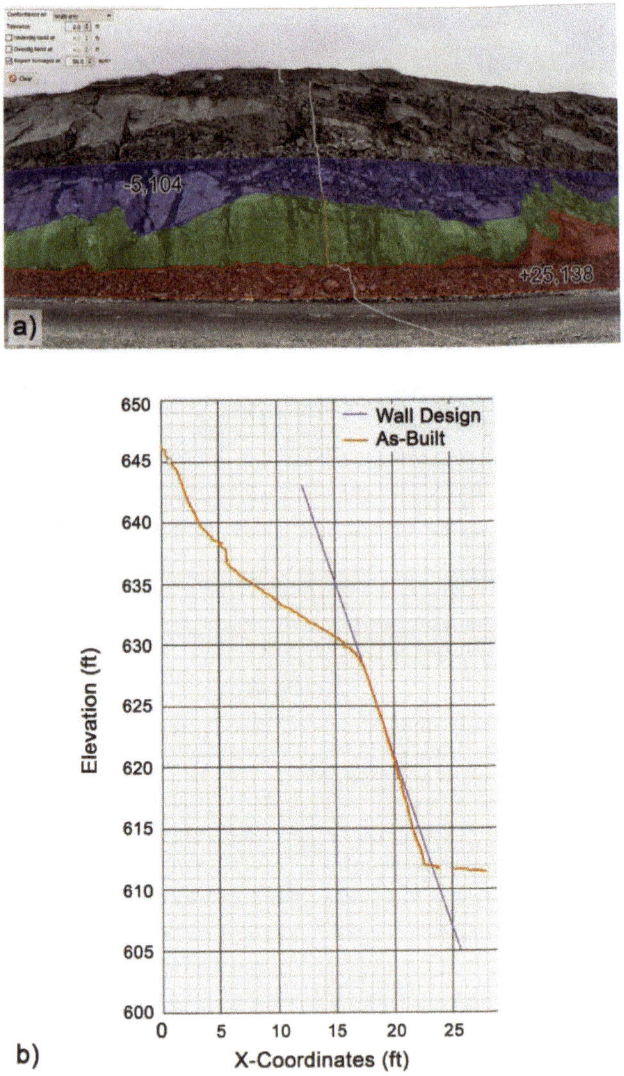

Fig. 4.47: (a) Perspective 3D view of over-excavation (blue) and under-excavation (red) measured using laser scanning data, and (b) as-built *v.* design cross-section view of the same slope showing variance in toe and crest positions. Note that the significant apparent under-excavated area shown in (a) is due to placement of a toe-berm that is not shown in the cross-section (b) (source: A. Goncalves).

4.7 Application: groundwater monitoring

4.7.1 Overview

A groundwater monitoring network that is integrated into the slope performance monitoring system is required for all mining operations, regardless of whether the mine workings extend below the water table. In addition to mine development and engineering considerations, groundwater monitoring is a permit requirement in virtually all regulatory jurisdictions worldwide. At any given mine site, the groundwater monitoring program may encompass:

- site-wide groundwater monitoring for general hydrogeological characterisation, environmental monitoring and groundwater protection;
- groundwater monitoring around individual site facilities;
- monitoring to assess the general mine dewatering program;
- specific local-scale pore pressure monitoring of the pit slopes; and

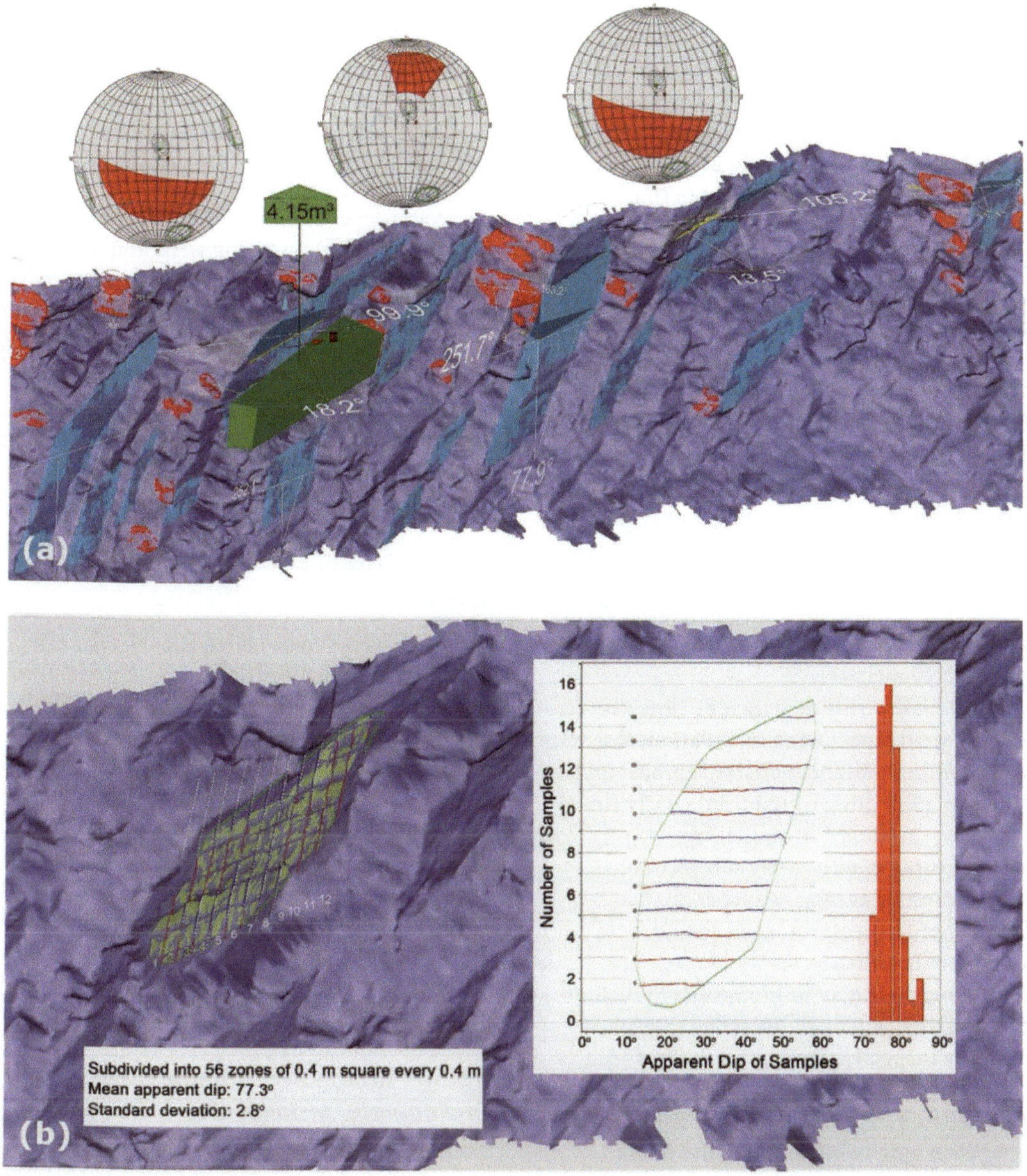

Fig. 4.48: (a) 3D joint set orientations, block volumes and kinematic analysis for a rockfall study. (b) Joint waviness analysis (source: B. Haugen).

- monitoring of the site water balance and external discharge of excess water.

For most mining operations, an understanding of the site-wide groundwater conditions is an essential component of the pore pressure analysis for the open pit slopes. Therefore, a combination of wider-scale general groundwater monitoring and local-scale specific pore pressure monitoring at depth within the pit slope is usually important. The monitoring system is integrated to track the effective area of influence of dewatering wells and to understand how the different formations in the pit slope respond to pumping and drainage.

It is important that the monitoring program evolves as the knowledge-base for the mining operation increases. The goals of monitoring programs often change with time in response to new hydrogeology or geotechnical data, improved or updated geology interpretations, or because of changing mine plans. Every borehole drilled in the pit area should be considered a potential candidate to be completed as a piezometer.

4.7.2 Types of monitoring installation

A groundwater monitoring network may include the following types of monitoring.

- *Precipitation*: most mines operate an on-site precipitation gauge. Additional precipitation gauges near the open pit may be necessary for sites located in areas of steep and variable topography or with convective precipitation. At some sites, monitoring of short-term (hourly) precipitation may be important for helping to interpret piezometer responses. In cold climates, temperature monitoring may also be important to assess the timing of snowmelt.

- *Individual dewatering wells*: this may include a comprehensive monitoring plan involving flow rate (cumulative and instantaneous measured during the daily inspection) using an in-line flowmeter; downhole water level monitored daily (or more frequently to track the pump duty cycle); wellhead pressure measured daily; wellhead power (line-to-line volts and amps) inspected daily; water chemistry field parameters (TDS, pH) measured monthly; and full water chemistry sampling, collected quarterly.
- *Multiple wells or well point systems*: monitoring of the discharge from individual wells may be combined to record combined flow rate (cumulative and instantaneous measured during the daily inspection); combined water chemistry field parameters (TDS, pH) measured monthly; and combined full water chemistry sampling, collected quarterly.
- *Horizontal drain holes*: this may include flow rate at the end of each drill pipe during installation; flow rate at the end of drilling; water chemistry sampled 24 h after completion of drilling (if required); daily flow rates for the first seven days after drilling; weekly flow rates for two to three months after completion; monthly flow rates for as long as access can be retained; and shut-in pressures, as needed, where shut-in valves are installed.
- *Individual sumps*: these may be monitored to include flow rate (cumulative and instantaneous measured during the daily inspection); water level, measured daily or automatically if tied to the pumping system; pumping pressure, measured during the daily inspection; power supply inspected daily; water chemistry field parameters (TDS, pH) measured monthly; and full water chemistry sampling, collected quarterly.
- *In-pit seeps*: this may include visual or photographic mapping of pit wall seeps; visual inspection of any runoff areas, paying particular attention to the potential for surface water runoff to enter tension cracks; flow measurement of prominent seeps, typically inspected monthly or quarterly; water chemistry monitoring of prominent seeps for field parameters (TDS, pH) sampled monthly for selected seeps as needed; and full water chemistry sampling, collected quarterly for selected seeps as needed.
- *Vertical or down-angle drain holes*: monitoring is typically limited to down-hole water level on an as-needed basis.
- *Water levels or pore pressures*: these may be measured in boreholes that are temporarily left open (with or without liner pipe); standpipe piezometers or observation wells (single or multi-level); monitoring wells (for groundwater chemistry); grouted-in vibrating wire piezometers (VWP); inactive (offline) pumping wells; and accessible nearby observation wells and/or pumping boreholes on adjacent land.

Water-level monitoring should be carried out in inactive (offline) pumping wells and, depending on local circumstances and access considerations, any surrounding third-party observation wells and pumping boreholes. Initially, monitoring networks should be planned to characterise both the regional-scale and site-scale hydrogeology. Monitoring networks necessarily evolve and become more focused after the commencement of mining and when a greater understanding of the hydrogeological regime is attained.

4.7.3 Integration of hydrogeology and geotechnical monitoring

The ability to manage large datasets has led to better integration of hydrogeology and geotechnical monitoring programs. There are ever-increasing amounts of data illustrating the relationship between pore pressure and slope deformation. This is facilitated by the relatively small diameter of the VWP sensors and cables now available, which leaves space within the borehole for additional instruments such as inclinometer casing, TDR cables, accelerometer arrays and borehole extensometers. These are described in the following subsections, noting that the frequency of pore pressure readings is often modified to match the monitoring frequency for geotechnical instruments.

4.7.3.1 Combined vibrating wire piezometers and inclinometer installations

Inclinometer systems typically comprise casing segments that are coupled together with a sealed cap on the end. The annulus between the borehole wall and the casing is typically grouted with a cement-bentonite mixture. Depending on the relative size of the borehole and the inclinometer casing, it may be possible to install one or more VWP sensors in the annular space between the borehole wall and the outside of the inclinometer casing.

The sensors, cables and PVC or steel grout tremie pipe are secured (taped) to the outside of the inclinometer casing as it is lowered into the hole (Fig. 4.49). The annulus is grouted from the bottom up using the tremie pipe. The grout encapsulates the VWP sensors and stabilises the inclinometer casing (Fig. 4.50). Depending on the depth of water in the hole, it may be necessary to fill the inside of the inclinometer casing with water to weigh it down. For deep installations, it may also be necessary to grout in stages, in which case more than one tremie pipe may be required. In boreholes subject to

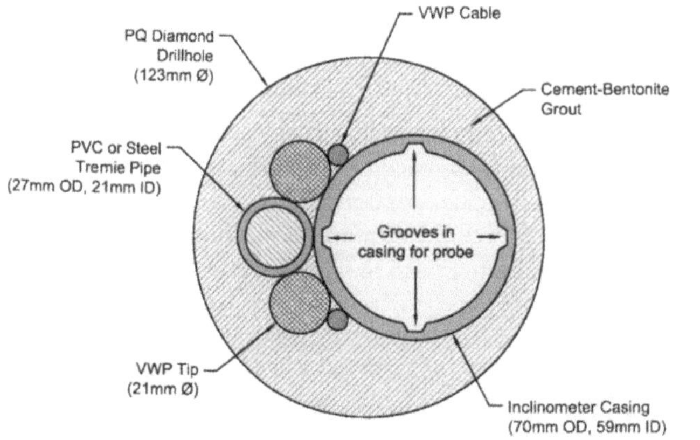

Fig. 4.49: Bundled inclinometer and VWP installation (source: Beale and Read 2013).

caving, it may be necessary to install the assembly through drill rods or steel casing, in which case it may be possible to use the rods or casing to inject the grout and eliminate the tremie pipe.

Small-diameter diamond drill holes (e.g. HQ size – 96 mm diameter or smaller) are generally not large enough to accommodate combined inclinometer and VWP installations; consequently, larger-diameter (e.g. PQ size – 123 mm diameter) diamond drill holes or reverse circulation (RC) holes are usually required. Most conventional inclinometer systems are installed in vertical or near-vertical boreholes. However, provided the borehole is stable, or can be stabilised with casing, installation in inclined holes is also possible.

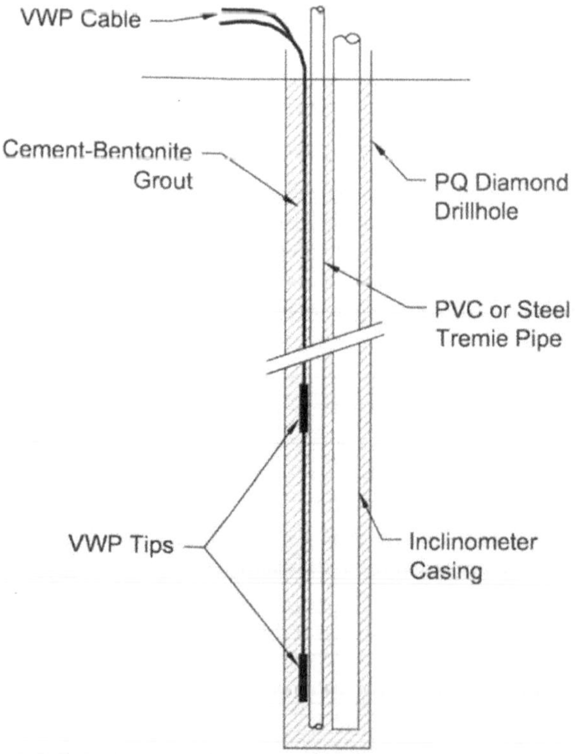

Fig. 4.50: Profile of inclinometer and VWP installation (source: Beale and Read 2013).

4.7.3.2 Combined vibrating wire piezometers and other geotechnical instrument installations

Several other combinations of geotechnical instrumentation can also be installed with VWPs. These include TDR, accelerometers and extensometers. In the case of TDR systems, the only downhole component is a coaxial cable. TDR can easily be installed with VWPs or with a combined inclinometer/VWP system, in which case the TDR would provide an inexpensive redundancy for the inclinometer. TDR/VWP bundles (without the conventional inclinometer casing) can also be installed in smaller-diameter drill holes.

Borehole accelerometer arrays can be bundled with VWPs in boreholes in the same way. This is illustrated schematically in Fig. 4.51. The accelerometer array consists of a series of mutually perpendicular accelerometers that are spaced evenly along a flexible or articulated cable. This cable can be lowered down a casing or conduit in a borehole (portable system) or can be permanently grouted into a borehole. Signal cables communicate the tilt of each accelerometer group to a readout box at the collar. Once processed, the data can be used to define the inclination (tilt) of the borehole in a similar manner as for a conventional inclinometer, and repeated readings can be compared against a baseline to detect deflections or distortions of the borehole over time. Like TDRs, they can be read remotely and provide real-time monitoring.

Borehole extensometers can be single or multi-point, and may be constructed from solid rods or wire cables. In a single-point extensometer, the end of the rod or wire is anchored in the bottom of the hole, and the rod or wire extends to the collar of the hole. Multi-point extensometers have multiple rods or wires anchored at various depths in the borehole. Deformations of rod extensometers are measured against a reference plate at the collar of the hole. In the case of wire extensometers, a variety of deformation measurement techniques may be used, depending on the system design. The key requirement of a borehole extensometer is that the

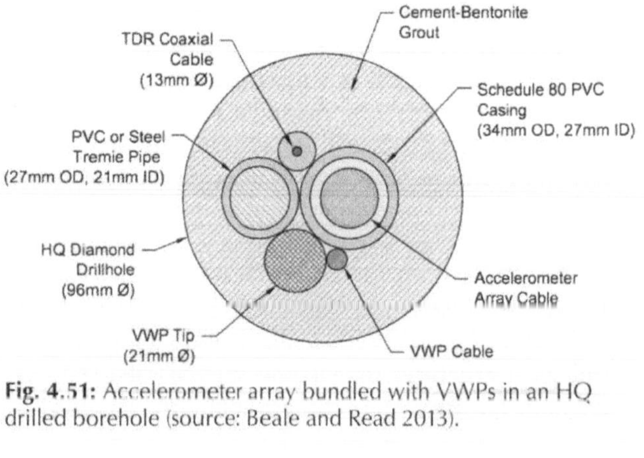

Fig. 4.51: Accelerometer array bundled with VWPs in an HQ drilled borehole (source: Beale and Read 2013).

portion of the rod or wire that extends between the anchor and the collar of the hole must be free to move. Consequently, the hole cannot be fully grouted unless the rods or wires are encased within a conduit.

Single borehole extensometers can be bundled with conventional inclinometers, provided there is sufficient space to accommodate both the rod/wire and the conduit. In this case, extensional and shear displacement could be monitored. Likewise, borehole extensometers could be bundled with VWPs and/or TDRs, subject only to the available annular space in a given borehole.

4.7.4 Presentation of hydrogeology monitoring data

As the ability to combine various types of monitoring data for analysis improves, there is increasing interest to integrate time-series plots of output from radar, prisms, TDRs, inclinometers and/or extensometers with pore pressure hydrographs, pumping hydrographs, timing of horizontal drain installations, and precipitation data. Hydrogeology data are usually presented in various formats, including:

- groundwater level or drawdown maps;
- maps of airlift yields from pilot holes;
- maps of horizontal drain flows;
- water chemistry maps;
- hydrographs of piezometers and observation wells;
- hydrographs showing well yields or drain flows with time;
- graphs showing changes in water chemistry parameters with time; and
- hydrogeological (hydro-geotechnical) cross-sections.

4.7.4.1 Water level maps

A base plan map provides the 'fixed' information onto which the monitoring results are added. For active mine sites, where the mine topography and mine plans are constantly evolving, it is necessary to progressively update the topographic contours and surface geology on the base map.

Groundwater-level maps may show past recorded levels, currently recorded levels, predicted levels indicated by modelling for some future date, or past and predicted future changes in levels over a defined time period. To aid data interpretation, it is normal that all elevations (such as groundwater heads) should be plotted relative to the mine datum. In some instances, maps showing depth to water below the ground surface may also be useful. Plotting of pore pressures onto maps is rarely done, as the pore pressure depends on the specific location (and elevation) of the piezometer sensor.

Ideally, all groundwater maps should show geology, geological structures, topography, surface water features and mine facilities. Maps that are contoured should also show the actual data points, as there is often little value in showing groundwater contours without illustrating the controlling features. Contoured maps not showing the individual data points may be misleading (e.g. where groundwater levels become strongly 'stair-stepped' across geological structures or between geological units). There will often be a sequence of maps which may show either total groundwater head at several 'snapshot' points in time and/or change in static head or pore pressure over a given time period.

Vertical differences in head can be represented on maps by colour-coding data from vertically separated piezometers. Strong vertical head gradients are associated with many pit slopes; in some cases, the vertical gradients may become larger than the horizontal gradients. Therefore, for an active mine site, the use of groundwater maps may be more applicable for presenting the site-scale data rather than the sector-scale data for which cross-sections are more useful. Points of hydraulic stress are usually included on the groundwater maps (Fig. 4.52). These are most commonly pumping wells, where the flow rate from the well may be added to the map. Flows from horizontal drains and prominent seeps can also be added where appropriate. Natural hydraulic stresses may be important in wetter climatic zones. Maps showing the distribution of horizontal drain flows provide an excellent way of rapidly reviewing drain hole results. The trace of the drain hole is shown on the map and is colour-coded to show the initial flow rate of the drain (Fig. 4.53). The position of the first water strike in each drain hole can be added to the map to provide an indication of the depth of water behind the slope.

4.7.4.2 Piezometer hydrographs

Hydrographs of piezometers are an essential component of any hydrogeological analysis. Multiple records should be included on the same plots to allow correlations to be made. The hydrographs should be annotated with as much pertinent information as possible to allow the influence of external variables to be determined and to facilitate an understanding of the principal factors that influence piezometric levels in each sector of the pit (Fig. 4.54).

Piezometer plots are usually grouped by their physical location (sector) within the pit, but may also be grouped by hydrogeological unit. Records from multiple piezometers in each sector, particularly vertically discretised piezometers, should ideally be included on a single plot (Fig. 4.55) provided the amount of information does not become overwhelming. If it does, several plots may be required. Plots of vertically discretised piezometric data against time, with the piezometers 'hung' on the plot at the appropriate depth, may illustrate the way in which vertical heads are changing with time. They can also indicate the effects of vertical compartmentalisation.

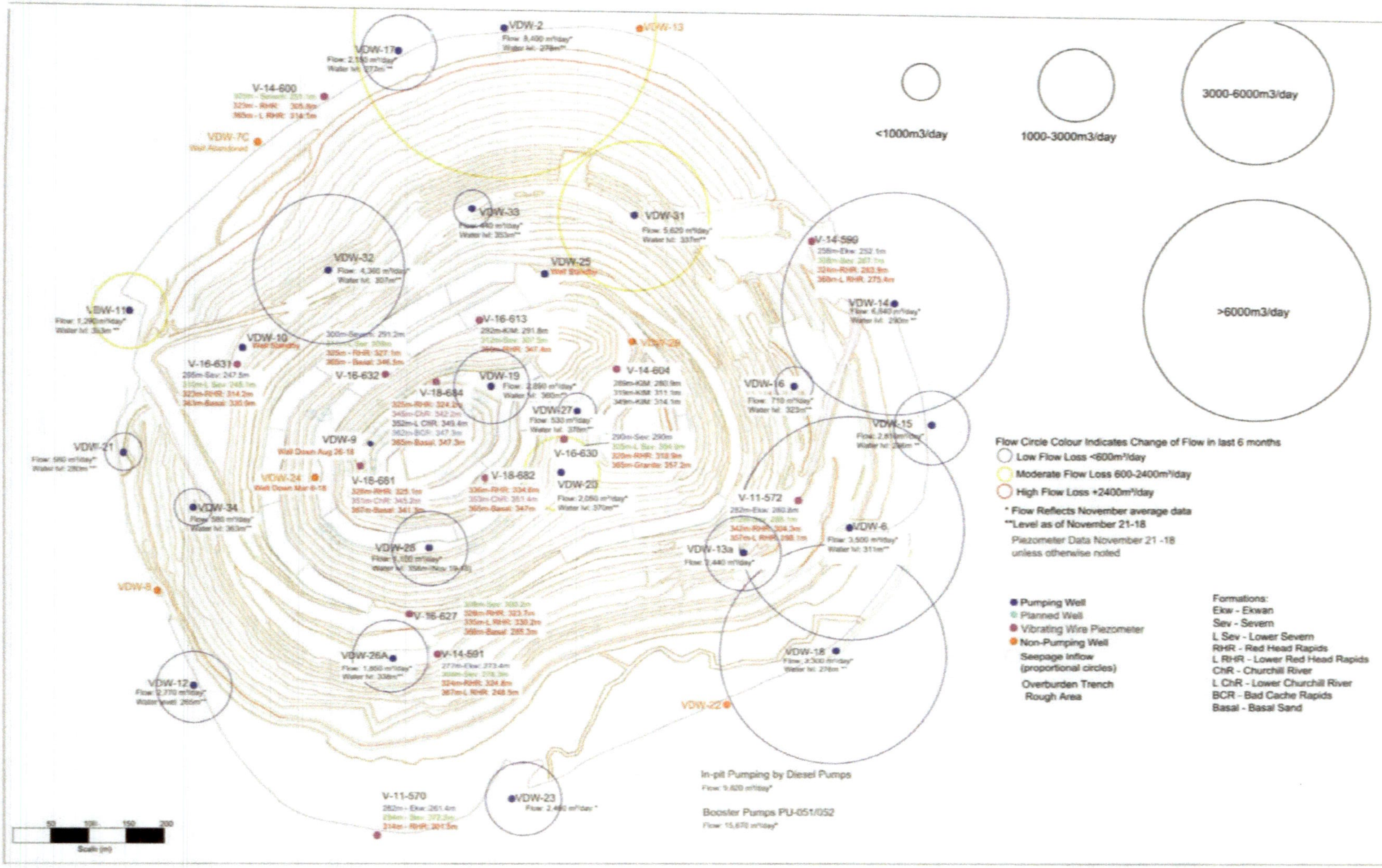

Fig. 4.52: Groundwater map where the pumping rates of individual wells are represented by the size of the circles and where multi-level piezometer data are colour-coded. This site has layer-cake stratigraphy and an absence of structural control, so it is not necessary to add the geology to the base map (source: M. Desjardins).

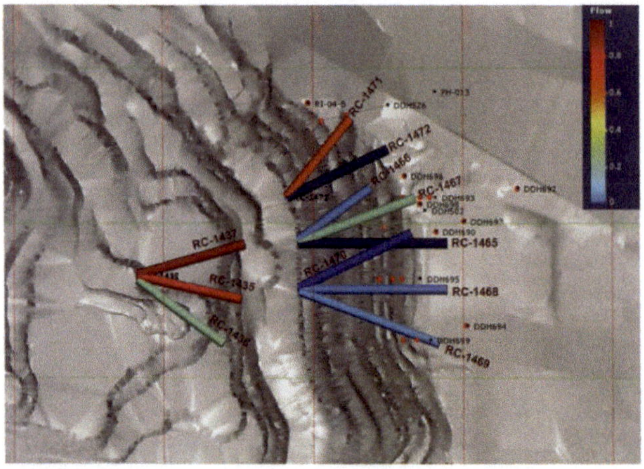

Fig. 4.53: Map illustrating horizontal drain hole results (source: G. Beale).

It is usually important to include the pumping rate on the piezometer hydrographs for each key well, for all wells in a particular sector, and/or for the total pumping rate at the site. Where appropriate, the piezometer hydrographs can also be annotated with horizontal drain flows, normally by sector but sometimes with the total flow from all drains. It is useful to annotate the hydrographs to show the start-up (or shut-down) of key wells or the timing of various horizontal drain programs. A histogram of daily, monthly or accumulated precipitation (depending on the time scale of the hydrograph) can be useful for identifying important recharge events or for identifying longer periods of enhanced recharge. In some circumstances, correction of the data for atmospheric pressure or other variables may be required to illustrate observed changes with time in the piezometric levels.

4.7.4.3 Flow rate hydrographs

Flow measurements from seepage faces, in-pit sumps, horizontal drains or pumping wells are an integral part of any monitoring program. Graphs showing flow rates from pumping wells or drains with time are fundamental to any hydrogeological analysis. Integrating the plots of the flow rate with hydrographs showing piezometric responses is common practice.

Pumping well installations from each sector of the pit may be included on a single plot, together with the total flow rate from all the wells in the sector. The plots of flow rates from wells may also be grouped by their hydrogeological unit. Plots of horizontal drain flows with time can be useful for determining whether a slope sector is draining and depressurising or whether there is ongoing recharge. Multiple drains from each sector of the slope may be included on a single plot, with the total flow rate from all drains in the sector shown on the plot. Where there is strong foliation or structure, the yield of the holes may show a strong correlation with drilling orientation.

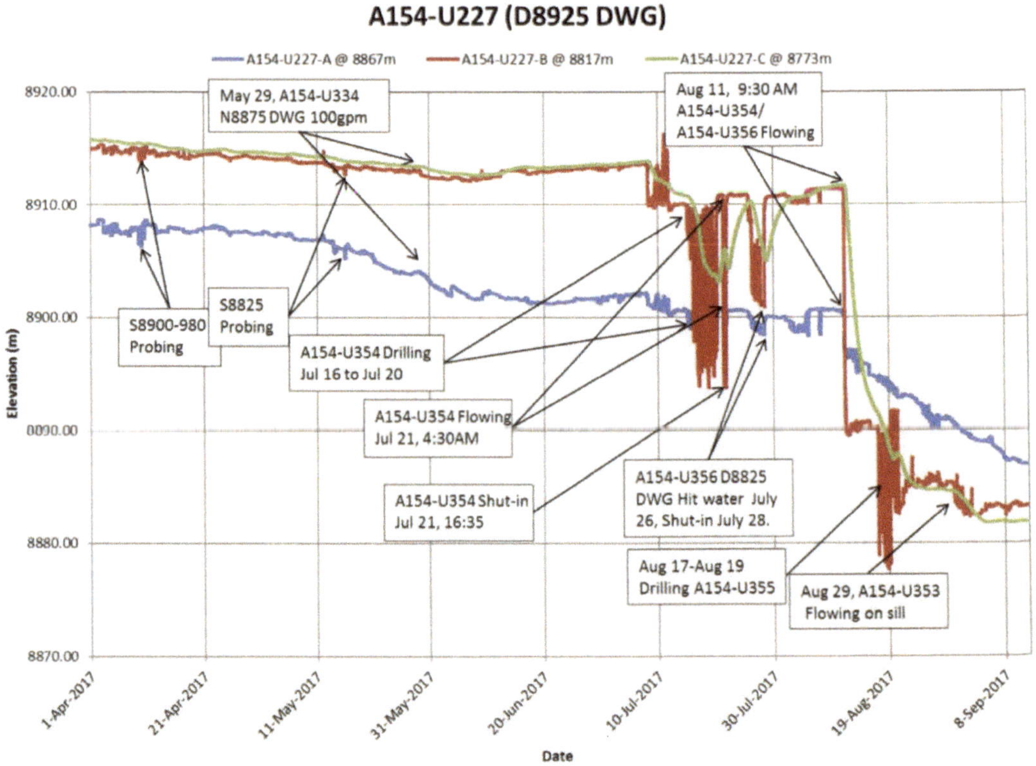

Fig. 4.54: Example of an annotated piezometer hydrograph (source: R. Quevedo and J. Sovka).

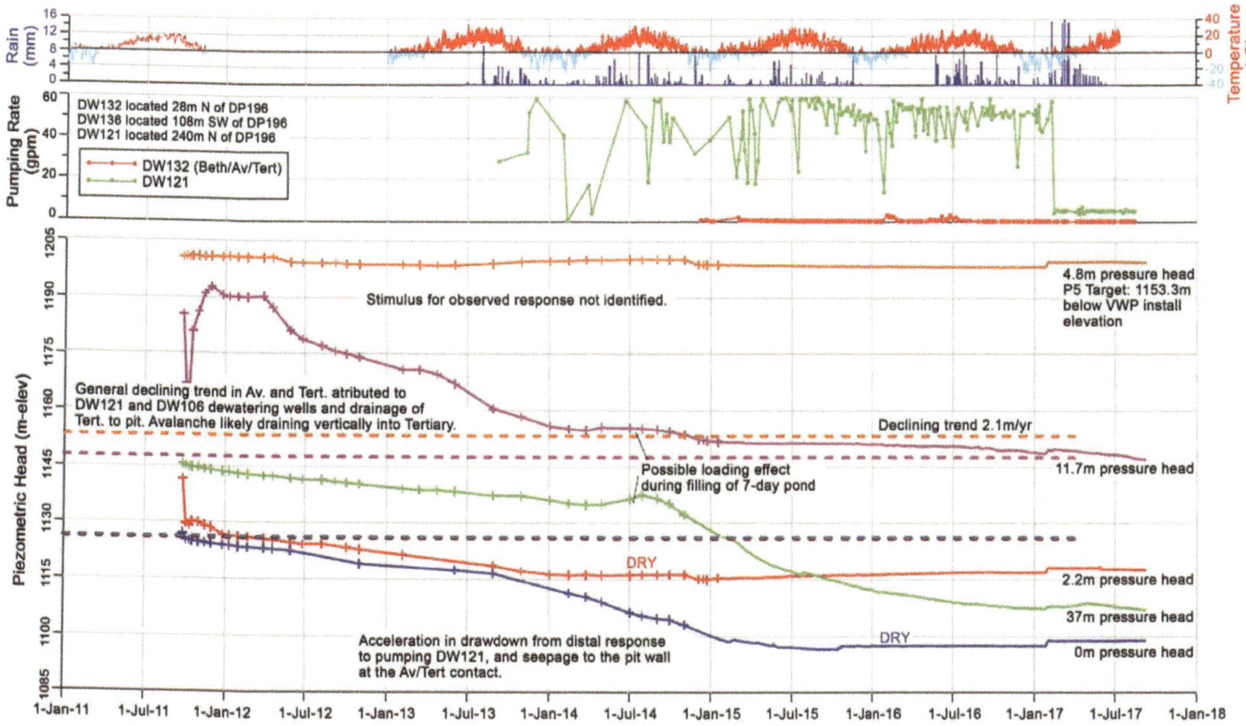

Fig. 4.55: Example of plotting multiple well hydrographs correlated with pumping, horizontal drain flow, rainfall and air temperature (release of water due to snow melting) (source: M. Veillette and S. Fortin).

4.7.4.4 Cross-sections

Sector-scale cross-sections are the basic component of any pit slope hydrogeology analysis. It is common practice for the cross-sections to include lithological, alteration and structural information. It is usually beneficial to make the hydrogeological cross-sections coincident with the geotechnical design sections. The sections may be annotated to include geotechnical information as well as hydrogeological information (Fig. 4.56).

Historical and current piezometer information can be plotted onto the sections to show the phreatic surface or to show 2D groundwater head (or pore pressure) values for a given time period, based on vertically discretised piezometers and/or changes of the phreatic surface or head distribution with time. The sections can also be annotated with known fracture or groundwater-flow zones and other pertinent features.

It is likely that the boreholes or data points will not lie exactly on the line of the section, requiring data to be projected from nearby measurement points onto the section. When this is done, the points should always be clearly identified and, if there are anomalies, the data that are furthest from the section line should be the first to be questioned. The lateral 'search distance' for adding piezometers to the section is typically 100–200 m, but may be greater where the density of data is low or where the geology of the slope is relatively consistent away from the section lines. It is usually of benefit to annotate the distance of the piezometer from the section line.

4.7.4.5 Composite data presentation

The integration of the geological and hydrogeological data onto maps and cross-sections is essential for providing meaningful interpretation and building the conceptual model. It is common practice to integrate plots of water level, pumping rate and chemistry onto a single display to show how the various measured parameters are linked. Every effort should be made to compile combined plots of piezometer data with deformation monitoring data. There is an ever-expanding database from the worldwide mining industry to illustrate the relationship of pore pressure and slope movement for various hydrogeological settings. Figure 4.57 shows the before and after effects of 12 horizontal drain holes on the slope movement monitored by radar.

4.7.5 Laying out the monitoring program

The factors typically considered when placing monitoring sensors on and near the pit slopes include the:

- overall groundwater flow system with the nature and location of the groundwater recharge sources;
- need for monitoring in each defined hydro-stratigraphic unit or compartment;

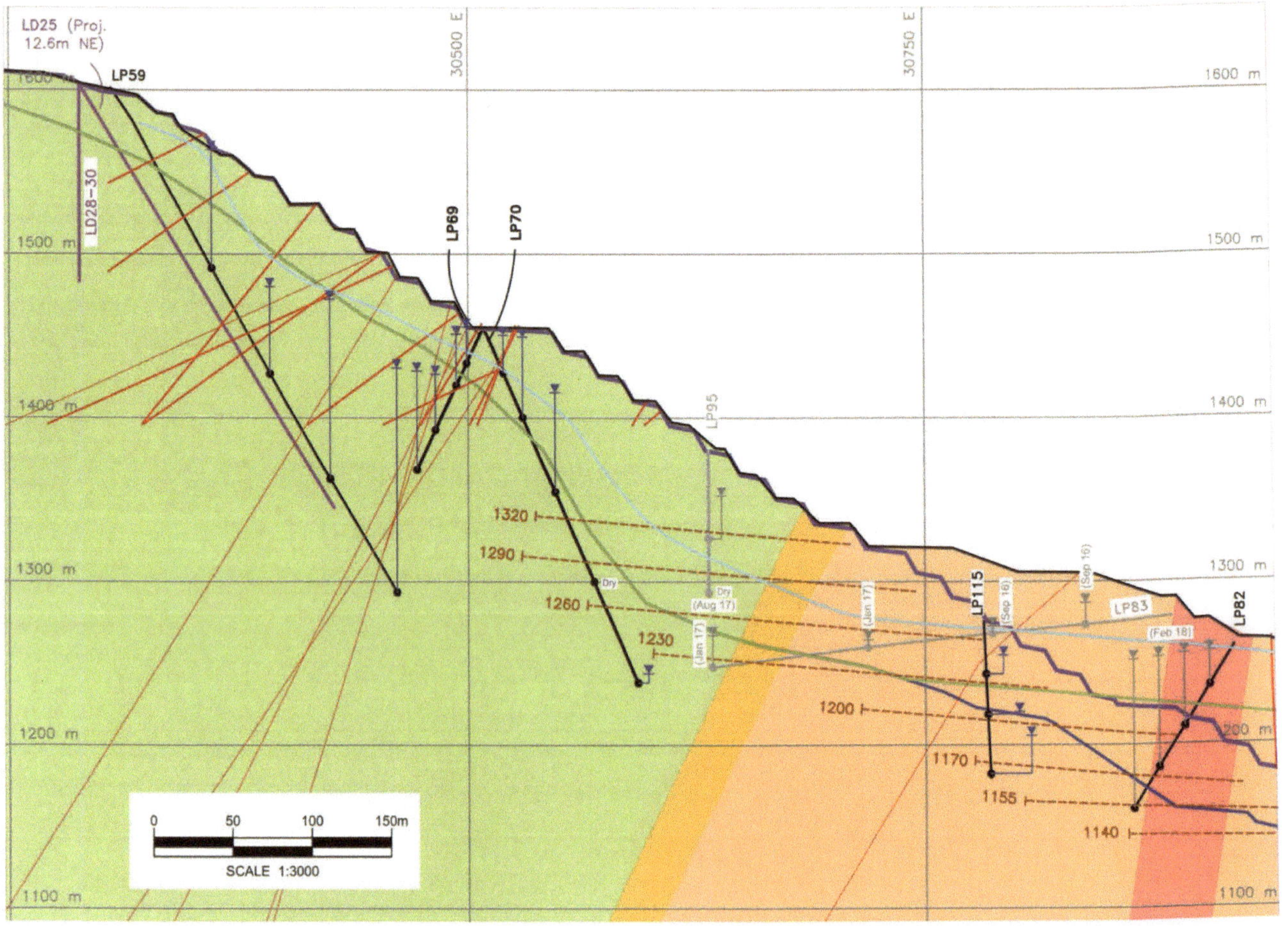

Fig. 4.56: Example of a detailed hydrogeological cross-section showing current and future pit slopes; current horizontal drains and location of water strikes; planned future horizontal drains; and vertical, angled and horizontal piezometers (source: M. Veillette and S. Fortin).

- requirement to monitor in weathered and non-weathered rock;
- need to place observation points to ensure general dewatering targets are achieved;
- slope sectors that need to be depressurised over the coming two-year period;
- need to monitor depressurisation in the various slope sectors, and the need to meet any defined targets for slope depressurisation;
- benefit of placing piezometers on or close to the geotechnical design sections;
- areas of weaker materials and sectors that have a higher risk of instability;
- ability to protect the installation from damage during mine development; and
- ability to retain access during mine development.

For a general open pit dewatering system, it is common practice to have 10–30 piezometers in service within and around the pit to monitor the overall behaviour of the water table. The exact number may depend on the scale of the mine, the scale and required timing of the dewatering operation, and the variability of the geological materials. The piezometers would be distributed at accessible locations within the pit and near the pit crest. In addition, for specific pore pressure monitoring and if the achievement of pore pressure targets is critical for a given slope sector, it is not uncommon to maintain five to 10 multi-level piezometer installations for a single sector.

Where appropriate, piezometers can be installed in angled and vertically drilled holes, or holes at multiple orientations from a common drilling platform (Fig. 4.58), to provide pore pressure information at various distances and depths behind the slope. For active pushbacks, it is obviously better to install the new piezometers behind the crest or on step-outs above the advancing pushback, rather than below it. The piezometer holes can be angled so the sensors are close to the slope (Fig. 4.58).

Many mining operations successfully mine around and preserve monitoring points, particularly where smaller mining equipment is being used. However, where there is a need to obtain critical pore pressure information and there are no other alternatives, it is sometimes necessary to install a sacrificial observation well or piezometer that may

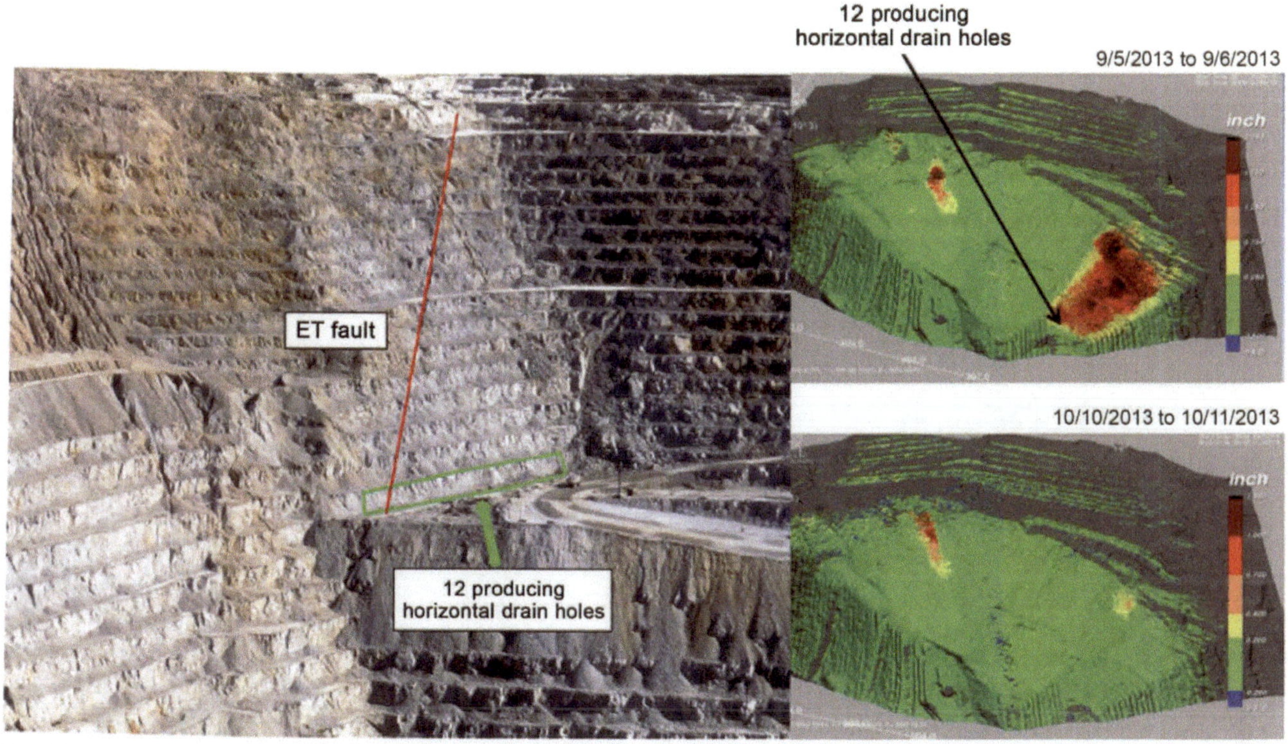

Fig. 4.57: Effect of horizontal drain holes monitored by radar (source: S. Borron, T. Ashinhurst and J. Dowling).

only last for several months before being destroyed by the mining operation.

4.7.6 Monitoring of transient pore pressure

Transient changes to the pore pressure distribution may occur due to several factors, including:

- blasting (very rapid response, potentially milliseconds, possibly creating water hammer effects in strongly jointed rock, or deformation and strain softening in weaker rock);
- deformation or failure of the slope materials (often a rapid response, potentially in minutes following a deformation triggering event);
- changes in rainfall patterns or a strong rainfall event (often causing an hourly or daily response);
- seasonal changes (often causing a daily to monthly response); and
- mining-related deformation of the slope (potentially on the scale of a week in areas of rapid mining, but generally longer-term).

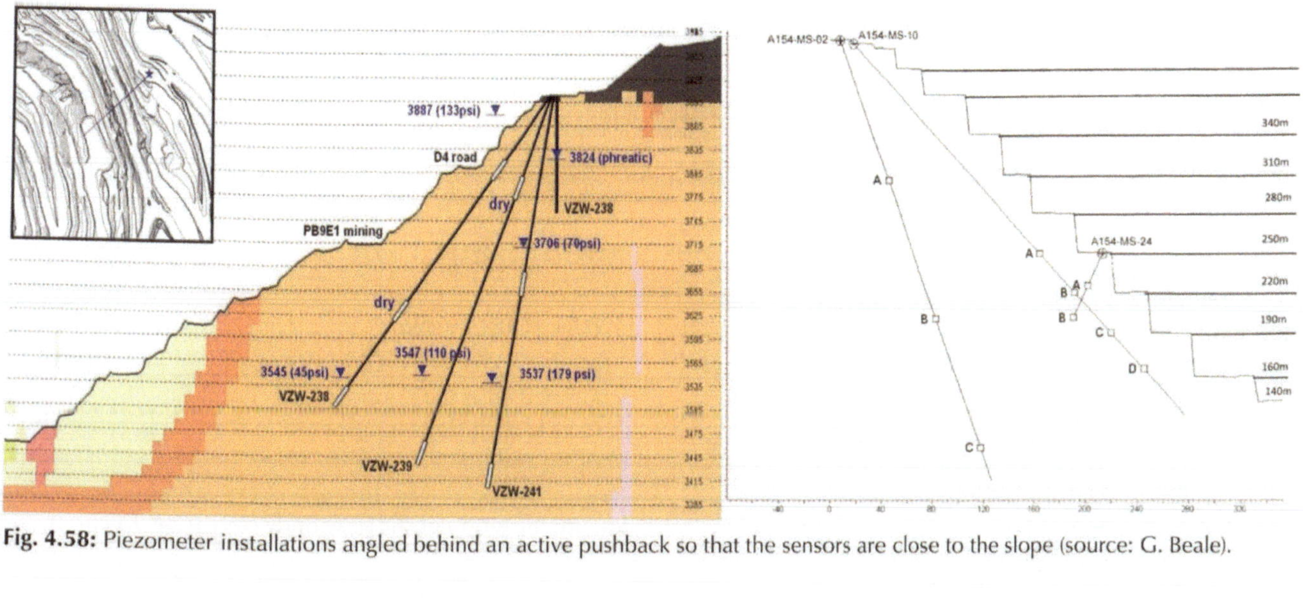

Fig. 4.58: Piezometer installations angled behind an active pushback so that the sensors are close to the slope (source: G. Beale).

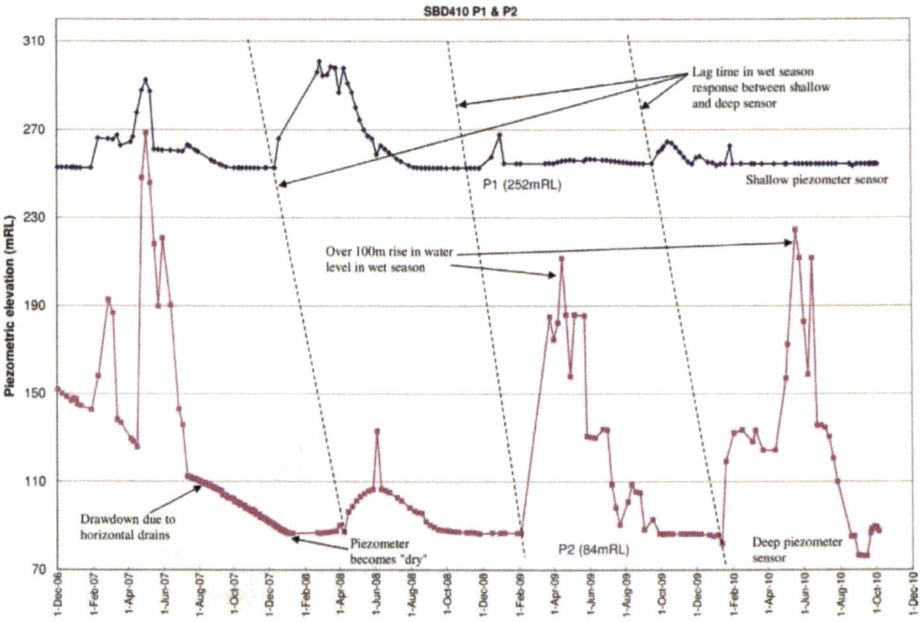

Fig. 4.59: Example of transient seasonal pore pressure responses in 'dry' piezometers (source: Beale and Read 2013).

Instability in weak rocks, wedge failures in jointed/faulted rocks or toppling in foliated rocks can be triggered by transient pore pressure caused by recharge and from storm events. There are many mines where much of the observed instability occurs at a relatively shallow depth in the slope, where confining stresses are low and where damage of the rock mass is greater. An increase in pore pressure at shallow depths typically has a greater influence on effective stress than equivalent pore pressure increases that occur deeper in the slope.

In settings with high rainfall or rapid melting of snowpack and where the consequence of instability is high, it may be appropriate to install shallow piezometers above the water table to monitor transient pore pressures in rock slopes that are normally drained. Because instability is frequently triggered by rainfall (or snowmelt) events, many operations install piezometers above the phreatic surface that are 30–60 m deep, with up to five grouted-in VWP sensors in a single borehole to assess the transient pressure profile. Piezometers that go dry as the drainage program progresses should be retained if there is potential for pore pressures to re-establish or if there is seasonal re-saturation and therefore a transient pore pressure response due to seasonal recharge.

Although VWP installations are generally preferred for monitoring transient pore pressure changes, pressure transducers installed within standpipe piezometers can also be connected to dataloggers, programmed for the desired data collection frequency, to monitor the transient response. The data collection frequency can be set to milliseconds if monitoring around a blast is required. Conversely, if transient trends over longer periods of time are the main objective and immediate responses are not important, daily or weekly readings may be adequate.

Figure 4.59 shows a dual-level VWP installation in a single borehole where the phreatic surface was below the location of the shallow sensor at the time of installation and subsequently fell below the location of a deeper sensor. The shallow piezometer shows transient positive pore pressures for up to three months each year, with the pore pressure rising from 0 to over 40 m head (~400 kPa). The deeper piezometer typically shows a greater transient response for up to four months each year, with the pore pressure rising from 0 to over 150 m head (~1500 kPa). There is typically more seasonal response in the deeper piezometer than in the shallower piezometer, and a time lag can be observed between the shallow piezometer and the deep piezometer as the seasonal recharge front moves downward through the rock mass.

4.8 Application: microseismic monitoring

4.8.1 Overview

Microseismic activity recorded in open pit slopes may be associated with the initiation and mobilisation of a sliding surface in a pit slope in response to fracturing caused by blasting and unloading of the slope, or after an instability has already developed, through the progressive decoupling and mobilisation of the slide body over time. In either case, the following benefits may be anticipated from a microseismic monitoring system:

- an indication of when particular mining activities are affecting the stability of the slope and by how much;
- an indication of when a particular geological structure is becoming seismically active;
- identification of previously unknown geological structures that may have an impact on stability; and
- early warning of slope instabilities.

The following case histories provide examples of these applications.

4.8.2 Case study: Sunrise Dam gold mine – Cleo open pit

The Sunrise Dam gold mine is a large open pit and underground mine located 730 km north-east of Perth, Western Australia. When the Cleo open pit was mined, the plan required leaving a mafic unit bullnose on top of a large healed shear zone. A microseismic array was installed as part of the performance monitoring strategy to study the rock mass response to mining. Although the mine plan later changed, eliminating the bullnose problem, the array provided invaluable insight into the rock mass response to mining to address the question raised and discussed by Lynch and Malovichko (2006), 'What causes open pit slope fracturing: blasting, support rock being broken, or removal of rock?'

It has been interpreted that microseismic activity recorded in open pit slopes is related to mining rates near the slopes. Again, this could be in response to fracturing caused by blasting, movement of broken rock to the toe of the slope and/or by unloading-induced stress redistribution. Insights answering this question were gained in data recorded at Sunrise Dam when two blocks totalling ~100 000 m^3 were blasted and removed directly below a slope being monitored with a small microseismic array (Fig. 4.60).

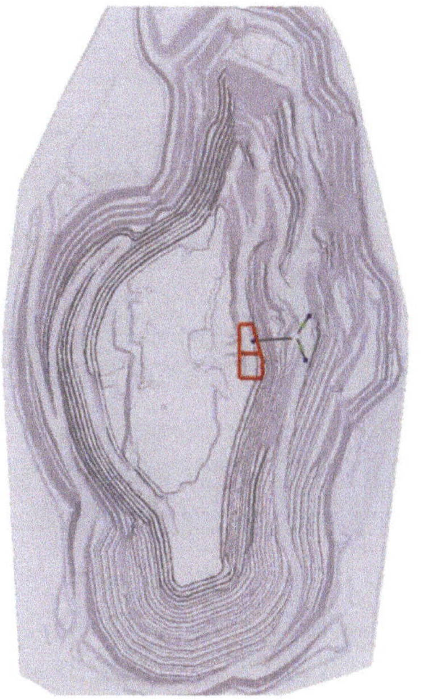

Fig. 4.60: Plan view of the Sunrise Dam gold mine (North at the top), indicating location of the mining area beneath a part of the eastern slope being monitored with a microseismic system. The two blocks mined, outlined in red, were blasted and loaded in April and May 2005 (source: Lynch and Malovichko 2006).

The cumulative number of seismic events recorded, processed and accepted during this period is shown in Fig. 4.61. A small jump in seismic activity was evident

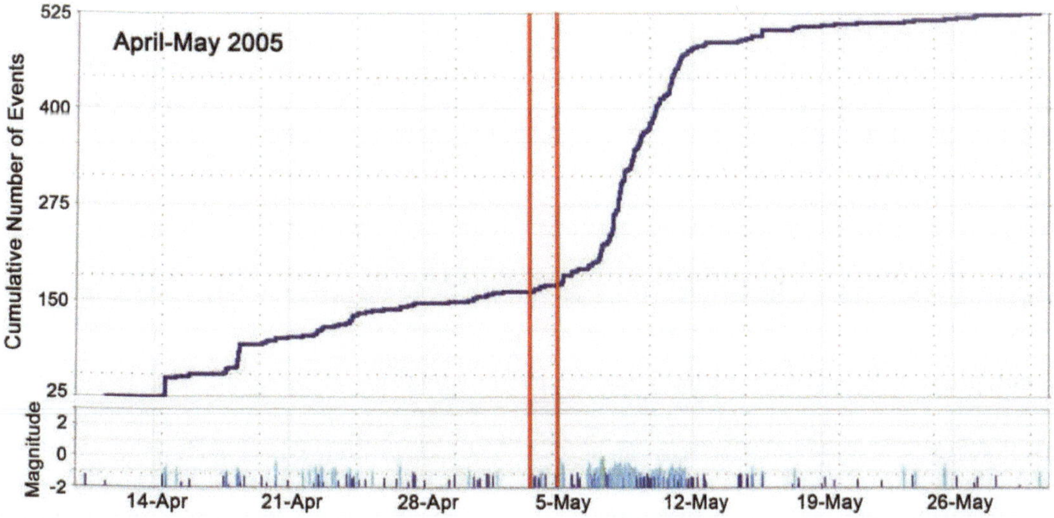

Fig. 4.61: Cumulative number of recorded seismic events as a function of time. The timing of the two blasts at the bottom of the slope (targeting the two blocks highlighted in Fig. 4.60) are indicated by the vertical red lines. The increase in events days after the blast demonstrate that the microseismic activity is not related to the transient loading of the dynamic elastic wave from blasting (source: Lynch and Malovichko 2006).

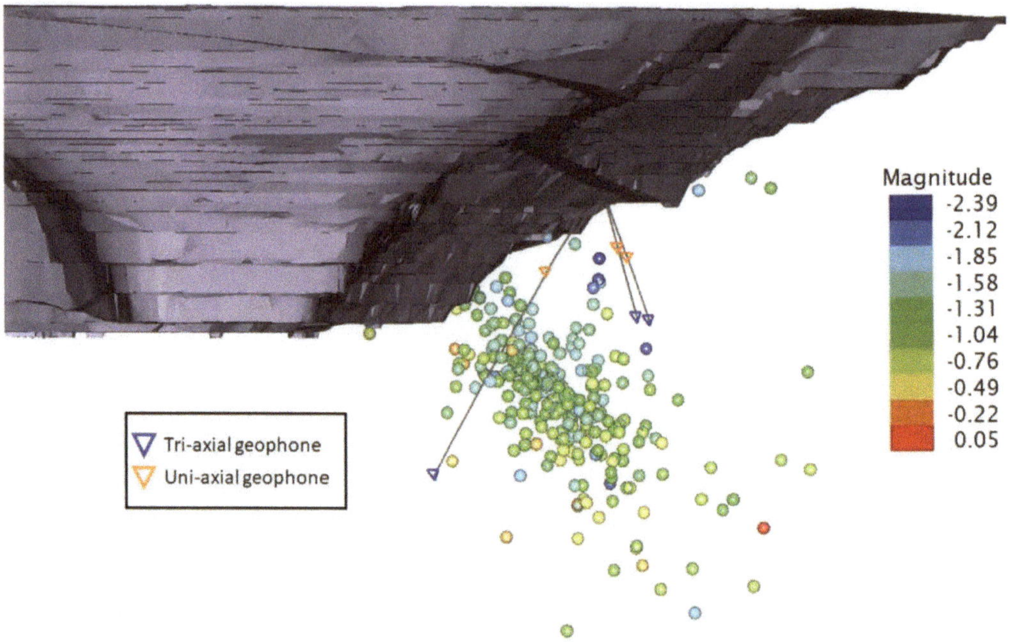

Fig. 4.62: Side view of microseismic events recorded during the seismic response to unloading and slope relaxation/deformation after blasting. The local magnitude scale is also shown (modified from Lynch and Malovichko 2006).

following the second blast, but the significant increase in microseismic activity was observed two to seven days after the last blast. This alone indicates that transient loading due to the dynamic elastic waves of the blast was not causing any significant fracturing within the slope.

It was further observed that stress changes caused by unloading the toe of the slope played a dominant role in the generation of microseismic activity. It is possible that the slope was relaxed before the first blast and absorbed some stress changes before fracturing commenced. However, the short time period between cessation of unloading and the drop in seismic activity indicates that this slope was not in a significantly fractured/weakened state and that stress changes were quickly distributed throughout the rock mass, as happens in an elastic medium.

Another important finding from this monitoring was that most of the events occurred beneath the floor of the pit. A major shear structure running roughly sub-parallel to the slope and 50 m below surface through the middle of the seismic array appeared to be seismically quiet. A side view of the seismic event locations during the period of unloading and resultant high seismic activity (Fig. 4.62) shows that the bulk of the located fracturing was taking place beneath the level of the pit floor. This suggests the shear zone was not mobilised by mining activity. This, combined with the absence of any surface movements, pointed to the relative stability of the shear structure.

4.8.3 Case study: Navachab gold mine

The Navachab gold mine is located near the town of Karibib ~200 km east of Swakopmund in Namibia. After a small slope failure at the mine in early 2000, faint popping noises were observed during the re-entry period after blasting. Preliminary seismic measurements were conducted in August 2001, and a small permanent system was installed in February 2002 for routine monitoring of microseismic activity behind the eastern highwall. Findings from a study by Lynch *et al.* (2005) are summarised below.

To better understand the origin of the recorded microseismicity, a simplified elastic numerical model of the eastern Navachab wall was constructed using FLAC2D. The purpose was to determine if a zone of tensile strain was present and whether fracturing of confined intact brittle rock could occur at Navachab. Such brittle fractures would be predominantly tensile in nature and could be recorded as seismic events (Stacey *et al.* 2003). However, the magnitudes of the modelled shear and tensile strains, assuming an isotropic homogeneous rock mass, indicated that brittle fracturing of the intact rock alone is not likely. Instead, the fact that microseismicity was being recorded in the area of interest pointed to the presence of geological structures as playing an important role. The major structures in this part of the Navachab pit were a bedding-parallel joint set that dipped roughly parallel to the slope surface at 45–55°. This coincided with the most likely stress-induced fracture orientations and thus provided a credible explanation of the mechanism for the recorded seismic events.

A plot of cumulative seismic events recorded in the Navachab eastern slope from April 2002 to January 2004 shows two distinct periods of increased microseismic activity: from October 2002 to January 2003 and from June

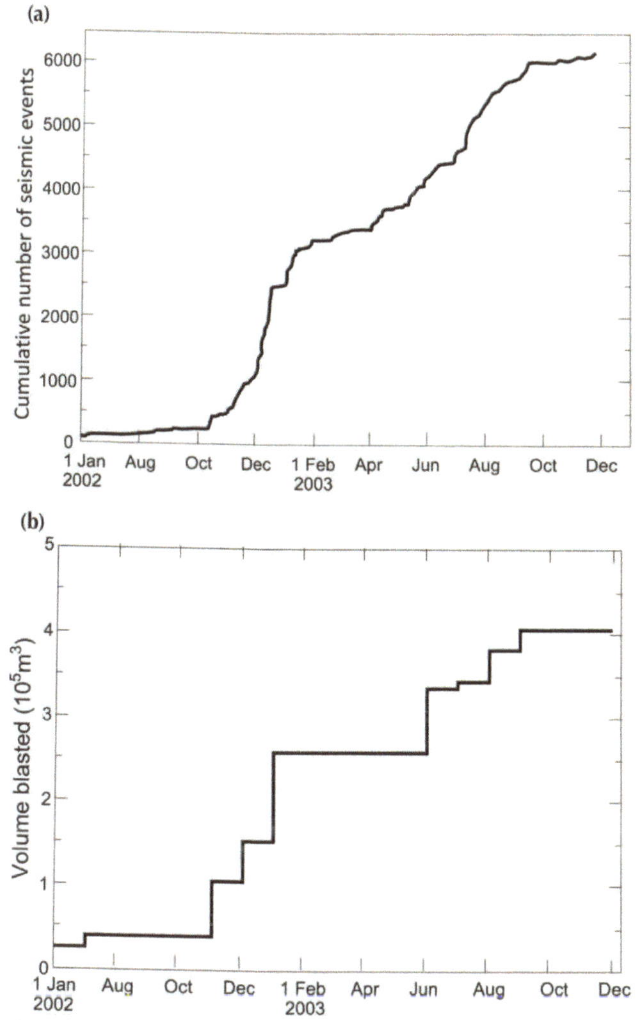

Fig. 4.63: (a) Cumulative number of seismic events recorded at Navachab between June 2002 and December 2003. (b) Cumulative volume of rock blasted and extracted from the pit bottom from June 2002 to December 2003 (source: Lynch *et al.* 2005).

to September 2003 (Fig. 4.63a). The general form of this plot is very similar to the cumulative volume of rock excavated from the bottom of the pit during the same period, shown in Fig. 4.63b. This close correlation between fracturing activity in the slope and mining at the toe of the slope is reasonable. Rainfall, typically between November and March, does not appear to have influenced fracturing to the same degree.

In early December 2002, the anomalous movement of a cluster of prisms used to monitor slope movement in the area of interest, shown in Fig. 4.64, was identified. As such, a correlation between slope movement and microseismic activity was investigated. Since the scalar seismic moment and source location are estimated for each seismic event, a standard expression for the static displacement field may be used (Aki and Richards 2002). The average radiation pattern is computed assuming fracture orientations such that the wall moves outwards, never inwards, with each

Fig. 4.64: The east highwall of the Navachab open pit. Prisms indicating anomalous movement are in a cluster located below the older bedding failure shown (source: E. Hamman).

event. No assumptions are made regarding specific mechanisms of the seismicity. After a scaling constant to account for free-surface effects and slope geometry was calibrated from the survey data, the inferred co-seismic displacement for any point on the highwall surface may be computed from the seismic data. Contours of relative slope movement are shown in Fig. 4.65 for each of the last six

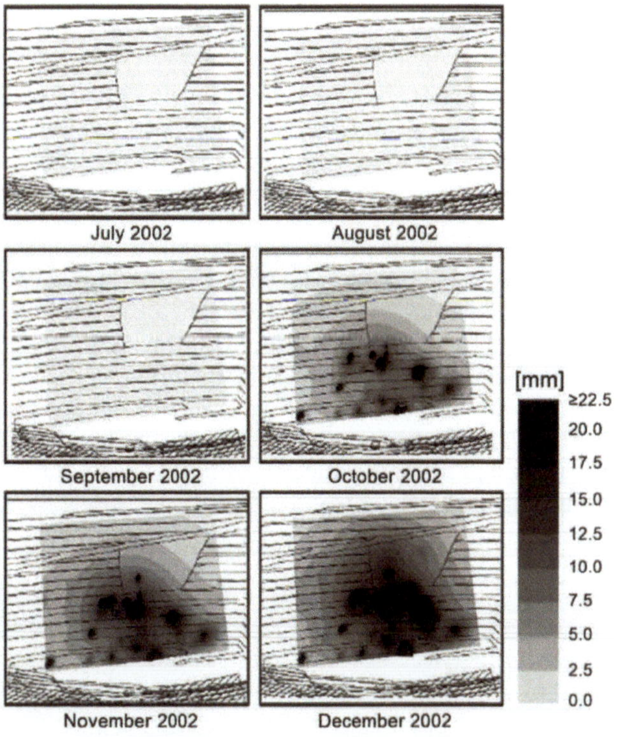

Fig. 4.65: Surface movements inferred from microseismic data recorded during the months of July to December 2002 (source: Lynch *et al.* 2005). See Fig. 4.64 for the location of a previous bedding failure relative to the location of the microseismic activity.

months of 2002. During the period July–September 2002, the maximum inferred surface movements were less than 1 mm, and so no contours were drawn.

The spatial relationship between the location of the seismic activity data and the area where the slope was observed to be moving (through bulging at the toe) was in close agreement. Another interesting result was the advanced indication of slope movement provided by the microseismic data, which in this case was ~45 days. This early warning can be seen clearly by computing inferred surface movements from the seismic data for a single point (coinciding with a geodetic target) and comparing this with the actual recorded surface movements at the target, as plotted in Fig. 4.66. The time lag is interpreted, at least in part, to be the result of the weakening introduced into the slope by the presence of large- and small-scale discontinuities, as well as mining-induced changes in the state of stress.

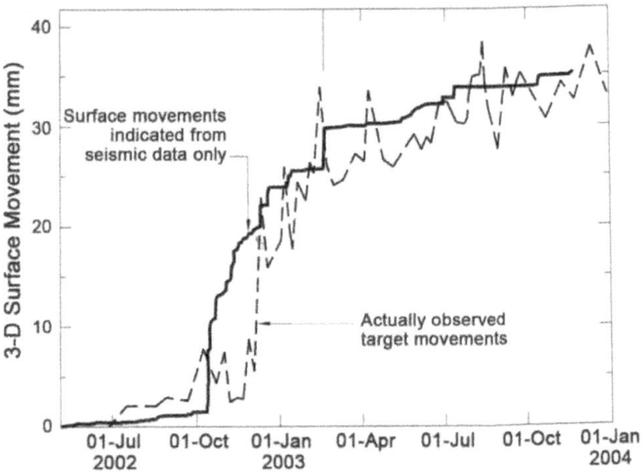

Fig. 4.66: Inferred surface movements at the location of a geodetic target computed from microseismic data (solid line) compared with the actual survey data (dashed line). These plots suggest an earlier warning of 45 days provided from the microseismic data compared to the warning given by geodetic prism measurements (source: Lynch *et al.* 2005).

4.8.4 Case study: Rössing uranium mine

The Rössing uranium mine, one of the largest and longest-running open pit uranium mines in the world, is located near the town of Arandis in the Namib Desert, 70 km from Swakopmund, Namibia. The mine site geology is characterised by highly metamorphosed rock subjected to intense deformation. The lithological units have uniaxial compressive strength (UCS) values ranging from ~90 MPa to 210 MPa and contain geologic structures composed of bedding and brittle deformation structures (faults and joints). Historic slope failures have been characterised as brittle, influenced by fault and shear structures.

Microseismic monitoring was introduced at the Rössing mine to provide early warning of brittle fracturing of the rock mass as a precursor to instability. Experience indicates that seismic data are able to indicate regions of deformation some 30–45 days before these movements are detected by surface monitoring instrumentation. The system's potential to locate major structures was particularly attractive to Rössing, given the complexity of the geological setting at the mine and the challenge in developing a structural understanding using conventional mapping methods.

The microseismic system at Rössing comprises a total of eight stations located as shown in Fig. 4.67. Each station has two or three geophones installed in boreholes at depths of 10–250 m. Combinations of uniaxial and triaxial geophones are installed at all stations. Data gathered from the geophones in the field are wirelessly transmitted to a monitoring computer located in the mine office and then transmitted via internet to speciality consultants for data analysis and interpretation. Daily reports are prepared on the health status of the stations, and monthly reports are compiled on the data analysed during the month, with alerts to the site if a significant event is detected any time during the month.

Data received and studied between 2013 and 2016 showed that a cluster of seismic events was recorded from 2013 to early 2014. The events were found to be localising in proximity to known major geological structures on both slopes (Fig. 4.68). Most of the events were located within a rock bridge between the south slope and a weak lithological unit (Fig. 4.69), helping to provide important insights into the deformation mechanism and kinematic controls.

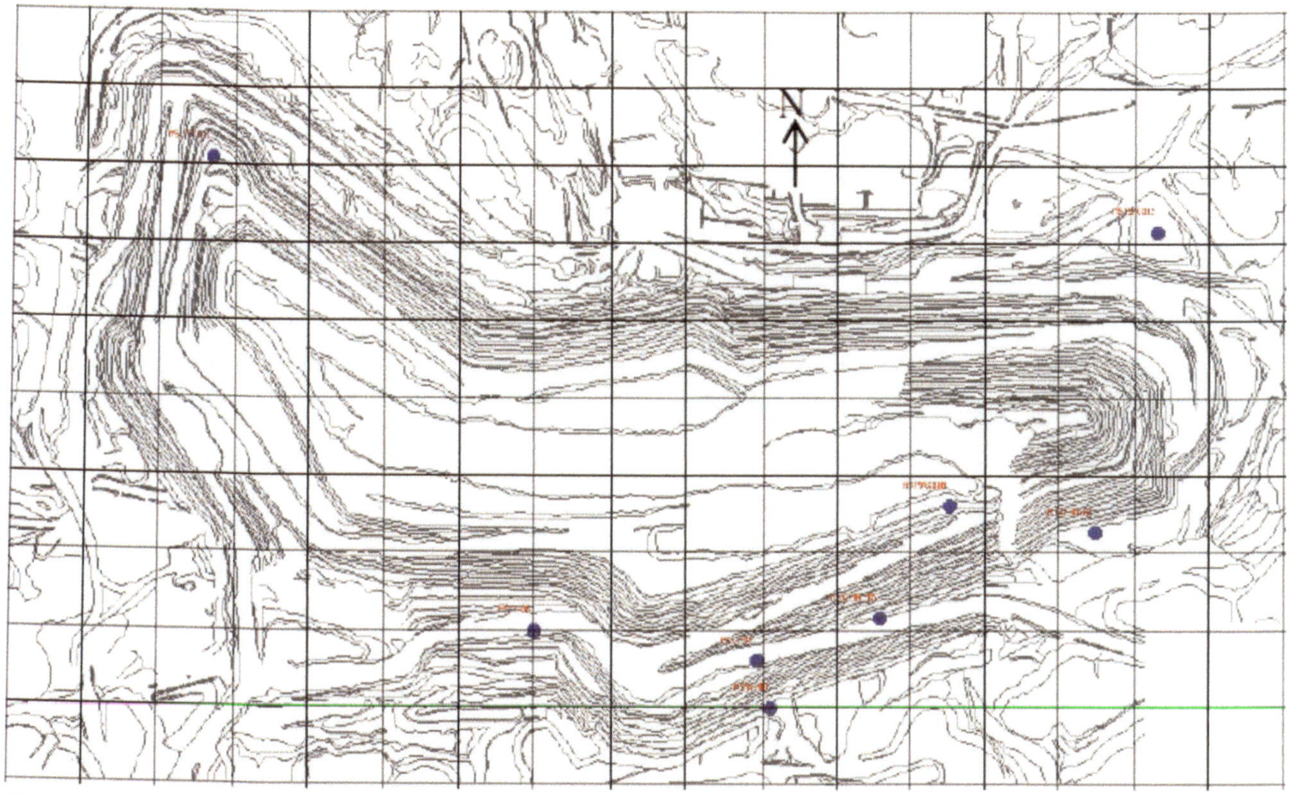

Fig. 4.67: Microseismic monitoring array at the Rössing uranium mine in Namibia (source: A. Creighton).

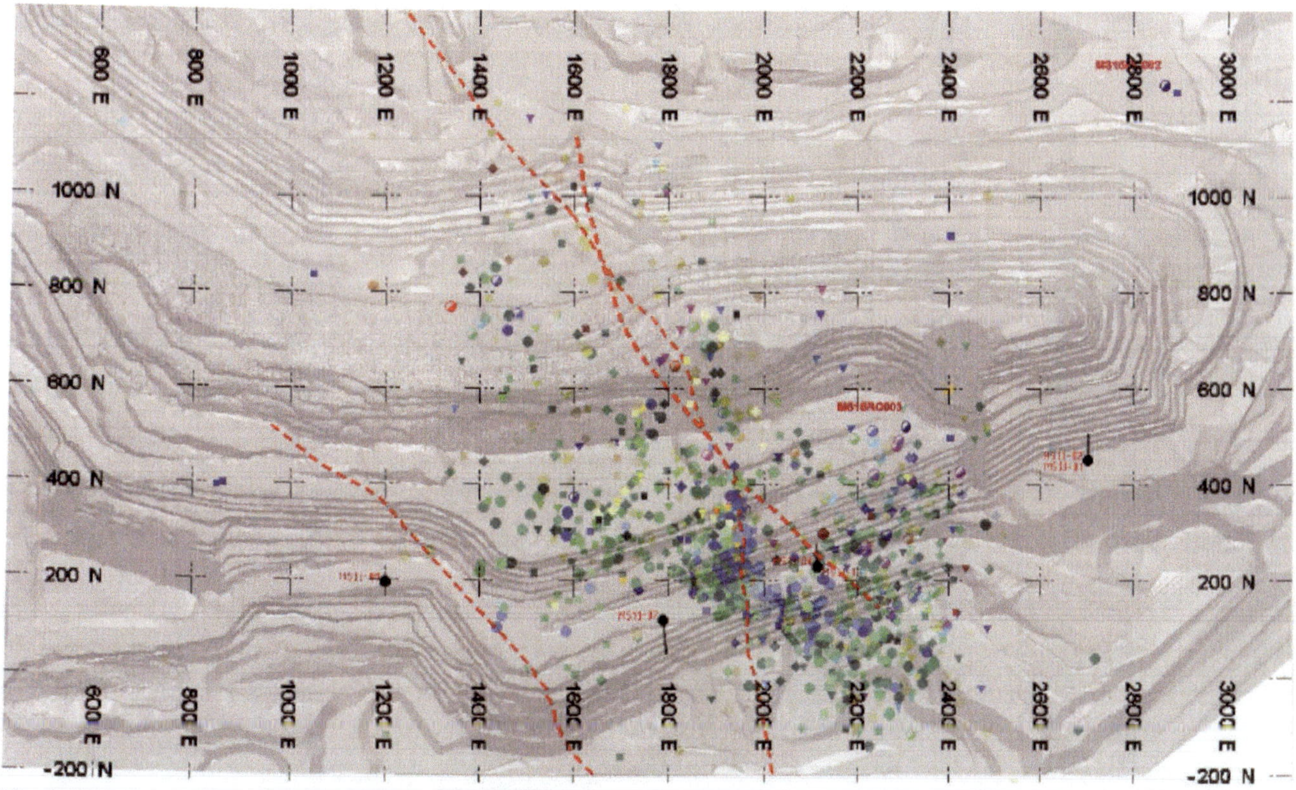

Fig. 4.68: A cluster of microseismic events recorded at the Rössing uranium mine between 2013 and early 2014, relative to a network of mapped faults (red dashed lines) (source: A. Creighton).

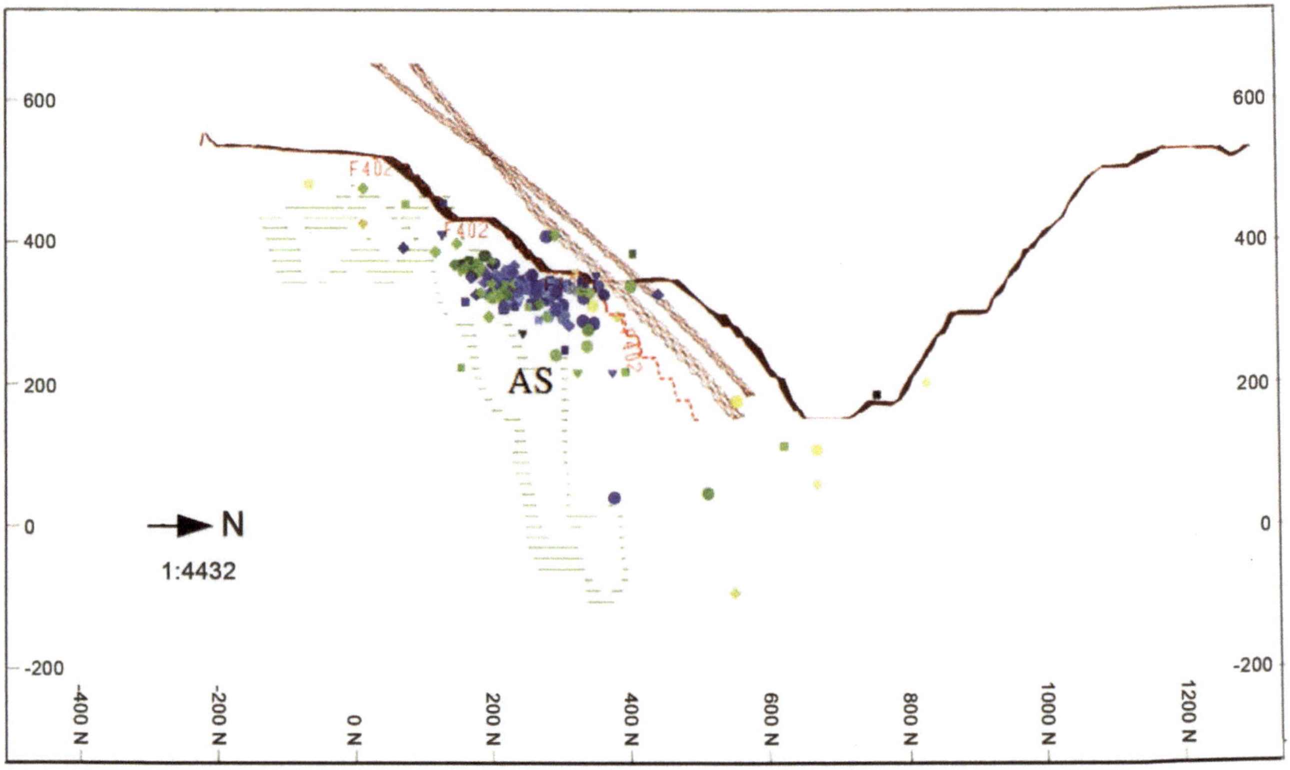

Fig. 4.69: Cluster of seismic events in the rock bridge between an amphibolite schist (AS) unit and the slope (source: A. Creighton).

5

SYSTEM SUPPORT AND OPERATION: PRACTICAL ASPECTS

David Ball, Johan Bergé, Niccolò Coli, Derek Hrubes, Lorenzo Leoni, Graeme Major, Michael Ness, Warren Newcomen, David Noon, Cliff Preston, Peter Saunders, Eric Schwarz, Robert Sharon, Peter Stacey

5.1 Mine operational considerations

5.1.1 Overview

The preceding chapters of the *Slope Monitoring Guidelines* have comprehensively developed the 'why, what and how' for building and using slope monitoring for mine safety, slope and mine design, and geotechnical risk management. The rest of the Guidelines focuses on putting the principles and theories into practice. The success of even the best-thought-out and -implemented monitoring system depends entirely on who uses it, and how and where it is used. The effective use of geotechnical monitoring is critical to modern mining, not only for the obviously necessary safety benefits but also because the ability to monitor and interpret slope stability conditions elevates mining from a chance endeavour to a reliable business venture.

A typical mine slope engineer understands that the value of geotechnical monitoring to the mining enterprise exceeds its cost by several orders of magnitude. However, it is important to recognise that the large efficient open pit mines of today are built by fiscally conservative managers who are particularly skilled at scrutinising any and all extravagance, and maximising value from all money spent. There are challenges with the development of sophisticated monitoring systems and the associated demands for additional skilled personnel required to manage them. In view of the increasing need to improve mine safety and to ensure short- and long-term economic performance as open pit mines increase in size, depth and service life, the resulting challenges should be seen as opportunities to validate the benefits of monitoring and to reinforce the overall geotechnical process at the mine.

The use of geotechnical monitoring for active mining has evolved from the reactive observation of failure to current sophisticated techniques of proactive early detection of developing instabilities, far in advance of what visual observation can do. The trend has been towards increased warning time, precision, confidence and capability; however, this has resulted in increased costs, both in terms of system costs and maintenance and the onsite personnel to use them. A new mine or an existing mine that lacks adequate geotechnical monitoring must develop a philosophy and plan for monitoring that may tend to mimic the evolution of a monitoring program, by starting slowly and gradually increasing commitment to monitoring as it proves its value to the operation. While it is still possible to use primitive methods such as a lath crack-gauge to manually measure and plot growth in tension crack width to visualise the failure acceleration curve and predict the time of slope failure, it would be a serious mistake to delay implementing an appropriate monitoring system while experimenting with insufficient techniques. In today's increasingly risk-averse mine environment, only complete commitment to appropriate geotechnical monitoring will achieve acceptable objectives of mine safety and economic performance. The following discussion on practical aspects of geotechnical monitoring in the active mine environment assumes an adequate initial commitment to achieve appropriate objectives.

5.1.2 Mine operational work environment

The two key responsibilities for the mine slope engineer working in an active mine environment (in order of priority) are:

1. *to protect the mining team and operation* using inspection and monitoring to mitigate geotechnical risks; and
2. *to protect the mine enterprise* by improving excavation performance and reliability, by adapting geotechnical designs to the conditions encountered while achieving mine design and operational objectives.

Both responsibilities are handled with appropriate use of slope monitoring, in data-driven processes, which befit the business of mining. The objective of slope performance monitoring in an operating mine is to achieve continuous awareness of the state of stability, then to interpret that stability data to know whether slope stability is improving ('regressive failure'), worsening ('progressive failure') or steady-state. Slope engineers use their expertise and this continuous awareness to control personnel and equipment exposure to geotechnical hazards and, in combination with an appropriate mine design, safe mining practices and effective communication, to mitigate geotechnical risks to the mine operation. Specifically, displacement monitoring (a proxy for stability state) is used to detect and then to ascertain the location, extent and urgency of stability-related hazards, to control the potential exposure of mine personnel and equipment to the impacts of those hazards.

The fundamental formula for geotechnical risk is:
geotechnical risk = p_h (geotechnical **hazard**) × p_e (**exposure** to that geotechnical hazard)

Geotechnical risk can be safely managed by the reduction (mitigation) of these hazards or by control of exposure to hazards (i.e. consequences), or a combination of both.

'Geotechnical hazards' is meant to be an all-encompassing term describing conditions or events that could negatively affect the safety or profitability of mining, but is specifically related to conditions where geological materials are not stable or self-supporting. Specific risks for personnel and equipment include being struck or buried by rockfall, rockslides, landslides or debris flows; if situated on unstable ground, there is risk of loss of support and collapse, or failure.

In this application, performance monitoring is the last line of defence in an entire system used to mitigate geotechnical hazards. Other key elements of geotechnical risk management include:

- appropriate mine slope design (e.g. minimum acceptable design safety factor);
- preservation of the catch bench minimum width;
- ensuring excavation equipment can reach bench crests;
- correct shovel mining practices for high-risk slopes (e.g. operator cabs kept away from wall); and
- blanket control of personnel access to highwalls (e.g. exclusion zones and field-level risk assessments).

Continuous monitoring and evaluation of displacement trends are used to predict when potential instability hazards become active geotechnical hazards that warrant triggering exposure control actions to protect personnel and equipment. Displacement trend analysis is also used to determine when personnel and equipment can safely re-enter a zone and resume normal operations after being removed from that hazard zone. Safe re-entry depends on whether the tracked instability progresses to failure, or regresses (restabilises) to displacement rates below the trigger level.

Operating mines also typically monitor blasting energy, excavation volume and locations, meteorological conditions, pore pressure and seismic conditions, and evaluate the influence of these factors on stability. When relationships are identified, the slope engineer has entered the realm of predictive slope stability. This work is the science and engineering of mine geomechanics. The same monitoring data that are used to control exposure to geotechnical hazards also serve as the best quantitative information to evaluate the geotechnical performance of the mine excavations, and therefore the mine designs, design parameters and assumptions. Clearly, the information gained contributes greatly to the value of the mining endeavour but production-focused mine operations personnel with boots on the ground may still ask:

- how does the slope engineer expect to contribute positively to the mining process (typical of a mining team that has not experienced serious slope failures and views geotechnical engineering as a hindrance, not necessarily a benefit); or
- how much slope engineering effort is needed to achieve mining objectives (typical of a mining team that understands the importance of geotechnical engineering)?

To clarify the need for active participation of geotechnical engineering in the mine, it is appropriate to consider that the long-established purpose of engineering in mining is to turn an unpredictable endeavour with uncertain value into a safely manageable, profitable and sustainable business, and that slope engineering is a critical component of the overall mine engineering effort. Open pit mining today is faster, deeper, steeper and less flexible than most historical open pit mining experience. What was historically achieved by good fortune, easier conditions and higher risk tolerance must today rely more heavily on appropriate engineering to achieve expectations of reliable design, safety, production, environmental stewardship, fiscal performance and social licence.

There is not, and should not be, much latitude in the application of diligent monitoring for personnel safety in mining; the mine must have the capability of measuring and tracking the state of stability in those areas where instability could affect personnel safety. Beyond that, the question of how much slope engineering effort is needed typically depends on two key factors.

1. *Local mining conditions and the adopted strategy for mining.* Aggressive mining conditions, with steep pit slope angles throughout the mine, in complex rock masses, with high production rates and low flexibility will require a high-level geotechnical team (experienced leader, adequate staff skills and numbers), and 360°, 24 h/day, seven days/week monitoring coverage. Less aggressive mining of high-quality rock masses, with more design flexibility, might be managed with only weekday engineering staffing with moderate experience and 24/7 monitoring coverage of active mining areas and infrastructure managed using Trigger Action Response Plans (TARPs).
2. *Key objectives of the management team and stakeholders.* The objectives of protecting the mining team and the corporation from geotechnical risk are top priorities, but sometimes more is required. If, for example, a mining company's reputation and social licence to operate might be jeopardised by even a well-managed slope failure, then the company might employ conservative, simple, consistent and unmistakable trigger levels to activate response plans to mitigate even low-risk events, and operating decisions could be made according to prescribed movement conditions requiring minimal geotechnical interpretation. The consequences of this approach could include excessive mine area closures, lower productivity and increased mining costs. If, on the other hand, an operation had a solid reputation and stakeholder confidence regarding the management of instabilities, and desired to maximise safe production, then the program would be more complex and less consistent because every instability would be tracked and studied carefully (i.e. 24/7) by engineers and technicians to ensure evacuations are appropriate and timed correctly. Both cases provide similar safety risk mitigation; the latter is more complex and expensive to implement but will result in fewer production interruptions, better decision confidence and higher operational cash flow. In practice, the geotechnical experience and skills required for the latter approach are sometimes difficult to obtain, and significant effort is required to build and maintain committed long-term geotechnical teams.

There are legitimate arguments to be made for every level of geotechnical effort; it is important that choices are appropriate and that consequences are clearly understood by all stakeholders. If a complex, sophisticated geotechnical management approach is chosen for the sake of safety and economics, then it is critical that highly skilled geotechnical personnel and adequate monitoring equipment, software and resources are in place and operating correctly to implement that program. If monitoring equipment fails or highly skilled technical personnel are not on duty, then the affected mine area may have to be closed until those deficiencies are remedied. This clearly requires a substantial commitment from upper management for the chosen program.

The most critical component of the geotechnical management program is the skilled and experienced personnel necessary to make the program perform reliably. Skilled geotechnical personnel can be difficult to identify and recruit, so most mining companies develop them internally. Retention of highly skilled geotechnical personnel may be more successful when the specialist is an integral part of the mining operations team, as opposed to an 'internal consultant' supporting role. Many times, so-called supporting roles are associated with terms like 'non-producing' or 'necessary overhead'. These labels can undermine effectiveness, limit career growth and challenge job satisfaction and commitment – outcomes which challenge the critical objective of ensuring that skill and experience levels are adequate to manage conditions encountered and to achieve the objectives of the enterprise. All mine decision-makers need an accurate and shared understanding of the association between the depth of the geotechnical program, capabilities and personnel required.

5.1.3 Stability in the mine operational environment

5.1.3.1 Defining and interpreting stability

In mining, slope performance success criteria do not necessarily strictly require stability. This consideration differentiates mine slope stability from civil construction and most other slope stability requirements. Instead, open pit slopes, or highwalls, are considered successful when they have performed their function to facilitate:

- the safe exposure and planned extraction of ore;
- timely access and haul road connections to adjacent or subsequent mining operations;
- negligible or manageable rockfall, sliding or other instability hazards during their service life; and
- the completion of each pushback without significantly destabilising subsequent mining.

Excavated slopes are considered successful when ore is exposed and extracted as planned, and instabilities that occur are tolerated provided they are managed safely and do not impact scheduled production or ore recovery. The reasons for this engineering function-based approach are both technical and financial.

1. Rock mechanics design and construction of engineered slopes is unique compared to other engineering disciplines in that rock engineering is the only one that is based on a partially understood geological material, rather than a well-understood engineered material. Therefore, rock engineering designs incorporate

geological, model and parameter uncertainty and thus are probabilistic, and all construction is 'learn as you go' in nature as the understanding of the materials and structural conditions evolves. Even extreme conservatism in design cannot guarantee 100% stability.

2. Mined (excavated) pit slopes are structures that typically have a short service life. During that lifetime, instabilities can be tolerated as long as robust performance monitoring protects personnel and equipment from the associated risks.
3. Many open pit mines operate in a continual state of economic challenge, where additional stripping may eliminate profitability. The ability to increase overburden stripping to improve slope stability (reliability) is limited and often economically unfeasible.

The action of mine development first creates sloped terrain and then continuously undercuts and deepens it, thus reducing relative stability. Open pit mines that develop below the water table behave like very large water wells. Because pit bottom operations require, at a minimum, removal of standing water, some inherent partial slope depressurisation will occur even in the absence of active dewatering measures. In this environment of concurrent destabilisation and partial depressurisation, instability displacement trends can be more difficult to interpret than simple classical or natural slope failures. Even continuous true vector displacement data tend to be a historical record of an accumulation of many smaller accelerating/decelerating displacements, with destabilisation potential driven by blasting, excavation and transient pore pressures. Restabilisation potential is driven by the depressurisation of the dilating rock mass (as the moving slope deforms) and slope self-buttressing from the buildup of displaced material at the slope's toe.

Therefore, in this complex 'destabilising' mine operating environment, instead of directed attention to identify a single unique acceleration curve, tracking becomes a comparison of each new developing acceleration curve to previous non-progressive failure curves to predict whether or not the latest acceleration will be the one leading to failure. Also, complex displacement profiles typically require visual inspections to interpret/confirm the nature of acceleration curve shapes, rather than simple displacement rate triggers in TARPs. Any changes in the size and boundary conditions of an instability must also be updated to reliably interpret individual displacement curves (for different monitoring points) and to identify potential hazard impact areas, should the instability accelerate to failure.

Under these conditions, even simple displacement profiles are not always easy to interpret. Frequently, slope engineers use engineering judgement to decide whether an accelerating event is likely to progress to failure, or regress to reduced velocities due to dilation and self-depressurisation so that normal mining operations can continue. This is not an easy task, and it would be impossible to interpret correctly all of the time. Where normal mining continues under a moving highwall, the slope engineer must monitor the developing displacement trend closely to ensure it remains in steady-state or accelerating at a minor rate. The slope engineer must be slightly conservative and ready to direct operators to evacuate the mining face, even for short periods, to enable evaluation of slope displacement behaviour under non-stimulated slope conditions. If rockfall protection is relevant, it is common to mine susceptible areas during the daytime only to enable visual observations and to provide a reasonable period of slope relaxation.

At a South American copper mine, challenging steep brittle slope conditions led to the creation of the position of a Slope Superintendent, who was specifically responsible for inspecting, field dressing and signing off on every slope section before work could commence below that bench. The unique feature that made this procedure successful was that the superintendent bridged the gap between engineering and mine operations, with authority and responsibility in each area.

Even without that level of effort, there is no substitute for an actively engaged slope engineering team that works directly with equipment operators and supervisors to manage slope development safety. Seasoned shovel operators, who have effectively been studying mine geomechanics for decades by observing the reaction of slopes to excavation, can be important contributors to the process of understanding slope behaviour to support operational safety and to fit mine slope designs to actual conditions. Personnel familiarity and frequent communications engender mutual trust, solid teamwork and safer operations.

5.1.3.2 *Rockfall and small-scale instability*

Most displacement monitoring in open pit mines is focused on the detection of bench- or inter-ramp-scale planar or wedge translation (sliding) that might lead to failure. There are cases of displacement monitoring and radar acceleration curves that provided good usable warning for non-translational rock instabilities, such as toppling or piece-wise collapse of very steep or overhanging slopes, apparently displaying averaged change over each pixel area proportional to the rate of loss of particles (rock mass disintegration producing rockfall) from that pixel area. The occurrence of similarly accelerating adjacent pixels warns of increased rockfall activity and provides delineation of rockfall sources that is useful for estimating potential downslope impact areas. However, as described previously in these Guidelines, ground-based (GB) radars are not generally reliable for predicting individual rockfall events because of the small relative size of rockfall versus pixel size and the higher relative particle velocities, and because the short warning period does not enable effective mitigation.

Experience shows that rockfall hazards can represent the most significant geotechnical risks to the workforce in an open pit mine. Operational activities which reduce rockfall risk, such as bench face scaling, catch bench cleanup and perimeter blast-hole drilling, also create significant exposure of mine personnel and smaller equipment to these hazards. Rockfall risk potential to shovel and haul truck operators can be largely mitigated by machinery design which places the operator cabins high and back from the face, outside the rockfall trajectory envelope. Dozers and other large equipment have various forms of rockfall protection, such as thick steel canopies. Remote-controlled equipment can eliminate operator exposure in cases where rockfall potential is considered to be an unacceptable risk.

Mining a highwall induces stress changes, and it may take time for the soil or rock mass to come to equilibrium. Where the soil or rock mass is weak or highly fractured, the bench face could unravel or fail, particularly during the initial 24 h after the face is excavated. Failure of the bench face could impact one or more benches above the freshly excavated ground, thus inducing a rockfall condition that may impact more than a single bench. A small-scale rockfall event can occur quickly and without warning. Since rockfall and bench-scale instability can occur suddenly, using instrumentation to detect small-scale movement in time for mine operators to avoid the incipient hazard can be challenging. An example is presented in Section 4.2.9. Mining companies are starting to study time-dependent behaviours of rock masses, such as chemical degradation, which may lead to revived rockfall even over short operating life-times.

Mining operations manage rockfall risk with procedures directed at limiting personnel access to the hazard, often referred to as standard operating procedures (SOP). Although equipment used to doze and excavate blasted material provides the operator protection from exposure to rockfall hazards, there are limits to the degree of protection that mine equipment can provide. In areas exposed to rockfall hazards and/or with potential for a bench face failure, the slope monitoring aspects of the SOP may include the utilisation of spotters, lights for excavation at night, barricaded exclusion zones, daylight mining only, increased frequency of inspections that may include drones or specialised surveys, and performance-monitoring instrumentation for safety-critical monitoring. Techniques that utilise photogrammetry, laser scanners and radar for the detection of small-scale movement in near real-time can be effective, but must be implemented with extreme caution due to equipment vulnerabilities and the speed with which rockfall can initiate (see Section 4.2.9). Limitations of such equipment include sensitivity to dust, rain, snowfall, changes in atmospheric conditions and deployment of instrumentation in strategic locations where they can be effective and not intrusive to the mining operation. Utilisation of instrumentation for the detection of rockfall and small-scale instability can present a risk by creating a false sense of security if mine operators trust the instrumentation to warn them in time to clear the area before the rockfall or failure event occurs. Rockfall protection is not designed for large rock blocks, such as room-sized boulders that develop substantial momentum. Where the dislocation of large rockfall particles are observed, or possible, special precautions should be taken to ensure potential impacts to equipment or personnel are mitigated.

The following considerations can improve the SOP for scaling and cleanup at the highwall toe for the purpose of mitigating the potential for rockfall incidents involving personnel and mobile equipment:

- recognition of potentially hazardous ground that requires a specific SOP for mine development;
- increased utilisation of drones for survey of as-built benches and bench faces, and rigorous search and detection of incipient cracks and other obvious and exceptional deformations;
- utilisation of light plants where needed to improve visibility at night;
- utilisation of spotters in areas determined to be potentially hazardous and where spotting can provide usable warning;
- recognition of the spotters' attention span to ensure they are effectively utilised;
- inspections and signoff by a qualified person before scaling and cleanup is approved; and
- experimentation with instrumentation that can potentially be used to detect small-scale incipient deformation and rockfall potential (e.g. photogrammetry, laser scanners and radar).

It is important to recognise the limitations of instrumentation for detecting sudden small-scale, potentially dangerous rockfall and bench-scale slope failure events. It is also important to understand that once adverse hazardous blocks dislocate and start to slide, even in small amounts, the radar monitoring for incipient early movement has done its job of warning, and subsequent movement warning times may be too short for effective evacuation. These brittle-type smaller-scale instability risks may not be reliably or adequately mitigated by the current state-of-the-art monitoring technology. Practical remedies for these cases are not necessarily robust, but have included:

- recognition and barricading of known hazard zones for special mining treatment, such as adequately sized and protected excavators and remote-controlled equipment;
- close examination of precisely located (relative to operating equipment) individual radar pixel changes from scan to scan (instead of relying on trigger limits), coupled with direct immediate instructions to move operating equipment away from impact zones at a moment's notice;
- use of specifically trained spotters to warn of signs and indications of early movement; and
- frequent or continuous real-time inspections by experienced slope engineers who are aware and knowledgeable of local ground conditions.

5.1.4 Mine operations: economics of stability

When mining proceeds as expected and cash flow is in accordance with the feasibility study predictions, the operation is considered to be successful. When unexpected instabilities occur that interrupt cash flow, the operation is said to be experiencing 'geotechnical' difficulties that are temporarily preventing success. If difficulties persist for long periods, questions are directed at the operation's geotechnical team, such as, 'why wasn't this instability foreseen?' or 'why was the mine designed for failure?'

The straightforward answers to those challenging questions can be found in *Guidelines for Open Pit Slope Design* (Read and Stacey 2009). However, the answers are not always satisfying when expectations and reality do not agree. In these cases, the slope engineer's job can be quite challenging. For practical purposes, it is not at all unusual to experience significant slope failures in operating large open pit mines, even when the pits are excavated in accordance with reasonably comprehensive slope designs. Slope instabilities are occasionally caused by deviations from design but more typically are the result of localised rock mass weaknesses, or the exposure of previously

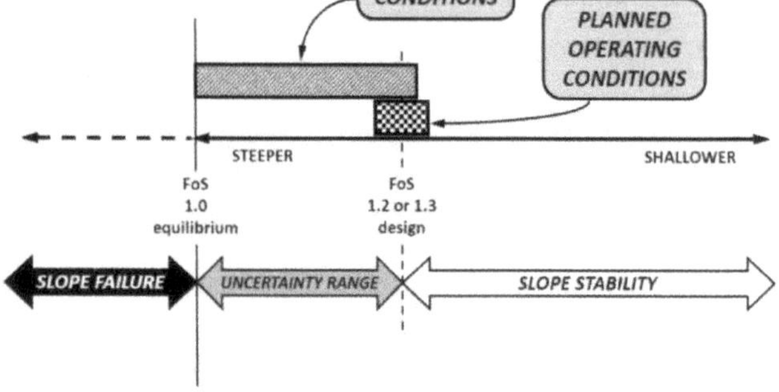

Fig. 5.1: Contrast between planned operating stability conditions and actual operating stability conditions (source: M. Ness).

unknown structural weaknesses, that were not accounted for in the slope design. It is not possible for even comprehensive slope designs to account for all the geological variability that will occur in mining, so modern mining operates in a probabilistic level of conservatism between two boundaries, as shown in Fig. 5.1.

The occurrence of instabilities generated by mining does not invalidate the mine design, but local experience suggests ways to effectively improve the design(s) to increase reliability as the operation matures. Performance monitoring data should be used to optimise mine design details to minimise production interruptions by conforming designs to actual conditions encountered. Typically, the more challenging problem for the slope engineer is to build or sustain the mining culture that understands, and works with, minor instabilities on a regular basis as part of the mining process. In this approach to slope stability management, the slope engineer must examine data for subtle signs indicating the potential for progressive slope failure. An easily overlooked concern is the potential for undercutting and destabilising an otherwise well-managed instability, leading to progressive expansion and upward migration into a larger instability. These conditions may require detailed analysis and engineered mitigation efforts.

Open pit mining experience suggests that designs could be made more reliable by reducing the slope angle, but in practice unanticipated slope failures of significant scale and consequence in mining have occurred on excavated slope angles as shallow as 12°. The wholesale flattening of slope angles quickly encounters the negative economics of an increasing stripping ratio. Slope monitoring programs are designed to manage risk associated with excavation of marginally stable slopes that are designed with a degree of uncertain performance reliability. The slope engineer's stabilising efforts are generally best served by attempting to control highwall damage from blasting, ensuring effective timely depressurisation and ensuring slopes are excavated in accordance with design recommendations.

An interesting and chronic misconception among many mine operators is that rock slopes should not be shallower than the angle of repose of broken rock, or that extensive blasting of wall rock to rearrange structural weaknesses should result in an isotropic rock mass where slopes will perform adequately at the angle of repose. The concept is theoretically possible only with extreme oversimplification of the problem, and in reality has never been satisfactorily demonstrated. It is not unusual for a failed or strongly blasted rock mass to retain weakness planes (potential failure surfaces) in roughly the same orientations as the surfaces instigating the original instability. In the context of slope monitoring, the problem is the suggestion that some rock masses would not need to be monitored since they are resistant or impervious to failure. It is appropriate to evaluate stable slopes with regard to their potential for instability, but, since it is still impossible to identify and characterise all the possible adversely oriented weaknesses in the rock mass, it is impossible to preclude the need for stability monitoring.

Almost every empirical study of enterprise value versus open pit slope angle, for every type of deposit, shows that the steepest obtainable slope angles generate the maximum net present value (NPV) of the mine. Since there will always be significant uncertainty in structural conditions and material strength of pit walls, stability of the steepest possible slope angles should be predicted probabilistically – sometimes it will be wrong. The most economic and effective slope angle process involves pushing slope steepness to the maximum manageable angles that can be developed safely and still allow detection and adequate warning time through displacement monitoring to mitigate stability hazards, with regular (periodic) design adjustments to fit designs to observed slope behaviours.

The slope engineer needs to accommodate mine planning and production objectives to contribute effectively to the operation's decision-making, based on the performance monitoring data. In practice, most operations use sensitivity analyses to evaluate a large number of economic factors that influence profitability, including the effect of slope angles. This information enables the slope engineer to determine the areas or sequences of mining that are most sensitive to slope angle changes, which areas are less sensitive, and where mine economics can tolerate more conservative designs or flexibility in the event that an instability occurs and remedial stabilisation measures are necessary. The slope engineer can use this understanding of ongoing stability, based on monitoring data, to assist the operating team to achieve development objectives in three different cases.

1. *Maintaining pushback value:* in areas where planned slope angle achievement is critical to mine economics, the slope engineer can pay close attention to small changes in slope behaviour that suggest adjustments in blasting or depressurisation are required to minimise the potential for instability, thereby reducing the potential need for slope angle adjustments or step-outs.
2. *Maintaining overall stability:* in significant exposures of weak rock sensitive to undercutting or over-steepening, and where failure might significantly impact revenue generation lower in the pushback, close attention to incipient instability generation could assist the slope engineer to determine where and when a slope design adjustment might be required to preserve a delicate equilibrium in weak rock above the active mining area.
3. *Optimise/field fit the pushback:* in areas less sensitive to the slope angle, the slope engineer might track stability to recommend opportunities for deferring and leaving waste in place (until the next pushback) or for local steepening to obtain future advantage lower in the pushback (i.e. mining to conditions with active feedback).

Guidelines for Open Pit Slope Design (Read and Stacey 2009) provides a probabilistic slope design procedure in which costs to mitigate potential slope failures should not exceed the costs to manage a failure's impact, multiplied by the probability of the failure occurring. Less guidance is available in dealing with the economics of unexpected slope instability in active mining, but the principles are the same. The slope engineer should be sensitive to the hourly or daily cost of lost production in each major mining area. Occasional closure of mining faces to mitigate geotechnical risk may be necessary, but even partial closures may significantly impact revenue generation. If a true cost of instability is required, a reasonable approach is to consider loss in the value of production during a closure period as a lost product sales opportunity (deferred sales value, debited according to the time value of money). The prudent slope engineer considers the practicalities of daily production when planning, then implements risk mitigation not for the purposes of moderating conservative slope design according to the cost of mitigation, but to be an engaged mining team member who clearly has taken all factors into account when making decisions regarding closure of unsafe areas.

Although every closure decision is also a significant business decision, appropriate risk mitigation/safety-related actions are taken without regard to cost. Appropriate and timely recommendations for area closures (and re-opening) are of great value to the mine and an important responsibility of the slope engineer. With each success, the slope engineer engenders more trust and authority, and implementation of recommendations becomes easier. A proactive slope engineer will champion flexible mine plans with alternate available mining faces and be prepared for potential instabilities that could result in significant economic impacts. The slope engineer is a geotechnical expert who interprets both empirical and consultant slope study information, helps mine engineers design flexible plans that take advantage of geotechnical conditions, and conveys risk potential for slope designs that are signed off and executed. When instabilities occur, the mine slope engineer must be the qualified and authoritative voice that contributes to development of a rational and defensible remediation plan.

The job of slope engineering may seem rather onerous, but it should be one of the most rewarding jobs at the operation. However, it is easy for the slope engineer, who is the in-house authority on stability, to sometimes feel isolated from design, operations and management personnel. Two remedies are suggested.

1. An early, open discussion with superiors and colleagues in the organisation to understand and agree upon the roles, responsibilities and capabilities of the slope engineer to know how to do the job and to prevent misunderstandings and misguided expectations. Typically, the slope engineer is expected to understand, explain and provide advice; the mine manager is expected to make the decisions; and the senior or shift superintendent is expected to execute the manager's decisions. This does not always happen. For example, in certain mining scenarios, the slope engineer is expected to make decisions and direct operators; when that is the case, a procedure that addresses the decision process needs to be defined and agreed upon.
2. A realistic view of stability and the geotechnical function in mining practice eliminates false expectations. The slope engineer is an applied scientist with special knowledge, skills and experience in geotechnical practice, capable of improving the quality of safety and business decisions made in the mine by taking advantage of the power of slope performance monitoring. The slope engineer is not a magician, soothsayer or fortune teller. The slope engineer is not at the mine to eliminate instability, but to help the mine identify and manage the associated risk. The slope engineer is a member of the team working to earn the maximum NPV of a mining enterprise by aiming for the optimum level of slope steepness versus instability. The optimal target is somewhere between complete stability and failure that compromises the objectives of the enterprise.

The slope engineer must collect and analyse all geotechnical data from existing instrumentation mainly because the hard-won status of 'diligent' can rapidly degrade to 'negligent' if a safety incident which was declared to have occurred without warning is later revealed to have been identifiable in the data beforehand. Every monitoring instrument used directly in TARPS, or otherwise used to make operating decisions, must be capable of providing warning time sufficient to act upon. It is likely that there will be instrument data that are useful to understand (i.e. investigative purposes) but that are not useful for warning purposes. In these cases, the data must be segregated to prevent inadvertent use, in the event that they may not be used to ensure personnel safety.

As slope engineers gain experience relating acceleration curves to a progressive failure condition, it becomes easier to identify acceleration curves that have a high probability of leading to failure and to ensure TARP effectiveness. However, there will be cases where slope displacement trends may look like impending failure but regress significantly later the same day; at times, confidence is elusive. With certain rock types, there is no adequate substitute for long-term direct experience with local instability events and their monitoring data to aid reliable decision-making. It is best to consider that the consequences of halting operations based on uncertain observations, while unpleasant, are less than the consequences of dismissing an acceleration event that later causes a significant loss. In performing, or defining, the slope engineer's role, it is useful to remember the following characteristics of the slope management challenge in active open pit mining.

1. *Mining is active destabilisation*: open pit mining steepens slopes to a less stable condition. Stability is a moving target and time-dependent. The result is a life-of-mine working environment subject to unplanned instabilities largely but not completely due to inherent uncertainties, mitigated by monitoring and appropriate slope design modifications.
2. *The high cost of exposure control demands correct interpretations, decisions and actions*: in mining, the safety of personnel is the highest operating consideration above all other risks, including economics. However, controlling personnel/equipment hazard exposure can partially or completely stop mining operations and impact economic return. The mine slope engineer needs reliable, timely and high-quality monitoring data to make reliable interpretations of stability conditions, to correctly assess hazards and potential impacts, and to contribute to good mining business decisions. Sometimes this capability requires time spent on detailed evaluation of the data; sometimes it requires more expensive, superior quality or increased quantity of instrumentation.
3. *The evolution of open pit mining*: open pit mining has evolved from designs with flatter slope angles for phase production cuts transitioning to steeper angles only for final walls, to steep slope angles on every cut. Modern open pit designs are steeper and deeper, with faster sinking rates. To address these increasing challenges, slope monitoring programs require similar increases in capacity and quality to achieve risk management objectives. In some cases, slope monitoring efforts are staffed full-time 24/7 – there are dedicated slope engineers and technicians on every mine shift. At times, seasoned mining personnel need to be convinced that conditions have changed and old paradigms are no longer adequate.
4. *Traditional thinking and false risk logic abound in mining*: confidence based on long personal mining experience without significant incidents is not necessarily relevant for changing mining conditions, nor is the longest mining career an adequate sample duration for high-consequence, low-likelihood instability events.

However, human nature tends to think it is. The slope engineer sometimes has the burden of not only providing correct and timely recommendations but also of convincing recalcitrant miners, who believe they know better, to heed recommendations they consider to be conservative.

5. *The slope engineer is an expert advisor and mining team member*: the mining team is impressed with correct, timely information that helps them achieve goals. If the slope engineer is clearly a mining team member, every recommendation will be utilised correctly and the entire mining enterprise will appreciate the importance of good slope management and good mining practices implemented to achieve success.

An important consideration related to the economics of mining involves the purchasing of slope monitoring equipment, which can be contentious. The selection of monitoring equipment and integrated data management systems must not be based on purchase price alone, but on quality, reliability and vendor support. For example, the slope monitoring profession has extensively tested retro-reflecting prisms: for long life and reliable long-distance shots, only high-quality, copper-metallic coated, sealed, mounted prisms should be used. Sealed prisms prevent interior moisture that dramatically shortens prism life. In quality installations, labour to install prisms can exceed the purchase price of the materials, so cheaper prisms (e.g. those made of three glued mirrors) are especially disappointing when they become unreadable and need replacing in a short period. In the case of vibrating wire piezometers (VWP), consider that they are being placed in drill holes that cost $400 000 to $1 million each – the cost of instrumentation is almost insignificant. The variation in quality and cost between VWPs themselves may not be significant, but they should be ordered correctly for application, including the ability to resist the pressure of the heavier-than-water grout column during installation. The most critical factor is to avoid unreliable downhole splices; it is highly recommended to order VWPs with cables of sufficient length that they can accommodate any length drill hole. Any excess cable can be cut off after installation.

While there may be some truth in the philosophy that 'we make money by saving money', assuming that correct instruments are being purchased with either a safety justification or economic analysis, there are few, if any, real cost-saving options. It is generally better for a slope engineer to be practical, successful and committed to best practice, rather than to seek opportunities to reduce costs by cutting corners. Slope engineers are unlikely to receive accolades for saving $100 000 on a monitoring program, but those who solve problems and generate millions of dollars in cash flow improvements do stand out. Choose wisely: slope engineering and associated risk management are not extra costs to be minimised, but they are opportunities to turn a good business into a safer and more prosperous one.

5.1.5 Instrumentation layout around an open pit

5.1.5.1 Historical perspective

Instrumenting an open pit mine is both simple and complicated. The concept of putting fixed instrumentation on and in the ground to detect and track movement is simple. The relationship between the distribution of instruments (locations) and likelihood of success in detecting instabilities depends on several factors and is complicated. The key objective should be to use instrumentation (type, quantity and layout) to satisfy both the safety and business risk mitigation needs of the operation. For a typical large open pit operation, these needs can be summarised into four monitoring strategies, ranked from bare minimum to maximum effort.

1. *Existing instability risk management*: the slope engineer already knows the ground is moving; instrumentation is focused on characterising the state of instability and a TARP is created to manage the instability. After the TARP has been triggered and the slope has failed, continued monitoring provides data to understand the restabilisation required for re-entry and resumption of mining. Historically, this was done with extensometers, crack mapping and prism surveys or global navigation satellite systems (GNSS) for obtaining total vector information. Today, GB radar is more commonly used for tracking movement, but total vector monitoring still must come from prisms or GNSS. This is sometimes called tactical monitoring of known problems.
2. *Detection of new instabilities*: the slope engineer wishes to be the first to detect new instabilities before operations reports rockfall and sliding. Continuous scanning over entire highwall regions using radar or laser scanners to detect and track incipient instability can be performed confidently without knowledge of instability history. Some slope engineers tend to focus scanning on areas of known or suspected weakness, but most prefer large-area surveillance to detect any and all incipient instabilities, because there may be significant unknowns not yet visible or identified. The priority is to scan any areas where instability could impact operations. Before radar was in common use, these areas were instrumented with prisms installed in a regular pattern, with concentration along highwall crests and the crests of haulage ramps. This approach provided valuable information upon detection of larger instabilities, but may have missed small instabilities that occurred between adjacent prisms. With sufficient

radar or LiDAR coverage, it is possible to survey movement on all pit slopes. This is sometimes called global strategic monitoring.

3. *Enhanced surveillance of stability state*: this is the same process as standard detection, using a deformation monitoring method that includes visual inspection, laser or radar scanning, and other tools used to yield the most comprehensive surveillance of the stability state. In this case, satellite radar (InSAR) is also used to observe areas adjacent to open pit crests and selected open pit slopes appropriate to coverage by InSAR. Unlike standard detection of new instabilities, which looks for movement above a certain trigger level, enhanced detection seeks to identify ground with any anomalous or differential movement to delineate boundaries between displaced weakened ground and ground that has not deformed. The benefit to enhanced stability surveillance is the maximum warning time and level of understanding of stability that can practically eliminate potential for surprise instabilities. Radar data are subject to additional under-sampling (long sample periods over some extended time) to obtain the most detail for very slow (early) movement areas, as needed.
4. *Predictive instability risk management*: the displacement monitoring activities are used as summarised above, but with the added objective of understanding what is driving each instability. This methodology is used to continually assess the performance of the geotechnical designs/design parameters and to modify them and subsequent designs when needed to better fit the actual conditions. Collective data assessment includes instability drivers such as production (excavation) by location centroid, precipitation, surface water, groundwater, seismicity, blasting/vibration energy and any other relevant information that can develop true cause and effect relationships, which can then be used to predict future stability conditions. This level of management provides direct feedback to modify future slope designs as needed.

This list of strategies is ranked from reactive minimum basic effort to proactive maximum comprehensive effort, which is also proportional to cost, personnel requirements, scope of coverage and level of effort. Most large open pit mines will use some part of each strategy, but the comprehensive list may be misleading in terms of benchmarking a specific slope engineering effort. The expectations, philosophies and capabilities for slope stability monitoring vary between mining companies. However, the chosen monitoring strategy should be appropriate for local conditions. It may be useful to review a simplified history of pit slope monitoring for context when evaluating a specific system and procedures.

Methods of monitoring, as well as motivation, have changed considerably over the years, and with those changes have come equally significant changes in expectations from slope stability monitoring programs, as follows.

- Initial mine slope stability monitoring was as simple as routine visual inspections, mapping cracks and displaced ground, with manual measurements across cracks tracked over time to derive displacement curves. Crude displacement vectors were analysed across crack mating surfaces. Definition of moving ground involved geology and cracks, and it became evident that slopes provided some warning time before failure. However, predicting when a failure would occur with any certainty was not an expectation.
- By the 1970s, manual survey point markers were installed on moving ground so that surveyors could obtain vector displacement data over time using steel chains or stadia rods. Another method of measuring point locations in a mine was to occupy the monument with a theodolite and perform a resection, where angles to multiple distant backsight monuments were taken and the only possible position was crudely derived by multi-angulation. In the late 1970s, the survey world was introduced to the WILD Tachymat, a theodolite with integrated electronic distance measurement (EDM) otherwise known as a 'total station'. Soon, retro-reflecting prisms were installed on monuments to be manually read by this instrument. Surveyors could shoot prisms (without the need for a rodman) every day if there were not too many, but they were hard to find at night. The idea developed to pre-emptively place prisms on non-moving ground in the hope that, if instabilities developed, the surveyor might obtain data without having to access the moving ground. With quantification of movement, time-displacement curves could be plotted to study slope instability behaviour. With better data, the slope engineer could try to estimate whether a failure might occur and, if so, estimate when. With engineering judgement and significant experience, some estimates were very good.
- By the 1980s, manually read wireline extensometers were used to provide real-time displacement data for active instabilities to plot an acceleration curve and, finally, to predict time to failure. These were subsequently improved into telemetered data units so that displacement plots could be generated automatically in the mine office. One well-placed extensometer could be used per instability. Even though only one dimension of velocity was measured (so really it was speed, not velocity), the resulting time versus displacement plot represented the acceleration well enough to predict slope collapse to within a day.

As such, an expectation developed that if instabilities were identified early enough, the ongoing displacement curve could show when failure was imminent. Two more instrumentation advances were developed in the mid- to late 1980s that further revolutionised slope stability monitoring. The Global Positioning System (GPS or GNSS) was introduced in the 1980s, but its use for slope stability monitoring did not occur until the mid-1990s. During this time, automated or robotic total stations (RTS) were also introduced, allowing regular unmanned collection of prism measurement data day and night.

- By the 1990s, RTS were adopted in mining, allowing continuous remote measurement of prisms. This increased the prism network of up to ~100 prisms that could be managed manually to 500–1000 prisms that could be monitored automatically. Prisms are one of the few sources for good-quality 3D vector data, but time averaging is needed to smooth out considerable noise from angle measurements. For short-term information, many slope engineers plotted the EDM distance component only (much less noise, higher precision) to graph a proxy for cumulative displacement and to assess acceleration. Fixed GNSS stations were the other method for obtaining high-quality 3D displacement data. Long-term averaging produced sufficient precision to measure very small movements, but, due to noise in the short term, GNSS was not well-suited for rapid warning of small movements. Crack mapping still provided instability boundaries, within which multiple prism or GNSS sensors tracked movement. Instability-specific wireline extensometers continued to provide the most real-time data for warning, but automated prism monitoring provided 24/7 alarming capability so that dispatchers could be alerted to take action based on alarms or TARPs. The evaluation of acceleration curve shapes was extended in attempts to reliably predict time of collapse using multiple data sources (e.g. the inverse velocity method; Rose and Hungr 2007).
- By the 2000s, GB radars revolutionised pit slope monitoring. For the first time, detailed remote coverage of entire slopes was possible and, more importantly, data could be collected without installing instruments on the slope. Quick and easy delineation of moving versus non-moving (stable) ground facilitated better data interpretation and correlation with causes. Radars provided relative measures of ground profile changes for a limited but useful band of movement speed, which were interpreted or modelled as slope displacements. Monitoring expectations sometimes exceeded capabilities. For many operations adopting higher standards of safety and performance reliability, mine slope stability monitoring changed from prudent to essential. The challenging concept of 100% remote monitoring coverage emerged. In the early years, all radar systems were hampered with power supply problems that limited machine availabilities. The concept of 'no monitoring, no mining' began to evolve as logic caught up with practice, placing extreme and sometimes unattainable demands on radar systems.
- By the 2010s, service providers were routinely utilising satellite InSAR to acquire data, processing them and reporting results for the mining industry. With scan cycle times of initially 24 days reducing to four days or less at the time of writing, InSAR could be utilised to detect very small movements with high precision on horizontal surfaces over wide areas of overburden, waste dumps, leach stockpiles, pit crest areas, tailings dams, haul roads and other infrastructure, that was not previously possible. Data from satellite InSAR were found to be very useful in combination with displacements measured around pit crests via other ground-based instruments.

5.1.5.2 Considerations for monitoring instrument layout

The starting point for consideration of slope stability monitoring instrumentation layout in open pit mining is dictated by purpose and by the specific monitoring goals at a mine site. Instrumentation layout consists of five principal elements:

- direct sensor installations (e.g. prisms, GPS, extensometers, tiltmeters, VWPs);
- indirect sensor station locations with associated ground coverage areas (e.g. GB radars and laser scanners with their locations delineated on a map to determine overlap and coverage);
- mobile indirect sensor coverage areas (e.g. InSAR, aerial photogrammetry);
- associated terrain information (e.g. stable instrument platform locations, road access) plus appropriate current and future topographic surfaces (e.g. current surface, mine plans); and
- monitoring support infrastructure (e.g. power lines, substations, transformers, fibre optic lines, repeaters).

Layouts for initial configurations are overlaid on current topography, and updated layouts are adapted to annual mine plans for the five-year business plan. It is useful to examine layouts yearly for the life-of-mine to better understand long-term instrumentation and associated infrastructure needs. These layouts are used for planning and to generate take-offs for budgeting. They are normally part of a mine site's Ground Control Management Plan. Although each mine's monitoring requirements will be unique, typical examples of open pit slope monitoring instrumentation layout are summarised below for several monitoring scenarios.

5.1.5.3 Instrumentation layout for existing instabilities

For tracking active or potentially active movement, the key objective is to be continuously aware of the stability state of the target region: whether it is accelerating (decreasing stability), decelerating (increasing stability) or steady-state. Providing it can be done safely (see discussion below when this is not possible), monitoring instruments are placed within the region of instability, typically favouring ground that would most directly impact active operations. It is common practice to make unstable ground off-limits to all personnel, with the possible exception of geotechnical staff who are tracking movement and surface conditions, understand the risks and are qualified and approved for the task. Even then, access to displacing ground with its deceptive appearance of stability should never be taken lightly. Monitoring moving ground typically involve the following approaches and suggested features.

1. *Wireline extensometers* with data telemetry (manual markers on wire allow visual tracking in the field):
   i) simple, reliable, robust, near real-time displacement data, graphed to detect and measure acceleration trends;
   ii) raw data are robust, correction for thermal expansion of the wireline is possible in theory, but difficult and not usually necessary;
   iii) typically one monitoring unit per instability, installed on relatively stable ground near the crest of the instability with the anchor pin in the movement zone;
   iv) tripod-mounted pulley at the edge of stable ground allows wire alignment as close as possible to actual movement direction. More distant placement of the instrument beyond the crest of the instability may be necessary for safety, but requires greater wireline length and associated data noise; and
   v) if needed, and with caution, they can be used in reverse – the instrument is located at the toe of instability with the anchor pin fixed uphill in moving ground converging on the instrument.
2. *Prisms monitored with RTS* with data telemetry (previously installed or installed with caution after movement detected):
   i) total vector movement (time-averaged smoothing of atmospheric, diurnal and angle noise);
   ii) track and plot EDM distance for relative line-of-sight (LOS) precision in near real-time for acceleration trends;
   iii) short-term measurement may not represent trends, due to variations; and
   iv) preferable to have several instability-centred prisms representing maximum rate and fringe prisms to clarify movement boundaries.
3. *Fixed GNSS monuments* with data telemetry (previously installed, or installed with caution after movement detected):
   i) good-quality total vector movement data with insignificant noise, requiring ±4 h processing with averaging (not for real-time monitoring);
   ii) reluctance to put expensive units in jeopardy – sometimes better used uphill or adjacent to moving ground to warn of instability growth approaching roads or facilities;
   iii) when multi-path (interference) error is prevented, GNSS provides very reliable, accurate, long-term data and a single well-placed unit can be representative of large moving areas; and
   iv) in some locations with steep slopes, GNSS units installed deep in the pit may not be within sight of a sufficient number of satellites to provide good data – it is advised to test satellite visibility before installing.
4. *GB radar monitoring* focused on the targeted moving zone as a sub-region of the highwall with sub-region-specific alarm settings (LiDAR may also be appropriate for geo-referenced tracking of specific changes):
   i) alarms set specifically for displacing ground to track displacement rate and search for extent and growth of the movement zone;
   ii) should achieve acceptable coverage of the entire instability, good signal geometry and no significant interference from shovel or other masking; and
   iii) displacement trend graphing when taken as a significant component of total vector movement.

When mining operations have procedures that preclude any personnel from accessing displacing ground, including predicted potential runout areas, installation of instrumentation in the moving ground is not allowed and all monitoring will be remote. Operations without such policies address hazard potential through a risk assessment process, where access to moving ground is considered only with appropriate precautions and monitoring. Access onto accelerating ground should never occur. Access onto meta-stable ground should be avoided directly after significant rainfall, snowmelt or earthquake activity.

5.1.5.4 Instrumentation layout to detect new instabilities

It is not possible to guarantee stability in slope design, due to variability and uncertainties inherent in geologic materials. Therefore, the operation's risk management program must include the capability to detect and track new unexpected instabilities in real-time as a component of the mine development plan. The slope monitoring system specifically designed and implemented for identifying new instability areas should always include regular visual inspection to assist the interpretation of monitoring data from instrumentation and as a check to ensure that completely unanticipated conditions are not missed. The two basic approaches regarding coverage area are either complete coverage of all mine and relevant stockpile slopes with instrumentation overlap for completeness and cross-validation, or selective coverage of all slopes above and below all active mining areas, access roads and other mine locations that personnel might access. Areas not covered by instability detection monitoring should either be demonstrably stable areas subject to periodic inspection, or physically closed to access by personnel and equipment.

Experience at a mine site for several pushback cycles might suggest certain geotechnical or monitoring domains and/or wall orientations that are more prone to failure than other areas. In this case, partial monitoring supported by a risk-based approach to system design that is reinforced by relevant development experience may be sufficient for the more stable areas. The following strategies are appropriate for the detection of new or incipient instabilities.

1. *Prisms monitored by RTS* with data telemetry based on systematic installation of prisms as new ground is exposed:
   i) installed for total movement tracking as required for distinct trend verification, including EDM tracking for short-term smoothing of angle noise and atmospheric disturbance;
   ii) some prisms concentrated near known instabilities, but other prisms placed on regular spacing along benches (50–75 m horizontal spacing and staggered placements on every other or third bench is common industry practice);
   iii) GNSS can be used as a substitute for or augmentation of prism monitoring; and
   iv) selected areas that are not easily monitored by RTS can be monitored by manual survey of prisms.
2. *GB radars*, either real aperture radar (RAR) or synthetic-aperture radar (SAR); LiDAR may be a potential substitute for GB radar in some applications:
   i) broad area surveillance monitoring with focus on certain active mining areas considered to be at risk;
   ii) large modern open pit mines use three to five radars for complete coverage, located on the crest or in the upper third of the open pit (depending on slope geometry and incidence angle, might have difficulty seeing certain slopes and/or top to bottom of the highwalls); and
   iii) most radar software packages have options for integrating multiple radars into one comprehensive view, with overlap management, coordinated alarming and movement vector enhancement (ability to improve on the LOS limitation).
3. *InSAR coverage of the mine area* augments prism and GB radar coverage of open pit slopes by adding incipient movement detection from pit crests outward of the open pit (also applicable for waste rock dumps, stockpiles, mill sites, heap leach facilities, tailings facilities and other important infrastructure over the entire mine property):
   i) initially added to provide site-wide stability monitoring over relatively flat surfaces to best advantage, and to monitor distant excavated slope and natural landslide regions that are less suitable for GB radar where incidence angles are marginal for signal return;
   ii) limited usefulness for pit slope monitoring due to incidence angle on relatively steep terrain, although some corroboration of GB instrument data is possible where dual-look satellite data from both ascending and descending orbits yield better coverage on steeper ground; and
   iii) service providers can contribute to reporting of ground movement trends and changes by providing periodic interpretive reports following satellite data acquisition (with or without the requirement for mine site support).

Other monitoring systems that warrant consideration include utilisation of LiDAR, or photogrammetry that could replace GB radar in some applications. Other devices worth consideration include borehole strain gauges, VWPs and additional devices that are addressed in Chapters 3 and 4.

5.1.6 Considerations for monitoring station locations

Monitoring stations have historically been structures designed and used to mount and house RTSs. However, they are now commonly used for a variety of purposes, including housing for SAR units, monitoring communications equipment and other electronics, lookouts and field work stations.

Stations should be strategically placed at stable locations that have the longest life (to minimise relocation, although most stations are built to be readily moved) and optimal coverage of the highwalls. For convenience and wherever possible, they should be located near power lines and access roads, and where they can transmit and receive strong signals. They should be located far enough away from haul roads and active mining to limit ground vibration and to avoid damage from fly rock.

Since open pit mine designs change when reserves, slope parameters and/or metal prices change, all monitoring installations, and pit facilities in general, are temporary. A monitoring station can vary from a sophisticated 'permanent' structure designed specifically for geotechnical monitoring, to a structure repurposed from mine dispatch or a trailer designed for rapid relocation. It is uncommon to encounter monitoring radar or RTS installations that have remained in one place for more than five years.

Most modern radar units are mounted on trailers with stabilising jacks to make them temporary fixed monitoring stations. Long-term pit crest SAR units are often mounted on large, stacked concrete barrier blocks for a rigid installation. Solid bases insulated and isolated from the structure are required to maximise RTS data quality and repeatability. Historically, RTSs have been fixed to a solid block of reinforced concrete ~1 m thick and sized to match the station footprint. Others are skid-mounted to improve mobility. The RTS monument can be a 150–200 mm diameter heavy steel pipe embedded in the concrete near the front of the block or a concrete column poured in a cylindrical tube fitted with a flat metal platen and tribrach bolt embedded in the top. A more basic alternative consists of stacked concrete barrier blocks placed by a loader. Where foundation concrete is used, the structures become a semi-permanent base for the instrument.

5.2 Equipment protection and reliability

5.2.1 Overview

Slope monitoring systems must be designed to work reliably and on a continuous basis in extreme environments. Open pit mines are located in some of the driest places on Earth, such as the Atacama region in Chile, to the wettest regions like Papua New Guinea, Colombia and other tropical locations. Some operations are subject to extreme cold, such as in the Canadian arctic, or hot seasonal temperatures and large night-to-day temperature changes, especially in high-altitude environments. Extreme wind environments produce unique challenges in some regions, such as near continental divides. Monitoring equipment must be rugged enough to survive and function for months under harsh working conditions such as in regions with significant winter snow accumulations, since there may be no access for long periods. Slope monitoring results are affected by snowpack, falling snow and rain, fog, dust and other impacts of changing atmospheric conditions that require data corrections. The primary benefit of a reliable monitoring system is not simply to minimise maintenance, but being able to continue mining because the stability state is being tracked and understood, versus having to stop production while the slope engineer lacks supporting data.

Most instrumentation utilised for surface deformation monitoring is designed to operate while exposed to the expected range of weather and climatic conditions, temperature, moisture and dust at a particular mine site, while others require shelter and protection from the elements. For some instruments, enclosures are designed and utilised to ensure maximum reliable performance, to extend service life and to minimise maintenance. Enclosures utilised effectively by the mining industry vary from simple home-built wooden shacks to roomy and robust heating, ventilation and air conditioning (HVAC) equipped sea containers. Instrument stations, or enclosures like several described below, have traditionally been used to protect RTSs and more recently for laser scanners, SAR units and other purposes.

5.2.2 Enclosures

The instrument shelter that houses the total station is an important aspect of equipment protection to consider for prism monitoring. The purpose of the shelter is to protect the RTS instrument from strong wind, dust, rainfall, extreme temperature changes, blast fly rock, tampering and theft. This prolongs and improves the performance and service life of the instrument and, to some extent, the quality and value of the data collected. The instrument shelter can be expanded to a slope monitoring field station and observation post with remote office capabilities. It can facilitate field meetings with mine operations personnel in inclement weather. The enclosure shown in Fig. 5.2 is a deluxe instrumentation shelter that includes a large air-conditioned work area, radio antenna, security fence and viewing area safety fence. Below the floor is a poured concrete cylinder that supports the station monument steel pipe that protrudes directly up into the front area of the

Fig. 5.2: A deluxe instrument shelter and monitoring station at a large open pit mine in South America, featuring a 180° horizontal view and elevation that provides a crest-to-toe view of the facing highwall. The structure is 4 m wide, 3 m deep and set 2 m above the ground surface (source: M. Ness).

Fig. 5.3: An example of a conventional instrument shelter at one of the Debswana operations. The metal cage is designed to protect the shelter from blast fly rock (source: Debswana Mining Co.).

Fig. 5.4: A small portable shelter for an RTS (source: M. Ness).

shelter. The elevated instrument monument and shelter are designed to improve visibility to the pit floor, allowing the station to set a comfortable distance from the highwall crest. At the floor level, a pair of lifting eyes on each side makes the structure relocatable for the next pushback. Not shown in the photograph are the power lines, transformers and lightning protection grounding for power, plus a separate lightning protection grounding grid for the field station and data cables.

When designing the instrument shelter, construction material that will not affect the accuracy of the measurements should be used. Figure 5.3 shows a conventional and efficient instrument shelter. Standard glass allows the total station to target any beacon or prism in its LOS with minimal hindrance from the shelter and impact from refraction. The use of special E-glass or other heat-reducing glass should be avoided since these significantly reduce the transmission of infrared (IR) laser distance measurement devices. Typically, acrylic, polycarbonate and other clear plastic windows also block IR, so these should also be avoided (there is considerable variability within even the same types of plastic; testing could reveal specific types that can transmit IR sufficiently). Glass with a thickness of 3 mm or less has minimal impact on the accuracy of measurements, and the errors can be adjusted using a tested formula. Tint and shape also matter; clear flat glass has the least impact (Afeni and Cawood 2010). If needed, the shelter can be protected from fly rock using metal cages, as shown in Fig. 5.3.

The small portable RTS shelter shown in Fig. 5.4 is utilised at a copper mine in Arizona, USA, to provide multiple protected monitoring locations that are faster to procure, build and install at minimal cost. The shelter uses thin regular glass sloped at 15° from vertical for maximum performance. It has filtered air inlets below the base and vents at the top where warmed air can exit passively, maintaining an interior temperature close to ambient, even in the heat of a southern Arizona summer. According to the operation, the units cost about US$4000 to build compared with US$20 000–60 000 for a full-size shelter. These small portable units are easy to move and thus improve instrument availability. The tribrach used for this station is attached to a steel platen with a centre mounting bolt embedded in the reinforced concrete column. A flexible conduit attached to the column carries the power and signal cables to an outdoor-rated electrical cabinet, which typically will contain instrumentation to operate the RTS and collect (buffer) the data, and radio equipment to transmit data back to a receiver and server in the office. Some units also have solar power units, solar controller and gel batteries to supply power independent of mine power or generators.

An important function of the instrument enclosure is to control differential expansion and tilting of the steel pipe monuments onto which the RTS are mounted, by shading them from direct sunlight. Without shading, one side of the pipe heats, causing differential expansion and a tilt of the top platen. The effect on distance measurement is minimal, but angle measurements can be significantly affected. One method of shading is to wrap the steel monument in reflecting insulation. As an alternative, some operations use reinforced cast concrete pillars. Small portable shelters like the one shown in Fig. 5.4 can be mounted on large concrete barrier blocks which, due to their mass and lower thermal coefficient, virtually eliminate concerns from differential heating.

5.2.2.1 Climate control and lightning protection

Instrumentation providers that support the mining industry continue to improve technology to extend the extreme limits of temperature and humidity under which their equipment will function reliably. Indoor and protected installation is typically the most efficient solution against heavy snow, storms, lightning and strong wind forces, especially for long-term applications since even rugged systems will deteriorate with exposure for extended periods.

RAR-type GB radar (those with a wide parabolic antenna) cannot easily be installed in an enclosure, for practical and physical reasons. RAR is therefore designed for reliable performance under extreme climatic conditions. A limitation for RAR is that the presence of strong wind acting directly on the parabolic dish can introduce vibrations that render radar measurements noisy and sometimes unusable. Improvements to the system include a more compact and durable design, and reduction of the dish surface area by introducing holes through the parabolic antenna. SAR-type GB radar is characterised by a smaller surface area and fewer moving components, improving radar monitoring suitability for high-wind environments (up to 150 km h^{-1} for mobile units and higher for semi-permanent installations).

Another issue for high-precision monitoring systems working in a harsh environment is the presence of dust. Very fine quartz dust can penetrate the most rugged and sealed systems and cause serious damage to mechanical systems with moving parts. In addition to effective dust protection, a regular cleaning and maintenance program is essential to protect monitoring equipment against dust-related damage.

Lightning has destroyed monitoring equipment at many locations, so it is important to install lightning protection correctly where lightning represents a risk. Design details for lightning protection vary from site to site, so proper design and installation should be done by the mine electrical team. Figure 5.5 shows an example of lightning protection of a small instrument station at a North American mine. The devices at the top of the poles are multi-point dissipators, called spline ball ionisers, which not only shield the shack from lightning strikes as a last resort but also act to discourage strikes by dissipating ionised clouds around conductive structures. A proper grounding array is attached to the poles and the shield wire, but isolated from all other grounding. In addition, the power and signal cables entering the station have lightning charge isolators to mitigate the occurrence of power surges.

Fig. 5.5: An example of a small RTS shelter showing lightning protection structures. A key element of lightning protection is the grounding grid buried as a large wire array located away from the shelter. A short section of low berm for improving the view angle is visible in front of the shack (the barrier fencing to prevent access to the low berm section was temporarily removed for picture clarity) (source: M. Ness).

5.2.2.2 Power supply

Large storm events frequently interrupt electrical power to the mine through lightning-induced surges and wind-induced line-slap. This can result in the monitoring system being down when the slope engineer has the greatest need for real-time data. In addition to lightning-surge problems, mine power is notoriously dirty power, with the capability to trip circuit breakers and damage instrumentation. In the aftermath of electrical storms, pit reticulation personnel prioritise repairing damage and restoring power to production equipment. Restoring circuits and refuelling generators for remote monitoring equipment is typically a low priority.

The first demands for significant electrical power in slope monitoring arrived with portable field radar units, each equipped with batteries charged automatically by small to medium generators. The practicalities of refuelling, servicing and otherwise operating these onboard generators proved challenging to geotechnical staff who often lacked the necessary expertise and time. The high equipment availability and reliability needed for full-time monitoring was impossible to obtain without significant effort on the part of geotechnical staff, vendors, mine maintenance personnel and others. To address increasing demand for reliable power, most operations positioned radars at locations where line power (with step-down transformers) was available, and radar batteries were charged with line power. While an improvement, this did not completely resolve challenges with reduced reliability, particularly since the power lines themselves sometimes served as conduits for damaging surges.

Many operations are evolving to use isolated solar power systems to directly charge radar batteries as well as supply power for instrument stations and remote monitoring sensors. Even in rainforest environments, mine sites can operate on solar power if enough panels are used. In one interesting example, unattended remote solar-powered GNSS units periodically dropped offline then resurrected themselves repeatedly for more than 10 years without a battery change or other maintenance. When used in conjunction with proper lightning protection, reliable monitoring installations are possible.

Several vendors provide power trailers for securing reliable solar- and wind-generated power to charge onboard battery banks and to sustain monitoring equipment power. This type of reliable, almost uninterruptible power is essential when operations must have active monitoring to allow access and production.

At a high-altitude mine in Colorado, USA, subject to deep winter snow that restricted access, the operator purchased small compressed natural gas thermoelectric generators (no moving parts) for RTS and SAR stations that could operate all winter on a single large tank of liquefied petroleum gas.

5.2.2.3 *Data transmission/telemetry*

The open pit mining environment presents several challenges for data transmission and telemetry, but also has the advantage of well-planned and -developed communications infrastructure that services mine offices, crushers, conveyors, dewatering pumps and controls, and electrical service for equipment. Wireless telemetry, particularly Wi-Fi, is common in open pit mining due to the prevalence of dispatch and production monitoring equipment. Power and even fibre optic networks are also commonly available, particularly at the pit boundaries.

Typically, the open space and clear LOS opportunities of open pit mines lend themselves well to wireless telemetry. Large open pit mines up to 4000 m across greatly exceed the few hundred metre capabilities of current Wi-Fi technology, but this can be solved by deploying multiple access points (APs), either attached to permanent infrastructure or in the form of equipment trailers that can be located around the pit to provide coverage in problematic areas. Coverage can be improved by deploying APs on equipment such as haul trucks as mesh access points (MAPs). Mesh technologies allow for extension of the network by routing and handing off subscriber connections as the equipment moves around the pit.

The design for wireless communications coverage becomes complex in large mines due to the large numbers of APs and MAPs. Like any complex system, it is only as reliable as the weakest link, so the possibility of the network being partially or completely down should be considered in the design of the data collection systems. This means either designing and deploying a separate higher-reliability data telemetry network using wireless technologies specific for data telemetry, building redundancy into the existing network so it is more robust, or ensuring the data acquisition gear has the ability to store data during network outages. Once the network is restored, data can then be collected from onboard memory. For simple instrumentation, such as VWPs or time-domain reflectometer (TDR) cables, specialised dataloggers with onboard memory are suitable. For more sophisticated equipment such as RTS or radars, industrial PCs may be required at the monitoring station. It may be desirable for the slope team to have one robust field laptop with all operating software on it, so that if data are needed but the network is down, the slope engineer can use the laptop to take over the monitoring unit temporarily from the field (mine) location.

Reliable power supply must be provided. All automated instrumentation systems should have backup power available, regardless of line power availability that can support multiple-day outages. In solar recharged battery systems, a minimum of seven days operating time on the battery alone is highly recommended. In cold-climate environments, even longer run-times should be designed, potentially for an entire winter season in latitudes where solar energy may be poor for several months at a time (see Section 5.2.5.2).

Power system monitoring should also be included in the design. Loss of power is the most common failure of wireless infrastructure. Electrical transients, usually from lightning, are the next likely cause of failure. Proper grounding and lightning protection are essential. Exposed antennas should have lightning arrestors installed in-line, with suitable grounding, either by a ground rod or ground plane (see Section 5.2.2.1). To prevent accidental destruction of monitoring components, the location of devices should be part of mine planning and discussed regularly during mine operations meetings. When mine plans call for removing or termination of equipment, plans should be in place to replace the function. Despite the lack of moving parts, solid-state circuits erode over time, especially power circuitry where energies are much higher.

5.2.3 Preventive maintenance requirements

Maintenance requirements for slope monitoring systems vary depending on the environmental conditions. However, the goal is always the same: to have extremely reliable monitoring systems. Slope failures can occur in the middle of the night, during a strong storm event or when the mine power is out (e.g. due to a lightning strike), and thus instrument durability, reliability and the maintenance program should be designed to handle such conditions. For a slope engineer, there is nothing so reassuring as arriving at a slope failure under those conditions and finding that the generator on the radar is still working and the radar is continuing to scan the slopes, as the RTS in its enclosure continues surveying on backup power.

Maintenance for monitoring systems involves the following types of components (with the most logical responsible party listed in parentheses):

- GB radars, RTS, laser scanners (vendor maintenance contracts);

- generator or power pack units for equipment (site technicians, vendor maintenance contracts);
- line power, transformers, light plants (mine electrical maintenance);
- radio data communications systems (mine IT group, system integration contractor);
- prisms, extensometers, manual monitoring (mine slope engineers and technicians);
- vehicles (mine light vehicle shop);
- computers, servers, office software (mine IT group); and
- geoscience, scanner, mine planning software (software licence agreements through mine IT group).

Computers, servers and software needed to operate the monitoring system and manage data require special consideration for licensed software, virus software and operating system patches and upgrades. Special maintenance plans for these, agreed with the mine IT group, will prevent system crashes and reboot problems for machines that are meant to run monitoring 24/7. Many monitoring headaches can be prevented with the purchase of high-end monitoring computer/server hardware and software exactly as specified by the system vendor, but this sometimes requires considerable support by the slope engineer. Systems that run seamlessly are rare, and represent excellent value for cost. A reliable accounting system that pays software licence fees and other system vendor maintenance charges on time is essential to ensure high monitoring availability.

Generator power for all radars was problematic in the early years, and consequently most mines have become proficient in providing alternative power (generally using line power or portable light plants supplied by the mine electrical team). Diesel generators require scheduled maintenance to ensure power on demand, but the backup power capability of the radar should be maintained, even with other primary power sources in place. Maintenance of generator engines is similar to that of other engines: they need care and regular maintenance per the manufacturer's recommendations.

Maintenance for the measuring components of a slope monitoring radar typically involves cleaning and lubrication of moving parts (rotating mechanisms for RAR, linear mechanisms for SAR), plus careful inspection of cables, bumper pads, exposed grommets and gaskets, and close inspection for signs of leakage or damage. The use of enclosures for SAR improves its life and minimises maintenance.

Vendors of high-cost equipment generally offer high-cost maintenance agreements, especially when those agreements are expected to support equipment availabilities of >98%. In the case of 'no mining without monitoring', high radar availabilities are essential, and mines will pay for high-cost maintenance agreements for reliable service. There have been cases of vendor technicians having extended stays of up to six months to ensure high radar availabilities at remote mines. The only reasonable alternative to acquiring more reliable systems is to purchase spare radars so they can be swapped out quickly to minimise mining downtime. Service agreements are structured to reflect the level of service and availability required, typically naming available plans Gold, Silver and Bronze. If mines have knowledgeable electro-mechanical slope engineers and technicians, it may be possible to save on service agreements. Experience shows that most mines eventually pay for high-level maintenance so the slope engineers and technicians can complete their assigned duties efficiently.

5.2.4 Equipment calibration and service

Onsite (if possible) or offsite instrument calibration (and maintenance) per the manufacturer's recommendations is essential to achieve data accuracy and quality, and to ensure that equipment is working as designed. Radars have complicated and specific function tests and calibration requirements beyond the skills of most geotechnical teams, so periodic vendor visits to the mine should be scheduled to perform these functions. Another calibration driver is the documentation requirement of health and safety certification programs, such as OHSAS 18001 (now changed to ISO 45001). Even if requirements of these programs are less prescriptive, standards will insist on good calibration record-keeping.

Many of the slope monitoring methods have evolved from standard surveying techniques, including the use of traditional survey error management techniques (turning direct and inverted angles, error distribution between starting and closing backsight shots in a group, etc.). These are used to manage survey error in the robotic survey process and to check whether calibration is needed outside of normal schedules. Techniques for maximising survey reliability and accuracy vary, and experimentation can assist in the development of procedures that work best for each mine environment. Calibration check procedures from user manuals should be implemented any time there is a concern about data quality, with a record of the results forming part of the instrument usage log. Using the monitoring software to check for control point repeatability from scan to scan, or day to day, is the most straightforward, effective and easy way to assess the need to recalibrate. Vendors should provide procedures for machine self-checking and should track the condition of equipment.

When instruments must be removed from site for an extended period for calibration, the operation requires spares or agreements for loaner equipment to fill the gap. Maintenance agreements and budgets must consider the additional instrumentation necessary for continuous reliable performance monitoring.

Maintenance and calibration documentation are important components of data validation and quality assurance, so these documents must be tracked, kept current and stored permanently in at least two separate locations. Documentation supporting the installation of VWPs that are permanently installed in grouted boreholes is particularly important. Common issues of data confidence with VWPs include incomplete calibration records and getting cables switched or confused. Every borehole instrumentation installation should be fully documented with serial numbers that are checked and rechecked during the installation process.

5.2.5 Considerations for cold weather operations

For open pit mining operations located in cold climates, the challenges in implementing and maintaining an effective slope monitoring system relate to extreme seasonal variation in surface conditions due to snow and ice, and use of technologies that were developed primarily for milder climates. This requires innovative solutions to keep equipment functional in accordance with operation requirements and the site's geotechnical risk management program.

5.2.5.1 Cold weather operations

The following factors affect slope monitoring system performance in cold climates:

- cold temperatures can prevent the instruments and the power supply from working;
- fog, snow and ice can either disrupt measurement or cause false alarms; and
- limited hours of daylight and low luminosity affect the way in which cold weather operations must monitor the slopes.

These are discussed in the following sections.

Temperature

The main factor affecting monitoring systems in cold weather operations is low temperature. In cold climates, the average temperature may be below 0°C most of the year. During the coldest months, the average temperature can decrease to below -20°C, as indicated in Fig. 5.6.

Most battery-operated instruments are designed to work between -25°C and +45°C. This limitation is mainly linked to battery performance, which is designed for an ideal working temperature of approximately +20°C. When a battery is used at a lower temperature, less current is released until a point is reached when the battery cannot deliver enough current to keep up with the demand.

Low temperatures can also potentially impact the mechanical parts of the monitoring equipment. Grease is more viscous in cold climates, so moving parts are not lubricated effectively. This can lead to premature wear and ultimately to component failure. In addition, some technology functions cannot operate in a cold climate. For example, at below -20°C a field computer screen will not turn on and a fibre optic cable is too brittle for practical utilisation.

Fog and white-out conditions

In cold climates, it is common for a mining operation to experience fog caused by cold air passing over warmer water or moist land. The site may also experience periodic intense snowfall conditions, as indicated in the bar chart in Fig. 5.6. In blizzard conditions, snow already on the ground can become windblown, creating white-out conditions. In fog and white-out conditions, the reduced visibility can affect the monitoring system function and performance.

A mining site comprising both open pit and underground operations may experience further challenges that impact its monitoring system performance. As shown in Fig. 5.7a, when positive warm airflow from

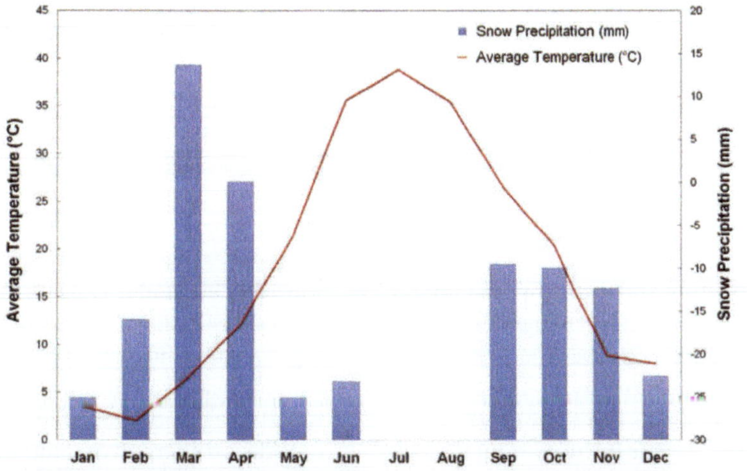

Fig. 5.6: Average temperature and snow precipitation at Diavik Diamond Mine, Northwest Territories, Canada in 2018 (source: Diavik Diamond Mine).

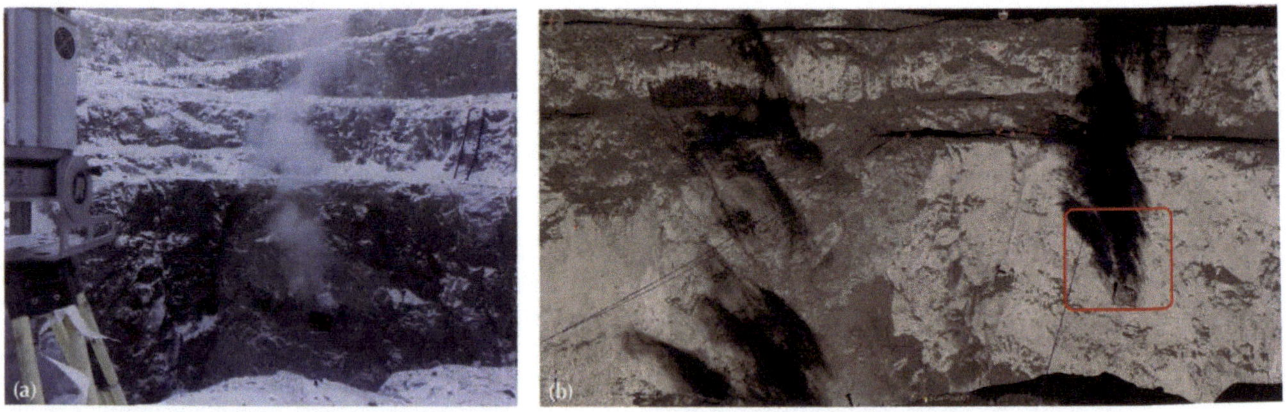

Fig. 5.7: (a) Warm air exiting a sublevel retreat opening at the underground/open pit interface and condensing to create fog. (b) Laser scan trial through the fog plume, showing an area with no collected data in black (source: Diavik Diamond Mine).

underground workings interacts with cold temperatures outside, the moisture condenses, creating a plume of fog that can impact monitoring results.

Technologies that use lasers (such as the EDM of an RTS and terrestrial laser scanners (TLS)) are only capable of monitoring with an unobstructed LOS. When the area of interest is not visible because of fog or white-out conditions, monitoring outcomes are affected. In the case of TLS monitoring, changing the range (Pulse Repetition Rate) of the laser has little effect on the beam's ability to penetrate fog. However, increasing the angular resolution slows the scanner enough that more reliable data may be collected due to the increased likelihood of the laser beam finding a gap in the fog cloud. However, this results in a much higher scan time, which can be problematic for critical tactical monitoring. Moreover, the density of the fog and its proximity to the desired target can have a significant impact on the scanner's penetration ability, as shown in Fig. 5.7b.

Ice and snow accumulation and depletion

Ice and snow accumulation can significantly impact the effectiveness of a monitoring system. It can mask areas of monitoring interest and obscure targets, such as prisms. Management of ice and snow accumulation requires good knowledge of the areas likely to be covered by snow. Adequate mapping of seepage areas where ice will build up during winter is important when planning the location of monitoring sensors, as well as the placement and height of prism installations.

Snow and ice accumulation and depletion also impact slope radar monitoring. Radars detect changes in slope geometry, which in cold climates can be related to accumulation or depletion of ice and snow. This is interpreted by the radar computer as a slope deformation towards (accumulation) or away (depletion) from the radar sensor, which can trigger alarms.

When installing downhole instruments in permafrost, water in boreholes can freeze, preventing the installation of equipment including VWPs, thermistors, inclinometers and TDR cables. Ground ice formation can kink and snap borehole instrumentation cables.

Low luminosity

Cold weather operations located at extreme latitudes with limited daylight or luminosity during winter, in combination with poor weather conditions, can make effective inspection and visual monitoring problematic. The reduced daylight hours also limit the time window to execute the task. Consequently, field inspection schedules may need to be changed, and the limited time may translate to less detailed inspections. Low luminosity also affects field cameras, and light stands should be installed where required.

Lack of light onsite can also lead to an outage of monitoring systems which rely on solar power. Solar panels need to be periodically cleared of snow to minimise these impacts.

5.2.5.2 Cold weather protection

Operations in cold climate regions require monitoring instrumentation to be protected from cold, snow and ice. This section provides examples of the protection required for specific slope monitoring instruments.

Slope radar (RAR and SAR systems)

Various service providers have been resourceful in finding ways for their radar systems to function in cold conditions, through enclosures and winterisation. The examples in Fig. 5.8 show different types of enclosures that provide a suitable heated environment to ensure 24/7 critical tactical monitoring using slope radar units. If the radar unit is outside, an alternative solution is to winterise the critical parts by adding insulation and heaters. The battery and

Fig. 5.8: Examples of protection for slope radar units. (Top) A SAR system scanning in a heated building through a special foam membrane. (Bottom) A RAR system scanning through a special plexiglass window (source: Diavik Diamond Mine).

generator can be kept in an adjacent heated building, as shown in Fig. 5.9.

Robotic Total Station

RTS systems must be enclosed in a building that provides heat and protection from dust, rain, ice and snow. Electrical heaters and fans are also required to prevent humidity from fogging the windows. Any apertures in the building, such as holes drilled for power and communications cables, and GPS units, must be caulked to avoid leaks. The building itself must have a functional geometry that prevents dust, rain, snow and ice from adhering to the glass, as illustrated in Fig. 5.10. The windows must be made of special glass, with a solid pane of consistent thickness and optical properties.

The pillar on which the RTS is installed is drilled through the concrete pad that supports the building with some

Fig. 5.9: RAR unit with winterised (insulated and heated) critical parts. In the background, the battery and generator are kept inside a heated container (source: Diavik Diamond Mine).

clearance, so that movement of the building due to ice heave in the fill below does not affect the absolute location of the total station. The pillar must be anchored into bedrock below any fill and below the permafrost active layer.

5.2.5.3 Prism design

Prisms used for slope monitoring should be customised to minimise buildup of snow and ice in front of the glass, which would prevent measurement. The overhanging edge in front of the glass of the prisms should be reduced (i.e. trimmed by a machinist), as shown in Fig. 5.11.

Prisms must be located where there is no seepage, as ice will build up during the winter and can prevent

Fig. 5.10: An example of an RTS functional building. The geometry of the building prevents snow and ice accumulation on the window, while a fan and heater keep the inside warm and dry to avoid condensation on the glass (source: Diavik Diamond Mine).

Fig. 5.11: The alteration of the overhanging edge surrounding the glass on the prism significantly improves monitoring throughout winter (i.e. there is less snow and ice accumulation). (Top) Before cut. (Bottom) After cut (source: Diavik Diamond Mine).

measurement. Ice buildup can also move and/or damage prisms, making monitoring impossible after only one winter.

5.2.5.4 Borehole instruments and dataloggers

Permafrost must be taken into consideration for cabled borehole instruments. During the grouting phase, the installer must not grout to the top of the collar. Usually, the casing is not grouted. After grouting, the ungrouted portion of the borehole is filled with glycol, and the collar is plugged. This prevents the cables from snapping due to casing movement caused by freeze–thaw action in the active layer. A steam truck may be required to re-open boreholes and wells for instrument installation and subsequent measurements.

Dataloggers for any borehole instrument must be installed in a heated cabinet. If a solar panel is used as a source of energy to charge the battery, an alternative source (such as line power or diesel generator) must be in

Fig. 5.12: Example of a solar-powered heated cabinet for a TDR datalogger. In winter, mine power is connected to the battery to provide sufficient power for operation of the datalogger and heater (source: Diavik Diamond Mine).

place, as low luminosity and limited daylight hours will not provide enough power during winter. An example is shown in Fig. 5.12.

5.2.5.5 Satellite InSAR

Satellite InSAR deformation monitoring has proven to be a valuable addition to terrestrial monitoring methods. However, under cold conditions, there are some limitations. LOS monitoring with InSAR is not possible when an area of interest is covered by snow. An effective solution to this limitation is achieved with the installation of a network of corner reflectors, which can act as prisms. However, even with good results from the corner reflectors, only single points are being observed, and the spatial advantage of InSAR over conventional survey techniques such as RTS and GNSS, is temporarily lost during winter. The reflectors have the same limitation as prisms and need to be clear of snow.

In regions with ice-rich permafrost, seasonal displacement effects can be observed in InSAR monitoring results: settlement occurs in summer due to a loss of volume in the active layer, and uplift occurs in winter due to volume gain. In certain cases, the InSAR service provider can assist to isolate seasonal displacement from long-term geotechnical displacement.

5.3 Program staffing and organisation

5.3.1 Overview

The most important component of any slope monitoring program is the personnel who install, utilise and manage the system. No amount of automated machinery can replace the human part of the performance monitoring equation. The human brain is the smartest and most rational computer, and the human eye is the best measuring instrument known. However, slope monitoring and management is a demanding full-time job that must overcome human limitations, including the potential for distraction, boredom with routine tasks, a limited attention span, and complacency. The qualities and capabilities of any monitoring system increase dramatically with the experience of the personnel at the mine running the system. Having high-quality, experienced, capable slope engineers and technicians is critical to success.

5.3.2 Organisation of the monitoring program

Some level of slope management can be provided by experienced consultants who are not on the mine site full-time, particularly during initial development. However, once the threshold of employed onsite mine engineering staff and technicians is crossed, building, training, retaining, replacing and supporting a full-time geotechnical team must be a systematic and continuous process.

The basic criteria for sizing and organising the slope management team include:

- skill and experience levels sufficient to manage the types of slope instabilities and slope crises possible at the mine;
- the team's ability to complete their slope monitoring-related work every day; and
- sufficient numbers of personnel and vehicles to visit all mine areas daily, or as required.

Considering the discussion on open pit monitoring history in Section 5.1.5.1, management and stakeholder expectations play an important role in the makeup and specific responsibilities of the slope team. Every mine slope management team has a senior person with requisite experience as leader and mentor. At some mines, this would be at a chief engineer level, but classifications vary. Full-time staffing requirements are determined by the size, complexity and risk management objectives of the operation. Most teams are expected to hire and train interns for the summer or, if they live nearby, interns who could work part-time during the school year. Some of these interns will be hired after they graduate to work full-time for the mine, either in the same discipline or related ones.

Qualified and knowledgeable technicians perform an important role in slope engineering. Especially in a business environment experiencing rapidly evolving communications, system automation and related electronics technology. Technicians do the basic and fundamentally important work that the monitoring program is built on. They also tend to be permanent employees who carry institutional knowledge of the site and are less mobile than professional or management staff. Technicians may not have university degrees, but they have the training, certification and, most importantly, the interest, curiosity, experience and practical skills necessary to maintain monitoring systems for reliable performance.

5.3.3 Skill sets and competencies

A monitoring skills matrix is intended to be a dynamic platform used by effective organisations to assist in developing desired skills for site personnel directly or indirectly involved in performance monitoring, by tracking specific tasks an individual is certified to perform and identifying where additional training may be required. The example monitoring system skill set matrix in Table 5.1 matches personnel categories to the required skill sets determined by the organisation for simplicity, but it can easily be expanded to include individuals.

The monitoring system skill set matrix promotes workforce flexibility and redundancy of skills availability by focusing on multi-skill tasking of onsite staff. In instances where site personnel are not available or capable of performing certain skills, consultants are often relied upon to provide qualified and experienced support. Suggestions for skill set requirements are summarised in Table 5.2.

5.3.4 Training for engineers and technicians

Training for slope engineering should be considered a career-long pursuit. In general, the more slope failures a slope engineer or technician has experienced, the better they understand how rock masses behave upon excavation, and the better they will be able to respond in the event of an emergency. Continuing education should include visiting other operations to examine a variety of slope issues and how they are dealt with, and periodic attendance and participation at conferences, short courses and university-level professional development courses where feasible.

Another critical aspect of slope engineering education includes periodic, specific training and certification in sophisticated slope monitoring instrumentation such as radars or similar equipment. Vendors are eager to supply training courses and provide certification. Because it is critically important to link the physical behaviour of slopes and rock masses with signals displayed on the radar output, sharing the experience of key slope instabilities within and even outside the operation are important responsibilities of the slope management teams.

5.3.5 Outsourcing: pros and cons

Full-time, dedicated, experienced, onsite and capable geotechnical teams are the ideal option to manage risks by providing a continuous presence at the mine site with

Table 5.1: Example of a monitoring system skill set matrix that defines requirements for the management of the slope performance monitoring system by members of the mine engineering and system support team

	PERSONNEL CATEGORIES						
SKILL SET	Technician	Surveyor	Dispatcher	Engineer	Manager	IT	Mine ops crew
Skill 1: System selection and layout	Yellow	Yellow	Grey	Green	Yellow	Grey	Grey
Skill 2: System principles	Yellow	Yellow	Grey	Green	Grey	Grey	Grey
Skill 3: System components	Green	Grey	Grey	Yellow	Grey	Yellow	Grey
Skill 4: System deployment and/or installation	Green	Green	Grey	Yellow	Grey	Red	Grey
Skill 5: System setup and configuration	Red	Grey	Grey	Yellow	Grey	Grey	Grey
Skill 6: System operation	Green	Grey	Green	Green	Grey	Grey	Grey
Skill 7: System diagnostics and troubleshooting	Green	Grey	Grey	Yellow	Grey	Grey	Grey
Skill 8: Data handling and management	Yellow	Grey	Grey	Yellow	Grey	Green	Grey
Skill 9: Data analysis and utilisation	Grey	Grey	Grey	Green	Grey	Grey	Grey
Skill 10: Alarm thresholds	Grey	Grey	Grey	Green	Grey	Grey	Grey
Skill 11: Communication and response protocols	Green	Green	Green	Green	Green	Green	Green
Skill 12: Documentation and reporting	Grey	Grey	Grey	Green	Green	Grey	Grey

Green → Full competency in skill is required.
Yellow → Familiar with skill, but require help and guidance.
Grey → No experience required for skill, personnel do not perform this task.
Red → A training deficiency that should be resolved.

site-specific knowledge and access capabilities. Training, certification and experience at operating and maintaining all of the complex performance monitoring instrumentation is essential, in addition to geotechnical expertise. For temporary or permanent cases of insufficient geotechnical staffing, or insufficient time for staff to become proficient with complex monitoring systems, an appropriate strategy may involve delegating activities that are critical but less purely geotechnical in nature, such as support from IT staff, or contracted monitoring integration services.

Remote monitoring, including remote operation of GB radars, is an external expert service that provides review and interpretation of monitoring data to ensure consistent data collection, management and use – from setup to response and notification of triggered alarms. Connection to the radar units from a remote monitoring location can be as 'real-time' as similar connections from the radar to the mine office. Given the focused expertise of the service provider, remote monitoring can provide the earliest possible warning based on identification of incipient instability hazards and careful tracking of existing hazards – with the potential to minimise surprises and the associated negative impacts on production caused by unforeseen instability events. Further advances in radar interpretation expertise may at least partially offset the lack of physical presence at the mine.

5.4 Interaction with operations

5.4.1 Overview

Since direct, continuous protection of mine operations personnel is a primary purpose of performance monitoring, the monitoring 'process' must be tailored to satisfy operational needs. The slope engineer must consider monitoring from the perspective of the operator who may be exposed daily to the geotechnical hazards; that is, the workforce wants and needs clear, simple, consistent and timely information (not data) about the stability of their work area and an assurance they are safe. A key characteristic of TARPs, summarised in Section 8.5, reinforces these same qualities, since TARPs were historically developed to allow non-technical pit dispatchers to implement safety controls at night and on weekends, when the slope engineer is not available.

The critical responsibility to provide mine operations with warning of, and mitigation controls for, all geotechnical hazards has three desired outcomes:

Table 5.2: Monitoring skills summary

Skill	Content	Implementation	Outcome
1: System selection /layout	Types of monitoring systems, capabilities and limitations, performing geotechnical and geological assessments, performing siting/location analyses	Internal experience-sharing, information from and presentations by manufacturers, research, feedback from industry colleagues	Proficiency in determining the appropriate monitoring system(s) to meet site requirements, evaluating factors that influence performance, performing location selection
2: System principles	History, physics, operating principles, applications, limitations	Manufacturer training programs, product manuals, expert guidance	Understanding of key system principles and theories
3: System components	Hardware and software components associated with the system	Manufacturer training programs, product manuals, expert guidance	Understanding of key system hardware and software components
4: System deployment/ installation	Site selection, deploying and/or installing system, geo-referencing	Manufacturer training programs, product manuals, expert guidance	Ability to efficiently and effectively deploy and/or install monitoring system
5: System setup/ configuration	Setup and configuration of a system from pre-determined criteria, calibration, baseline testing, determining data validity	Manufacturer training programs, product manuals, expert guidance	Ability to set up and configure system
6: System operation	Hardware, software, communications, data management operation	Manufacturer training programs, product manuals, expert guidance	Understanding operation of the system components and how data are distributed and managed
7: System diagnostics/ troubleshooting	Diagnostic and troubleshooting tools and procedures, available resources, contact information	Manufacturer training programs and/or maintenance support, product manuals, expert guidance	Ability to follow basic diagnosis script and resolve expected problems, as well as the protocol for elevating to manufacturer support
8: Data handling / management	Network communications, database management, data validation and reliability, archiving	Site-specific knowledge, internal experience-sharing, product manuals, expert guidance	Ability to facilitate automatic and manual collection, transfer, validation, and backup of data at routine intervals
9: Data analysis / utilisation	Data visualisation, analysis, trends and interpretation; failure prediction, inverse velocity, acceleration trends, forward and back analyses	Formal education, manufacturer training programs, product manuals, expert guidance	Ability to perform complex data analysis, interpretation and decision-making
10: Alarm thresholds	Alarming methods and types, alarming theory, case history and databased analyses, determination of data 'noise', level of monitoring detail, time windowing	Site-specific knowledge, internal experience-sharing, formal education, manufacturer training programs, product manuals, expert guidance	Ability to determine applicable alarm types, settings and thresholds
11: Communication / response protocols	Distribution of alarms and notifications, responsible parties, protocols (TARPs, etc.), evaluation	Site-specific training, internal experience-sharing, expert guidance	Distribution of alarms to appropriate personnel for response
12: Documentation / reporting	Periodic system/monitoring status reports, baseline configurations and change tracking logs, monitoring plans, project organisational charts, instrumentation maps, decision flowcharts, evacuation procedures, communications lists, alarm review and signoff forms, system and visual inspection forms, geotechnical reports, operational reports	Site-specific training, internal experience-sharing, expert guidance	Ability to determine applicable monitoring documentation and reporting needs, and develop appropriate reporting intervals

- every operator, every time, will understand, accept and take appropriate actions to mitigate geotechnical risks;
- each warning communication will be a mentoring opportunity to cumulatively improve the geotechnical skills of the operating team; and
- the wording of warnings will inform operator safety, not undercut nor substitute for any operator's own awareness of conditions and individual safety responsibility.

In practice, sound risk management logic and reasonable care are sufficient to manage the technical framework and details of geotechnical risk communications to mine operations.

Successful communication is a complex process requiring active involvement from all participants in the process. A large part of the process is not even about the technical information being conveyed, but rather how it is conveyed. Safety information must be communicated clearly and unambiguously. A useful analogy may be for the slope engineer to think how they would like to receive serious and complicated medical information from their doctor. A patient might expect from a physician:

- *whole picture view*: understand the real problem, know what the patient needs to know and to decide;
- *useful information*: no data, no boilerplate, no caveats – best interpretation without useless info;
- *consistency*: honesty, explain changed interpretations, logical approach – create no reason to doubt;
- *expertise*: most knowledgeable, best informed, experienced – especially on this specific problem;
- *familiarity*: know patient, sincere concern, invested in outcome (like the patient themselves); and
- *attitude*: focused, available, dedicated, confident but not condescending.

Shift superintendents and mine operators might have similar wants when being advised by the slope engineer about potentially life-threatening safety risks in their operating areas. For example:

- convey critical information pertinent to any problem succinctly and ensure that it is understood;
- equipment operators must evaluate risk information by judging the slope engineer's apparent competence, motivation and credibility, and by tracking their consistency and engagement with the operating team;
- do not underestimate the knowledge of experienced mine operators and operations supervisors, who may understand local geotechnical hazards as well as, or better than, the slope engineer;
- when helping to describe a particular slope hazard to the shift crews, focus on that issue, or the opportunity to effectively communicate the information is lost; and
- be fully aware of a risk that is being communicated to the workforce – as part of inspections, spend time at the operating face with operators on a regular basis.

Very few people are natural-born communicators. However, since communications about risk will rank among the most important communications of a person's career, the responsibility to overcome all obstacles to communication cannot be ignored. Fortunately, most problems are eliminated when the slope engineers simply present their genuine self without embellishment – honest, sincere and available.

5.4.2 Control centres

The control centre of most open pit mines typically has been 'mine dispatch', traditionally a small relocatable structure perched on the pit crest at a good overlook. When slope engineers began managing telemetered data from various sources, it made sense to put a computer/monitor in the dispatch building to apprise the dispatcher of any slope conditions that might require attention. Some operations have a full-time engineer or technician observing telemetered monitoring data and information alongside the mine dispatcher, so that the dispatcher can get the geotechnical information first-hand.

Some operations have large integrated monitoring systems built around mine dispatch, with various screens for mine operations (dispatch), mine maintenance, geotechnical engineering and ore flow, in a master control centre. These may have one or more walls of large monitoring and remote camera display screens with several rows of computer desk stations arranged like a small auditorium. The idea is to assure timely face-to-face communications between critical personnel from all disciplines in the same room, as well as to provide a venue for discussion of concerns or even crisis situations with upper management and other key stakeholders.

If a mine lacks a dispatch or control centre, it is still important to have a dedicated space at the operation to accommodate slope stability instrument monitoring screens and personnel to observe them, write reports and communicate with the mining team. Most operations have provisions for offsite geotechnical personnel to use their phones or computers to connect to the monitoring servers and view the same types of information remotely. Facilitating the accessing, compiling, plotting and reviewing of monitoring data increases the likelihood that it will be done regularly and consistently, and decreases the likelihood that pertinent data will be overlooked.

5.4.3 Regular communications

The subject of communications, including risk escalation and formal reports, is addressed in Sections 8.5 and 8.6. This section examines communications with a focus on operational considerations. Although not always desirable or convenient, regularly scheduled meetings and reports are necessary for the exchange of essential information between relevant staff and other stakeholders. Mining company cultures vary on the necessity and intensity of meetings, but most operations tend to have fairly brief and efficient meetings, and informational reports are likewise concise. The key venues that typically facilitate sharing of slope performance monitoring information include the following.

- *Daily/pre-shift safety meetings (tailgate meetings)*: the geotechnical team conducts brief, documented, daily safety meetings to collaborate on current activities and to share and resolve any concerns. At times the slope engineer may be called to help with pre-shift safety inspections for operations groups. They should always take advantage of these opportunities to facilitate communications.
- *Daily published slope monitoring reports*: concise summaries of ongoing instabilities, usually delivered

by email, showing location, conditions, access limitation and other impacts to the operation. A mine plan indicating location of the instability and associated/potential impact areas is the main graphic; radar DTM images are sometimes used. In some cases, a table, listing each instability, relevance (pushback or haul road affected), trigger level, current rate and whether the instability is accelerating, decelerating or steady-state can be useful for decision-making throughout the shift. This report is catalogued to become the history/verification of slope monitoring and communications.

- *Current hazard map*: this comprises a large-scale map, or set of DTM views, locating and describing all active hazards that are being tracked, their extent and areas of potential impact, a photo (if relevant), description of each hazard and/or failure mechanism, and mitigation instructions. It is typically posted in the main mine office, crew lineout areas, covered viewpoints, slope engineer's office and mine meeting room. The purpose is to keep personnel current on the locations of hazards and how they are being managed. Keeping the map updated is critical (weekly, monthly or as needed). Such a visual display is helpful to mine inspectors and other external stakeholders as an example of slope management diligence.
- *Daily mine planning and/or operations meeting*: most operations have a short daily meeting, commonly at an open pit overlook, with operations, mine engineering, survey, geotechnical engineering and geology staff to review the general plan for the day, discuss new issues and/or clarify concerns and directives. This is an opportunity to communicate specifics about slope management and to share and potentially resolve any concerns from mine operations regarding geotechnical issues.
- *Weekly mine meeting*: this is the principal information-sharing meeting for most operations, including mine operations, maintenance, engineering, survey, geotechnical and geology staff. The slope engineer typically prepares a short presentation to cover details about specific geotechnical concerns and discusses projects, slope stability status, instructions and recommendations to operations. Other slope stability updates might cover dewatering progress, and drill-and-blast management for stability and fragmentation. The slope engineer may also advise on issues, problems or questions from other presenters, including changes to short-range plans and preparation of contingency plans for uncertain problem areas. Information is shared regarding night, weekend and on-call staffing for the geotechnical team.
- *Weekly mine-mill meeting*: this meeting may be a venue for contributing to mill feed risk discussions that involve slope stability factors, and for discussing issues of blast damage versus mill feed fragmentation.
- *Monthly reconciliation/forecast meeting*: some operations conduct a meeting with the general manager and other site management to review the month's performance; others produce written reports. Either way, the slope engineer should provide the slope stability portions containing factual information.
- *Annual mine budget meeting*: most operations prepare budgets and forecasts at least yearly for review and approval by corporate financial and human resources teams. These reviews are typically based on production rates, sinking rates or other relevant factors. The slope engineer will have a slope management plan that supports the same periods, which permits development of a realistic geotechnical budget. Personnel, capital expenditures, operating and other costs should be estimated for every period. The geotechnical budget should include monitoring equipment, software, equipment training and maintenance fees, personnel by category for new positions and attrition replacement, consulting services, and hydrogeology drilling and management costs that pertain to slope stability.

Meetings should be kept concise to focus attention and maximise information-sharing directed at achieving slope stability objectives.

5.4.4 Operational support for monitoring

Senior management, IT, operations, maintenance, engineering, survey and geology teams are sometimes called on to support open pit slope performance monitoring. The types of operational work or services specifically supporting the monitoring function vary considerably from mine to mine, but generally fall into the following categories:

- *earthwork construction*: instrumentation installation sites, protection berms, access roads;
- *power supply and radio communications*: line power drops/transformers, mesh network repeaters, lightning protection, grounding, radios;
- *civil construction*: overlook guard rails, equipment shelters, offices, concrete structures/barriers;
- *operations coordination and communications*: mine radios/radio assignments, mine central dispatcher;
- *mine mobile equipment and maintenance*: field vehicles and equipment, service of generators and instruments;
- *engineering and survey*: engineering design, survey, laser scanning, drone operations, GNSS base for real-time kinematic GNSS survey or monuments; and
- *safety*: health and safety oversight and reporting, mine regulator (e.g. Mine Safety and Health Administration) interaction, annual refresher training, overall mine site security.

Many of these items involve administrative tasks that occur in a bureaucratic and priority-based environment, requiring the slope engineer to function in the administrative environment, which includes anticipating delays, foreseeing challenges and managing the process for positive results. This may require developing the relationships necessary to conduct this business efficiently.

Occasionally, the installation or even the existence of monitoring equipment in the mine causes production interruption, either by partially blocking haul roads or by requiring valuable support equipment time. If the instrumentation is required to facilitate mining, then it is a high priority since production would be lost if the equipment is not installed and working properly. Negative responses can be minimised by creating a culture where the instrumentation belongs to the workforce, developing an installation and communication plan in advance of the installation event, and assuring operators that such instrumentation is needed to manage risk and to keep them safe. Fitting instrumentation into the mine environment can be a challenge, but most challenges are overcome with proper planning, communication and budgeting, so that each installation becomes a managed project.

The slope engineer should anticipate all installation needs in advance of construction and operation. Details of the instrumentation project that are well-conceived, scheduled and in the budget go much more smoothly than a last-minute emergency purchase and install. The best scenario for success is an aware and diligent mine operations team that installs its own monitoring equipment in a manner that maximises its objectives. That way, the minor production interruptions associated with the installation and management of the monitoring system are seen as part of the cost of doing business. Once the monitoring equipment is running, pre-shift meetings that allow all operators to see the equipment and monitoring results create a culture characterised by active sharing of information.

6

DATA ACQUISITION, TRANSMITTAL, VALIDATION AND MANAGEMENT

Phil de Graaf, Erik Eberhardt, Alex Neuwirt, Fanie Wessels

6.1 Introduction

This chapter describes the collection, transfer and management of the data generated by monitoring systems. This includes mechanisms to acquire and transfer data, understanding data integrity challenges associated with errors and limitations of data sensors, as well as recognising what instrument specifications need to be considered to achieve the monitoring objectives. Examples of managing data uncertainty are also presented.

The increased types and volumes of real-time monitoring data now possible bring new challenges in managing data flow and converting data into information upon which to base informed decision-making. Data represent a set of discrete (unprocessed) signals or observations about events (e.g. vibrating wire piezometer (VWP) pore pressure readings for a month), which are converted into information by providing context, aggregation or organisation of processed and validated data, which make decision-making easier (e.g. VWP pore pressure plots for a specific geological unit showing daily rainfall and dewatering pumping rates). This information in turn is converted into knowledge through insight and understanding of how it applies to achieving monitoring and decision-making objectives (e.g. supporting a decision to increase pumping capacity at a defined future date). Clear focus and understanding of the monitoring objectives are essential to determine the nature and volume of data that are required for supporting the operation's risk management program and the relevant decision-making process.

6.2 Data communications

6.2.1 Overview

Mining environments present numerous challenges for reliable communications to support system management and data collection. There are essentially three elements of data communication requirements for monitoring mine slopes. The first element is the acquisition of slope performance monitoring data. That is, measuring some physical response, such as deformation or pore pressure, using instrumentation designed for that purpose. The second element is a mechanism for communicating the performance measurement from the monitored slopes to the local mine network. Communicating from the operation to a network operations centre or corporate operations centre, which may or may not be located at the mine, is the final element. These communications may include transfer of data as well as notification of alarms and alerts.

This section will address the general challenges of communications in the mining environment as well as specific technologies that may be utilised for each communications application. The most significant challenges are summarised as follows.

- *Geographical area*: typically, in open pit mining, a large geographical area must be serviced and is often situated in remote locations with limited existing communication infrastructure. While the open pits themselves may be modest in area, the supporting area of the mine can be quite extensive. For example, a tailings impoundment located some distance from the mill and even further from the mine requires performance monitoring. Another example might be groundwater extraction or groundwater remediation in the form of pumping or injection wells that can extend for many kilometres beyond the boundaries of the open pit itself.
- *Topographical characteristics*: since mineral deposits are commonly found where significant geological or tectonic plate action has occurred, mines are often located in mountainous regions where topography works against wireless or wired solutions. It works against wireless systems in that line-of-sight wireless is limited; by the same token, wired systems are more

challenging to install over uneven terrain. Access can be difficult for installation of wireless repeaters or bases, or for the installation of copper or fibre optic hard-wired communications. Furthermore, the mine area is a dynamic environment, requiring expansion and flexibility of the monitoring system.

- *Rapid technological change*: communication technologies are becoming more sophisticated with each passing year. Miners are making significant investments in communications, largely driven by advances in production management systems such as route planning, running autonomous vehicles and automation of data integration. These systems require real-time tracking of virtually all movable assets on the mine site including shovels, haul trucks and loaders, together with other types of equipment. Personnel may also be monitored and tracked, with alarms triggered by systems such as Geofence if someone inadvertently enters a restricted area such as a hazard zone. Fixed assets such as crushers, conveyors and mills also require monitoring and wireless systems. Add to this real-time kinematic systems for accurate positioning information, and the wireless space becomes very cluttered. Often these systems operate in different wireless domains; however, significant power being broadcast in one domain will affect others.
- *Competing vendors*: there is little coordination among vendors regarding how communications systems operate. Even technologies which are similar (e.g. spread spectrum radio) and based on standards such as IEEE 802.11 (i.e. WiFi, of which there are close to two dozen versions) are often found to be completely incompatible due to this lack of coordination. Vendors are not generally incentivised by customers to work together and often use their proprietary systems and methods to create 'vendor lock', whereby once an investment is made to utilise a particular system, it is very expensive to switch to another vendor.
- *Local regulations*: wireless standards vary from country to country. Each country regulates its own use of wireless spectrums. While there is some coordination for commonly used spectrums, such as cell modems or WiFi, use of other parts of the spectrum can vary considerably. This creates challenges for vendors, as spectrum that is available in one country may not be available in another. Licensed or approved spectrum is often the most reliable; however, the availability of a certain spectrum can vary within a country depending on geographic location (i.e. a frequency licensed at one mine may not be allowed at another).
- *Environment*: systems providing data are generally located outdoors and are often fully exposed to hostile operating environments. This presents numerous challenges to the reliable operation of the communications systems. Lightning wreaks havoc with systems, as wireless repeaters or base stations generally have to be located at the highest points surrounding a property, making them vulnerable. Other environmental challenges include water (usually in the form of precipitation), heat, cold, solar radiation and low and high humidity, all of which conspire to degrade and destroy electronics, connectors and wiring. Figure 6.1 shows two challenging environmental conditions including sub-arctic conditions affecting prism optics (DeBeers' Gahcho Kue Mine in Canada's Northwest Territories) and an aggressive windswept coastal setting for VWPs (Namdeb's Southern Coastal Mine on the Namibian Atlantic seaboard). Fly rock generated by blasting operations can also damage equipment.
- *Power*: remote sites usually require self-power, in the form of extended-life batteries or rechargeable batteries charged with solar (most common) and/or wind energy. In locations with a high number of sunny days (more than 300), solar is an excellent source of charging power; however, this is not common in many mining locations. Even in locales with excellent solar or wind recharging, battery sizes and the power systems in general are still limited compared to equipment powered off main lines (mains power). A radio that consumes 2 A continuously with a 24 VDC (volts DC) supply should have a 24 VDC-chargeable power system with over 300 AHr (ampere hours) of storage, meaning over 600 AHr of 12 VDC battery storage. A typical car battery is rated at around 100 AHr. Many sites that have mains power are subject to power outages due to their remote locations and other challenges associated with providing the line power.
- *Cost*: strictly speaking, mining is focused on the efficient and cost-effective excavation and processing of materials. As such, communications systems can be a difficult sell, as their benefits may not always be tangible. How can systems of one design be justified over another design? What will be the tangible benefits of a network design that supports 1 Gbps (billion bits per second) *v.* 100 Mbps (million bits per second) *v.* 10 Mbps? The cost will vary considerably across these different bandwidths.
- *Theft and vandalism:* most mine properties are secured against unauthorised entry, which greatly limits theft and vandalism. However, even secured mine properties suffer from theft and vandalism. Equipment is often powered by batteries recharged with solar panels. Solar panels are generic by nature and are easily removed from installations, so it is not unusual for them to be taken from site. Equipment installed at properties located in developing nations can be particularly difficult to keep secure, due to the premium placed on

Fig. 6.1: Examples of instrumentation located in challenging environments. (Left) A hoarfrost-obscured prism located in arctic conditions at the DeBeers Gahcho Kue Mine in Canada's Northwest Territories (source: J. Mason). (Right) VWPs and telemetry box installation located in an aggressive coastal setting at Namdeb's Southern Coastal Mine on the Atlantic seaboard (source: H. Ileka).

batteries and solar panels for home electricity. Often, elaborate protection schemes or 24 h security is required to keep equipment secure.

The following sections summarise some of the basic technologies currently employed in open pit mining to facilitate communications. For example, Fig. 6.2 illustrates some of the telemetry solutions that may be deployed in a typical mining environment, including but not limited to ethernet/fibre networks, WiFi, radios and microwave links.

6.2.2 Hard-wired systems

Numerous hard-wired systems have been developed over the years to support data communications; however, only a

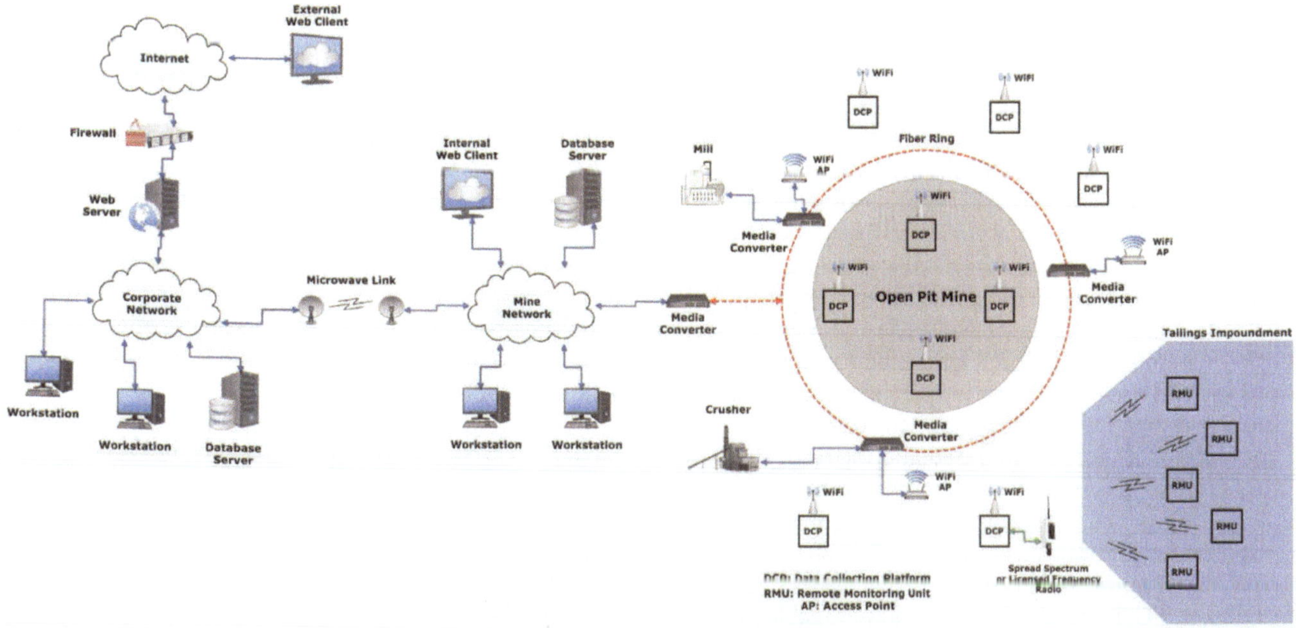

Fig. 6.2: Typical multi-faceted mine telemetry solution including WiFi, hardwired (ethernet/fibre optics), radio and microwave link communications (source: A. Neuwirt).

few have found use in mining. The two systems that have found success are ethernet and single-mode fibre optic.

Ethernet comes in several data transmission speeds: 10BaseT, 100BaseT and 1000BaseT are the most common, where the number prefix represents the bits per second in millions of bits per second (e.g. 100BaseT means 100 million bits per second). Several cable categories have been developed to support the various speeds; the most common is category 5 (Cat5), rated for 100 MHz. Other types include Cat5e, Cat6 and Cat6a. All ethernet varieties are rated for a maximum distance of 100 m, which limits its use in remote telemetry. In practice, longer distances are possible and repeaters or boosters can be used to extend the distance, but running the cable is often problematic. The primary advantages of ethernet are the availability of connections due to its use in computer networking and the ability to network numerous devices on the same cable.

Fibre optic is also popular largely due to its use and availability in long-distance network links that are generally installed around a mine site to support plant and other infrastructure communications. Two types of fibre optic cable are utilised: multi-mode and single-mode. Multi-mode cable is less expensive and the electronics are less expensive, but the distance is limited to ~2 km, so it is often used within buildings or structures. Single-mode fibre is more expensive but supports much higher data rates (up to 60 GB/s) and greater distances (up to 100 km without amplifiers/boosters and thousands of kilometres with boosters/amplifiers). Fibre optic cables typically consist of multiple strands (up to 48 are common), so if remote telemetry locations pass near a fibre cable, it is generally a good idea to break out a fibre for connecting the equipment.

6.2.3 WiFi

The term 'WiFi' is generally considered to be a play on high fidelity (hi-fi), as wireless fidelity. It refers to a class of wireless communications that fall under the IEEE 802.11 standard and consists of a protocol for encoding internet protocol (IP) packets for wireless transport or Wireless Local Area Network (WLAN) technology. Today, hundreds of WiFi solutions are available, driven largely using WLAN adapters in personal computers and personal electronic devices. Use of WiFi in mining is becoming more and more common, verging on standard for certain types of communications. It offers compatibility across many electronic device types and widespread support in the computing field. However, it has several important limitations including distance, bandwidth and channel capacity. Manufacturers have responded by developing versions of IEEE 802.11 systems with higher power and at alternate frequencies from the original 2.4 GHz standard. WiFi is now available in the 900 MHz, 3.65 GHz, 4.9 GHz, 5 GHz, 5.9 GHz and even 60 GHz range. Each base frequency has several bands or channels to provide for separation of communications for devices operating in the same band. For example, the 2.4 GHz frequency has up to 14 channels or bands.

6.2.4 Radio communications

Numerous radio systems have been developed by many vendors. Unlike the standards of IEEE 802.11, hardly any standards exist for the transmission mechanisms of these systems.

The 1960s to 1980s saw the development of numerous radio systems for low-power wireless data telemetry. Systems varied in sophistication and reliability. The earliest radios used licensed bands well-suited to wireless communications, such as the 150–174 MHz band. The advantage with licensed bands is that much higher power levels are allowed, licensees are guaranteed use of the band for their region, and in many countries enforcement options exist in the form of government licensing agencies. In the USA, the Federal Communications Commission provides licensing and enforcement. The disadvantages with licensing are that rules and regulations vary from country to country and that they generally require a study of local frequency usage and application fees, along with the delays associated with dealing with governmental agencies. The delays can be particularly problematic when rapid deployment of monitoring equipment is required.

As signal processing methodologies, electronics and processors improved throughout the 1980s and 1990s, the use of unlicensed bands became more attractive. While the power of unlicensed bands is generally limited to 1 W (watt) or less, the improvements in signal processing technologies, such as the technique of hopping among numerous narrow-band ranges within the larger unlicensed bands, dramatically improved the reliability of communications. Radios with modern frequency-hopping techniques with only 1 W of radiated power can now communicate reliably over great distances – up to 50 km with directional gain antennas and in some cases up to 100 km.

While licensed band radios still have application in long-distance or otherwise challenging (obstructed) telemetry applications, most remote telemetry can now be supported reliably with radios operating in one of several unlicensed bands. The most common is 900 MHz, followed by 2.4 GHz.

6.2.5 Other communications systems

Some sites are so remote or without adequate infrastructure that alternate communications systems must be considered. More common types are as follows.

6.2.5.1 Microwave

Microwave is used extensively by telecommunications companies for providing digital backhaul over distances as great as several hundred kilometres. Areas of thousands of square kilometres are supported through networks of microwave repeaters.

One challenge with microwave systems is the high frequency of transmission and curvature of the Earth. The high carrier frequency demands unobstructed views, which can be complicated by the curvature of the Earth over significant distances. This requires antennas to be mounted on the highest possible topography, such as mountain tops. Mountain tops are by nature difficult to access. In remote areas there may be no roads, which means flying in towers, buildings and equipment by helicopter. Due to the high-power nature of microwave transmission, line power is generally required or significant off-grid systems must be deployed, in the 25–50 kW range. By comparison, a typical home solar system is ~10 kW.

However, once a microwave link is established to a remote area, local radio networks using low-power, lower-frequency telemetry radios, cellular or WiFi may be deployed to provide access to monitoring resources near the microwave link location.

6.2.5.2 Satellite

Several types of satellite solutions are available for remote telemetry. We focus on the two common systems available for mining applications.

Broadband global area network (BGAN)

The BGAN provides worldwide data telemetry services up to 492 kbits/s using several geostationary (35 680 km) satellites, as shown in Fig. 6.3. The network originated in the late 1970s when the International Maritime Satellite Organization (INMARSAT) was formed to provide communications options to the maritime industry. Latitudes up to ± 75° are supported, although it is not recommended for use beyond ± 70°.

At the time of writing, the most common radio module for BGAN is the 9502, manufactured by Hughes. It offers IP bandwidth of up to 488 kbits/s, but latencies due to the time required for signals to travel to geostationary orbit and back can be as high as 2.5 s. Data collection equipment must be able to handle the relatively high latency of the communications. Packet-type protocols with configurable response times function best. Multiple IP devices can be connected to a single transmitter.

Iridium

The Iridium network consists of 66 active low-Earth orbit (~775 km) satellites used for worldwide voice and data communications, illustrated in Fig. 6.4. Iridium services were originally launched in 1998. At the time of writing, ~95 satellites have been launched to maintain the network. They can usually be seen in the night sky as satellite flares or short-lived bright flashes of light.

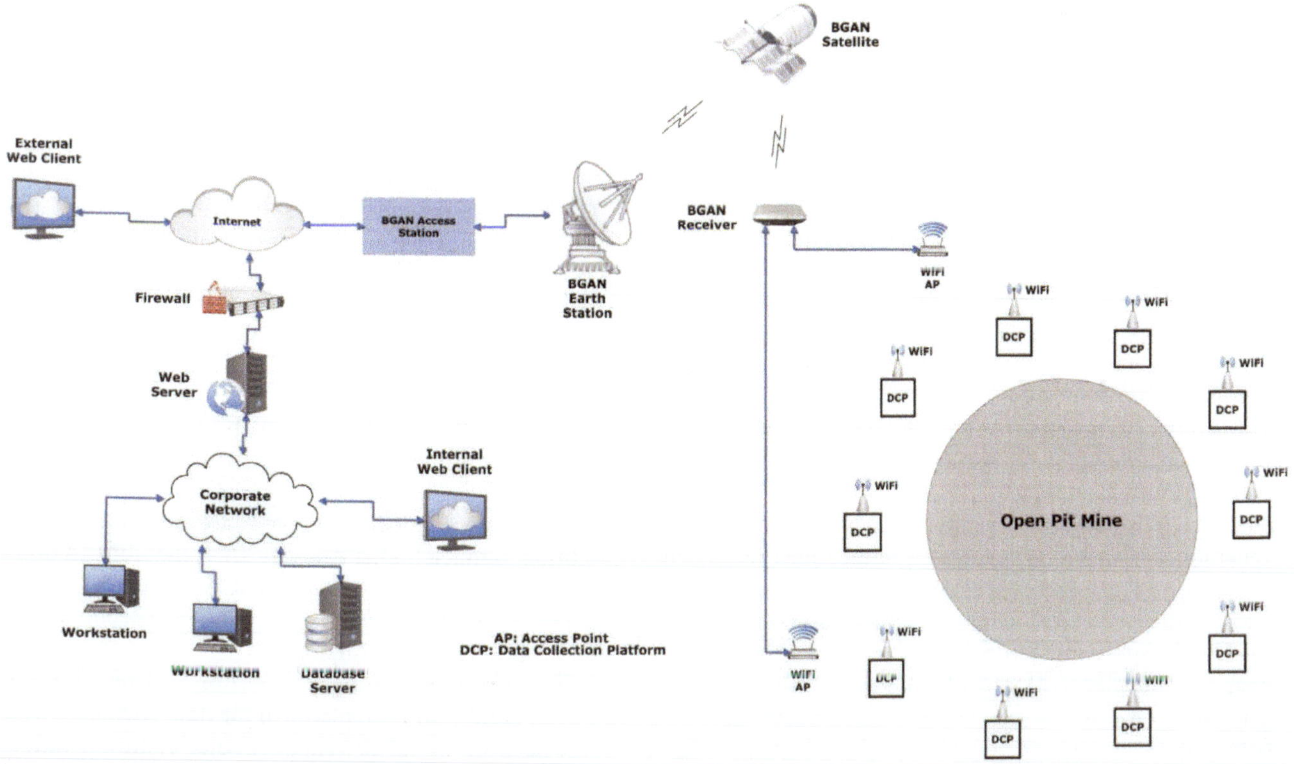

Fig. 6.3: Illustration of a BGAN system (source: A. Neuwirt).

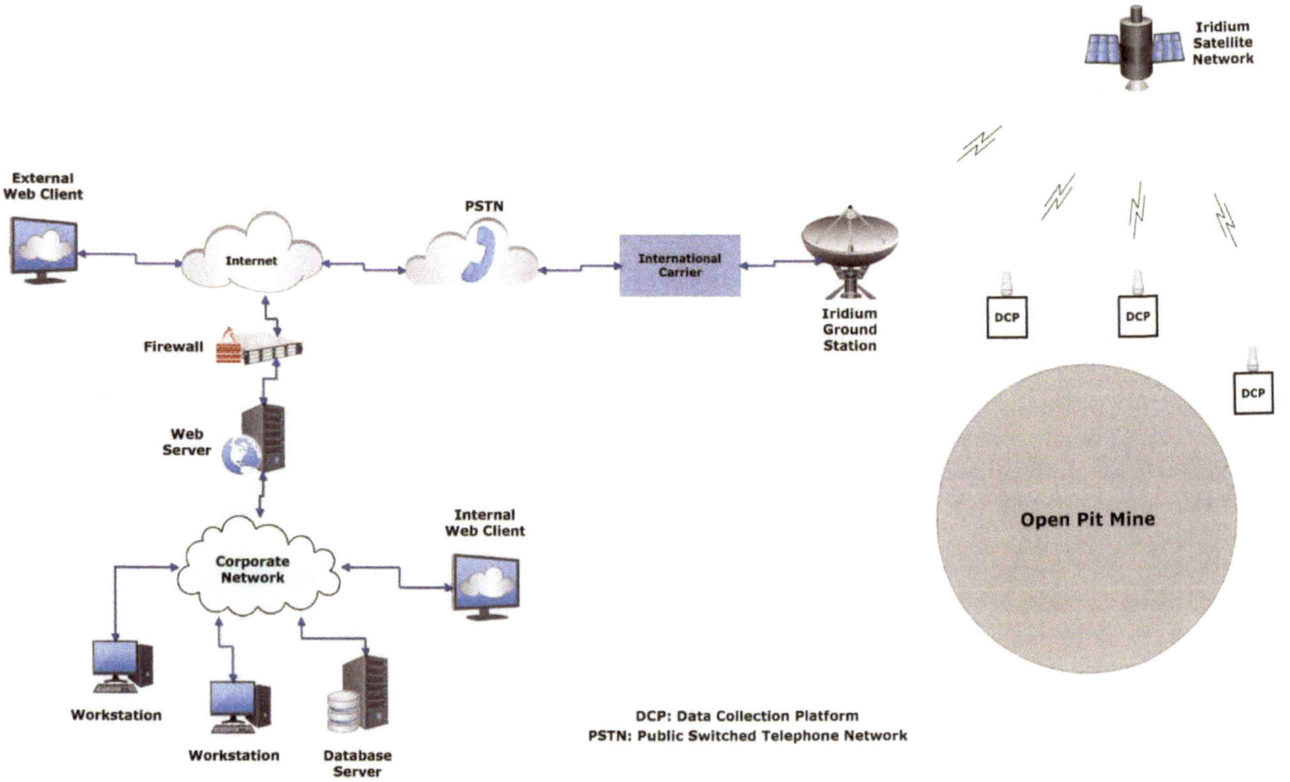

Fig. 6.4: Illustration of an Iridium satellite communication system (source: A. Neuwirt).

For data telemetry, several versions of '9522' series modems are available, including an original equipment manufacturer (OEM) version (the 9523) that has been integrated into products from several manufacturers. The current version of the 9522 is the 9522B, available from Iridium.

Data telemetry rates for Iridium are up to 2400 bits/s, so it is not suitable for high-speed data transmission. Latency is improved compared to BGAN; however, it can be up to 2 s due to routing through multiple satellites. IP services are not generally supported. Several service providers offer telemetry plans; the type of plan and costs depend on the provider and type of plan purchased.

6.2.5.3 *Cellular technologies*

The cellular era began in 1979 in Japan with the launch of the first commercial cell phone network by Nippon Telegraph and Telephone. This first-generation (1G) analogue technology has been followed by successive generations with added digital communications and much higher bandwidth and connection capacity. At the time of writing, there are over 7 billion cell phone accounts worldwide, approximately the same number as there are people on the planet. Most deployments are third-generation (3G) and, in more recent years, fourth- and fifth-generation (4G and 5G). Use of cellular networks for data telemetry became commercially viable with 3G technology, which was released in 2000. In 2010, 4G technology made its appearance, providing full IP-based communications, data bandwidth of up to 1 Gbps and full WiFi support. As the technologies evolved, vendors produced specialised data modems for wireless data telemetry use. Several vendors such as Cisco have developed multi-band, multi-purpose modems such as the '800' series routers/modems, which combine multi-band WiFi/4G/LTE in a single package. These types of technology integrations are encouraging more sites to adopt WiFi/4G infrastructure.

The technology continues to advance, with 5G technology readied for release by 2020. This combines the best of WiFi (low-latency, high bandwidth) with the best of cellular technology (high security, mobile channel switching, high capacity, high reliability), as shown in Fig. 6.5. The 5G technology is often described as being able to properly support the Wireless World Wide Web (WWWW). Multi-gigabyte speeds are supported in addition to much higher capacity and reliability. In other words, all the current limitations of WiFi and 4G technology are hoped to be solved with the deployment and adoption of 5G technology.

If the claims regarding 5G are substantiated, it offers mine operators an ideal platform for data telemetry. The same network would also support voice communications, video services, machine communications and LAN/WAN

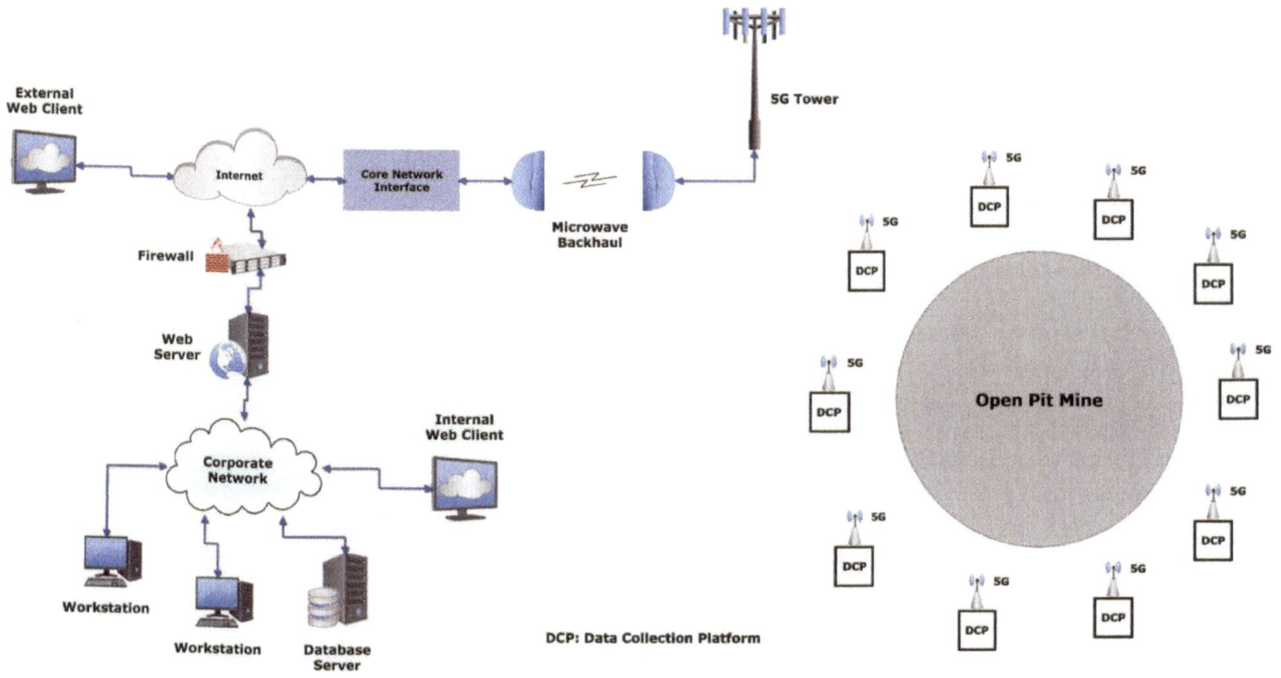

Fig. 6.5: Illustration of the 5G network (source: A. Neuwirt).

network access. This potentially offers significant cost savings, as deploying infrastructure for a single wireless network will always be more cost-effective than the several that are typically required at present. Further, much of the land-based networking, such as fibre or ethernet infrastructure, would become unnecessary.

6.2.6 Network requirements

The backbone of any telemetry system is the data communications network. Process and business operations require the design and installation of LAN and WLAN topology. Typically, the physical infrastructure consists of one or more fibre rings or loops around a mine property, with strategically placed terminations to provide hubs for wired and wireless ethernet communications. The fibre rings are generally multi-gigabit with multiple fibres to support all site communications, including process system communications, business computing (including internet access), cameras and other security apparatuses, and phone or other voice communication systems. In the past, implementation of mine-wide communications was an afterthought to mine design; now, design and implementation of mine-wide communications networks are undertaken in tandem with the development of the mine itself.

Leveraging this infrastructure makes good business sense: the investment has already been made, and bandwidth load needed to support remote telemetry is generally minimal. Where an IP camera might require 1 Mbit bandwidth per second, the bandwidth of an entire remote telemetry system might be 1 Mbit per day.

6.2.7 Limitations and cautions

Reliable communications systems are critical to mine operations, particularly in relation to collecting data from remote locations to provide an effective and reliable risk management system. If a slope is failing and there are monitoring devices which detect the onset of instability, the system will only work if the data can be communicated reliably and quickly to decision-makers.

Historically, the culture of mining has been resistant to investing resources in these types of networks. Mine operators have typically focused their resources on increasing production (e.g. it has been easier to sell them a US$5 million haul truck than a US$500 000 radio network). However, improvements in the efficiency of fleet management, facilitated by the reliable and rapid collection of data from equipment, and the communications requirements of autonomous equipment, have encouraged mine operators to invest in communications technologies. Improvements in remote monitoring equipment have similarly encouraged mine operators to invest in these technologies to improve their risk management systems. The improvements in monitoring equipment have led to higher demands on the communications network. For example, collecting data from a single remote telemetry station monitoring a piezometer installed on an excavated slope requires only a few bytes to be transmitted per hour, but collecting data from a ground-based (GB) radar or camera can require several million bytes per second.

It is critical that the communications network design be organised across all groups at a mining operation. Due to similar spectrum licensing schemes adopted worldwide, radios from different vendors for different applications will commonly use the same frequencies. The protocols for communications may be different, but there is only so much radio frequency (RF) bandwidth. Two radio networks using the same frequency will essentially be fighting with each other to reliably exchange their packets. Ideally, radio networks for different applications will utilise different frequencies, or a single network will be built (such as WiFi) to support the multiple applications. Arguably these challenges are easier to manage during the operational phase, when resources are more readily available to implement, support and manage these systems. During mine construction and closure, these systems are often overlooked and receive attention late in the development process or when something goes wrong. Early planning during the mine startup operational readiness and pre-closure phases will ensure that the monitoring system objectives are achieved.

Due to compartmentalisation, it can be difficult to coordinate among different groups at a mine; however, failure to do so can mean that no network operates reliably. Every mine should have a wireless oversight group, which coordinates all network deployments. This group should have RF technicians skilled in the use of spectrum and network analysers so that the wireless space may be profiled before the deployment of equipment or used for troubleshooting reliability issues. If a mine is in an urban area, there will likely be numerous wireless systems nearby, including those used by wireless phones, walkie-talkies, WiFi and cellular. Understanding the wireless space is critical to maintaining a reliable wireless telemetry network.

6.2.8 Future developments

As with all technology, change is the only constant. This has been seen with communications systems. In the early days, licensed band radios were commonly required. This gave way to unlicensed band spread spectrum technology, and the latest iteration utilises more efficient and longer-distance WiFi. The advantage of WiFi is that multiple types of communications are supported, since transmission control protocol/internet protocol (TCP/IP) is a somewhat generic medium for the transmission of data. As more devices adopt TCP/IP support, they can leverage that single network for communications.

Further improvements to WiFi technology are anticipated, including enhancements in the signal-processing algorithms and receiver sensitivities to allow longer distances, lower power, higher bandwidth and lower cost.

Cellular technologies are continuing to improve in range and bandwidth. Data telemetry options are currently available for cell networks. However, due to limited coverage in remote areas, they find limited use in mining. This is likely to change soon, particularly with the release of 5G cellular technology.

6.3 Data management and handling

6.3.1 Overview

We live in an age of rapidly increasing information creation. By some estimates, ~2.5 quintillion bytes of data are created each day. This pace of data creation is accelerating. To find specific information in this deluge of data, ~5 billion searches are conducted each day using search engines; ~75% using Google's search tools, the rest by others including Baidu and Bing. This is in addition to searches conducted on local computers and/or networks. It is becoming more and more difficult to find information of interest. New storage and search methods must be developed to allow productive use of this growing mass of data.

The data explosion has arrived in the mining market. Dispatch systems are continuously collecting vehicle health and position information; production systems such as crushers, mills and plants are generating thousands of data points every second; and geotechnical monitoring systems have become much more sophisticated over the last couple of decades. Prior to the arrival of electronic measurement and storage systems, data were collected by hand and often plotted on graph paper; an example of a time-series of several crack meters plotted by hand over a five-year period is shown in Fig. 6.6.

Electronic instruments have been used in mining for ~80 years. For example, the Carlson strain and crack meters and vibrating wire strain gauges both date from the early 1930s. With the advent of battery-powered electronic measurement and storage devices in the 1980s, the frequency at which measurements could be made increased substantially. Development of various wireless telemetry solutions in the latter part of the 20th century allowed for automated measurement and data collection. Advances in laser scanning, satellite-based radar and GB radar in the last two decades have dramatically increased the types and quantity of slope performance and mapping data that may be collected.

6.3.2 Data storage requirements

The first consideration in designing a data storage platform is the quantity of data being collected. Use of laser scanning, GB radar or photogrammetric systems will generate significant quantities of data.

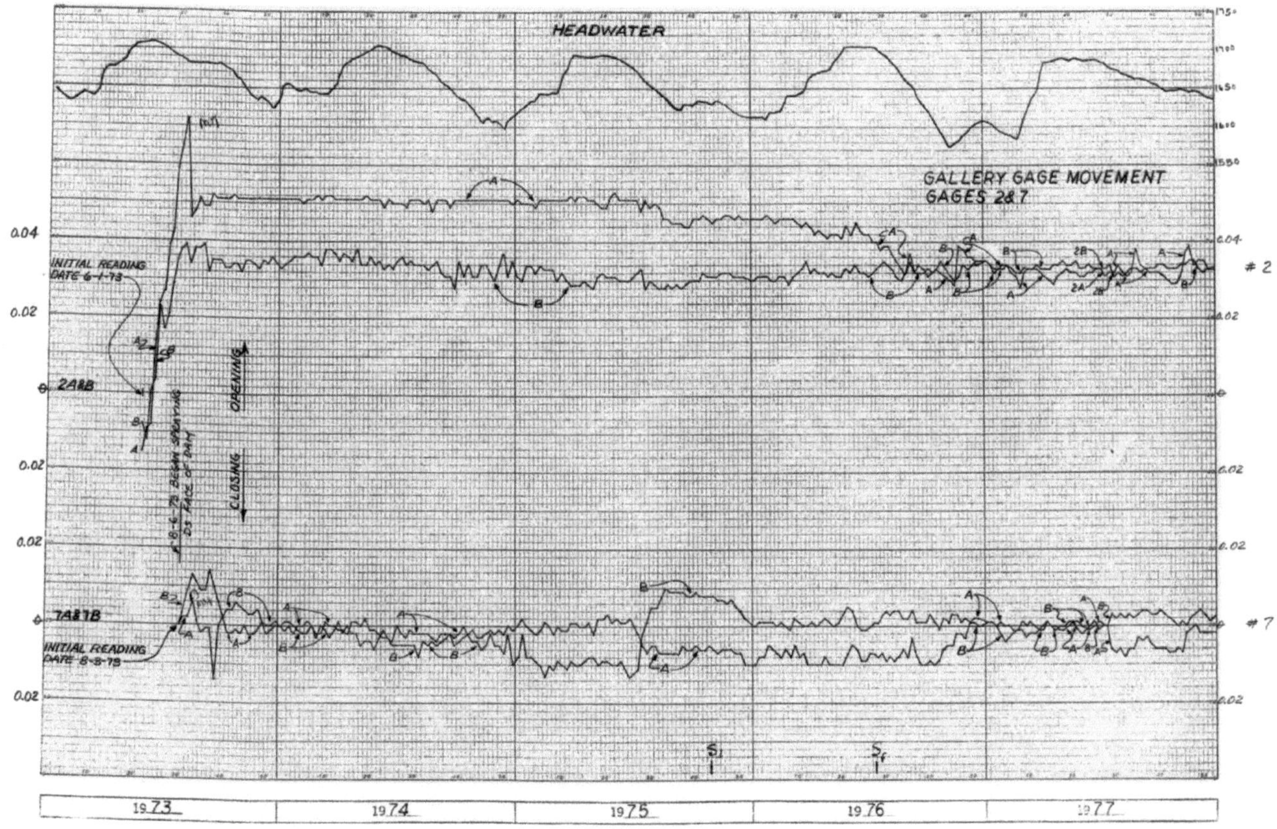

Fig. 6.6: Hand-plotted crack meter displacement graph from the 1970s (source: A. Neuwirt).

Even more important is the storage allocation for data backup. Partial backups are often archived daily, with full backups on a weekly or perhaps monthly basis. Where full backups are conducted on a weekly basis, the data storage requirements are multiplied by 52 for a year. Provisions should be made for archiving the backups to a secondary data storage array or backup facility.

Consider the example of a mine monitoring system which consists of two weather stations with five instruments each, 200 piezometers, two RTS units monitoring a total of 600 prisms, a single GB radar, laser scanner, InSAR and 200 photos captured per day. The values for this example are summarised in Table 6.1, which provides an estimate of data storage requirements for such a monitoring system. These values are approximate and subject to variation depending on types of equipment and frequency of data collection.

6.3.3 Technologies

As with most technologies, there are several options for the design and implementation of data management systems. Careful consideration must be given to the storage system, not just in terms of capacity, but also performance. All database systems, which form the basis of all integrated data management platforms, require high-performing platforms for the best result.

It is important to understand the difference between a block device and filesystem. Most are familiar with the filesystem method of data storage: users have files such as spreadsheets, text documents, presentations and photos and retrieve them from their computers. However, database systems require a block device storage method, where blocks of information are written and read to the storage system. A typical database system does not store the entire file of the database in memory; rather, blocks of the database are written/read at high speed. This type of system provides much higher reliability than filesystem methodologies as, in the event of a power outage or other disruption, any database corruption would be limited to a single block. Depending on the database system, blocks can vary in size from 2 Kb to 16 Kb.

Database systems include the use of transaction logging. Even in the event of a corrupted block, the transaction logged can be used to identify that specific block and rollback the database. Three types of storage systems are illustrated in Fig. 6.7.

- *Directly attached storage* (DAS): a block device which is directly attached to the host computer (i.e. local hard drives or USB storage devices). A filesystem must be configured on it before it can be used; for a Windows

Table 6.1: An example of typical data storage requirements

Equipment	Scanning interval	Each scan (bytes)	Daily storage (bytes)	Weekly storage (bytes)
Weather stations	15 min	100 b	10 Kb	70 Kb
Piezometers	15 min	2 Kb	200 Kb	1.5 Mb
Robotic total stations	15 min	30 Kb	3 Mb	21 Mb
Ground-based radar	15 min	5 Mb	480 Mb	3.36 Gb
Laser scanner	Daily	50 Mb	50 Mb	350 Mb
Photogrammetry	Daily	1 Gb	1 Gb	7 Gb
Satellite-based radar	Weekly	1 Gb	N/A	1 Gb

For this example, ~12 Gb of data will be collected on a weekly basis. Over the course of a year, that equates to ~625 Gb. If four weeks of backups are retained, then drive space of ~2.5 Tb is required. Fortunately, we live in an age of inexpensive drive storage, where 2 Tb solid state drives (SSD) can be purchased for less than US$500.

Server, it is typically the new technology filesystem (NTFS).

- *Network attached storage* (NAS): a filesystem delivered over a network. It is well-suited for systems with significant file storage requirements, but not recommended for database systems.
- *Storage area network* (SAN): a block device delivered over a network. It is well-suited for database systems, but is considerably more expensive and complex than NAS or DAS.

If the mining operation does not have the resources for a SAN-based storage platform, then the use of DAS is suggested. NAS should be avoided, except for archiving of database backups and general file storage. It can be a challenge to convince IT staff to abandon NAS for database servers due to the limited IT resources of most mining organisations.

6.3.4 Data reporting

The wide range of geotechnical data types and formats requires numerous data reporting methods. Tabular reports, such as the VWP example shown in Fig. 6.8, were popular some years ago as a simple method of summarising data from a given instrument, or group of instruments. The usefulness of this reporting method is limited as, typically, many pages must be reviewed to understand trends. If the magnitude of changes in data is small, it becomes impossible to visually identify them.

A key consideration of any reporting engine is the ability to refactor or recalculate data, present several different types of data to understand correlations, and calculate useful metrics. For example, the instrumentation report in Fig. 6.9 provides a statistical summary of data, in addition to the last values recorded.

Reporting can include status information that is colour-coded to highlight the operating state of the instruments. Figure 6.10 provides an example which summarises deformation data from a series of prisms and indicates the status of each prism in the far-right column.

Reports may highlight the data values based on defined limits. Figure 6.11 illustrates this principle – the shortening of time-domain reflectometer (TDR) cable lengths is highlighted. This allows for quick identification of potential risks.

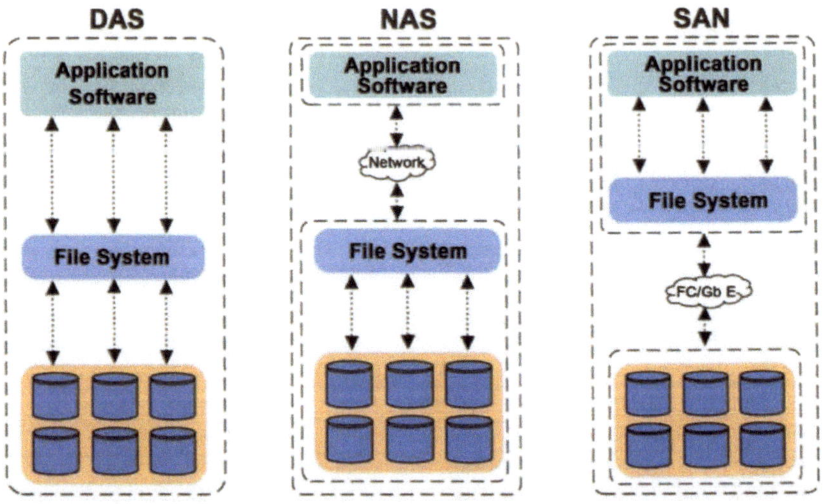

Fig. 6.7: Schematic comparing the three types of storage systems – DAS, NAS and SAN (source: A. Neuwirt).

VW Measurements

RECORDED	D512@VW_Barometer_Temp	D512@VW_Barometer_D	D512@VW_Barometer
6/10/2018 0:00:00	56.11	9601.483398	32.37
6/10/2018 0:15:00	55.31	9601.985352	32.35
6/10/2018 0:30:00	54.22	9602.810547	32.31
6/10/2018 0:45:00	52.95	9603.750977	32.27
6/10/2018 1:00:00	52.05	9604.520508	32.23
6/10/2018 1:15:00	51.55	9604.928711	32.21
6/10/2018 1:30:00	50.84	9605.450195	32.19
6/10/2018 1:45:00	49.86	9606.137695	32.16
6/10/2018 2:00:00	49.00	9606.696289	32.14
6/10/2018 2:15:00	48.32	9607.155273	32.12
6/10/2018 2:30:00	47.83	9607.545898	32.10
6/10/2018 2:45:00	47.93	9607.78125	32.09
6/10/2018 3:00:00	48.20	9607.645508	32.09
6/10/2018 3:15:00	48.47	9607.37793	32.10
6/10/2018 3:30:00	48.61	9607.260742	32.11
6/10/2018 3:45:00	48.59	9607.173828	32.11
6/10/2018 4:00:00	48.31	9607.285156	32.11
6/10/2018 4:15:00	47.63	9607.675781	32.09
6/10/2018 4:30:00	46.89	9608.078125	32.07
6/10/2018 4:45:00	46.31	9608.568359	32.05

Fig. 6.8: Example of a tabular summary report (source: A. Neuwirt).

Instrument Report

Name	Units	Total Readings	Invalid Readings	Valid Percent	Last Reading Value
W1A#PZ_W1A	feet H2O	2861	0	100	86.87
W1B#PZ_W1B	feet H2O	2861	0	100	87.01
W1C#PZ_W1C	feet H2O	2861	0	100	47.50
W2#PZ_W2	feet H2O	2861	0	100	13.87
W1A#PZ_W1A_Temp	celsius	2861	0	100	9.84
W1B#PZ_W1B_Temp	celsius	2861	0	100	9.52
W1C#PZ_W1C_Temp	celsius	2861	0	100	8.32
W2#PZ_W2_Temp	celsius	2861	0	100	9.67
VW_Barometer_Temp	fahrenheit	2861	0	100	85.32
VW_Barometer	feet H2O	2861	0	100	33.45
Relative_Humidity	percent relative	2861	0	100	37.048
Ambient_Temp	celsius	2861	0	100	28.29
Solar_Radiation	Watts	2861	0	100	208.21
Barometer_mBar	millibar	2861	0	100	969.12
RainFall_In	inch	2861	0	100	0
Wind_Speed	miles per hour	2861	0	100	3.71
Wind_Direction	arc degree	2861	0	100	208.58
Battery_Volts	volt	2861	0	100	13.71
Panel_Temp	fahrenheit	2861	0	100	91.58

Fig. 6.9: Instrument report with statistical summary. This type of reporting could also include aggregates such as the maximum, minimum and averages over a given period (source: A. Neuwirt).

Dispatch East Weekly Report

July 09, 2018

Number of Active Prisms: 108

Prism ID	Last Reading Date (dd/mm/yyyy)	Slope Distance (m)	EDM_CRD (mm)		Velocity (mm/day)	Bearing (deg)	Plunge (deg)	Status
			Weekly Delta	Current Reading				
CTZ_103	09/07/2018	1368	1.95	341	0.28	310	7	Active
CTZ_104	09/07/2018	1364	1.75	337	0.35	153	69	Active
CTZ_106	09/07/2018	2307	-1.95	-3	-0.40	131	36	Active
CTZ_107	09/07/2018	1895	-1.45	-2	-0.21	155	-2	Active
CTZ_108	09/07/2018	1907	-0.25	0	-0.04	334	39	Active
CTZ_25	-							Inactive
CTZ_26	08/07/2018	2262	2.35	600	0.37	288	11	Active
CTZ_27	08/07/2018	2289	0.25	608	0.51	284	43	Active
CTZ_32	08/07/2018	2128	4.10	-33	0.64	161	75	Active
CTZ_40	08/07/2018	2128	4.25	-33	0.66	308	-2	Active
CTZ_42	09/07/2018	1786	1.50	-9012	0.22	326	-45	Active
CTZ_43	05/07/2018	1939	2.85	-48	0.93	314	12	Active
CTZ_44	-							Inactive
CTZ_45	09/07/2018	1786	1.10	-30	0.16	327	-24	Active
CTZ_46	09/07/2018	1778	1.40	-41	0.20	323	29	Active
CTZ_47	08/07/2018	1777	8.15	-41	1.17	284	43	Active
CTZ_48	09/07/2018	1786	1.45	-45	0.21	325	-40	Active
CTZ_5	09/07/2018	1372	1.80	-5	0.27	140	-13	Active
CTZ_50	09/07/2018	1913	1.35	-51	0.20	300	76	Active
CTZ_51	09/07/2018	1887	-0.05	-48	-0.01	325	-40	Active
CTZ_52	-							Inactive
CTZ_53	-							Inactive
CTZ_54	08/07/2018	1689	4.25	-33	0.61	326	-25	Active
CTZ_55	09/07/2018	1685	1.50	-38	0.22	327	3	Active
CTZ_56	09/07/2018	1675	1.20	-40	0.17	144	-71	Active
CTZ_57	09/07/2018	1691	2.00	-40	0.29	317	-30	Active
CTZ_58	09/07/2018	1736	-1.35	-45	-0.29	135	-2	Active
CTZ_59	09/07/2018	1792	2.65	-41	0.39	305	5	Active
CTZ_6	09/07/2018	1386	3.50	-6	0.58	318	-8	Active
CTZ_61	09/07/2018	1839	1.55	-26	0.22	146	78	Active
CTZ_62	08/07/2018	2394	4.25	-20	0.65	299	-56	Active

Fig. 6.10: Summary report with instrument status (source: A. Neuwirt).

TDR Daily Report

Date: 6/28/2018

Station Name	Location	Initial Date (mm/dd/yyyy)	Last Recorded (mm/dd/yyyy)	Initial Install Length (m)	Initial Reading (m)	Last Week Cable Length (m)	Last Recorded Cable Length (m)	Δ Initial Reading (m)	Δ Last Week (m)	Δ Daily (m)	Status Active/Inactive
DZJ4_02G_01	J4	12/5/2017	6/28/2018	558	425	426	426	0.0	0.0	0.0	Active
DZJ4_02G_04	J4	12/5/2017	6/28/2018	578	577	577	577	0.0	0.0	0.0	Active
DZJ4_02G_05	J4	12/5/2017	6/28/2018	555	555	557	557	0.0	0.0	0.0	Active
DZJ4_02G_08	J4	12/5/2017	6/28/2018	576	576	578	578	0.0	0.0	0.0	Active
DZ30_01GT_01	DDS#8	2/13/2018	6/28/2018	480	479	483	484	0.0	0.0	0.0	Active
DZ30_01GT_02	DDS#8	2/13/2018	6/28/2018	435	434	435	435	0.0	0.0	0.0	Active
DZ30_01GT_03	DDS#8	2/13/2018	6/28/2018	460	420	420	420	0.0	0.0	0.0	Active
DZ30_01GT_04	DDS#8	2/13/2018	6/28/2018	480	441	429	429	12.0	0.0	0.0	Active
DZ30_01GT_05	DDS#8	2/13/2018	6/28/2018	470	470	471	471	0.0	0.0	0.0	Active
DZ30_01GT_06	DDS#8	2/13/2018	6/28/2018	525	103	103	103	0.0	0.0	0.0	Active
DZ30_01GT_07	DDS#8	2/13/2018	6/28/2018	485	467	468	468	0.0	0.0	0.0	Active
DZ30_01GT_10	DDS#8	2/13/2018	6/28/2018	513	382	384	384	0.0	0.0	0.0	Active
DZ30_01GT_101	DDS#8	2/13/2018	6/28/2018	250	240	240	240	0.0	0.0	0.0	Active
DZ30_01GT_102	DDS#8	2/13/2018	6/28/2018	250	235	235	235	0.0	0.0	0.0	Active
DZ30_01GT_103	DDS#8	2/13/2018	6/28/2018	260	240	241	241	0.0	0.0	0.0	Active
DZ30_01GT_104	DDS#8	2/13/2018	6/28/2018	270	247	247	247	0.0	0.0	0.0	Active
DZ30_01GT_106	DDS#8	2/13/2018	6/28/2018	296	280	281	281	0.0	0.0	0.0	Active
DZ30_01GT_107	DDS#8	2/13/2018	6/28/2018	330	292	292	292	0.0	0.0	0.0	Active
DZ30_01GT_108	DDS#8	2/13/2018	6/28/2018	425	401	401	401	0.0	0.0	0.0	Active
DZ30_01GT_109	DDS#8	2/13/2018	6/28/2018	480	478	480	480	0.0	0.0	0.0	Active
DZ30_01GT_11	DDS#8	2/13/2018	6/28/2018	440	427	429	429	0.0	0.0	0.0	Active
DZ30_01GT_13	DDS#8	2/14/2018	6/28/2018	443	237	238	238	0.0	0.0	0.0	Active
DZ30_01GT_14	DDS#8	2/13/2018	6/28/2018	462	432	436	436	0.0	0.0	0.0	Active
DZ30_01GT_16	DDS#8	2/13/2018	6/28/2018	467	199	200	200	0.0	0.0	0.0	Active
DZ30_01GT_17	DDS#8	2/13/2018	6/28/2018	467	455	459	459	0.0	0.0	0.0	Active
DZ30_01GT_90	DDS#8	2/13/2018	6/28/2018	438	424	428	429	0.0	0.0	0.0	Active
DZ30-02G-101	DDS#13	2/20/2018	6/28/2018	545	602	600	600	2.0	0.0	0.0	Active
DZ30-02G-102	DDS#13	2/20/2018	6/28/2018	531	258	258	258	0.0	0.0	0.0	Active
DZ30-02G-103	DDS#13	2/20/2018	6/28/2018	516	470	468	468	2.0	0.0	0.0	Active
DZ30-02G-105	DDS#13	2/20/2018	6/28/2018	502	527	525	524	3.0	1.0	1.0	Active
DZ30-02G-106	DDS#13	2/20/2018	6/28/2018	476	99	99	99	0.0	0.0	0.0	Active
DZ30-02GT-201	DDS#13	2/20/2018	6/28/2018	466	499	495	495	4.0	0.0	0.0	Active

Fig. 6.11: Summary report with exceptions highlighted (source: A. Neuwirt).

As user demands and reporting systems have become more sophisticated, there is a trend towards combining tabular presentation of data, such as sensor identification, with graphical presentation. Figure 6.12 provides an example report depicting two types of instruments. On the left is a tabular summary of instruments measuring the status of water in a series of standpipes using a wet/dry sensor; the plot on the right shows current water level as measured by a pressure transducer.

The following capabilities must be supported by the reporting engine to effectively utilise the instrumentation data for risk or production management purposes:

- ability to easily configure date range of interest for one or more instruments;
- flexibility to select the measurement units and automatic conversion to other units (e.g. conversion of Fahrenheit to Celsius);
- ability to refactor or recalculate data flexibly. This includes not only simple math operations, but also more advanced math functions such as aggregates and historical functions;
- ability to easily create statistical reporting of data. This helps to identify developing issues with quality or reliability of data sources;
- ability to utilise data from different instruments in the same calculation. This requires that the calculation engine supports data that have unsynchronised time stamps;
- flexibility of format in terms of data presentation. Ideally this includes helpful text descriptions based on the data or on calculation results derived from the data. This requires the reporting engine to have scripting of programming support; and
- ability to provide forecasts from one or more data sources. At a minimum this includes trending functions, but could also include output from models.

6.3.5 Challenges

There are several challenges to deploying and maintaining the type of infrastructure best suited to handling large quantities of data and related high-performance

WET/DRY Status	
PZ_TH2_1_C	DRY
PZ_TH2_2_C	DRY
PZ_TH2_3_C	DRY
PZ_P_22_C	DRY
PZ_3_3_C	DRY
PZ_TH2_3A_C	DRY
PZ_TH2_2A_C	DRY
PZ_TH2_2B_C	DRY
PZ_TH2_1C_C	DRY
PZ_TH2_1D_C	DRY
PZ_TH3_12P1_C	DRY
PZ_TH3_12P3_C	DRY
PZ_TH3_12P4_C	DRY
PZ_TH3_12P5_C	DRY
PZ_TH3_12P6_C	DRY
PZ_Old_A_C	DRY
PZ_4_E_C	DRY
PZ_TH4_3A_C	DRY
PZ_TH4_4A_C	DRY
PZ_TH5_12P1_C	DRY
PZ_TH5_12P2_C	DRY
PZ_TH5_12P3_C	DRY
PZ_TH5_12P4_C	DRY
PZ_TH5_12P5_C	DRY
PZ_TH5_12P6_C	DRY

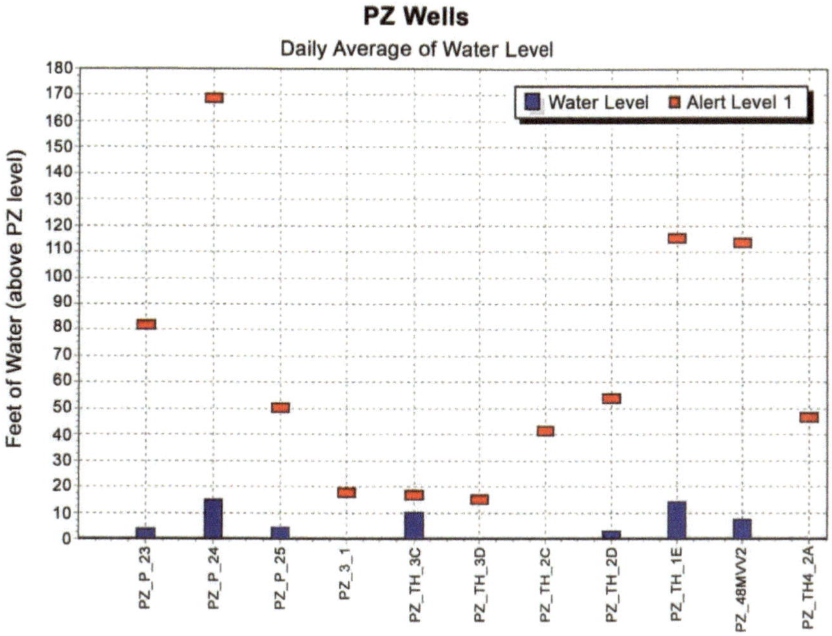

Fig. 6.12: Graphical plot of water level measured by a pressure transducer and a tabular summary of borehole status measured by a wet/dry sensor (source: A. Neuwirt).

requirements. The most significant challenges are summarised as follows.

- Deploying infrastructure can be problematic, considering the challenging working environment for the equipment. Also, priority in a mining operation is given to production equipment and production operations, so access to data management equipment and other resources may be limited.
- Maintaining equipment in the challenging and dynamic environments that characterise mining is difficult. Servers may be in less than ideal environments and exposed to water, dust, extreme and wide temperature fluctuations, vibration or noise, leading to premature machine failures.
- Mine operators are experts at mining operations, but there are often gaps in the IT domain.
- Electronic hardware and software are in a continual state of change. Simply keeping systems and software up-to-date requires substantial resources, and priority for resources may be given instead to mine production.
- The sheer quantity of information demands substantial storage and computing resources. The requirement for these resources is again competing with the systems that provide directly for mine production.
- The use of virtual machines (VM) has become common. These provide easier management of computing infrastructure by allowing multiple virtual servers to exist on a single build of physical hardware. However, in many cases, the use of VMs has introduced substantial degradation of computing performance due to the nature of the shared resources. Performance degradation is often particularly acute with storage systems, due to the shared nature of the connections and disks/drives.

There are also challenges of evolving needs for visualising and communicating the data. Simple tabular presentations of data are decreasingly useful or even no longer acceptable for proper data utilisation. Simple graphical presentations of data from a single instrument or type of instrumentation, that can be easily created with software such as Excel, are limited in their ability to convey meaning for multiple types of data in the quantities being collected.

The rise of database management systems, to more easily aggregate information and allow correlation of different data types and warehousing of the large quantities of data being created, has necessitated more thoughtful infrastructure design and more resources for implementation and support. If appropriate resources are not devoted to these systems, then much of the value of the investment in monitoring technologies will be lost as the data will have limited value for supporting operational and safety risk management.

6.4 Data validation and reliability

6.4.1 Overview

Slope performance monitoring is an integral part of effective slope management. It provides not only a safe working environment, but also important slope performance evidence for the calibration and optimisation of designs. The most important purposes of a slope monitoring program are to:

- maintain safe operational practices;
- provide advance notice of instability; and
- provide additional geotechnical information regarding slope behaviour (Sjöberg 1996b).

Monitoring is routinely adopted in the mining environment to manage the residual risk associated with uncertainty in design. Even the most carefully designed slopes may experience failure from unknown structures, unexpected weather patterns or seismic loading (Girard and McHugh 2000).

Eberhardt and Stead (2011), Marr (2007), Dunnicliff (1993) and Sullivan (1993) provide more detailed discussion on the objectives and planning of monitoring programs. A useful reference for the methods and monitoring requirements for open pit applications in relation to the type and scale of the mine slopes being considered is provided by Hawley *et al.* (2009).

Key requirements of a well-designed monitoring system include:

- early detection of potential incipient instability;
- communication of alarm exceedance;
- analytical functionality in assessing current conditions and estimation of forecast predictions;
- confirming physical characteristics (e.g. deformation, pore pressures, etc.) are within the expected or required performance and serviceable range; and
- identifying, for the purpose of data reliability, the critical elements which will ensure the success of the monitoring program.

The utilisation of geotechnical instrumentation does not involve merely the selection of instruments. It requires a comprehensive step-by-step engineering process beginning with a definition of the objectives and system design criteria, and ending with communication of the data and decision-making that can impact operational safety and performance. Each step is critical to the success or failure of the entire program, and the engineering process involves combining the capabilities of the instruments and people (Dunnicliff 1993).

The purpose of geotechnical instrumentation is to assist with addressing specific questions or concerns about the behaviour of the pit slopes in response to mining or construction activity. If there is no activity that can influence behaviour, there should be no instrumentation (Dunnicliff and Powderham 2001). Questions typically arise where a geological feature or design input is recognised to control slope stability, but the impact on slope performance is uncertain (known unknown), or around addressing areas of concern where reliable slope design input data are scarce (unknown unknown).

The following must be considered.

1. What are the objectives?
2. What is the risk and consequence of getting it wrong?
   i) Indicate impending failure (provide warning).
   ii) Reveal unknowns.
   iii) Evaluate critical design assumptions.
   iv) Control the excavation process.
   v) Provide data to help select remedial stabilisation methods.
   vi) Inform stakeholders.
   vii) Satisfy regulatory requirements.
   viii) Advance state-of-knowledge (Marr 2007).
3. What parameters must be measured and to what sensitivity and frequency?
   i) What are the expected characteristics of the movements (location, magnitude, vector, rate)?
   ii) When must the instruments be installed to achieve the objectives?
   iii) What physical attributes/specifications of the instrument are required to satisfy these objectives?
   iv) What density of monitoring is required to detect critical deformation (considering rock mass or structure variability)?
   v) What is the optimal location (and orientation) for installation?
   vi) What types of instrumentation are suitable?
4. What level of redundancy (if any) is required?
5. What technical and maintenance support are required?

Addressing these elements and questions will define the system requirements, including functionality and reliability, with more rigorous requirements depending on the significance of the consequence and risk potential being managed.

Broadly, three monitoring levels are defined (de Graaf and Wessels 2016):

- *critical monitoring*: normally undertaken in high-risk areas. It requires real-time monitoring and 24/7 alarming, and is mandatory while mining operations continue in the area;
- *diagnostic surveillance monitoring*: typically applied in moderate risk areas, but can be employed in any area. Monitoring may include real-time monitoring and 24/7 alarming; and
- *background monitoring*: the baseline monitoring that is employed on all pit slopes (specifically low-risk slopes). This monitoring is normally done with visual inspections and prism monitoring, and the frequency can vary. Although slope monitoring radar is not normally used in these circumstances, it can be

deployed where limited or no prism monitoring is available. It is expected that where radar is deployed, it will be fully geo-referenced and alarmed at all times.

The monitoring plan is not static and may change over time. The reliance on performance monitoring in the risk management plan may also change over time, due to changes in instrument response, such as a trigger alarm being exceeded, or change in risk exposure during mine development. Consequently, the monitoring system must have the capacity to evolve during mine development to address the expected changing requirements over the life of the project.

6.4.2 Data reliability

Data reliability is critical: poor-quality or inaccurate data can be misleading and are worse than having no data. The reliability of reported instrument readings is typically described by terms that are sometimes misunderstood or improperly used interchangeably, namely, 'accuracy', 'precision', 'resolution' and 'full scale'. These four inherent characteristics of monitoring devices and the data they generate are materially different.

The American Society of Testing and Materials (ASTM) has defined a standard for each of these terms in the field of engineering. Ideally, we would like to have instrumentation that produces reliable data with a combination of high resolution, accuracy and precision over a long measurement period. However, there is some trade-off between these attributes and cost implications.

Sensor specifications, such as resolution, precision, accuracy and full scale, help us understand the effective range, reliability and potential inherent error in resultant data. These elements must be carefully considered in selecting the appropriate monitoring device for a particular application. Arguably, in many mining geotechnical applications the relative measurement change or rate of change is a more important trigger than the precise absolute measurement at a point in time. Nevertheless, reading reliability must be aligned with the monitoring system requirements.

For example, in applications where brittle catastrophic slope failure could occur with only very small displacement as a precursor to sudden collapse, very sensitive monitoring devices operating over a small base length are required. In other applications, where substantial deformation is expected and can be tolerated (e.g. waste dump), a lower-resolution system with a large range of measurement may be suitable.

6.4.3 Resolution

Sensor resolution specifies the smallest measurement or change in measurement that the sensor can report on the readout scale. This is conceptualised in Fig. 6.13 using two targets with different resolutions, illustrated by the different number and thicknesses of the bands. The bullseye represents the true value. A 10 cm resolution instrument would report 1 m if it perceives the result to lie anywhere between 0.95 m and 1.05 m. This means that 1 m will be reported as long as the target is seen as sitting anywhere in that 10 cm range. Alternatively, if the sensor has 1 mm resolution, the bullseye would be seen as sitting between 999.5 mm and 1000.5 mm.

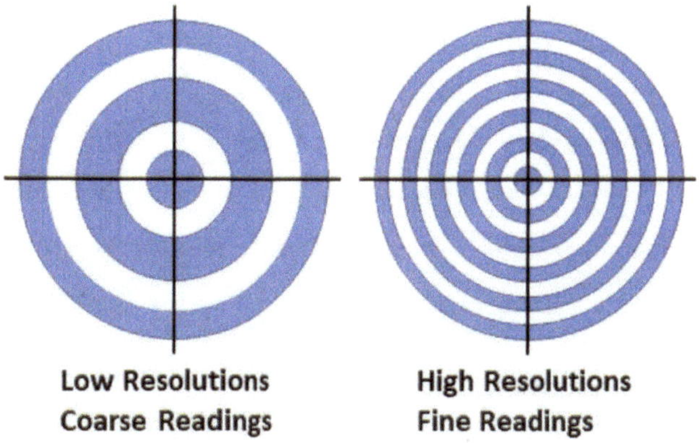

Fig. 6.13: Bullseye charts used to visualise relative resolution (source: MaxBotix).

6.4.4 Accuracy

Accuracy is defined as 'the closeness of approach of a measurement to the true value of the quantity measured' (Dunnicliff 1993). Accuracy is synonymous with the degree of correctness and is typically presented as the maximum allowable difference between the reported value and the actual value. Accuracy is evaluated during device calibration and is reported as a ± number. When a device is stated to be accurate to ± 0.5 mm, the measured value may vary by up to 0.5 mm from the true value; ± 1% RD accuracy means the reported value is within 1% of the true value. Often suppliers specify accuracy as a ± percentage of full scale (FS). An instrument with an accuracy ± 1% FS indicates a lesser accuracy than ± 1% RD, since the percentage applies to the full-scale range of the device rather than the measured value.

6.4.5 Precision

Precision is 'the closeness of approach of each of a number of similar measurements to the arithmetic mean' (Dunnicliff 1993). This is a measure of randomness of measurement, synonymous with reproducibility and repeatability. Similar to accuracy, precision is reported as a ± value. The greater the number of significant figures specified, the higher the precision, or the better the reproducibility. System precision of ± 0.25 mm does not necessarily indicate a system accuracy of ± 0.25 mm. It is

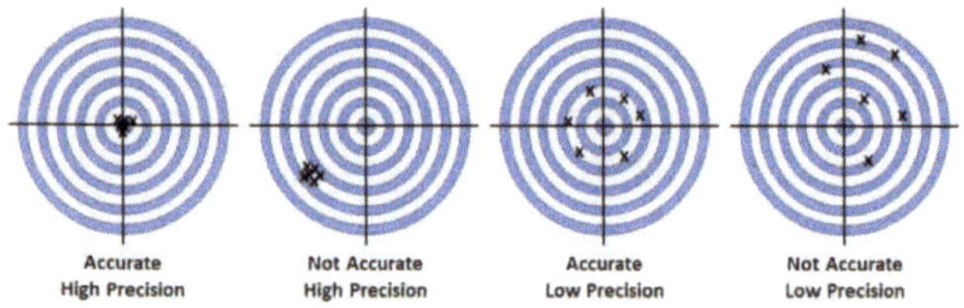

Fig. 6.14: Precision *v.* accuracy bullseye comparison (source: MaxBotix).

possible, for example, to have a system with 0.25 mm precision and, say, 2 mm accuracy.

The difference between accuracy and precision is illustrated in Fig. 6.14, where the target bullseye represents the true value. From left to right, the first image illustrates the ideal condition, where measurements are highly repeatable (precise) and close to the true value (accurate). In the second image, measurements are precise but not accurate – this may be representative of a systematic error, and the results may still be useful for identifying changes in trends or rate of change. In the third image, the measurements are not precise but, provided sufficient readings are taken, the average will be accurate. In the final image, measurements are neither accurate nor precise, which is unsatisfactory for most monitoring situations.

6.4.6 Full scale (range of measure)

In measurement devices, full scale is the maximum (or minimum) representable value and is analogous to the range of measure. Once a device has reached full scale, any further increase will result in an error known as clipping. Using the bullseye analogy, Fig. 6.15 shows similar measurement data at two different instrument full scale ranges. In the image on the right, half of the data are censored where it is clear that the full range is inadequate to capture the data distribution.

6.4.7 Errors

All measurements include some degree of error. Measurement error is defined as the difference between the measured value and the actual or true value. Understanding measurement error and sources of error are essential elements in planning and managing monitoring systems. Monitoring system errors can introduce uncertainty in interpretation and may result in significant mistakes.

Measured value = true value + measurement error

Three principal types of measurement error are recognised:

Error = gross error + systematic error + random error

Gross errors include unpredictable human mistakes, carelessness in reading instruments and recording and calculating measurements, or equipment failure. These outliers are typically significantly different from the true or expected value and are usually discarded when assessing the data.

Systematic and random errors are typically associated with measurement or instrument errors. Instrument error refers to the combined accuracy and precision of a measuring instrument or the difference between the actual (true) value and the value indicated by the instrument.

Systematic error, or bias, is a permanent deflection in the same direction as the true value. If identified and understood, it can be corrected or compensated.

Random error is associated with short-term scatter of readings around a mean value and cannot be predicted. Sometimes these errors are compensating and tend to at least partially cancel themselves mathematically. They cannot be corrected on an individual measurement basis.

In recognising and isolating the potential sources of error, they can be differentiated into controllable and uncontrollable elements. Effort must be applied to managing the most significant controllable sources of errors. Controllable sources of error include instrument selection, specifications, installation, maintenance and calibration. Table 6.2 summarises causes and remedies for typical measurement errors. It should be no surprise that adequate training, experience, due diligence and careful deployment of well-engineered monitoring systems with good support will significantly mitigate most error sources.

Reliability is the degree to which the result of a measurement, calculation or specification can be depended on to be accurate. This is intrinsically related to measurement error and the data validation process. System reliability is discussed in more detail below, in Section 6.4.10.

Dunnicliff (1993) suggests two primary controls to ensure performance monitoring reliability:

1. instrument factors, including:
   i) simplicity – inherent in reliability is maximum simplicity;

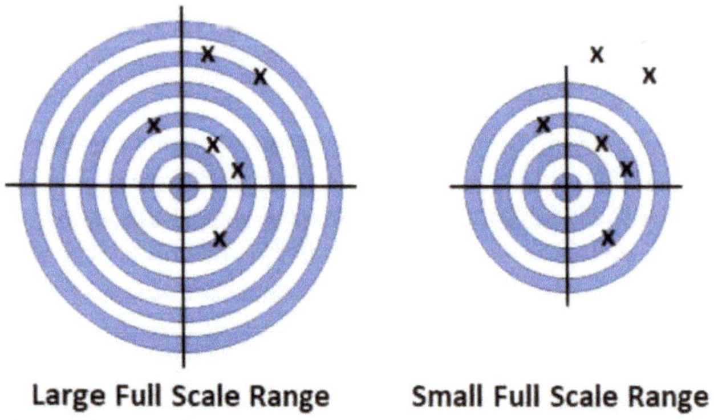

Fig. 6.15: Full scale bullseye comparison (source: P. de Graaf).

Table 6.2: Causes and remedies of measurement error

Type of error			Causes	Remedies
Gross error	Misreading Misrecording Computational error		Inexperience	Care, experience
			Training	
			Duplicate readings, data auditing	
			Dual observations	
			Checking against previous readings	
			Datalogger	
Measurement errors	Systematic error	Instrument error	Improper calibration	Use of correct calibration
			Loss of calibration	Recalibration
			Nonlinearity	Use of standards
				Use of consistent reading procedures
		Conformance error	Inappropriate installation details	Select appropriate instrument
			Instrument design limitations	Modify installation procedure
				Improve instrument design
		Environmental error	Weather	Record environmental changes and apply corrections
			Temperature	
			Vibration	
			Corrosion	Make correct choice of instrument materials
		Observational error	Variation between observers	Training
				Use of dataloggers
		Sampling error	Variability in the measured parameter	Install sufficient number of instruments at representative locations
			Incorrect sampling techniques	
	Random error		Noise	Correct choice of instrument
			Friction	Temporary elimination of noise
			Environmental effects (sudden changes)	Multiple readings
				Statistical analysis

Source: modified from Dunnicliff (1993).

ii) self-verification – failsafe built-in checking mechanisms, periodic instrument calibration; and
iii) durability in the installed environment – proven longevity to suit application.

2. human factors, including:
   i) thorough planning – logical and comprehensive engineering process;
   ii) initial calibrations and inspections – ensure instruments are functioning properly and to specifications before installation;
   iii) installation care – prevent inadvertent damage; and
   iv) regular maintenance and calibration – careful data collection (i.e. develop procedures), processing and interpretation required.

Accurate understanding and interpretation of data are required to facilitate the right decision-making at the right time in the right location. However, unidentified anomalies may result in suboptimal decision-making that may jeopardise the safety, environmental and economic objectives of a monitoring program. Therefore, it is vital to develop approaches to analyse the quality of the existing data. This must include identifying spurious error readings from inherent natural variability.

6.4.8 Managing data noise

Noise in data can mask true deformation, which can contribute to apparent changing trends of deformation and trigger false alarms, thus eroding confidence in the monitoring system. Data validation is the process of checking data to determine whether they are accepted or rejected based on established criteria. This involves critical review of the data to identify spurious readings. There is no universally accepted test for data validity. Data validation is integral to many fields in industry and research, where extensive temporal productivity and/or data quality are required to be analysed on an ongoing basis. Various approaches can be adopted. Ratliff (2003) provides a good overview of these; some are presented in the following paragraphs. They range from a cursory scan to detect extreme values or spot outliers, to detailed statistical evaluations. It is important that records are kept as to which readings are judged invalid or otherwise suspicious. No readings should be rejected without detailed review and documentation.

Manual or visual validation is by far the most common approach, but it relies extensively on expert judgement. This approach is useful for identifying extreme or outlying values which are noticeably higher or lower than expected, but to ensure consistency it should rely on predetermined criteria to define what constitutes an outlier. Validation criteria must be aligned with the monitoring requirements and objectives.

Obviously improved inherent system reliability will reduce data noise or flag poor-quality data. Some geotechnical data self-verification examples include:

- vibrating wire instrument data acquisition systems which have sophisticated built-in noise-filtering procedures to identify and eliminate unwanted noise (over and above the routine temperature corrections) to ensure high resolution output; and
- the prism monitoring system described by Audigé *et al.* (2017), which uses least-squares regression checks on overall data quality to determine whether data are acceptable.

Several statistical tools are available for the validation of data, such as the *t*-test, chi-square test and F-test. Some are integral to monitoring analysis software, but unfortunately are often underutilised. Sequential analysis is a powerful approach which allows a decision on acceptability of the data to be made with fewer readings; it can also, importantly, flag changes in trends. This includes Cusums and control chart techniques. The control charts concept, originating in the 1920s, is widely used in production and manufacturing. It has not frequently been applied in geotechnical monitoring, but it presents an opportunity for improved data validation and interpretation. This approach potentially provides a powerful tool for distinguishing the pattern of indeterminate (random) error or variation from the determinate (assignable cause) error or variation (Ratliff 2003).

This technique graphically compares the variability of all sequential data with the average or expected variability of small groups of data; that is, it compares 'within groups' variability *v.* 'between groups' variability. Control charts can be thought of as visual hypothesis tests.

When plotted on a run chart, such as the sequential time-series plot illustrated in Fig. 6.16, with upper and lower bound specification limits (typically ± 3 standard deviations) and automated flagging of variations, a control chart provides a first-pass indication of acceptability of data or changed conditions. If observations start to appear outside of these limits, we would suspect that the former process is no longer in control and that the data distribution has changed (Wild and Seber 2000). Automated flagging of variations from expected conditions can be identified on a control chart as follows:

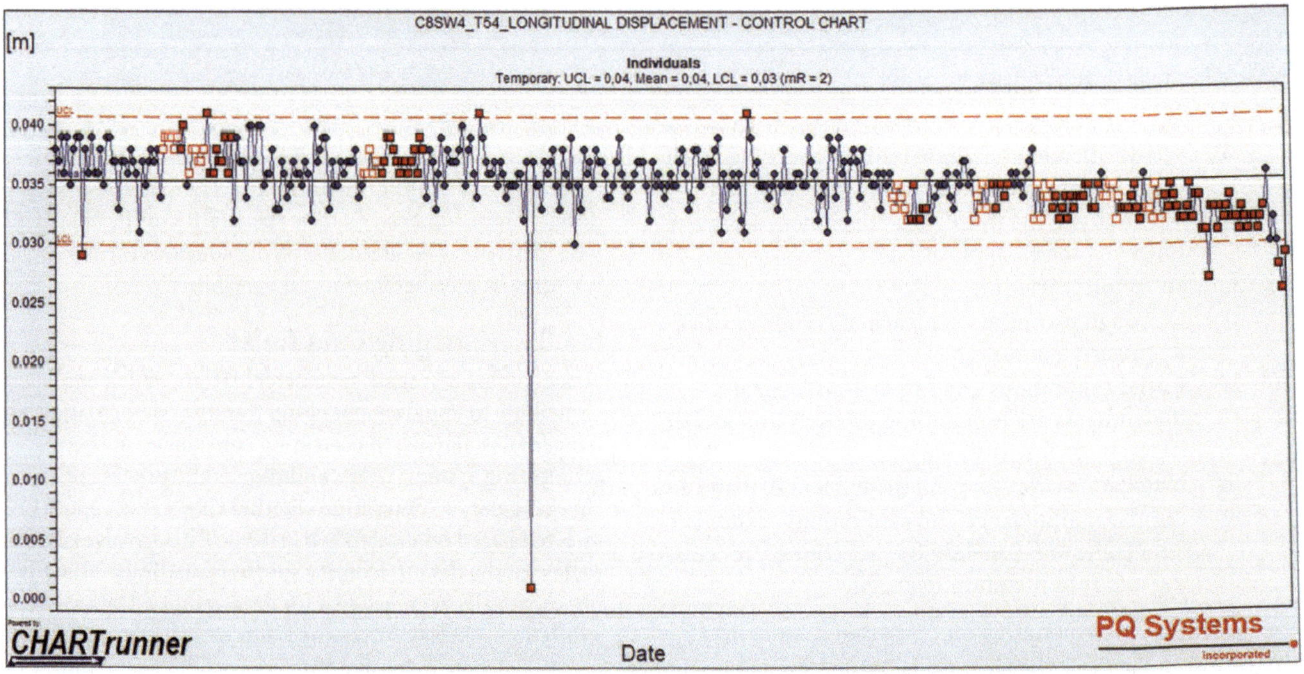

Fig. 6.16: Control charts can provide a first-pass indication of acceptability of data or changed conditions. Variations from expected conditions (i.e. variation from a statistically stable response) are flagged in red. Note the outliers and emerging changing trend (source: C. Visser).

- any point outside control limits (typically ± 3 standard deviations);
- run of seven points;
- in a row above or below the average;
- in a row in one direction, up or down;
- non-random patterns;
- too close to the average;
- too far from the average; and
- the occurrence of cycles.

A statistically stable process deviates about its centre in a random fashion. Non-random behaviour indicates that the process is not statistically stable over time. Gross outliers are a sign of an unstable process. This allows for rapid evaluation of system stability (Wild and Seber 2000). Data stability facilitates reliable retrospective analysis and better predictions. Control chart limits can also be used as alarm triggers for notification of a change in trend. These dynamic triggers are potentially useful secondary triggers.

Adopting appropriate statistical tests to flag potentially suspect data is a sound approach that can be utilised to remove subjectivity and efficiently review the often-substantial volumes of suspect data before the data are used for meaningful analysis. Unfortunately, few rigorous data verification features are available in most commercial geotechnical monitoring data software programs. Therefore, much of the current verification process is reliant on visual inspections. Marr (2000) suggests that data validation is an area that could benefit from additional research. Unfortunately, advances in data validation have yet to be made and it therefore remains an opportunity for further improvement.

6.4.9 Long-term trends and continuity of data

Long-term trends and data continuity are important for capturing the full deformation response of an excavated slope. When a change in the slope behaviour is observed, the historical data can be analysed to determine or interpret the reasons for the change. Further comparison to mining activity or other temporal data, such as rainfall, will contribute valuable insights into slope performance, such as:

- defining 'normal' or 'expected' rock mass behaviour;
- assisting with discriminating between noise and actual movement;
- validating appropriate trigger thresholds for hazard management; and
- identifying the sensitivity of pit slope movement to factors such as geology, mining rate, mining location relative to the slope, seismicity or rainfall, which can assist in improving pit slope design.

The following brief case studies illustrate comparisons of manual and automated prism data with slope monitoring radar data. Prism and GNSS monitoring have significant advantages in terms of long-term monitoring of slopes. With adequate redundancy and allowing for data losses due to minor rockfalls, blasting fly rock and dust, monitoring with prisms or GNSS facilitates a documented history of deformations that can be tracked over a period of several years. The length of time and typical wide coverage that prism and GNSS data offer are not currently viable for other monitoring methods.

Each case study also shows how the availability of historical slope monitoring data proved useful for developing a better understanding of slope behaviour, as well as providing insights into predicting future developments and facilitating effective slope management. It is important to keep the focus on early deployment of monitoring instruments and maintaining data continuity (even if only with manual readings). To achieve this, it is important to ensure that the system health (and all its components) is monitored. To ensure that all the data required for analysis are available and are reliable, the slope engineer needs to know that the system is working as intended.

6.4.9.1 *Case study 1: copper mine pit slope*

During a routine review of monitoring data from a large open pit copper mine, a change in the slope behaviour was observed. This was confirmed by visual inspection. Fortunately, historical manual survey data were available for review. Figure 6.17 shows the velocity trends for four months recorded for several prisms. The vertical red lines represent blasting events near the highwall; the response of the highwall movements is indicated by the velocity changes recorded for the prisms. Deterioration of the slope's stability state was evident from the increased velocities that follow the blasts, especially in the latter half of the data record where the minimum velocity to which the slope returned after each blast was higher than the previously recorded rate. The trend for monitoring pit slopes that are showing signs of instability is to increasingly rely on GB radar for safety-critical slope monitoring. This case study nevertheless shows that manual prism monitoring can be a valuable slope management tool.

6.4.9.2 *Case study 2: coal mine footwall slope*

An example of prism monitoring over a seven-year period of a footwall slope in coal measure rocks is shown in Fig. 6.18. The prisms were read manually; noise was in the order of ± 30 mm. As such, if the data are only looked at over a short period of time (e.g. the three-month window shown in the insert), it may not be possible to define a trend of movement. However, if viewed over a longer time-frame, the following observations can be made from the time-series trend:

- the initial data up to the end of 2011 indicated very low movement rates, in the order of 4 mm/month;

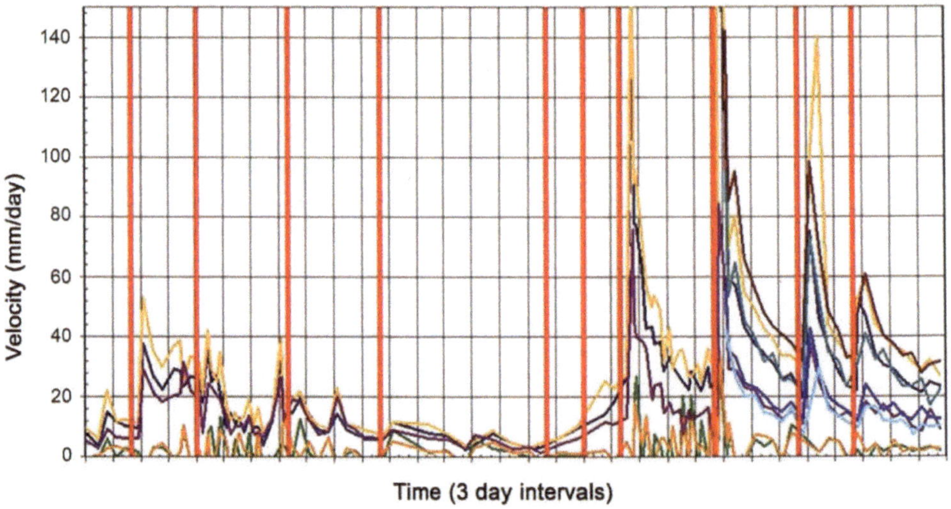

Fig. 6.17: Analysis of manually monitored prism data over a four-month period that shows the effect of blasting on slope movement. The red vertical lines delineate blast events. The velocity trends indicate increased deterioration of the slope's stability state over the monitoring period (source: F. Wessels).

- there were four periods (early 2011, mid 2012, early 2013 and end 2013) with clearly faster rates of movement, although these periods of faster rates were not evident in all prisms. The faster rates were similar through all four periods and nominally 40 mm/month;
- each period of faster rates was followed by a period of decreased rates similar to the initial rate of ~4 mm/month (shown as the black dashed lines); and
- there was a clear change in behaviour in early 2015, with significant change in rates by mid 2015 where rates exceeded the periods noted above.

This example highlights the value of maintaining and reviewing the history of slope movements over the long term. It is noted that the rates of movement before mid 2015 would have been unlikely to trigger operational responses. However, from an engineering perspective, they allowed an understanding of what would be considered the creep rate for the slope (4 mm/month) that should have triggered investigation into the causes of the increased rates as far back as mid 2011 and provided ample warning of the impending failure, which occurred towards the latter half of 2015. However, the

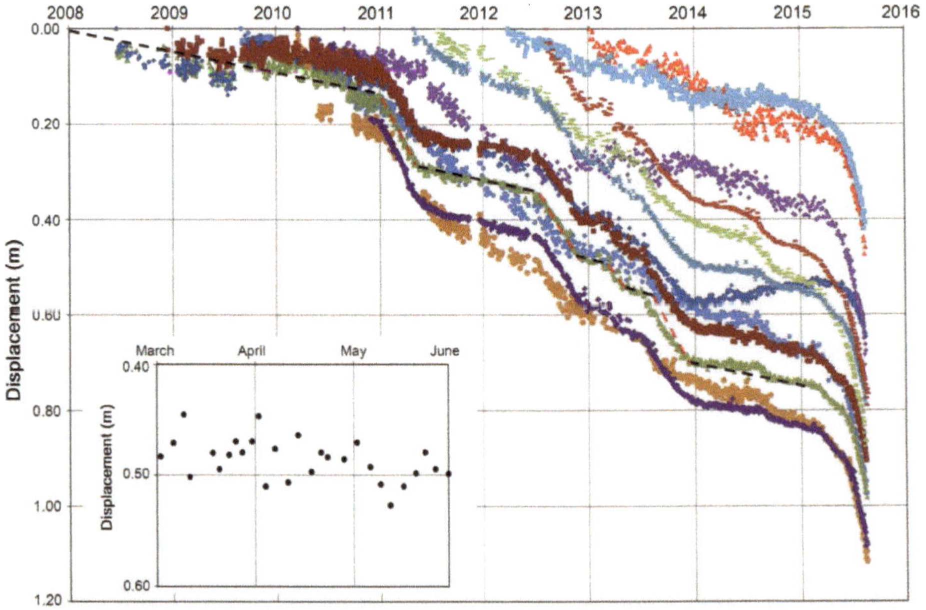

Fig. 6.18: Prism monitoring of a footwall slope showing clear long-term trends over a seven-year period. The inset graph indicates short-term variability and no discernible change in movement trend over a sample three-month period (source: A. Duran).

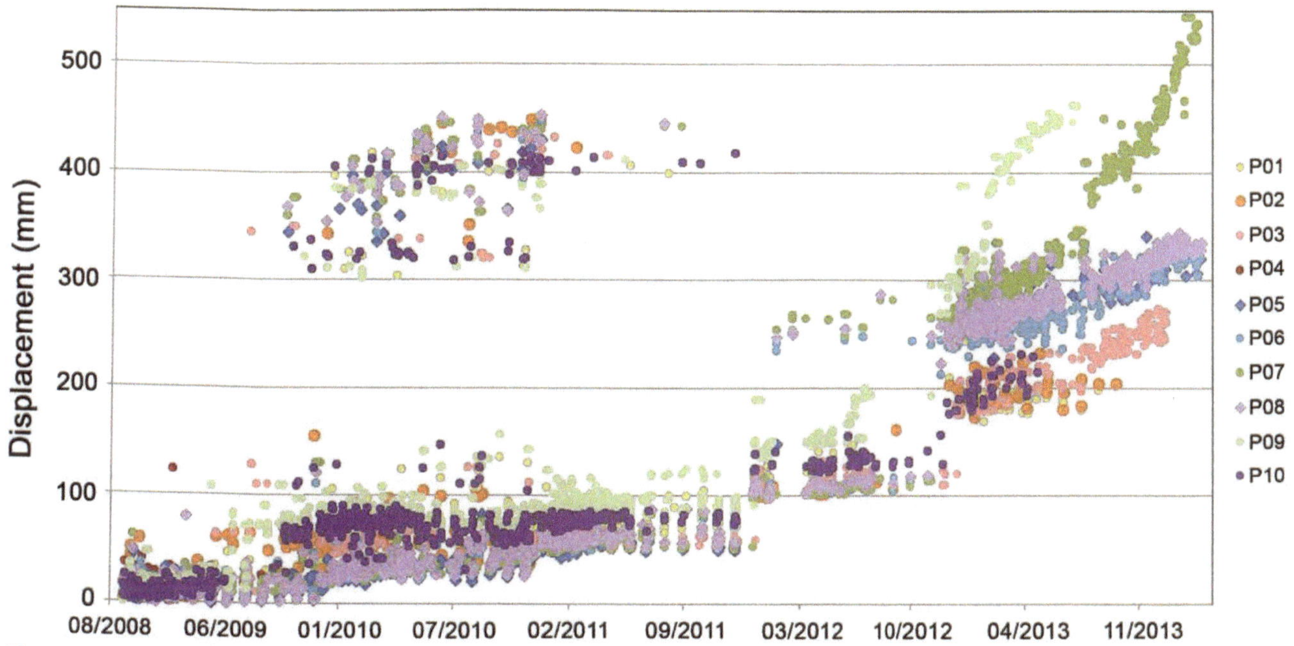

Fig. 6.19: Prism monitoring at a large copper porphyry mine over a six-year period from 2007 to 2013 (source: F. Weir).

opportunity to investigate the underlying causes of the faster rates was not taken. Hence, there is uncertainty about whether remedial measures could have been implemented and, if so, whether they would have succeeded in mitigating the instability.

6.4.9.3 *Case study 3: copper porphyry mine pit slope*

An example of prism monitoring from a large copper porphyry mine over a six-year period is shown in Fig. 6.19. The prisms were located on both sides of a small failure. They clearly demonstrated that the entire region of the pit wall was moving, not just the failure zone itself. The monitoring data indicated that this area of the pit wall began moving in late 2008, with the following observations:

- from 2007 to 2009, a notable jump in the data in the order of ± 100 mm was measured. The cause of this jump was not sufficiently investigated at the time. Several possible contributing factors were interpreted;
- prism data can require longer time-frames to identify trends;
- long-term continuing movement of the wall of nominally 2.8 mm/month occurred from 2007 to mid 2012;
- an acceleration event occurred in 2012 with an increase in rate to nominally 6 mm/month. The acceleration was related to mining at the toe and ongoing water management issues;
- a further increase in rate to nominally 22 mm/month occurred after mid 2012;
- anomalous readings (300–450 mm) for several prisms were recorded multiple times during 2010 and 2011. In the context of the longer-term trends, it is evident that these were erroneous, possibly due to manual reading errors; and
- one prism underwent a total displacement of over 450 mm before it was lost in a failure event.

This case study again highlights the value of maintaining and reviewing the history of slope movement over the long term. Prism data clearly demonstrated that the entire area of the pit wall, not just the small embedded failure, was moving. If understood and addressed early, a larger-scale failure could have been predicted.

6.4.9.4 *Case study 4: iron ore mine pit slope*

A structurally controlled instability occurred in this iron ore mine. A section of slope, ~120 m high and from ~200 m to 300 m in width failed, resulting in ~0.7 Mt of debris. The instability was controlled by a known but not well-modelled shale band and triggered by blasting near the toe. The historical automated prism monitoring data were reviewed as part of the back-analysis. Prism monitoring data were available for at least two years before the event. Total vector (North, East, Elevation) data from three prisms, covering this period, are plotted in Fig. 6.20. Notable blasts occurred in close proximity to the ultimate failure in November 2008 and December 2009; these blast events are shown in Fig. 6.21.

A review of the prism monitoring data indicated a brief period of increased velocities shortly after each blast event, particularly the November 2008 blast event. The increased

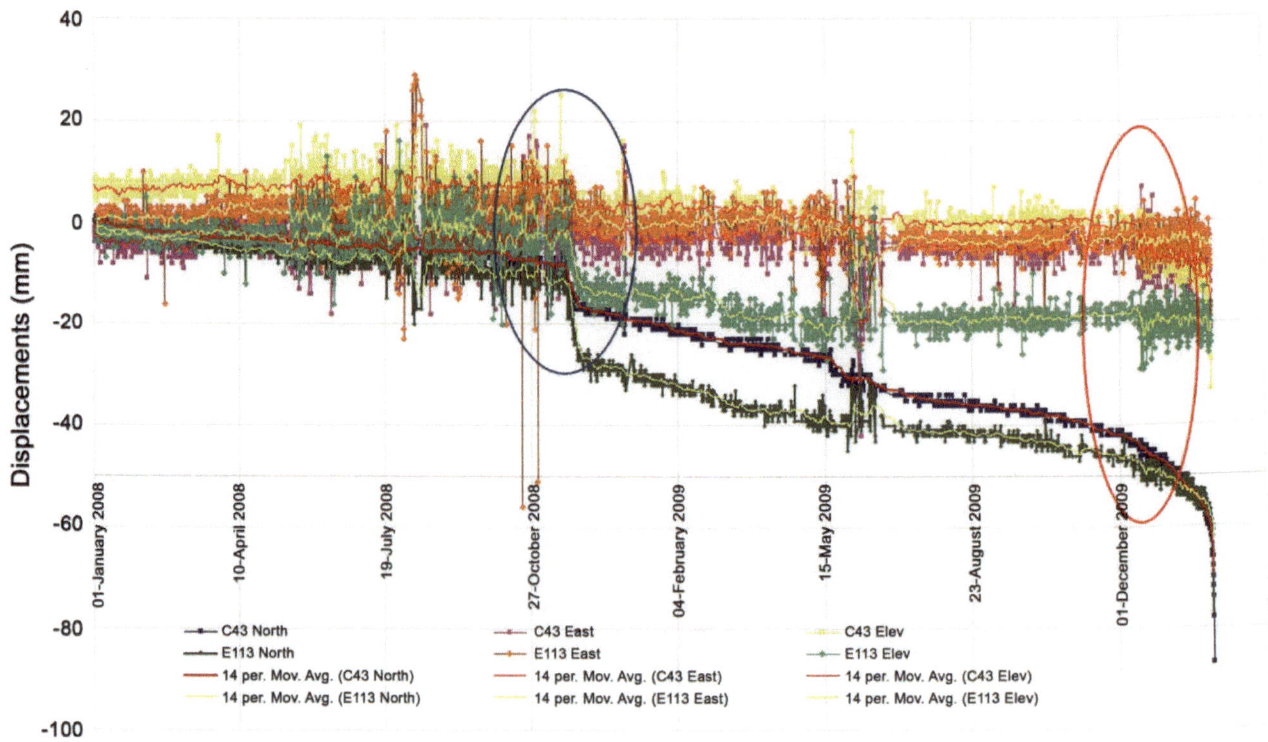

Fig. 6.20: Analysis of historical prism data indicated a sensitivity to blasting activities of a slope (notable blasts occurring in November 2008 and December 2009). The jump and increase in northing displacement data following the November 2008 blast occurred 14 months before the slope failure (source: F. Wessels).

velocity from the November 2008 blast event was recorded ~14 months before the slope failure.

In this case study, the historical prism data assisted in the interpretation of contributing factors leading to slope failure. It appears that the November 2008 blast event resulted in a reduction in the shear strength of the shale band. Failure onset was triggered when the bench of the toe of the slope, which provided a buttress, was mined out. This understanding resulted in improved blasting controls for the remainder of mining of the area.

6.4.9.5 Case study 5: dragline strip mine low-wall dump slope

In a dragline strip mining operation, the low-wall dump is a dynamic environment which poses challenges to conventional geotechnical monitoring. With the advent of slope radar monitoring, it became feasible to effectively monitor for displacements which may be a precursor to low-wall instability. The dragline profile, at the toe of a low-wall dump, comprises blasted overburden placed via a dragline in a loose state which can be at, or steeper than,

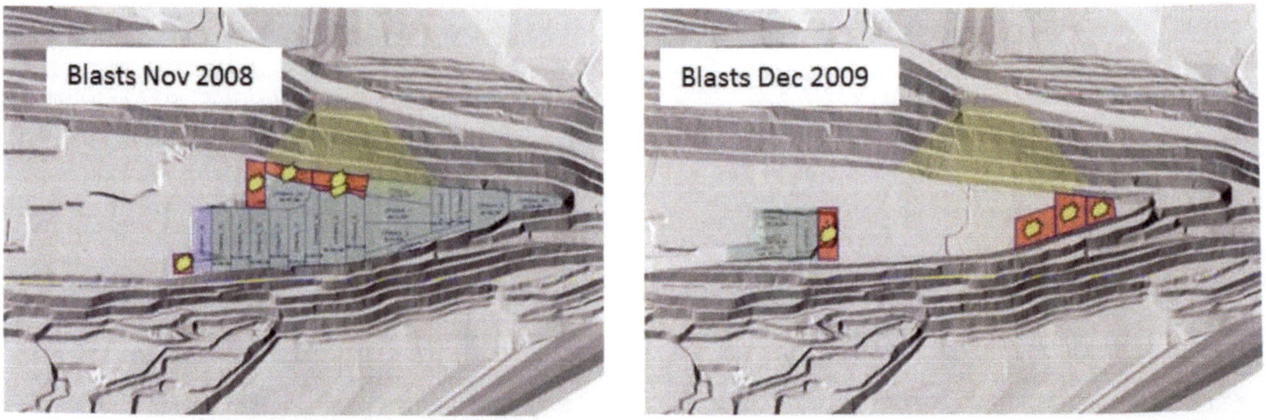

Fig. 6.21: Cross-referencing of mining activities with the monitoring data. The blasts on the left correspond with the initial increase in velocities recorded in November 2008. The blasts on the right triggered the second phase of increased velocities in December 2009 (source: A. Creighton).

Fig. 6.22: Pre- and post-failure photographs (upper and lower images, respectively) for a low-wall dragline mining operation (source: A. Creighton).

the angle of repose. The overburden is relatively recently placed and hence is subject to settlement. Vibrations from production blasting will result in further disturbance. Hence, displacement in a low-wall can be expected. The challenge at this coal mine site was to discern precursors of instability from background displacement trends.

In October 2016, the dragline profile and low-wall dump failed, involving a total displaced volume of ~38 million m^3. The failure extended ~1.3 km along the length of the mine and ~800 m behind the low-wall toe. Photos of the pre- and post-failure conditions are shown in Fig. 6.22. A summary of mining activities for the operation for the period 2013–2016 is shown in Fig. 6.23.

The failure was preceded by a similar event in July 2013, which had been interpreted as comprising shear failure on a relatively steeply dipping, very low shear strength, tuff layer within the floor of the pit, in combination with elevated pore pressures.

An extensive process of back-analysis and predictive stability analysis of the earlier event resulted in a recovery mine plan that prompted the operation to implement a monitoring strategy which involved slope monitoring radar. The monitoring strategy called for the radar to be deployed in a tactical manner, while coaling activities were undertaken. Tactical monitoring in the context of this case study refers to the radar being used to detect movement and to track and report that movement relative to a trigger action response plan (TARP). The radar was deployed elsewhere when active coaling was not taking place. Deployment of the radar was effective in detecting and tracking low-wall displacements, and the pit was evacuated a few days before the failure.

The geotechnical investigation into the 2016 failure was examined in some detail and concluded the following:

- the tactical use of the radar, as defined above, over successive episodes of mining meant that no one was looking at, or attempting to understand, low-wall performance relative to the predictive design acceptance criteria;
- combining and examining the raw radar movement rate data from the mining period July 2013–November 2016 showed that there was a direct correlation between movement rate, mining method and the geology which relates mining activity timeline, coal geology and radar movement rate;
- the long-term raw radar data, which were provided as a moving average velocity (expressed as mm/day) over a rolling 60 min period and a 24 h period, respectively, were plotted and examined with time for

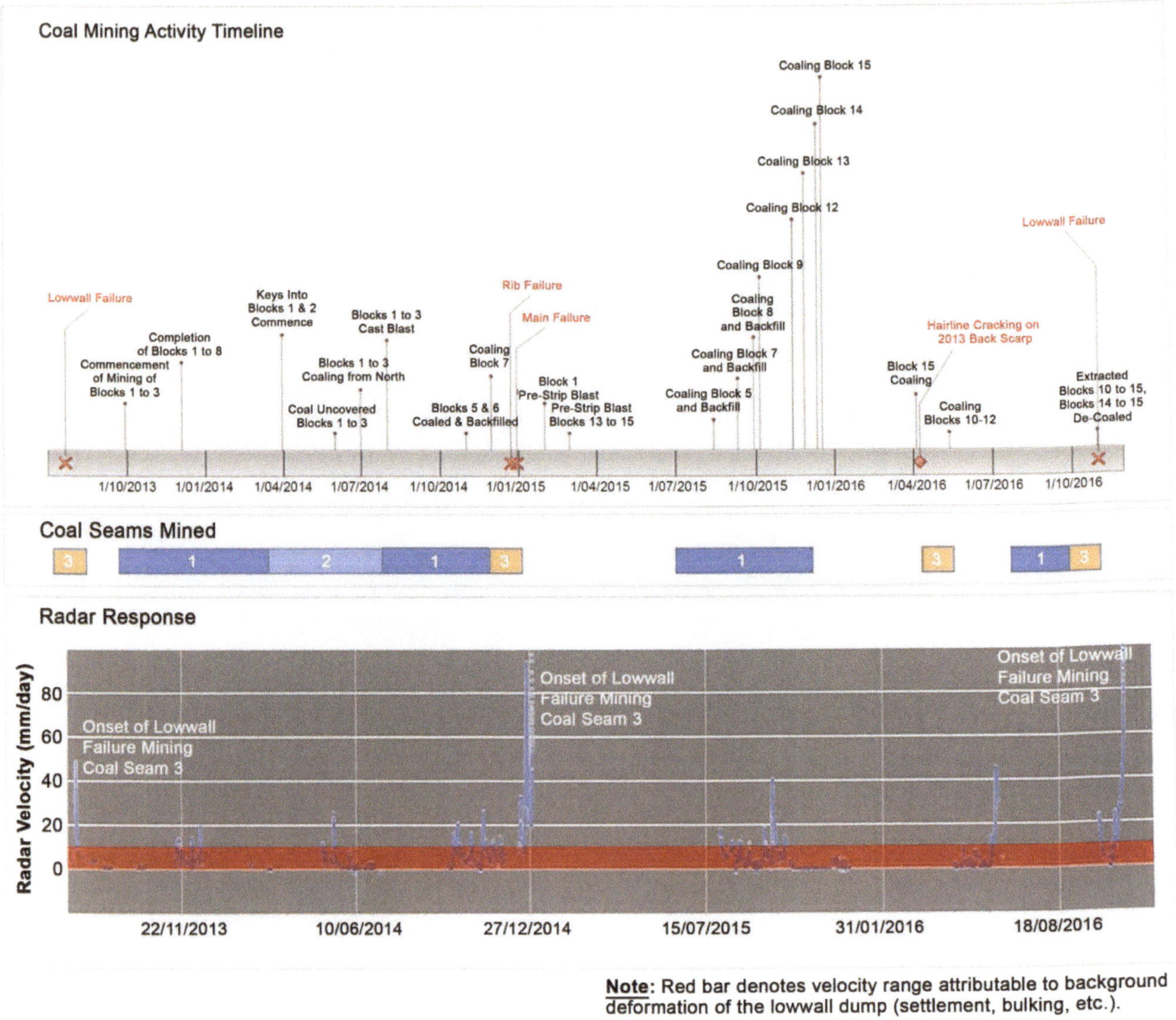

Fig. 6.23: Event time-series showing key slope performance, mining stages and radar velocity plots (source: A. Creighton).

the period 2013–2016. By comparing observed low-wall performance with the measured movement rate, it could be discerned that 'normal' or 'tolerable' low-wall behaviour was represented by a movement rate of up to 10 mm/day over a 24 h period (represented as the red band in Fig. 6.23). Movement rates which exceeded this threshold were anomalous and possibly indicated the onset of movement which could lead to instability; and

- this assessment also indicated that significant accelerating movement was experienced when mining Geology Type 3 material. If this had been recognised in the period leading up to the failure, it may have contributed to a change in mining approach or contingency planning which could have prevented the failure from occurring.

6.4.10 System availability

The full system health should be monitored and availability reported to stakeholders. This includes overall system health, performance of individual sensors, data communication from the field sensors written to the respective databases, and the alarm notifications to first responders.

System health demonstrates that availability is maintained and is functioning to predefined monitoring requirements. It is important that a set of key performance indicators (KPI) are tracked and reported on a regular basis. Not only will this provide confidence to stakeholders that monitoring systems are operational and effective, it will also assist to identify trends and weaknesses in the system. Some of the parameters that should be reported to

Radar Weekly Health Checksheet

RioTinto			
Radar Name	MSR048	Mine Site	TOM PRICE
Radar Location	SEP SOUTH WALL	Date	23/01/2015
Scan region	SEP NORTH WALL & STR3	Personnel	D. E.
Wall file name	RADAR-23/01/2015	Monitoring Type	CRITICAL
Reason for Monitoring area and Monitoring Type assigned			

At Radar	✓ / ✗	Details
Radar isolated with Cones / Signage	✓	
Pad condtion (evidence of erosion, washout)	GOOD	
Oil / fuel leaks / spills	NO	
Oil (level)	✓	ON EXTERNAL ✓
Coolant (Groundprobe only)	n/A	
Fuel (level)	✓	NEEDS
Dust (computer box)	NO	
Operation of computer screen & keyboard	GOOD	
Generator Hrs	✓	11669:24:44
Fire Ext/s (6 mths from tally plate: mth/year ie Jan15)	✓	
Hard drive (monthly download)	✓	
In office at PMP		
Age of data (last updated)	✓	
Age of wall file (restart after 1 month)	✓	
DTM / Photo alignment	✓	
Georeferenced	✓	
Alarms enabled to OC	✓	
Alarms tested	x	
Restart computer	✓	AM on 23/01/2015
Has radar been moved in past week?	x	

Details on any new fault, breakdown or standby in past week

ON EXTERNAL GENERATOR

Signed	[illegible]	Date	23/01/2015

Process: Scan file and email to Radar Officers by COB Friday each week (Filename eg:MSR047_yyyymmdd)
Radar Officers to review and provide to Radar Vendors

Fig. 6.24: Example of a daily radar checksheet (source: F. Wessels).

track the performance of the monitoring device and alarming are listed below.

- *Availability of sensors and monitoring devices over the reporting period*: this includes logs of downtime or outages, reasons for downtime (e.g. device problems), communication breakdown and network issues. Details of the downtime should be recorded to ensure the correct remediation actions are implemented. For example, slope radar performance should include details on generator hours, fuel status, maintenance checks, area scanned and current alarm trigger thresholds. An example daily radar check sheet is provided in Fig. 6.24, a weekly summary status dashboard for multiple radars is presented in Fig. 6.25 and a similar template for prism monitoring is shown in Fig. 6.26.
- *Currency of data*: it is important to confirm that the sensor has recorded and transmitted data on the expected frequency. This helps troubleshoot issues, such as whether the instrument is functioning and whether data have been transmitted and written to the database.
- *Alarm testing and evacuation trials*: slope monitoring alarm reporting processes should be checked and reported on a regular basis. This is to provide stakeholders, including mine management, with the assurance that monitoring alarms will be reported and that there will be a correct response of personnel to alarms. It is prudent to perform regular tests of alarm notification communications to confirm that the system is functioning, and to undertake evacuation

<u>Radar Status</u>

	Brockman 2 SSR067	West Angelas MSR047	Tom Price MSR048	West Angelas MSR076	Hope Downs 1 MSR077
Area monitored	BS2 Sth Wall	CEPN Sth wall	SEP north wall	CEPN Nth wall	NE3
Monitoring type*	Diagnostic	Critical	Critical	Critical	Critical
Alarming enabled	OC	OC	OC	OC	OC
Sensitive to mine plan**	Yes	Yes	Yes	Yes	Yes
Red Alarms	0	1	0	0	0
Orange Alarms	0	6	0	0	0
Grey Alerts	0	0	0	4	0
Weekly Availability	50	100	100	100	100
Use of availability (%)	100	99	100	100	96
Georeferenced	Yes	Yes	Yes	Yes	Yes
Weekly Checklist received	Yes	Yes	Yes	Yes	Yes
Next maintenance	TBA	19/01/2015	21/01/2015	19/01/2015	20/01/2015

*Critical: No mining without radar,	** Downtime can affect mining plan during the next week
*Diagnostic: Radar not a monitoring pre-requisite,	
*Background: Normal baseline monitoring	

- SSR067: Availability down as the radar was decommissioned during the week.
- MSR047: 6 Orange alarms due to weather events.
- MSR048:
- MSR076: 4 Grey alarms due to stabilisation and generator running too long events.
- MSR077: UoA down to 96% and Geotech alarms were due to stabilisation events.

Fig. 6.25: Example of a weekly summary status dashboard for multiple radars (source: F. Wessels).

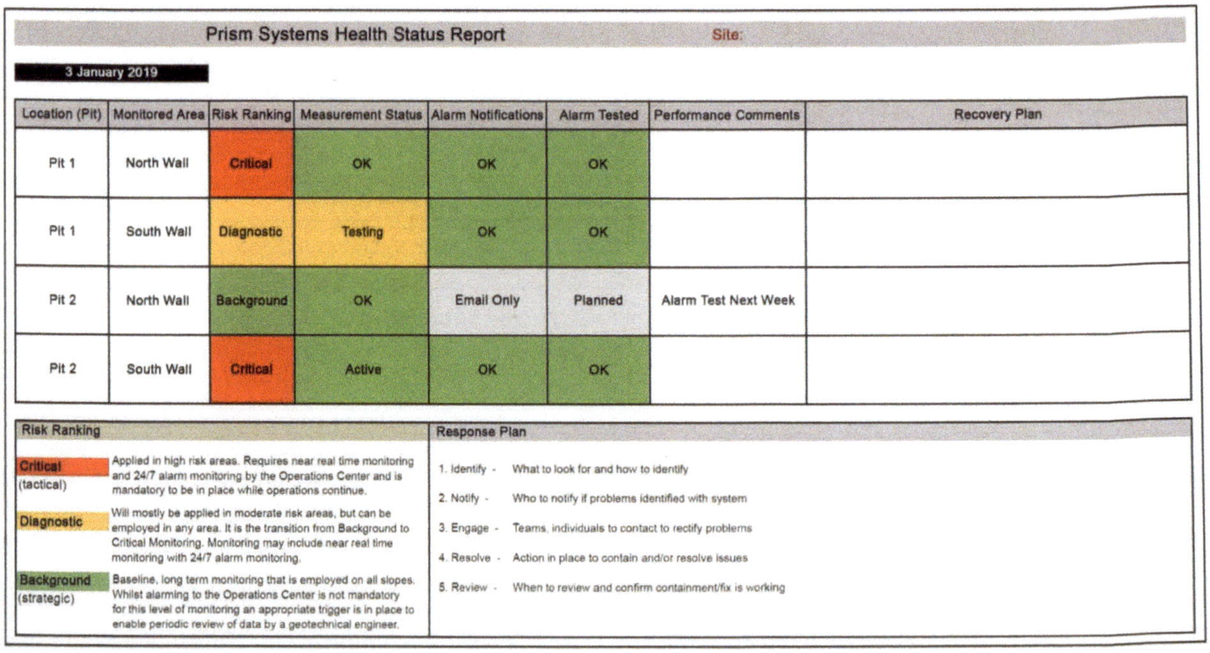

Prism Systems Health Status Report — Site:

3 January 2019

Location (Pit)	Monitored Area	Risk Ranking	Measurement Status	Alarm Notifications	Alarm Tested	Performance Comments	Recovery Plan
Pit 1	North Wall	Critical	OK	OK	OK		
Pit 1	South Wall	Diagnostic	Testing	OK	OK		
Pit 2	North Wall	Background	OK	Email Only	Planned	Alarm Test Next Week	
Pit 2	South Wall	Critical	Active	OK	OK		

Risk Ranking	
Critical (tactical)	Applied in high risk areas. Requires near real time monitoring and 24/7 alarm monitoring by the Operations Center and is mandatory to be in place while operations continue.
Diagnostic	Will mostly be applied in moderate risk areas, but can be employed in any area. It is the transition from Background to Critical Monitoring. Monitoring may include near real time monitoring with 24/7 alarm monitoring.
Background (strategic)	Baseline, long term monitoring that is employed on all slopes. Whilst alarming to the Operations Center is not mandatory for this level of monitoring an appropriate trigger is in place to enable periodic review of data by a geotechnical engineer.

Response Plan	
1. Identify -	What to look for and how to identify
2. Notify -	Who to notify if problems identified with system
3. Engage -	Teams, individuals to contact to rectify problems
4. Resolve -	Action in place to contain and/or resolve issues
5. Review -	When to review and confirm containment/fix is working

Fig. 6.26: Example of a weekly summary health status dashboard for prism monitoring (source: F. Wessels).

'dry runs' to ensure personnel are aware of how to respond when an alarm is triggered. These evacuation trials could be recorded and feedback provided in the routine monitoring report.

Monitoring device and system dashboards are essential to demonstrate system status and for the proactive maintenance of systems. If well-structured, these will help ensure that troubleshooting can focus on specific areas of malfunction. Specialised (multi-skilled) personnel are required to troubleshoot and resolve system issues related to the integration of hardware, software and communications. A master service agreement (MSA) with a service provider (if the required skills are not available internally), or internal service team, is essential. It is important to clearly define the MSA KPIs such as system availability, both for individual instruments/sensors and telemetry communications, as well as the maximum allowable duration of system outage. These system health indicators are a key input into the TARP. The MSA KPIs should be aligned with risk, where selective sensors monitoring high-risk environments may require higher availability and hence higher-priority support. Critical controls or critical monitoring require 100% availability (24/7); they need high levels of technical support and, potentially, system redundancy.

6.4.11 Data redundancy

Data redundancy is an important additional control or backup on the monitoring system. Two principal needs for redundancy are recognised:

- to verify or corroborate monitoring trends; and
- to provide a backup source of data should one system fail.

Ideally, data redundancy should be provided from different types of instruments. The premise is that if the physical manifestation of change is detected by different systems, then the change must be real. If prism, radar and wireline extensometer responses show similar spatial displacements, the collective data would strongly indicate that real displacement is occurring. The hierarchical layered approach to monitoring described by de Graaf and Wessels (2013) can be used to develop a monitoring program aligned with the risk to be managed, as summarised in Fig. 6.27. This also provides a basis for selecting which systems may require additional redundancy. However, redundancy comes at a cost with additional hardware, installations, maintenance and, importantly, substantial increase in the data generated, which then need to be analysed and managed. Redundancy provided by duplicate systems, such as two or more radars scanning the same slope sector, would typically be only considered for exceptional circumstances.

A balanced, risk-based approach is recommended. The number and types of instruments must be aligned with the expected modes of instability and appropriately resourced. The amount and type of redundancy requires careful consideration and must be aligned with the objectives of the monitoring program, including addressing potential risk and tolerance. Monitoring system scenarios are shown in Fig. 6.28.

Care must be taken to avoid the 'kitchen sink' or 'overkill' approach to monitoring, where multiple types and high numbers of instruments are deployed in the misguided belief that more is better (safer). This approach places a significant burden on the team to maintain the

Slope Risk	Monitoring Method	Monitoring Frequency	Access Required
Low Risk	Berm inspections (crack inspections)	Increasing monitoring frequency (some can be alarmed if appropriate telemetry & alert systems are in place)	Access to berms
	Manually read crack monitors or surface extensometers		Access to berms
Moderate Risk	Laser scanners		Access to berms (install/maintenance)
	Inclinometers		Access to berms (readings)
	TDR/VWP		Access to berms (readings)/Telemetry
	Automated surface extensometers		Access to berms (readings)/Telemetry
	Manually read prisms		Initial access to berms (install targets)
High Risk	ATS Prisms (Automated Total Station)	Requires near real time telemetry and alarms (i.e. Alert, Alarm, Evacuate)	Access to berms (install/maintenance)
Critical Risk (Safety Critical Monitoring)	Radar		Remote
	Automated surface extensometers with local or remote alarms		Remote

Fig. 6.27: Risk-based hierarchical approach to slope monitoring (source: de Graaf and Wessels 2013).

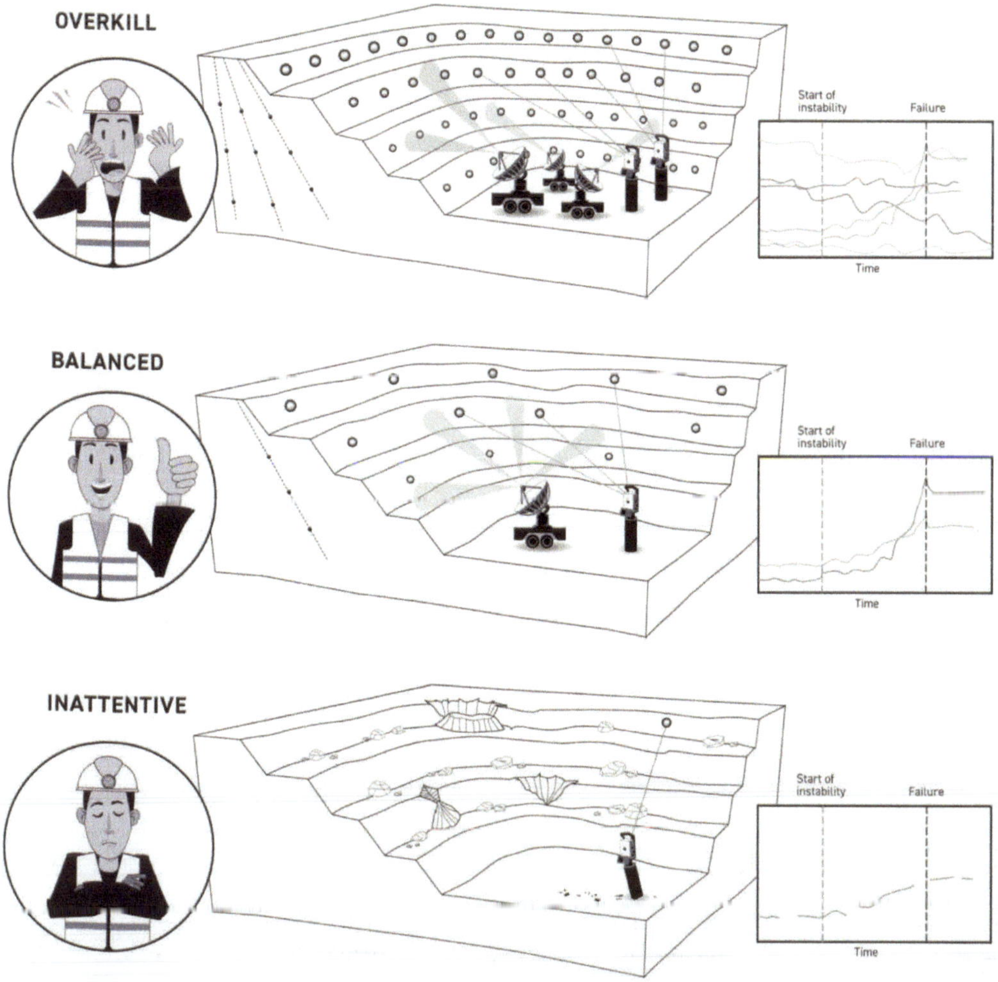

Fig. 6.28: Effective slope monitoring requires a robust risk-based approach founded on sound engineering judgement and a strong individual duty of care (source: P. de Graaf).

monitoring system, manage the substantial data flow, analyse the data in a timely manner, and interrogate and resolve any discrepancies between the duplicate systems. Data overload and resulting confusion can (and has) led to significant foreseeable events being missed. 'More can be less' as this approach has the potential to divert attention and/or delay critical response time. Practical considerations include which alarm system will be the primary trigger and how to manage the occurrence of false alarms. The availability of system technical support and monitoring program management must be commensurate with the increased requirements.

Unfortunately, instances of significant unforeseen (surprise) slope failures, some involving fatalities, continue to occur either:

- on well-monitored slopes where failure was not detected due to slope engineers being overwhelmed with data and not having the time to interpret or recognise the onset of instability from the data or from observation; or
- on poorly monitored slopes with inappropriately selected instrument types, insufficient and poorly maintained instruments, and inattentive staff who did not understand the significance of the data or their duty of care regarding coworker safety.

7

DATA ANALYSIS AND UTILISATION

Jorge Armstrong, Eric Audigé, Niccolò Coli, Erik Eberhardt, Chris Fagan, Paolo Farina, James Jung, Tatyana Katsaga, Lorenzo Leoni, Loren Lorig, Mike Ness, Alex Neuwirt, Cliff Preston, Marc Ruest, Eric Schwarz, Robert Sharon, Peter Stacey

7.1 Introduction

The focus of the previous chapters in these Guidelines has been on understanding what slope performance monitoring is, instrumentation types and how the various components work, capabilities and limitations, monitoring system development and application, and data management. Training of technical personnel who collect and manage monitoring data is an important component of slope monitoring and is covered in detail in Chapter 5. Details of communicating the results of data generated from slope monitoring and escalating risk potential are addressed in Chapter 8. The ultimate goal is to obtain and use data to better understand slope behaviour and predict future performance so that decisions relevant to geotechnical risk management, or optimisation of development opportunities for the operation, are well-informed. Unfortunately, there remain significant instances where terabytes of data are collected and not utilised to maximum potential, for a variety of reasons. Often, personnel onsite either do not have the time necessary to parse and analyse the data, or do not have the experience needed to make the best use of the data. Imagine the scenario where a major failure could have been predicted, if the appropriate amount of time was taken to compile, examine and understand the data recovered from slope monitoring. In real life worst-case scenarios, damage to equipment, injury to personnel and even fatalities could have been prevented. The only thing worse than not having data is having good data that are not utilised.

In large open pits that are well-resourced, and where competent geotechnical personnel are available for data analysis, there are often exceptional practitioners who have taken an interest in understanding the output from monitoring systems. From ground-based (GB) radar technicians and geotechnical practitioners with expertise in assessing movement trends to microseismic system specialists, expertise in understanding the story that data can tell is incredibly valuable to the operation. It can mean the difference between the loss of credibility of a geotechnical department by an operation weary of false alarms, or unpredicted disasters due to missed signals and a well-executed trigger action response plan (TARP) that prevented the loss of millions of dollars of equipment or even the loss of life.

Throughout these Guidelines, the appropriate treatment and communication of data have received considerable focus. In this chapter, concepts relevant to the analysis of data to best understand slope performance, assess the failure mechanism, identify risk areas, predict time to failure and provide relevant information for mine design are examined.

7.2 Basic data analysis

7.2.1 Overview

The variety of instrumentation utilised for surface and subsurface slope monitoring described in these Guidelines provide signals that report to data acquisition systems that are compiled, interpreted and analysed. The data come in various forms but are eventually interpreted as either point data or area data. Point data are measurements at a discrete, single location on a slope, while area data are amalgamations of many measurement points where data are contoured continuously over an area or volume for interpretation. Data can be available at a particular location in space (at a point, such as a prism mounted on a slope), one-dimensional (along the length of a borehole or line-of-sight, LOS), two-dimensional (covering an area, such as the ground surface), or three-dimensional (volumetric, such as a region of an underground mine). The richest data are clearly 3D where a full vector (magnitude and direction) can be analysed. However, the nature of the instrument can limit the data to 1D and 2D which, depending on the quantity and understanding of equipment limitations, is useful in its own right. For example, a practitioner may be monitoring a crack opening at the crest of a critical ramp, and important decisions can be made if the crack shows an opening trend. The following sections provide further discussion on these data types.

7.2.2 Point data *v.* area data

Displacement data provided by robotic total stations (RTS), global positioning systems (GPS), crack gauges and tiltmeters are examples of point data. The instrument or sensor is installed at a strategic and deliberate location, and provides a measure of movement for that specific location or 'point' over time, usually at a set measurement frequency. Because it is a point measurement, interpreting the meaning of the measurement is simpler than interpreting a point cloud. The discrete nature of these data is advantageous to long-term movement monitoring, as the data are easily processed using methods such as time-series displacement or inverse velocity plots for predicting time to failure.

Where the point data are obtained from an array covering a large area, both the point data and area data are effective at detecting small- and large-scale slope movement to provide alerts regarding important movement events to technical teams. GB radar systems provide area data and have the distinct advantage of providing fast feedback for assessing movement trends over large areas. Point data such as those produced by an RTS only provide movement information for specific locations where reflective prisms are installed; thus, movements between points can go undetected. However, RTS does provide total vector (or true movement) for each monitored point, whereas the radar is limited to movement along the LOS for the area under surveillance.

7.2.3 1D *v.* 3D

The 1D applications produce a movement magnitude over time between two points without any information indicating the orientation of true movement. It is useful for tell-tale applications where the practitioner is interested in the presence or absence of movement and possibly whether a particular threshold has been reached. A borehole extensometer is a type of instrumentation that produces 1D data.

Prism survey data from the total station theodolite produce the total vector (or true magnitude and direction of movement) whereas radar data only produce a relative LOS change in movement (or slope distance), where true movement is unknown or not certain in the absence of data validation. It is important to understand and account for the non-vectoral quality of radar data when applying such data to pre-established monitoring thresholds or TARPs. Prism data are more reliable than radar data for establishing and comparing measured total movement to TARPs because of their vectoral and discrete qualities.

7.3 Data visualisation

7.3.1 Overview

The way we choose to visualise data is extremely important, as it determines what and how information is communicated, often at a moment's notice. Creating effective data visualisation is part art, part science. Fundamentally, the user needs to understand the data behind any visualisation, but it is the artistic perspective of a realistic view that helps to convey the data effectively. In terms of slope monitoring, any type of reading that differs from the status quo needs to be instantly recognisable. Examples of advances made in mixed-reality visualisation are illustrated in Chapter 10.

For all aspects of data visualisation, a layer-based approach is recommended, whereby the user can overlay various datasets and choose to turn some on or off. This style of data visualisation is commonplace within geographical information systems (GIS); however, the adoption of GIS-based data management in slope monitoring software packages has been a relatively recent development. As the quantity of slope monitoring sensors and systems grows, the user must be able to easily switch and compare between one system and another. Some of the more common means of slope monitoring data visualisation are examined in the following sections.

7.3.2 2D visualisation

The 2D top-down plan view is the type of visualisation that most people intuitively feel comfortable working with; to most, it is a familiar representation of geographical data. It will therefore form one of the most important visual components of any slope monitoring system. There are many components to an effective 2D visualisation, and these may differ from site to site and organisation to organisation. Each site will have different geotechnical and hydrogeological conditions that will influence the information that is deemed most critical. However, some key considerations for an effective 2D visualisation are as follows.

- Data points and sensors should be clickable and link to information such as the time-history of data points, a journal of events associated with the sensor, and

Fig. 7.1: Example of a 2D visualisation showing integration of a variety of sensors with displacement isolines (source: Sixense-Soldata and Debswana, Jwaneng mine).

relevant documents. Simple analytical tools should exist for computing trend lines and statistical analysis of data points on demand. It is also important that data points and sensors with different data types are directly comparable, using graphs with multiple axes.

- Data points and sensors should be clearly represented with a severity scale of colours and/or size. For example, sensors having crossed pre-defined thresholds should evolve from green to yellow to orange and red, and/or grow in size.
- An effective visualisation should emphasise turning data into information. Typically, a vast amount of data are gathered from a multitude of sensors, and sometimes the sheer volume can be overwhelming. For instance, the grouping of all displacement data to produce a contour plot with displacement isolines, as shown in Fig. 7.1, has the ability to translate large amounts of data into a simple and easy-to-interpret representation of surface displacement.
- The visualisation should be layer-based, similar to a GIS system, whereby the user can turn different slope monitoring datasets on and off to customise the view.
- The concept of a time-line should exist in the visualisation, so that the user can switch between real-time and historical datasets. As the user scrolls back and forth through time, all elements shown in the visualisation should change accordingly.
- Where possible, a visualisation should contain a system health dashboard. A dashboard display of system health parameters is an effective way to synthesise large sets of vital information for a slope monitoring system onto one screen. Such parameters might include a summary of the different alarm severities by sensor type or zone. Rates of alarm acknowledgements, quantity of unacknowledged alarms and percentage of sensors not reporting data within expected times should be available and partitioned by sensor type or zone.

7.3.3 3D visualisation

Another effective way to visualise slope monitoring data is with a 3D view. A capable 3D system will have advanced navigation controls, whereby all three axes are rotatable, combined with varying zoom and pan. Not all elements of an effective 2D view will be applicable to a 3D view; however, there are many in common. By adjusting the view angle of an open pit's 3D model, some monitoring data may appear more clearly than they would have in a 2D visualisation. In particular, GB radar or laser scanning data that detect the movement of near-vertical faces of a pit wall will be easier to visualise when displayed in a 3D view. The display of in-ground instruments, such as multi-level piezometers, also benefit from a 3D view, where the 3D locations of piezometer transducers can be easily seen in relation to pit walls and other slope monitoring data.

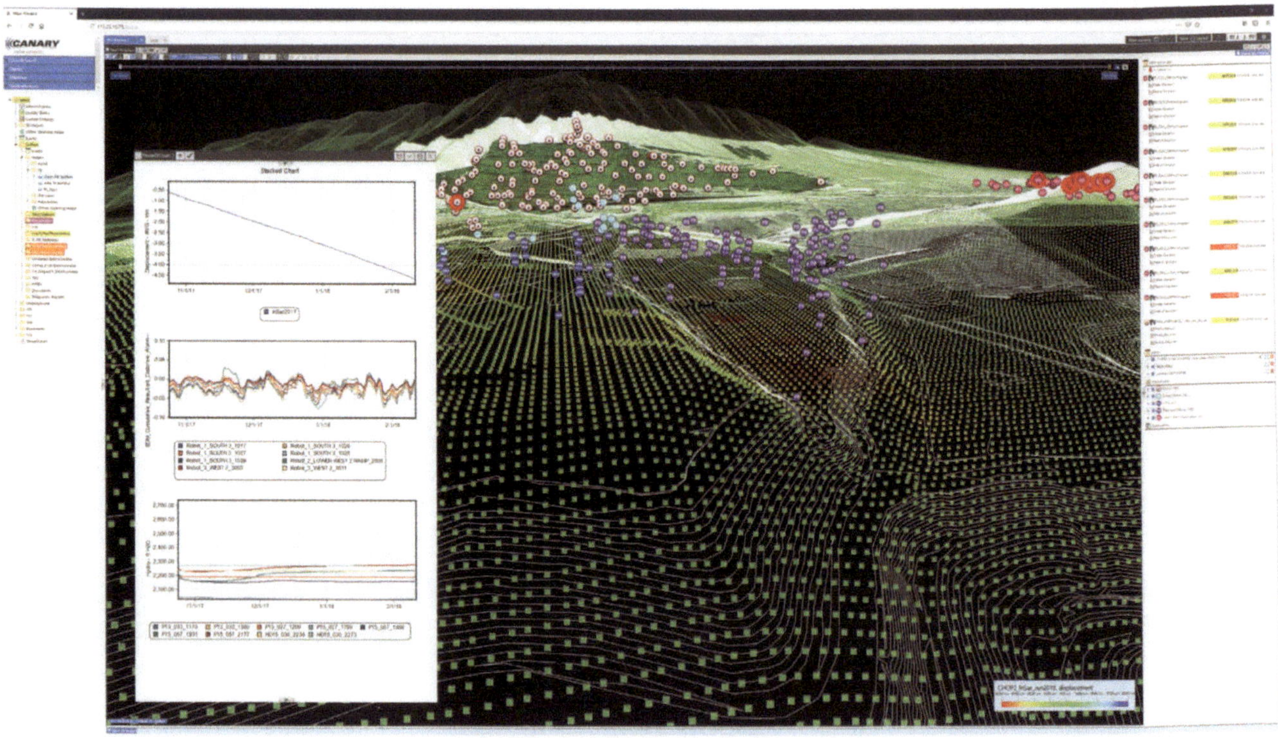

Fig. 7.2: Example of a 3D visualisation showing inset time-series results from data generated by satellite radar (InSAR), RTS (EDM) and vibrating wire piezometers (source: A. Neuwirt).

Effective 3D visualisation will allow the slope engineer to identify with spatial context the locations of various monitoring instruments on the 3D scene with access to associated data of interest on demand. Figure 7.2 illustrates 3D visualisation of multiple types of sensors with time-series displays of specific data within the field of view. To have an effective 3D visualisation, an accurate 3D digital terrain model (DTM) of the pit must exist and have been loaded into the slope monitoring software. In most cases, these models will reside with the survey or mine planning departments. The 3D model should be updated regularly as the datasets become available.

7.4 Integrated monitoring

7.4.1 Importance and utility of various monitoring devices

When designing the components of a slope monitoring system, each system, sensor or group of sensors should aim to answer at least one specific question or one specific problem (Dunnicliff 1993). A comprehensive slope monitoring system should consist of a broad range of complementary sensors and systems, including RTS and prisms, GNSS, GB radars, laser scanners, photogrammetry, InSAR and subsurface geotechnical and hydrogeologic sensors. Each provides a unique set of information that answers a specific question. Users of slope monitoring systems must clearly understand each sensor and know which question the readings for each sensor are seeking to answer.

Unfortunately, the data generated by these sensors and systems are often viewed in isolation on separate computers by different users. In the past, this has largely been the result of slope monitoring software package limitations, but a recent trend has been to integrate all monitoring data types into a single monitoring software platform. Changes in ground movement are most effectively detected through the analysis of multiple data types from varying sensors. Thus, it is important that data from the varying sensors are managed in a slope monitoring software package that has a strong emphasis on data integration.

By combining data from two different sensors, such as a piezometer and a prism, the slope engineer can gather a more holistic understanding in terms of data context. For instance, a spike in pore pressure measured by a piezometer behind a pit wall, in combination with a subsequent increase in rate of deformation at a nearby prism, is a series of events that can be interpreted in terms of cause and effect. Data validation and reliability of the interpretation are achieved when this observation is coupled with increased velocities measured by GB radar or InSAR. The key is to be able to quickly and easily tie events together from the piezometer, prism and radar.

Table 7.1: Reduction of fragmented monitoring software following the deployment of an integrated monitoring system

		Before	After	
Sensors	**Data acquisition**	**Visualisation**	**Visualisation**	**Analysis**
GB radar	Vendor	Vendor	Integrating Software	Vendor
3D laser	Vendor	Vendor		Vendor
Prisms	Vendor	Vendor		Integrating software
Continuous Operational Reference System (CORS)	Vendor	Vendor		
Precise levelling	Manual	Spreadsheet		
Tiltmeters	CSI datalogger	Loggernet		
Inclinometers	CSI datalogger	Loggernet		
TDR	CSI datalogger	Loggernet		
Weather station	CSI datalogger	Loggernet		
Piezometers	Speciality software	Spreadsheet		
	CSI datalogger	Loggernet		
	Vendor	Loggernet		
Divers	Manual	Spreadsheet		
Dewatering boreholes	Manual	Spreadsheet		

Rapid changes to data trends for slope monitoring sensors can require time-critical responses, where switching between software packages and computers for analysis is not ideal. The slope monitoring system must therefore be capable of displaying multiple sensors with varied data types side-by-side, and layered one on top of another, for direct comparison and analysis.

Another important benefit of system integration relates to the triggering and management of alarms. Having multiple systems, each with their own alarm management system, on separate computers is inefficient and possibly ineffective. The integration of all data types in a single location allows for the triggering and managing of alarms from a single point with embedded traceability. In this way, the overall process activating the TARP is streamlined.

An example of a site that adopted the concept of data integration for all its slope monitoring sensors is Debswana's Jwaneng Diamond Mine in Botswana. Table 7.1 provides a summary of the variety of slope monitoring sensors utilised at the operation and the associated number of software packages originally required to visualise and analyse the data before integration into a single data hub. In some cases, vendor software is maintained for more instrument-specific complex analysis, as the integrated system is sometimes not permitted to reproduce vendor intellectual property. Details of Jwaneng's monitoring system are summarised in Chapter 9.

7.4.2 Advantages of integration

The development of numerous systems for monitoring geomechanics issues in mining has necessitated the development of methods for integrating these various data sources. From a general point-of-view, if the data can be integrated, then a more complete picture of the risks of mining can be understood. There are several advantages of this approach.

- *Safe working environments*: injuries to personnel or damages to equipment are prohibitively expensive. Furthermore, in this age of large publicly traded mining companies, perception can be worth billions of dollars. If the public perceives a company to be operating without regard for the safety of its employees and infrastructure, the 'license to operate' will be impacted.
- *More predictable operating costs*: operational risks may be identified before failure, which will allow mitigation. The cost of failure has become so extreme that one significant failure could jeopardise an entire mining operation.
- *Improving monitoring return-on-investment (ROI)*: data integration provide the ability to compare disparate but related monitoring data to create or substantiate models of behaviour and risk. In other words, it is a classic example of 'the whole is greater than the sum of the parts'. Increasing ROI incentivises mine operators to invest more significantly in all monitoring technologies.
- *More efficient training*: reducing the number of systems that operators must learn and use reduces the cost of training and increases the likelihood of them using the tools properly to maximise their risk management capabilities. Also, mine operators are known to move personnel around a site and among properties, so

having a single integrated tool in use across an organisation means training investments are not lost when personnel are relocated.

- *Improving workplace attitudes*: employees seeing the investment made by their employer to manage risks of operations will respect their employer and often be more engaged and involved with operations. They will tend to better understand and appreciate their role in helping manage the monitoring system and work hard for the success of the system and the data it generates.
- *Improving stakeholder relations*: public markets can be punishing. Perceptions can cost mine operators millions, if not billions, of dollars. They can also delay approvals for new mines or mine expansions. Investments in technologies to integrate data can demonstrate to stakeholders the mine operators' commitment to operating safely and as efficiently as possible.

7.4.3 Challenges of integration

There are numerous challenges associated with the development and management of integrated mine monitoring systems. An analogous scenario is the development of safety systems in mobile equipment; for example, the number of safety systems in a car has increased over time through experience. Early automobiles did not have safety belts, crumple zones, safety glass or airbags, among other features taken for granted today. Systems are continually added to automobiles to provide more protection for occupants. New safety systems do not replace old systems: they augment them. Likewise in geotechnical applications, no single device provides complete risk management; rather, they are complementary in their ability to help manage risk.

Mine operators have always employed the observational method, using their eyes and ears to monitor their operations. The development of electronic technologies, including instrumentation, hardware and software, has led to automated methods to augment the continued observational approach. The challenge is that, unlike the automotive industry where the manufacturer must integrate all the safety technologies into one automobile, the mining industry is served by many industries. There are often multiple vendors for each type of monitoring product and technology. The technologies are complementary to each other; even within the same type of technology, there are capabilities and limitations associated with each vendor's product or service. However, since they are offered by different vendors, there is a decided lack of integration of the information produced by each type of risk management technology.

Vendors are not incentivised to work together. On the contrary, they often use proprietary technology as leverage to discourage customers from working with other vendors. As a consequence, customers often remain bound to a single vendor. Ultimately, the limitations of mine monitoring system integration are analogous to the limitations of an automobile's safety system that is not integrated. If the electronic system designed to detect a collision is not connected to the airbags which protect the passengers in the event of a collision, then the benefit of the two systems working together will be lost.

Geographical or geo-referenced software systems (e.g. GIS) are becoming popular for collecting and managing the various types of information produced by different monitoring technologies. However, the residual challenge is integrating all of these systems into the same coordinate system. Base maps may be in one of several coordinate systems, including systems developed by the mine itself. Data from InSAR and GB radars may associate with one coordinate system, mine drawings and plans may be in another system, and data from other sensors located on and below the ground surface may be referenced in another system altogether. Initial planning for an integrated system must include surveyors, mine planners and all other stakeholders involved with organising information for the mine.

Automated systems work very well to create large quantities of data automatically, silently and out of view. The challenge is that more data collection can make for more processing, analysing and reporting work. In designing a mine monitoring system, thought must be given to the types of data to be collected and the minimum frequency of collection that can be reasonably processed and analysed. It may appear counter-intuitive, but less is typically more in an automated system because more often means the volume of data is unmanageable, and hence the value is lost.

Technology is intrinsically complicated, and complexity only increases with time. This means that personnel skills must be continually updated to reflect the increasing complexity of the systems. Mine personnel must learn basic electronics, instrumentation, communications, database systems and software systems, not to mention acquire the expertise required to properly process and analyse data. The competitive nature of the commodities markets also means having to continually do more with less. Technology has helped in some ways, but has made other challenges more acute. It can become difficult for operators to make time to properly understand what the systems are doing and what the data indicate.

Finally, the economics of mining are highly cyclical and subject to many forces that can be global, national or local. Even when the industry is enjoying generous profits, budgets are often constrained by the need to operate more efficiently. It is often difficult to obtain approval for investing in risk management systems if the benefits to operations are not well-articulated or clearly understood. Integrating data types does not directly produce a product, and ultimately resources within mining companies tend to be prioritised and focused on production. Once management understands how risk management technologies provide more stability to operations and help mitigate expensive failures, budgets are not as challenged, even in times of difficult economic conditions.

7.4.4 Benefits of integrating various data types

Integration of data sources and types contributes materially to the advancement of mine monitoring technologies. Each solution has advantages and limitations, where combining the solutions mitigates the limitations and ultimately increases the value of investments made by the mine owner. At the risk of experiencing management system failure, mine operators must not underestimate the effort required to integrate data types. Careful planning should be made to provide the resources needed to achieve the intended benefits.

Several areas for continuing improvement with data integration include:

- the integration of radar and laser scanning point cloud data;
- the integration of high-resolution cameras into the performance monitoring process;
- the use of 3D navigation to advance understanding of monitoring data in relation to rock mechanics and auto-correlation mechanisms; and
- the ability to automatically identify trends and relationships between different types of data.

The following are common mine monitoring methods (described in greater detail in Chapters 3 and 4) that have been successfully integrated, with some detail on the benefits.

- *Hydrogeology*: measurement of changes in groundwater pressure allows the capture of depressurisation efforts. Correlations between slope movements, mine development and their influence on pore pressure distribution can be useful for interpreting the extent and efficiency of depressurisation efforts. Integrating flow data from pumping or leakage has proven to be useful in understanding the overall hydrogeology of the operation, including optimisation of the groundwater pressure management system.
- *Surface water*: measurement of surface water, either through rain gauges or collection weirs, allows the calibration of the water balance model. Infiltration of surface water is often a material contributing factor to slope instability; thus, assimilating rainfall data with local slope movements can facilitate an understanding of whether surface grading, water collection and diversion are working effectively to minimise operating risks.
- *Subsurface displacements*: discrete displacements below the ground surface may be determined either with time-domain reflectometry (TDR) cables or inclinometers that are installed/operated in boreholes. Slope monitoring instruments utilised to measure surface movements are not suited for identifying the limits of subsurface displacements. Correlations between surface and subsurface movements can be useful for evaluating movement progression and for defining and confirming failure mechanisms. Failure surface models can also be cross-validated with subsurface deformation data, thereby improving the reliability of remedial stabilisation plans and the effectiveness of remediation efforts.
- *Global positioning systems (GPS/GNSS)*: GNSS is a platform that can be rapidly deployed where slope movements are identified or anticipated. Other ground surface monitoring instrumentation can initially be used to identify areas of concern (e.g. InSAR), and GNSS may then be deployed for a closer assessment of movement and results confirmation. Once deployed, true movement (i.e. total vector) with accuracy up to 1–2 mm is possible. Total vector data improve the ability to determine the failure mechanism.
- *Robotic total stations (RTS)*: RTS provides true movement (i.e. total vector), including magnitudes and velocities. Scanning technologies such as radar, InSAR and laser scanning can be used to identify areas of movement, then prisms are deployed to provide more detailed information. A single station can be used to survey a large area; a range of up to 3000 m is possible with equipment from most vendors.
- *Slope extensometers*: several types are available, typically using a wire attached to a rotary potentiometer or wound around an optical encoder. While they only provide a single line of movement between fixed points, this information is useful when used in combination with scanning technologies to more accurately detail the extent and velocity of slope movement.
- *GB radar*: this technology has proven to be a popular and highly regarded management tool for excavated slope movement monitoring. However, the LOS movement limitation must be recognised and accounted for. It is therefore important to calibrate and validate radar data with geodetic slope movement data, such as RTS and GNSS, and with other sensor data, such as from slope extensometers, to ensure data and results are interpreted reliability.
- *InSAR*: InSAR has proven to be an excellent site-wide risk management tool for mine operators. It is limited in terms of the frequency of data collection, so it is not well-suited for dynamic (i.e. real-time) monitoring, but

it is well-suited to identify areas of concern which can then be monitored using other more dynamic technologies such as GB radar, laser scanners, photogrammetry, GNSS, RTS or slope extensometers.

- *Laser scanning*: this technology provides one of the highest-density surface monitoring solutions available and has the added capability for mapping structure and as-built slope geometries, as well as for volumetric measurements. However, the large datasets produced from laser scanning can be a challenge to integrate with other systems.
- *Photogrammetry*: photogrammetry is becoming an essential method to adequately characterise (map) and visualise geologic features in excavated slopes, in order to better understand displacement mechanisms. Photogrammetry's role revolves around its generally high relative measurement accuracy and its ability to accurately map geological structures and their orientations.
- *Microseismic monitoring*: this technology is recognised as an important tool for mapping stress-induced rock fracturing and fracture slip in rock masses. This method uses distributed geophones/accelerometers to capture seismic signals initiated by a fracturing event or disturbances on existing structures. Inversion of the seismic signals allows the source location and magnitude of the event to be evaluated and the fracture mechanism to be interpreted.

7.4.5 Importance of confirming data reliability

With the increasing volume of data generated by modern electronic slope monitoring systems and the increasing variety of sensors, there is a growing need to combine and integrate these data types in efficient, effective and innovative ways. It is important that geotechnical staff do not spend their time wading through piles of data, but instead focus on data interpretation, reliable results reporting, and addressing urgent actions as needed. A slope monitoring system must be capable of treating all data from different sensors in a collective unified way. For example, where multiple technologies and sensors are being used to measure vertical displacement data, all data from those sensors should be available in a single data hub for rapid compilation, data cross-validation and analysis.

Users of monitoring data must be aware of the limitations and potential traps of comparing data from different sensors that were put into service to answer different questions. However, with a good understanding of the limitations of each dataset, material value can be gained from an integrated approach to data analysis. An example of a potential trap is where GB radar measurements indicate movement on a highwall, but the corresponding prisms located on the same wall indicate little or no movement. This comparative observation is illustrated in Fig. 7.3. What is actually happening is that the radar is measuring ravelling of loose surface material, while the prism that is anchored 1 m into the highwall is indicating that the overall mass is stable. For the complete picture of this scenario, both the radar and prism data need to be examined in unison; no single sensor can tell the complete story.

7.5 Trend analysis

7.5.1 The importance of long-term slope monitoring

There are two main categories of pit wall surface deformation monitoring: point monitoring (prisms,

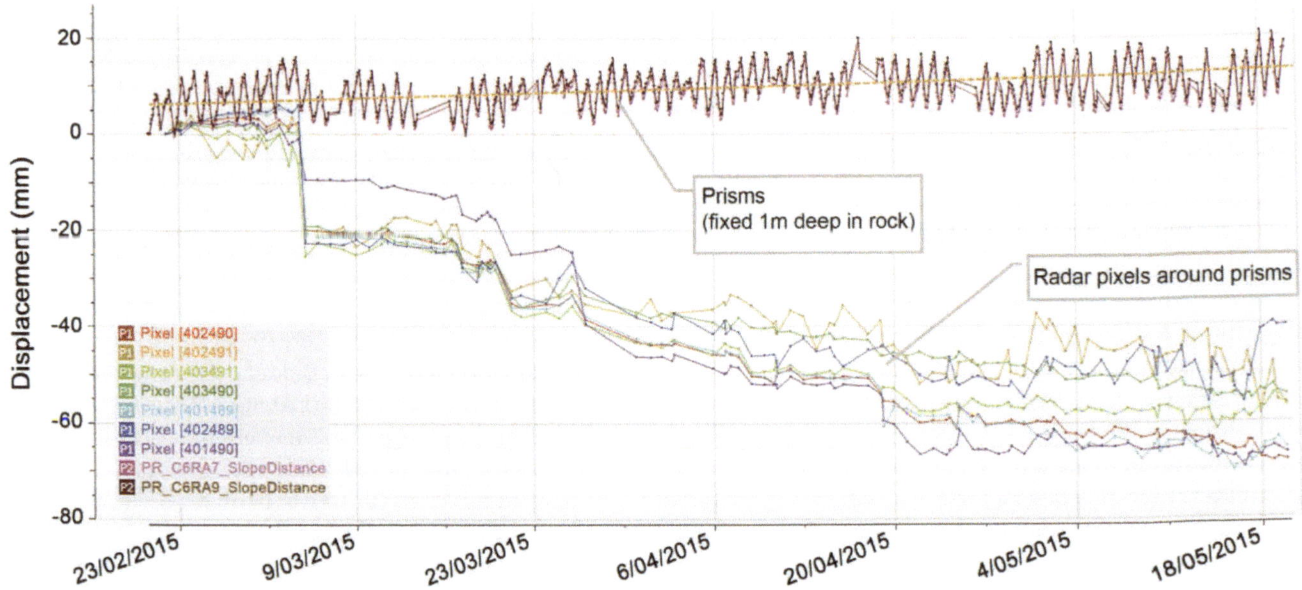

Fig. 7.3: Comparison of prism and radar displacement, where results appear to generate conflicting information about real slope movement (source: Sixense-Soldata).

extensometers, GPS) and area monitoring (laser scanning, photogrammetry, radar). The evolution of geotechnical monitoring technology for assessing slope stability issues in real-time has progressed rapidly in the last few decades. Recent efforts have largely focused on improving safety performance at mining operations and the ability to predict the onset of instability that could endanger personnel and equipment. While most area-based slope monitoring systems are designed to give operators near real-time warning of dangerous slope movement, they may not be well-suited for the assessment of longer-term monitoring of slope performance. To evaluate long-term slope performance, it is important to consider the expected long-term response of the slope to the excavation process.

A displacement monitoring system should be implemented as soon as practical during the early stages of mining and should be maintained throughout the life of the operation. A monitoring system may even be required beyond closure. Pre-excavation monitoring can provide important baseline data for subsequent stability monitoring, as well as to validate design assumptions and modify designs as determined from slope development experience. The long-term monitoring system should be robust enough to endure the rigours of the mining environment.

Geodetic surveying with prisms and/or GPS often provides the best and most practical methods for monitoring long-term movements, as these instruments deliver the most accurate measure of total displacement. Allowing for the LOS limitation, GB radar, satellite InSAR and LiDAR also produce data that can be used to assess long-term slope movements. The monitoring system selected should not rely on a single monitoring method. Multiple monitoring methods, such as prisms and radar, are beneficial for cross-validation of movement (data reliability) and to ensure a high level of availability, in the event that part of the monitoring system is not operational or is damaged by mining activities. This approach will maximise the potential for creation of a continuous long-term monitoring record.

Typical monitoring systems generate displacement *v.* time plots (time-series); a potential shortcoming is the uncertainty of how much initial slope movement has not been recorded. This may be due to operational difficulties related to the installation and maintenance of monitoring stations located near active operational areas, with resultant loss of the initial displacement record (Zavodni 2001). The value of understanding the relative stability of an excavated slope requires early installation of monitoring systems. The value of these early monitoring trend records is realised when a complete history of slope movement patterns is recorded.

7.5.2 Slope movement patterns: general

Pit slopes are expected to experience elasto-plastic responses in the early stages of mining. Steady-state slope movements of the rock mass typically occur in response to mining. These can continue long after mining is completed without visual evidence of loss in rock mass strength. As experience is gained with rock mass behaviour during the excavation process, mine operators can get a feel for the level of comfort when working in close proximity to the slopes. Movement rate thresholds are important to understand when managing risk, with most practitioners focusing on accelerating movement rates that may indicate the development of an incipient unstable condition.

When external forces (gravity) impose a stress on a pit slope, the rock mass responds by deforming (strain). Pit slope design practitioners have used the rock mass yield strength and elasto-plastic properties to define the onset of progressive slope failure (Zavodni 2001). Sullivan (1993) expanded the failure classification process by dividing pit slope movement patterns into three phases (Fig. 7.4). This

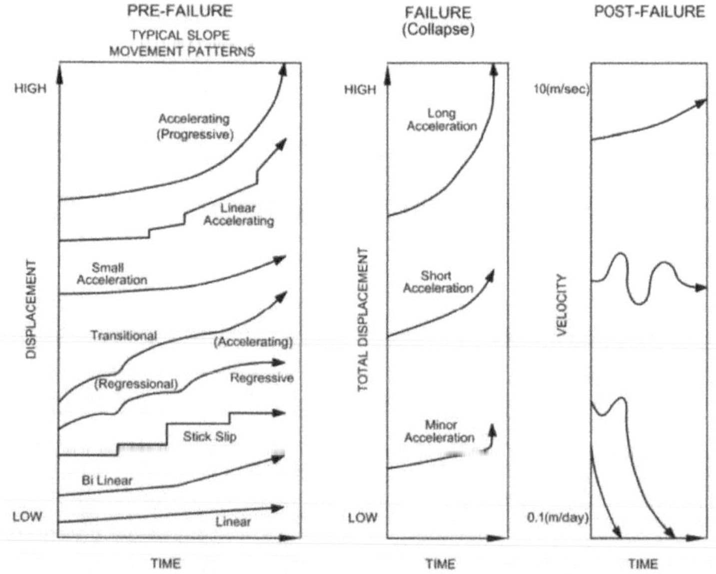

Fig. 7.4: Modified pit slope failure classification system (source: Sullivan 1993).

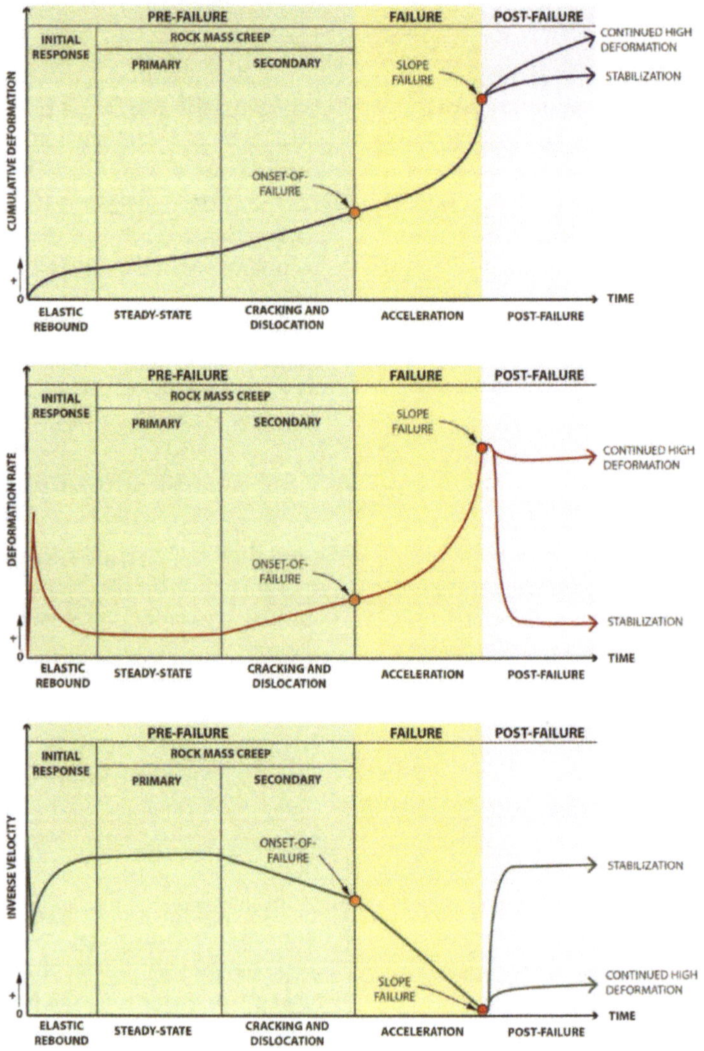

Fig. 7.5: Schematic plots of cumulative deformation *v.* time, deformation rate *v.* time, and inverse velocity *v.* time curves leading to failure (modified from Newcomen and Dick 2015).

allows practitioners to use more detailed descriptions to characterise the evolution of the slope deformation process from the onset of instability through failure and post-failure.

Newcomen and Dick (2015) generated a simplification of previous concepts to demonstrate a schematic of deformation *v.* time leading to slope failure, as shown in Fig. 7.5.

Many practitioners have noted that slope failures do not come without warning. Terzaghi (1950) stated that 'if a landslide comes as a surprise to the eye-witnesses, it would be more accurate to say that the observers failed to detect the phenomena which preceded the slide'. Similarly, 'slopes seldom fail without giving adequate warning' (Hoek and Bray 1974).

Call (1982) and Zavodni (2001) summarise the distinction between a theoretical and an operational slope failure. Theoretical failure is displacement beyond recoverable strain, if the rock is considered to be an elastic material. Operational failure is defined as when the rate of displacement is greater than the rate at which the material can be mined safely and economically, or the movement produces unacceptable damage to a permanent facility.

Assuming that slope movement trends can be projected forward and used to estimate the deformation response in the slope at some future point in time, the maximum deformation magnitude or rate thresholds proposed could be used to assist with contingency planning. Global experience demonstrates that the minimum deformation at failure varies substantially depending on local conditions (geology, groundwater, climate, excavation practice, geometry) and is also sensitive to the failure mode. Slope designers and those responsible for implementing the designs must have an early understanding of the potential failure mode so that proper protocols are in place in case the slope movement progresses towards a critical deformation magnitude or rate threshold. This highlights the importance of a good geotechnical data acquisition program at the mine (Newcomen and Dick 2015).

A range of movement thresholds, utilised to alert mine operators to a potential failure condition, should be developed based on local conditions and performance experience. Each slope has a unique combination of geometry, rock mass properties and structural setting. Threshold movement criteria are therefore site-specific, even slope-specific, at any given operation. Experience indicates that movement rate thresholds that are based on back-analyses of site-specific incidents will provide more reliable results than analyses using non-calibrated stress-strain models. An important part of the risk management strategy is to keep the workforce fully informed of potentially hazardous conditions at all times. Movement trigger levels should always be used in conjunction with visual inspections and careful analysis of movement trends.

7.5.3 Movement patterns: their importance and what it means to the slope engineer

Having an extended database showing a wide variation (in space and time, of displacement information, etc.) raises the challenge of how to properly query and visualise the data in order to identify information otherwise hidden within each set of records and measurements. For example, Fig. 7.6 presents comparable cumulative displacement images over a pit wall area monitored over a 20-day period. Both images represent the same dataset. Figure 7.6a uses a three-colour legend with regularly spaced classes of values between zero (in green, stable sectors) and 50 mm (in red, areas showing some displacement). From this image, it may be difficult to appreciate all of the information contained in the dataset and to take advantage of the wide range of velocities the GB SAR is capable of covering.

With a simple change in the number of colours in the legend and the category of velocities represented by each colour shown in Fig. 7.6b, the user can obtain much more information about relative deformation that has occurred on the highwall. Note that in Fig. 7.6b, the categories of velocities are not equally representative of movement ranges like those in Fig. 7.6a (e.g. 0 mm, 25 mm and 50 mm velocity categories) but have a logarithmic ranged distribution (0 mm, 25 mm, 50 mm, 100 mm, 180 mm, 480 mm and 750 mm). With this simple adjustment, it is possible to highlight information that was not available in the previous representation. Note, for example, the gradient of different velocity rates within the wide unstable region on the left side of the highwall, or the additional

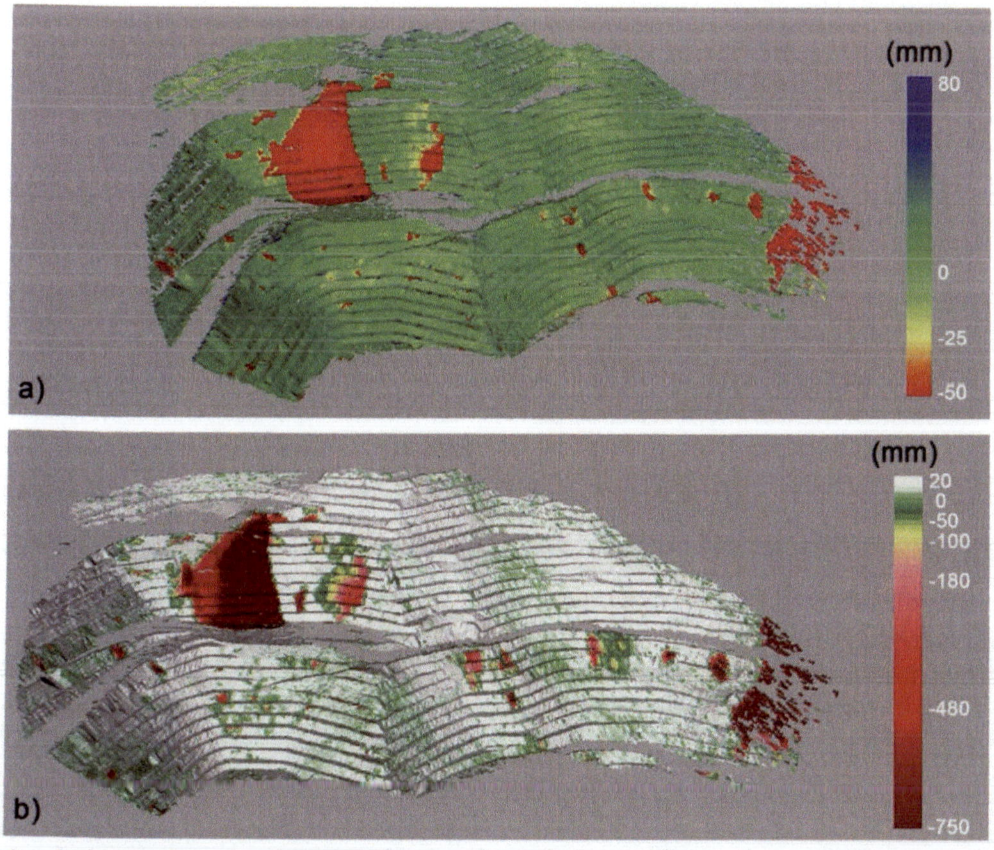

Fig. 7.6: Cumulative displacement images representing the same dataset but using different colours and different scale bars to exploit all the information recorded by the GB radar (source: IDS).

information about small displacements occurring in the right region of the highwall that were not visible in the previous image. Potentially valuable information may be made visible with a few simple adjustments to the scale and colour bar, while still querying the same dataset.

This simple example was developed to highlight how it is the slope engineer's responsibility to take advantage of all the tools available through the monitoring technology they are using, to extract all potentially valuable information within the stored dataset.

7.5.4 Estimating time-to-failure

As reviewed in Read and Stacey (2009), the interpretation of data to predict future responses depends on many factors, including mining activity and precipitation (rainfall, snowmelt), both of which can result in further accelerations, or the onset of freezing conditions which can slow movement. Several approaches to predicting slope failure are available in the literature. Commonly known is the inverse velocity method (Fukuzono 1985), which provides a simple means for estimating the time of slope failure under certain conditions of progressive movement, illustrated in Fig. 7.5.

The concept of inverse velocity is based upon observations from large-scale well-instrumented laboratory tests simulating rain-induced landslides in soil. The conditions simulated in the laboratory were considered to be characteristic of accelerating creep (i.e. slow continuous deformation) under gravity loading. When the inverse of observed displacement time rate (inverse velocity) was plotted against time, its values approached zero as velocity increased asymptotically towards failure. A trend-line through values of inverse velocity *v.* time could be projected to the zero value on the *x*-axis, predicting the approximate time of failure, as shown in Fig. 7.7.

Based on the back-analysis of slope monitoring data from several large pit slope failures, Rose and Hungr (2007) suggest that a linear trend of inverse velocity data could be applied to long-term trends as well as the short-term forecasts proposed by Fukuzono (1985). Figure 7.8 is a plot of inverse velocity *v.* time showing the trends of nine survey prisms located at various elevations on the south-east highwall of the Betze-Post open pit, over the six weeks preceding slope failure in August 2001. Filtering (data smoothing) of two-hour RTS prism monitoring measurement data was achieved by calculating six-day average (incremental) slope distance velocities to reduce the effects of instrument error in low-level velocity values. As displacement rates increased, a clear inverse velocity trend developed and began to converge at the failure time of 29 August 2001. Linear regression was applied to the inverse velocity values, which defined regression coefficients (R^2) of 99% for all nine prisms. The failure

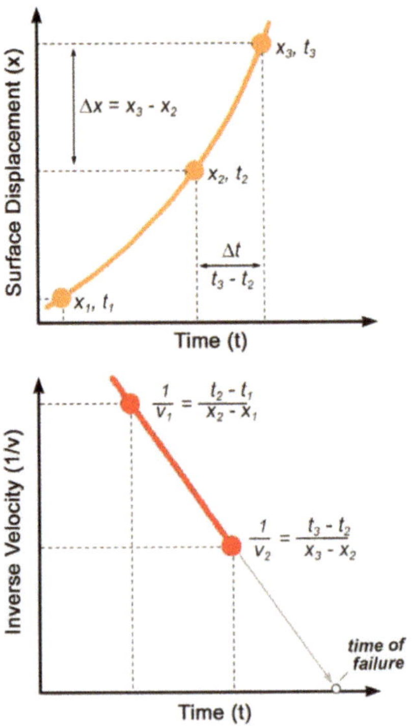

Fig. 7.7: Inverse velocity *v.* time relationships preceding slope failure (source: Eberhardt 2008).

occurred on the predicted date, encompassing an overall slope height of 550 m and an estimated 47 million t. The failure lasted several hours as a series of nested wedge failures and rock avalanches.

Two subsequent slope failure prediction case histories presented by Rose and Hungr (2007) utilised shorter time durations in data filtering of average prism and wireline extensometer data. Time increments were reduced to one or two days to avoid the observed divergence from the linear trends in inverse velocity data from the 2001 Betze-Post open pit case history in the days leading up to failure (Fig. 7.8). The appropriate amount of data filtering with the inverse velocity method varies with each application and depends on the type of monitoring system and the associated error or noise in the data. Thus, it should be recognised that, independent of slope failure mode and whether the displacement data are derived from crack meters across individual tension cracks or from RTS prisms covering an entire slope, the technique is applied in the same manner (Eberhardt 2008). To contend with this, Rose and Hungr (2007) recommend that data should be linearly extrapolated over varying lengths of time, looking for a consistent trend and noting and re-evaluating any departures from it, combined with observation of various controlling factors and the developing failure mechanism. As velocities increase and inverse velocities correspondingly decrease, reductions in the filtered time increment may be required to provide more accurate

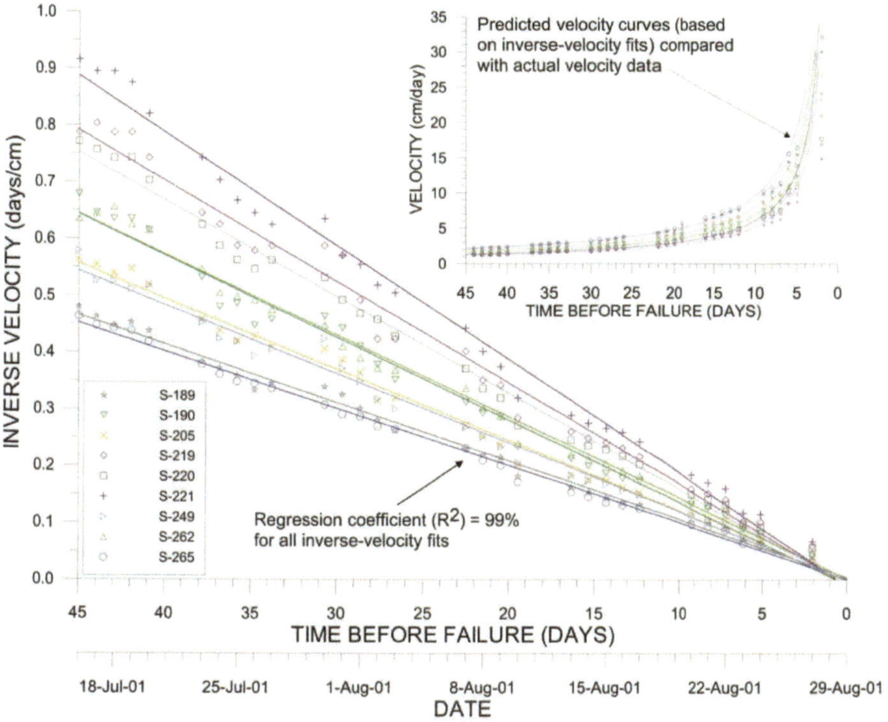

Fig. 7.8: Plot of six-day average slope distance inverse velocity and velocity (predicted curves *v.* actual values in the inset graph) *v.* time for nine prisms (time 0 was the observed time of failure) (source: after Rose and Hungr 2007; copyright 2006 with permission from Elsevier Ltd).

predictions of failure time (Dick *et al.* 2015). Thus, longer-term predictions are never entirely definite and monitoring needs to continue up to the point of failure. Additional limitations and cautions regarding the utilisation of the inverse velocity method for predicting the time of a slope failure are summarised in Chapter 8. Forecasting rock slope failure is a complex problem involving observations, analysis and experienced judgement. The inverse velocity method is a simple but powerful tool to aid the process in many cases.

7.5.5 Time-to-failure analysis adapted for ground-based radar

The use of GB radar for safety-critical monitoring provides several key advantages over point measurement data (e.g. geodetic prism monitoring), including monitoring of a broad coverage area and updating of repeat scans to track the distribution of slope surface deformations and their progression over time. This presents the opportunity to consider the full spatial coverage of a moving slope for a time-to-failure analysis instead of using only single monitoring points, as is conventionally done when applying the inverse velocity method (Fukuzono 1985) or similar methods like the SLOpe gradient (SLO) method (Mufundirwa *et al.* 2010). Relying on a single point or series of points can introduce uncertainty and ambiguity into the time-to-failure estimates owing to rock slope heterogeneity and complex failure modes that can then be misinterpreted; for example, a pit slope may collapse in a single event or fail retrogressively as several smaller events over time.

To take advantage of GB radar's ability to scan a full pit slope, Dick *et al.* (2015) developed a procedure to apply time to failure methods to radar data. The method was developed and calibrated using eight different open pit slope failures selected to represent a diverse sampling of different slope heights, geology and rock mass conditions. These led to the development of a real time time to failure analysis methodology involving a multi-pixel selection procedure and integrating the inverse velocity and SLO time-to-failure methods. The procedure is outlined in Fig. 7.9. It begins with the triggering of a GB radar alarm. Using the pixel that triggered the alarm or the pixel with the highest accumulated deformation, a grouping of pixels is selected that exceed 50% of the deformation increment of the benchmark pixel, for which the time-to-failure analysis is applied. This recognises that slope instabilities may or may not fail as one coherent mass. An example is provided in Fig. 7.10.

The triggering of an alarm will initiate the TARP. The time-to-failure analysis that follows applies a data filtering method to the 50% grouping of pixels involving the averaging of the deformation measurements over a moving time period to reduce the data noise and provide more

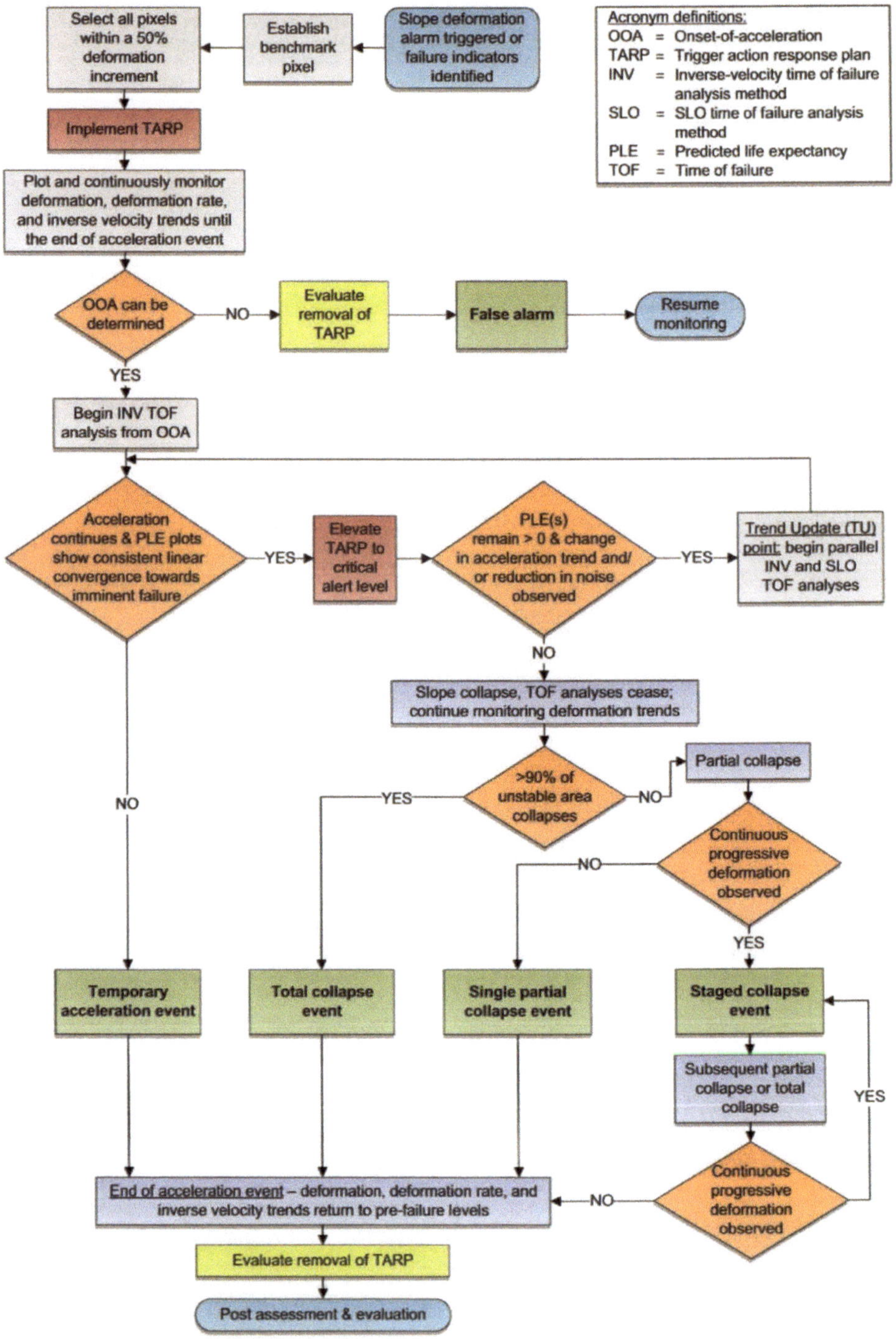

Fig. 7.9: Methodology for time-to-failure analysis for use with GB radar (source: Dick *et al.* 2014).

accurate and reliable results. This is used to determine the onset-of-acceleration (OOA), based on subtle changes in the slope velocity trend as the real-time radar data updates. Filtering over longer time periods provides less noisy plots, allowing the visualisation of long-term data trends and an earlier OOA (Fig. 7.10). Filtering over a shorter period, although typically noisier, shows short-term trends in the data. This data analysis is carried out by plotting the data as inverse velocity (Fig. 7.11). If a significant change in the acceleration trend and/or a reduction in data noise is observed, then an additional time-to-failure analysis is to be conducted in parallel with the inverse velocity analyses from the original OOA for additional comparison.

Both the inverse velocity and SLO methods can be applied in this real-time procedure, thus providing multiple time-to-failure predictions (or 'life expectancy' if

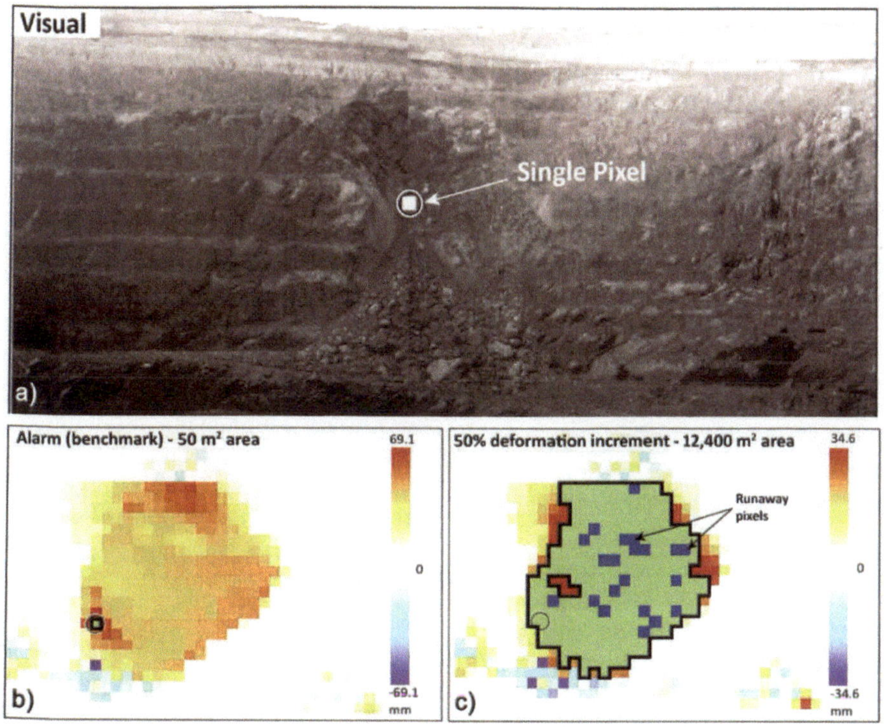

Fig. 7.10: Example of time-to-failure analysis developed for GB radar data, showing (a) pit slope failure analysed, (b) radar deformation map and the benchmark pixel that first triggered the alarm and (c) 50% deformation increment pixel grouping used for time-to-failure analysis (modified from Dick *et al.* 2014).

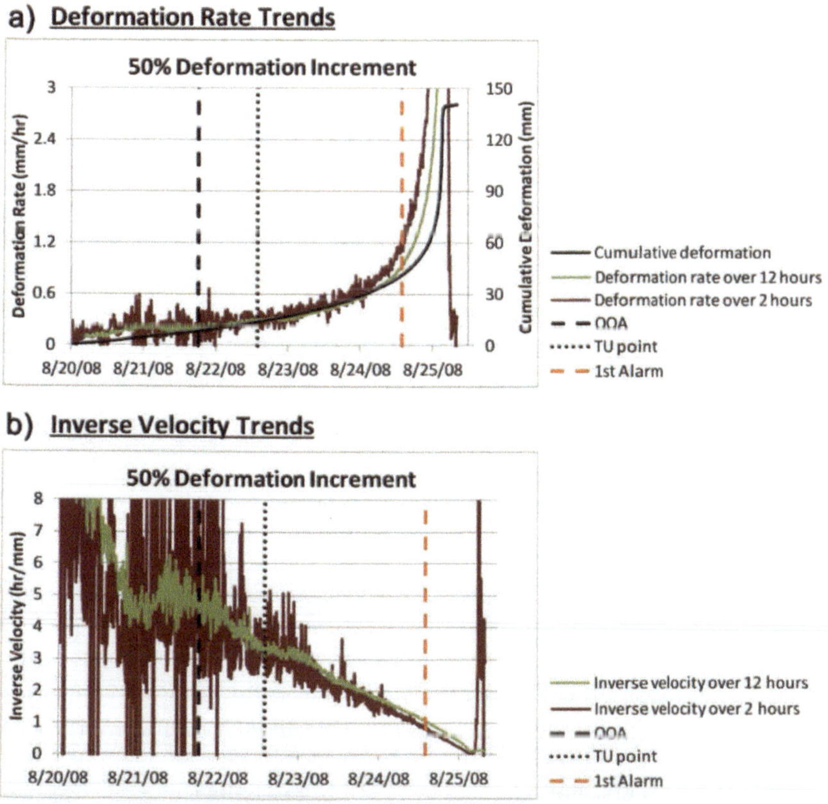

Fig. 7.11: Time-to-failure procedure for GB radar data using the example in Fig. 7.10, showing (a) real-time determination of onset-of-acceleration and (b) inverse velocity trends for different time-interval sampling (modified from Dick *et al.* 2014).

using the SLO method), allowing for different portions of the radar dataset to be utilised or omitted to help add confidence in the early warning decision-making. Only predicted time-to-failure with a converging negative linear trend is considered to be valid and used confidently. The procedure allows for consideration of different acceleration event outcomes (Fig. 7.9): (i) temporary acceleration, (ii) total collapse, (iii) single partial collapse and (iv) staged collapse. Overall, the procedure developed by Dick *et al.* (2014) offers a more robust, reliable and accurate set of results with the following key attributes:

- averaging the deformation measurements over multiple pixels provides smoother, less noisy trends resulting in improved time-to-failure analysis results;
- benchmark pixel data can perform well for partial collapse events as they typically coincide with the most critical area of the instability;
- analysing deformation measurements averaged over a large area of the instability is essential in understanding the overall time-dependent behaviour of the rock mass, especially for a staged collapse event; and
- actively comparing the benchmark and 50% deformation increment datasets provides redundancy in the time-to-failure results to help improve decision confidence.

7.6 Utilisation of monitoring data in the design process

7.6.1 Geotechnical reconciliation

7.6.1.1 Mine planning

Creation of long-term mine plans for an open pit mining operation is typically the responsibility of the long-range planning group, who convert strategic plans into workable mine plans, with support from the geotechnical team who ensure that design parameters and other relevant consideration, such as water management, are implemented correctly into the plans. The day-to-day engineering for most open pit mine operations is typically the responsibility of the short-range planning group, who field-fit the long-range plans, with support from the geotechnical team who support the geotechnical design basis, performance monitoring and design adjustments required, based on as-built performance.

The operations team then implement the daily and weekly plans, which are meant to convey the schedule (sequence), location and methods of mining during that period. This process includes safely excavating and delivering expected ore to the mill while developing the excavation shape (slopes, roads) in preparation for subsequent mining periods. It is rare that every detail of as-built ground conditions match expected conditions. Therefore, the field-fitting process typically involves changes to the planned excavation shape and sequence. The as-built data collection required for the short-range planners to understand changes made to design assumptions, and the associated impacts to production and to future mining periods, shares much with what the geotechnical team requires in order to understand the impacts to stability. It is logical and efficient, but not always common practice, for the short-range planning and geotechnical teams to work together to understand all the impacts to production and future development. The slope engineer has limited interest in differences between predicted and sampled metal grades, and the mine planner has limited interest in differences between predicted and observed geotechnical domaining; however, both have the same need to understand variations between the planned and actual excavation shapes. Laser scanners and photogrammetry are ideally suited to provide the necessary information. It is suggested that the reconciliation process be approached holistically.

Many open pit operations are migrating from traditional occupied point survey of mined crests and toes, to the use of laser scanners and photogrammetry, mainly because images generated from a distance are safer and more accurate. Usually, toe and crest areas are off limits to personnel, so offsets are estimated. While laser scanning can be used to quantify displacements and measure changing conditions, the scanners are not necessarily utilised for slope monitoring. However, the process overlaps sufficiently with monitoring to include laser scanning in these Guidelines as a slope management tool. At even the lowest scan resolution, mine planners get more data than they need, and slope engineers have enough to characterise bench face angles, catch bench widths and plan achievement. A higher scan resolution is useful for extracting rock structure data and can be used to develop a base 'map' for geology mapping or characterisation.

7.6.1.2 Slope design confirmation from monitoring data

The operation's slope engineer, or chief engineer, is typically responsible for ensuring that the consultant's geotechnical design parameters, including slope angles, single/double benching and domain boundaries, have been incorporated into the plan. The slope engineer also confirms that as-built shapes conform to the mine plan or, when deviations are necessary, the as-builts are suitable for actual conditions and that the deviation-altered mine plans conform to current geotechnical design guidance. The slope engineer is also responsible for communicating with the slope designer (usually a qualified geotechnical consultant) about the deviation, so that they can include it in future analyses and can decide if a visit to review options or provide further analysis is required.

Once the planned and as-built surfaces are in the mine planning model, there is an opportunity to evaluate how closely mining and as-built performance conform with geotechnical design intent. This may be facilitated by visualising the mine plan model in terms of a smoothed triangulation of the mine design (consisting of a dense triangulation of the wall scan data, faults/major structures, geological contacts, geotechnical/design domain boundaries and shot blasthole locations). Blasthole information can be examined by creating shot volumes by drawing perimeter lines around each blast at a distance of half of the design burden from the outermost drill collars, then extending them downward to the depth of blasthole drilling including the subdrill.

When these items are assembled in the mine plan model and rendered with partial transparency, they can be viewed together in 3D and evaluated together as a slope reconciliation to address the following questions.

- Did operations achieve the mine plan (the actual slope toe is close to the design toe)?
- Were design catch bench widths obtained with selected/achievable reliability?
- Was the entire blast volume excavated or are there areas of broken rock left on the highwall? Or did blasting contribute to catch bench loss (did blast patterns fit correctly)?
- Is bench toe cleanup adequate or is muck left that diminishes effective catch bench capacity?
- Compare the appearance of as-built bench faces for all adjacent (and higher) benches in the same geotechnical domain. Are they uniform in terms of crest loss and face roughness, suggesting correct domaining and mining?
- If more severe instabilities are exhibited irregularly, are they explainable using structure mapping or the location (fit) of trim row blast patterns? Or, according to their cut orientation, are they within the same domain?
- Do identifiable groundwater seeps correlate with bench crest/face damage, suggesting increased groundwater management efforts are called for?
- Is the mining effort building the pit highwall that was expected or are revisions to the mine plan or mining method needed?

Another component of the slope reconciliation process is evaluating whether conditions encountered during mine development match conditions anticipated by the model. Major structure and lithology boundaries are easily marked on 3D surface models derived from scanning (if available; scanning camera data help in geology mapping). The intersection points of all these features with the as-built surface shape become new triangulation lines used to reinterpret model shapes as needed for subsequent mining. The extraction of structural fabric data from detailed scans is not new, but improvements continue to be made in facilitating the process. At any point along the process, the scan with identified surfaces can at least aid the geologist and slope engineer to obtain good measurement data and accelerate the field mapping processes.

There is no doubt that additional uses for laser scanned data will develop, as well as capabilities for improved data that could allow in-the-office detailed visualisation of slopes that could rival visiting the slope in person. Experience suggests that eliminating regular in-person field inspections is a bad thing, but that using high-end technology to enhance data quality from those inspections, and to make them much more efficient and thorough, is a good thing.

7.6.1.3 Slope performance monitoring for design reconciliation

In principle, the design of inter-ramp slope angles is dictated by rock mass quality and/or the occurrence of persistent and adversely oriented (relative to the slope face) structures or weakness trends which have the ability to form planar or wedge instabilities, or toppling/caving zones. In practice, the design inter-ramp angle of many slopes that do not exhibit such characteristics of instability may be controlled by catch bench design and the achievement of catch bench width adequate to control rockfall and bench-scale instabilities. Procedures implemented to inspect and evaluate bench-scale features as a function of blasting, mining and slope geometry are described in detail in Chapter 11 of *Guidelines for Open Pit Slope Design* (Read and Stacey 2009). An important task for the slope engineer includes management of a continuous program to collect and analyse as-built highwall performance data to address the following questions.

- Are slope design objectives being achieved at the bench and highwall scales, or should the inter-ramp slope angle, or bench design, be modified?
- Are inter-ramp slope angles being achieved, but minimum catch bench widths are not? If so, does the scan data reveal clues about blasting, depressurisation or mining which might suggest that improvements in operational controls are needed to remedy the situation?
- Do inadequate slope geometry parameters and stability results suggest the need for more significant design changes? Is it appropriate to involve the design consultants to analyse potential adjustments to the slope design plan, control destabilising forces and provide room for drilling for instrumentation?

Overall mine or design domain changes to inter-ramp slope angles should be engineered, based on as-built slope performance and analysis of the data obtained from the scans and other slope monitoring. The use of scan data in this way might be considered pre-slope, or experience-based, monitoring in that an assessment of the performance of as-built slopes could be used to assess opportunities for positive modification of designs to control potential future instabilities and optimise the slope design.

7.6.2 Interpretation of the failure mechanism

With the implementation of a reliable monitoring system, slope failures do not occur without warning. Despite the large availability of monitoring data in modern open pit mines, negative events associated with slope instabilities continue to occur in surface mine operations, often to the surprise of the mine operator. One possible explanation involves the difficulties experienced by site staff in understanding the failure mechanism and making the connection with data measured by the slope monitoring system.

To design a slope monitoring program, the geotechnical constraints that justify the slope design should drive the selection of the parameters to be measured and, consequently, the slope monitoring methods to be used. Instrumentation utilised for real-time slope monitoring should allow the slope engineer to simultaneously achieve the objectives of short-term safety-critical monitoring and long-term investigative monitoring. It is worth noting that these two objectives are strictly connected. To be effectively used as an early warning system capable of predicting an impending failure, a monitoring program should be designed to take into account the most likely failure mechanisms in all geotechnical domains of the pit, including temporal evolution and spatial scale of potential instabilities.

Experience indicates that the approach of designing and evolving a monitoring system in terms of the potential failure mechanism that controls the slope design is seldom implemented. The reasons for this include:

- failure to identify the controlling potential failure mechanism that justifies the slope design (this can be specific to each geotechnical domain in the open pit or a global design deficiency);
- limitations in staff availability, or motivation, to compile and interpret monitoring records;
- inadequate compilation and interpretation of key geology/structural and hydrogeological data;
- poor-quality data management and/or loss of institutional knowledge due to staff turnover;
- lack of training on the use of the data for analyses;
- focus on safety-critical or reactive monitoring; and
- failure to fully exploit the informative content of the collected data.

Site geotechnical staff often limit their attention to performance monitoring to the alarms generated when data trigger user-defined thresholds. However, the definition of thresholds for alarming purposes should be the result of a process aimed at understanding the most likely risk scenarios associated with the potential slope instabilities that may occur during mine development. In other words, if the failure mechanism is not well-understood, it is unlikely that the slope engineer will be able to set up alarms that are representative of failure conditions.

When incipient deformation is detected from a sensor at the mine site, questions to be addressed by the slope engineer can be summarised as follows.

- What is the spatial extent of the unstable area in terms of surface, volume and possible run-out distance?
- What is the expected behaviour (brittle, plastic) of the material involved in the detected movement?
- What is the most likely temporal evolution of the instability (e.g. regressive, steady-state, progressive)?
- Eventually, in the case of progressive movement, when will the slope likely fail?

To address these questions, it is of paramount importance to understand the mechanism driving the detected instability. To achieve this objective, the operation requires resources to develop the methodology for data collection and for the interpretation of monitoring data needed to determine the underlying mechanism, thus enabling the estimation of risk potential. In many cases, especially for small-scale movements (bench-scale or multi-bench-scale), the slope engineer is tasked to assess risk associated with such movement, using only surface deformation measurements.

Monitoring data from subsurface measurements are much less frequently available, especially in real-time and in the early stage of movement development. This results in the need for interpretative methods aimed at enabling an inversion of the deformation field measured on the surface. The analogy applies to assessment of co-seismic displacements assisted with data from surface monitoring techniques, such as InSAR and GPS, used to interpret the static co-seismic rupturing mechanism and to locate the rupture surface geometry. Such an inversion could be done solely based on monitoring data or, for movements controlled by structure, by combining monitoring data with structural data.

For slope movements where displacement vector measurements have been recorded on the surface by RTS using prisms or by GNSS receivers, a possible approach to invert the detected surface deformations, thus supporting the interpretation of the mechanism, is the graphical method originally proposed by Carter and Bentley (1985), then generalised by Cruden (1986). Such a method is

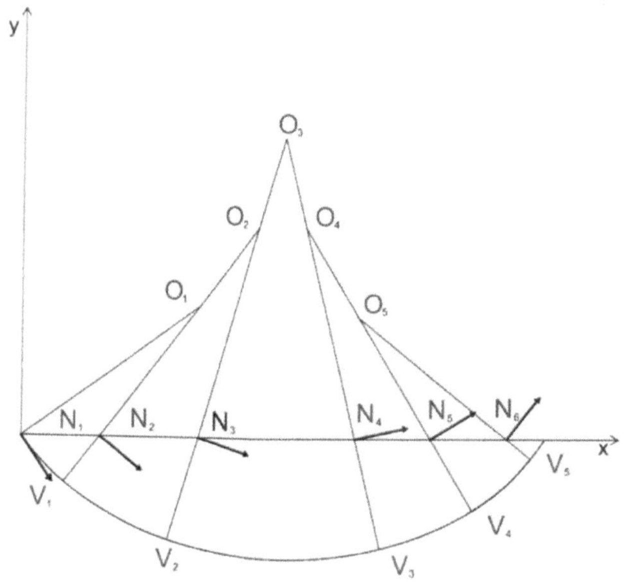

Fig. 7.12: Calculation of the sliding surface using the graphical method proposed by Cruden (1986) (source: P. Farina).

aimed at predicting the depth and geometry of the sliding surface of a landslide from surface displacement vectors.

In particular, the graphical method (see Fig. 7.12) proposed by Cruden (1986) starts with the drawing of normals to slip vectors N_1, N_2, ... N_n along a cross-section of a landslide. Then, centred at the intersection of two first-consecutive normals (O_1), an arc can be drawn starting from the location of the scarp slip vector. The intersection V_1 between the arc and the projection of the normal to the next vector represents the first estimated point of the slip surface. Then the surface can be extended by using an arc centred at the intersection between the normals to the second and third vectors, and so on.

The technique in its original form has been tested against results of laboratory experiments of slope failures over different slip surface geometries and against real cases of landslides for which investigation data were available (Carter and Bentley 1985), demonstrating the accuracy and reliability of the estimations.

Despite the usefulness of this technique in geological engineering practice, especially for the design of site investigations (e.g. locating inclinometers), the technical literature is limited to applying the method to landslide investigations. The limited experience-base may also derive from the graphical nature of the method that in the computer era has probably lost its original appeal. In some cases, especially for deep-seated landslides where the detection of the slip surface may be difficult using inclinometers, or for bench-scale movements where there is no time for detailed geological investigation, subsurface measurement is impractical or unsafe. As a consequence, surface movement is the only information available to interpret the kinematics of the landslide. Nevertheless, it can provide useful insights for reconstruction of the mechanism.

Recent advances of geodetic techniques and the availability of monitoring records from prisms and GNSS receivers in large open pit mines, with high spatial density and temporal frequency, increase the capability for the retrieval of precise measurements of ground motion over unstable slopes in real-time. The estimation of slip surfaces using surficial displacements can thus benefit from the use of these monitoring methods, in combination with selectively located cross-sections and highly accurate topographic surveys. Similarly, the application of the method has been extended to the use of displacement vectors obtained from satellite InSAR measurements by combining LOS measurements from ascending and descending orbits (Farina *et al.* 2007). It should be noted that these cases show how the availability of a large number of displacement measurements, despite increasing the quality of the resulting estimation of the rupture surface, can further complicate the use of the graphical method, making its implementation a laborious and time-consuming process (Carter and Bentley 1985).

7.6.3 Estimating failure limits from surface monitoring

An example of the implementation of such a method in an open pit mine is briefly described here. The E–W and vertical components of the InSAR measured displacement over an open pit mine, obtained from the combination of ascending and descending InSAR data over the pit (Fig. 7.13), were used to retrieve the displacement vectors aligned along cross-sections of two detected instabilities. The displacement vectors were then plotted along topographical cross-sections. The aforementioned graphical method proposed by Cruden (1986) has been translated into its analytical form, approaching the problem in terms of equations of simple algebraic equations of lines, circles and their intersections, then applied to the InSAR data. The resulting sliding surfaces, or displacement contacts, are shown in Figs. 7.14 and 7.15. The interpreted contacts were then used to optimise the position of inclinometers to confirm the presence of movements at the depth estimated by the method.

The proposed method is simply an estimation of the displacement contact based on assumptions of the presence of a single displacing surface and of discrete deformation between each available measurement (Mertie 1947). Where data from inclinometers are available, it is recommended to use them to validate the displacement surface estimated from surface deformation records.

In another example of utilising monitoring records for the interpretation of the failure mechanism, InSAR data acquired before the failure event were analysed to assist the

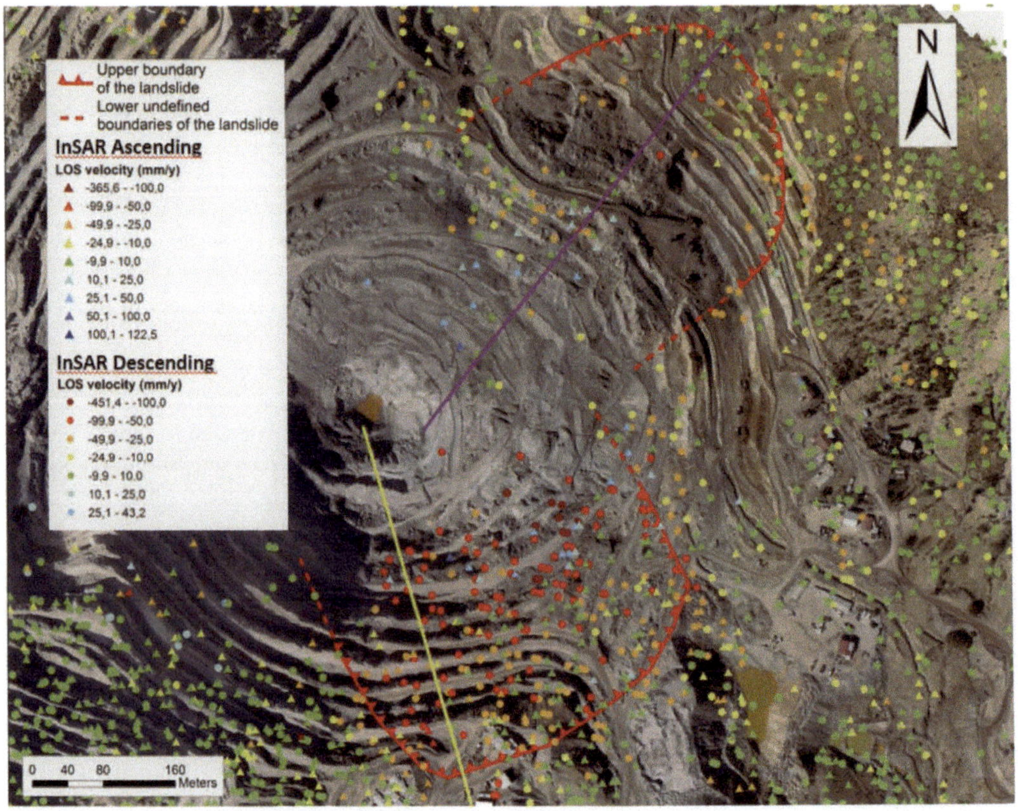

Fig. 7.13: Satellite InSAR analysis over an undisclosed open pit mine with boundaries of two moving areas (source: P. Farina).

back-analysis of a large slope failure that occurred unexpectedly in an undisclosed open pit mine. These data were used to characterise the actual size and temporal evolution of the event, and to identify whether precursors of incipient failure could be identified. The delimitation of the area where InSAR measurement points featured a clear accelerating slope movement trend confirmed the margins of the failure as mapped in the field. These measurements were then used to refine the estimation of the volume of the instability and, consequently, the potential run-out distance. As shown in Fig. 7.16, the movement area delineated with InSAR matches well with the failed area mapped in the field and with the comparison between the pre- and post-failure topography; it also includes most of the main accumulation zone of the failure, suggesting that the main failure surface may extend downslope further

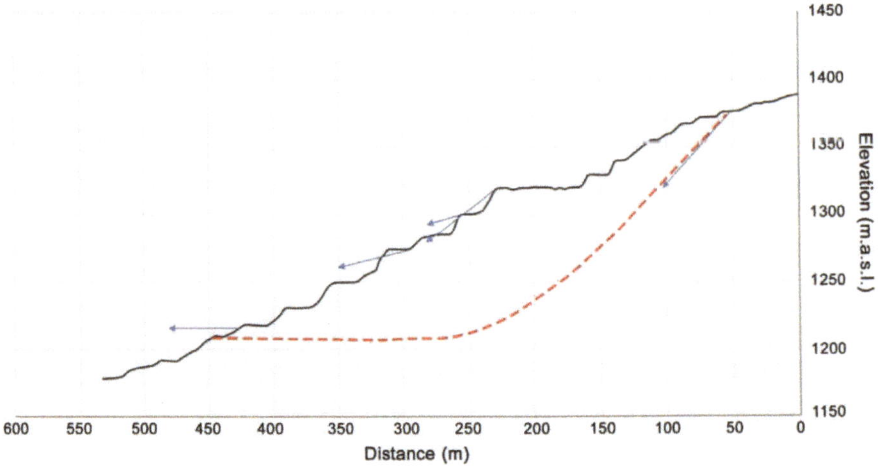

Fig. 7.14: Interpretation of a displacement contact in the NE sector of an undisclosed open pit mine along the violet cross-section of Fig. 7.13 (source: P. Farina).

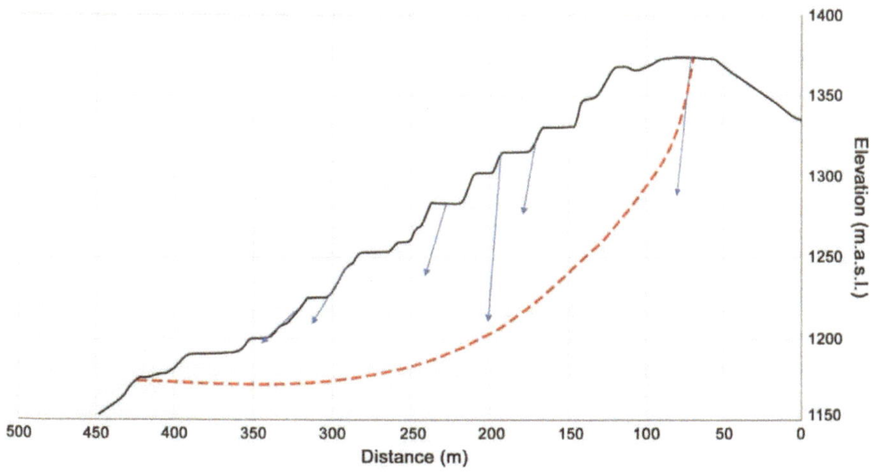

Fig. 7.15: Interpretation of displacement contact in the southern sector of an undisclosed open pit mine along the yellow cross-section of Fig. 7.13 (source: P. Farina).

than what could be inferred only from the change in topographic elevation.

Figure 7.17 presents the slope cross-section along the profile trace indicated in Fig. 7.16 before and after the failure. The extent of failure at the ground surface mapped in the field is also indicated. The point of intersection between the pre- and post-failure topography was preliminary, associated with the approximate boundary between the detachment and main accumulation zones. The extent of the failure zone was unclear due to uncertainty regarding the thickness of the failed material and how far the failure surface persisted downslope beneath debris run-out. The initial estimation of the failure volume (410 000 m^3) was based only on the amount of material included within the interpreted detachment zone with an average depth of 11 m.

The pink shaded area in Fig. 7.17 marks the cross-section length of the slope sector displaying accelerating slope behaviour, as defined according to the intersection of the profile trace with the limits of the white polygon in Fig. 7.16. Since accelerating slope behaviour is supposed to affect those areas where rock fracture propagation and strain accumulation is occurring (Kemeny 2003; Kilburn 2012), it can be inferred that the basal damage zone, and

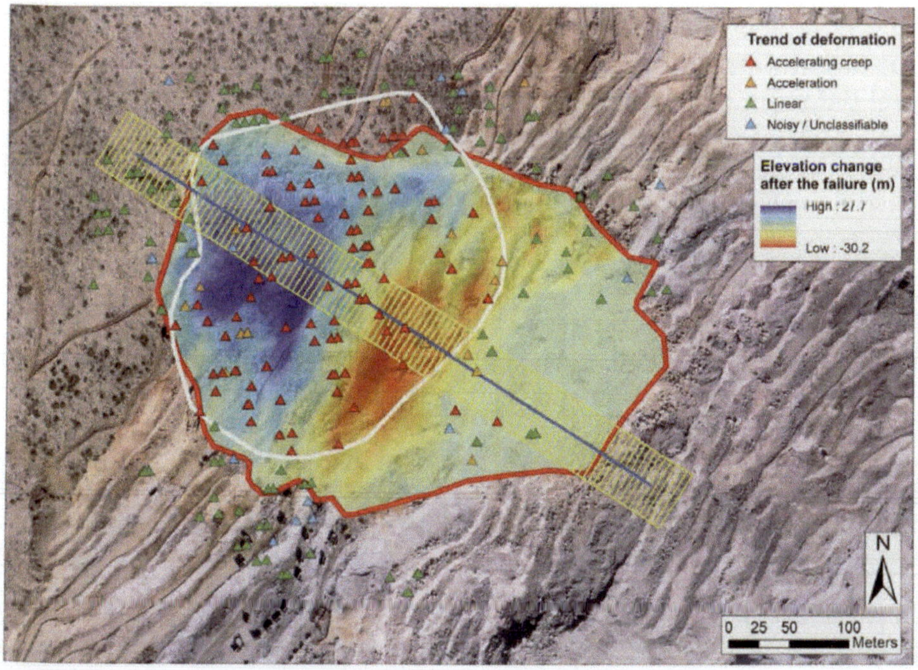

Fig. 7.16: Satellite InSAR data (ascending geometry) within and in proximity to the failure, classified according to the different deformation behaviour. The blue line represents the trace of a cross-section shown in Fig. 7.17 (source: P. Farina).

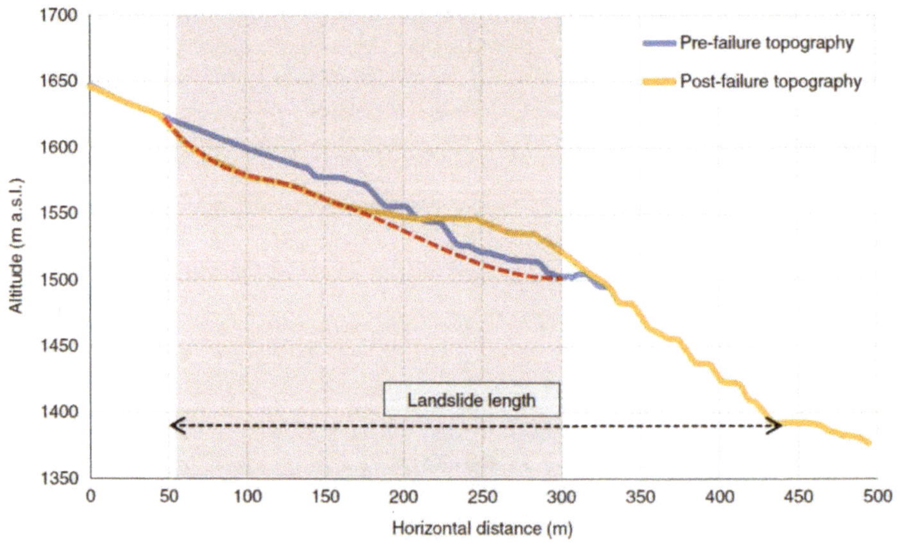

Fig. 7.17: Slope cross-section along the profile trace indicated in Fig. 7.16, where the pink shaded area includes the slope sector affected by accelerating slope behaviour, while the red dashed line indicates the interpreted extension of the basal damage zone (source: P. Farina).

consequently the potential detachment zone, extended downslope in the proximity of the lower boundary of the main accumulation zone (red dashed line in Fig. 7.17). Still, the estimation of the overall volume of the instability can be refined. Considering an extent of 61 000 m^2 for the accelerating slope area mapped in Fig. 7.16 and the average interpreted depth of the failure surface (11 m), the estimation yields a volume of ~670 000 m^3. This is more than 1.5 times the estimation that was derived from the comparison between the pre- and post-failure digital terrain models (DTM).

7.6.4 Interpretation of failure mechanisms from structural data and monitoring with radar

In a moderately jointed rock mass, slope failure is generally defined by the presence and orientation of discontinuities that act as planes of weakness within the rock mass, thus controlling the size and the direction of ground movement. For this reason, when dealing with structurally controlled slope movements, a method to support the interpretation of monitoring data could be to combine displacement records with structural data, especially when the monitoring system relies on GB radar. Recognising that movement measurements are LOS, data from radar provide only magnitudes, not directions of slope displacement, thus limiting their use for interpreting the 3D failure kinematics and the corresponding behaviour. This poses a challenge in the interpretation of the deformation data provided by the radar and in the setup of alarm thresholds that commonly assume that monitoring data provide true (total vector) movement, which could be significantly different from that measured by radar.

To reconcile the directional component of radar measurements with actual true movement, utilisation of a kinematic analysis approach was investigated to interpret true movement from LOS data. The term 'kinematic' refers to the study of movement, without reference to the forces that produce it. The term has been extended to the graphical analysis of block behaviour under its weight and considers only frictional shear strength along the discontinuity faces (Goodman and Bray 1976; Hoek and Bray 1981; Matheson 1983; Hudson and Harrison 1997). This kind of analysis is utilised to establish if and where a particular instability mechanism is kinematically feasible, given the geometry of the slope and discontinuities present. To identify areas prone to failure on an excavated slope, an analysis of the kinematic admissibility to potential wedges or planes that are exposed by the excavation is typically recommended. Kinematic analyses are useful for the first definition of susceptibility for the main mechanisms of instability, especially at the bench scale. These are identified by combining discontinuity dip and dip directions with local slope orientations. The main instability mechanisms investigated with this approach are plane failure, wedge failure, block toppling, flexural toppling and rockfall. An evolution of the kinematic analysis process was introduced by Casagli and Pini (1993) with the concept of the kinematic hazard index for each instability mechanism.

Based on the concept of the kinematic hazard index, it is possible, by using specific algorithms such as those included in the DiANA code (Gigli and Casagli 2011), to manage a great number of discontinuities with assigned friction angles together with high-resolution point clouds acquired by laser or photogrammetric systems.

Fig. 7.18: Example of 3D kinematic analysis identifying the sectors of the slope more prone to a wedge failure mode, compared with radar data acquired over the same area. The red boxes indicate areas where the radar detected significant movements corresponding to areas of strong kinematic potential for planar failure (source: P. Farina).

Intersection lines of discontinuity sets are determined by the method and together with the equivalent friction angle, based on the friction angles of the intersecting planes, potential kinematic geometries are determined. The analysis can be performed for specific slope orientations or for each point of a point cloud. Such a method has been called 3D kinematic analysis. The outcome of the 3D kinematic analyses are maps, by design sector, of the relative probability of each different failure mode (plane failure, wedge, direct toppling, flexural toppling, rockfall). The maps, one for each failure mechanism, show for each point of the point cloud values ranging from zero to a value associated with the relative probability of each specific failure mode.

The 3D kinematic maps can become an essential input for the interpretation of radar data, allowing the slope engineer to associate measured ground movements with a possible failure mechanism. By comparing radar displacement maps to 3D kinematic maps, similarly to the example shown in Fig. 7.18 for a sector of an open pit mine, it is possible to infer the most likely mechanism generating a certain instability. Moreover, from the 3D kinematic maps, it is possible to derive maps showing the most likely direction of movement of each kinematic block, thus enabling the estimation of the sensitivity to movement measured by radar to an estimation of true movement for each design sector of the pit. As a consequence, 3D kinematic maps can be used to refine the radar alarm thresholds, by correcting the LOS measurements with respect to their sensitivity to the expected direction of movement.

The described method, based on the quantitative combination of structural data in the form of 3D kinematic index maps with displacement records acquired by radar, represents an operational methodology to support the understanding of the relationship between observed deformations and structural stability controls. In addition, the output of the analysis improves the effectiveness of the slope monitoring program attributed to the input that provides improved selection of alarm thresholds.

7.7 Combining monitoring data with numerical modelling

7.7.1 Overview

Numerical models are widely used to analyse slope stability. They are capable of representing deformation and displacements, as the model is cycled through mining sequences. Pit slope monitoring data and observations are often used to calibrate numerical models to the best possible representation of actual conditions, which yields higher confidence in predictive results. Instrumentation data and observed conditions that can be utilised for calibration include:

- LOS prism displacements;
- vector data as derived from prisms, inclinometers, GNSS units and field measurements (e.g. slickensides or striations);
- TDR or inclinometer shear points at depth;
- cracking from slope deformation;
- radar images (InSAR) and data for spatial extents of deformation;
- microseismic data;
- collapse features; and
- pore pressure changes (in low-permeability material).

Utilising information from all these sources is often not possible. However, the more the model behaviour can be matched to multiple points of actual observed behaviour and mechanisms, the higher will be the confidence in the calibration. History points that can track displacements, tilt and pore pressure changes in the model should be fully utilised, placing them at the same locations in the model as the instruments are located in the pit slope, and setting them up to record the same information (e.g. horizontal displacement) as the instrumentation data being used for model calibration.

7.7.2 Numerical model basics

Geotechnical numerical models attempt to represent the mechanical response of a rock mass subjected to a set of initial conditions (e.g. *in situ* stresses and water levels), boundary conditions and induced changes (e.g. slope excavation). With numerical models, a 'full' solution of the coupled stress/displacement, equilibrium and constitutive equations is made. This can be carried out by treating the slope as an equivalent continuum, using methods like finite-difference or finite-element, where the role of discontinuities is treated implicitly as reducing the shear strength of the continuum rock mass; or by treating the slope as a discontinuum, using methods like the distinct-element method, where discontinuities are treated explicitly, allowing for kinematics through the dislocation and slip along fractures to be accounted for in the stability analysis. A detailed discussion of different slope analysis techniques including their strengths and weaknesses can be found in Stead *et al.* (2006).

Given a set of properties, the modelled system is found to be either stable or unstable. If an equilibrium result is obtained, the resultant displacements at any point in the rock mass can be compared with measured values. If an unstable/collapse result is obtained, the predicted mode of failure is demonstrated. By performing a series of sensitivity simulations with varying properties, the strength reduction factor (SRF) can be found. This factor demonstrates the margin of stability in a system. This method is typically done by reducing all strength parameters in the model that influence stability (rock mass, ubiquity and explicitly defined discontinuities such as faults). Partial SRFs are sometimes used to analyse a specific parameter or area, depending on the mechanism being represented and knowledge of intact and residual strength states.

Note that SRF is differentiated from factor of safety (FOS); SRF is specific to numerical modelling, whereas FOS is specific to limit equilibrium analyses. The two are not always equivalent, as numerical modelling is generally capable of more complexity than limit equilibrium, and matching mechanisms between the methods in complex rock slopes is difficult to achieve (noting that the failure mechanism is assumed as an input in limit equilibrium analyses, whereas in a numerical modelling, it emerges from the results).

There are several reasons why numerical models are used for slope stability studies:

- numerical models can model complex behaviour, including deformations, failure mechanisms, creep, dynamic effects and coupled hydromechanical modelling;
- numerical models can be extrapolated confidently to a wide variety of conditions in comparison to empirical methods, which are limited to the experience base (i.e. database) from which they are developed;
- numerical analyses can incorporate key geologic features, such as faults and groundwater, providing more realistic approximations of the interactions between these features and thus the behaviour of real slopes. In comparison, analytical, physical or limit equilibrium methods may be unsuitable for some pit slopes, requiring oversimplifying assumptions to represent key features, possibly leading to non-conservative or overly conservative solutions;
- numerical analyses can help explain observed physical behaviour, especially that being measured by slope monitoring instruments (e.g. slope displacement is a key output of a numerical analysis, whereas limit equilibrium does not calculate displacements); and
- numerical analyses can evaluate multiple scenarios to test the sensitivity of the results to different geological models, failure modes and design options, thus providing an important tool for managing uncertainty.

7.7.3 Using monitoring data in model setup

The first step in the modelling process is to assess the geology, rock mass properties, hydrogeology and geological structures in detail. Monitoring data can and should be considered as early as possible during model setup, as the data can help define the purposes of the analyses, conceptual picture of the problem (potential failure modes), model objectives (design or remedial stabilisation optimisation) and discretisation parameters. For example:

- prism, radar and LiDAR data can reveal the extents of mining response and areas of dilation above normal response to define model extents;
- downhole/subsurface data can show the extent of shear on a given structure. This can be vital when considering how to represent geology and structure in the model (i.e. explicit interface or a material zone); and
- vector data (from RTS and GNSS) can reveal the mechanism that needs to be represented in the model.

By considering monitoring data during model setup, key monitoring points to be used for model calibration can be built in as model prisms, model inclinometers and model piezometers to record the history of displacements and pore pressures as the model steps through the mine sequencing being simulated.

7.7.4 Using monitoring data in calibration

To achieve calibration, a set of criteria should be established with consideration of model limitations and realistic expectations of how well a model can represent reality. The monitoring tools themselves also have limitations in accuracy, and data used for model calibration should be well-vetted.

- LOS distance readings on prisms are often of the order of 0.5 mm/day (15 mm/month) for normal background rebound. A numerical model may not be able to reasonably match this in what it simulates as time is not explicit in the model calculation but treated implicitly by simulating incremental stages in the mine plan sequence. Thus, a mined volume might be staged in increments of several months or years; staging this in days or weeks in the model might not be computationally practical. Comparing or attempting to calibrate to natural rebound with low rates is often not achievable without adjusting elastic inputs, for which it is difficult to establish a defensible basis.
- Back-analysis of failures and even of stable slopes can play a significant role in the determination of material properties. Detailed records of the performance of excavated slopes and the initial stages of ultimate slopes can provide field-scale assessments of properties that can normally only be determined through small-scale laboratory tests during the feasibility and earlier stages of design. However, representing periods of higher than normal rates or collapse is much more achievable.
- Downhole measurements from inclinometers can yield 2D vectors and magnitudes. Model shear on a defined interface can and should reasonably represent vector direction and be within an order of magnitude of observed shear. Elastic properties on interfaces can have an impact on how much model shear is predicted.

Criteria should be outlined collaboratively with the modeller, practitioner and key stakeholders during model construction. Non-convergence criteria should be based on back-analyses of a collapse state whenever possible. The parameters to be adjusted for calibration purposes should also be agreed; typically, these are rock mass strengths and/or fault strengths. The chosen dials for adjustment and calibration should start from laboratory-derived properties that will ultimately be defensible. Final calibrated strengths should compare reasonably to laboratory data and field observations. A leading practical approach for defining calibration criteria is outlined as follows.

Matching behavioural trends observed at historical topographies:

- failure/collapse (Fig. 7.19);
- non-convergence in numerical models and FOS = 1.0 in limit equilibrium models;
- accelerating trends;
- increased deformation in numerical model relative to a more stable time-step;
- decelerating trends; and
- decreased deformation in model relative to a less stable time-step.

Matching observed deformations (via monitoring data):

- model matches LOS prism movements to about ±50% of magnitudes (Figs. 7.20); and
- deformation patterns show good correlation to radar data for spatial extents (Fig. 7.21).

Other observations:

- model is within same order of magnitude when compared to inclinometer data;
- model may over- or underpredict shear on beds at inclinometer intersections, but be within an order of magnitude; and
- model can represent the location of subsurface shearing from downhole installations such as inclinometer or TDR installs (Fig. 7.22).

Matching observed mechanisms:

- model movement vectors match well with observations;
- model yield state plots (Fig. 7.23) demonstrate shear on major structures involved in the mechanism – active slip condition if the structure is presented as an explicit feature or active shear condition if the structure is presented implicitly as a zone; and
- tensile and rock mass yield in areas with observed cracks (Fig. 7.23).

A state of equilibrium assessment is a fourth consideration in calibration against known slope conditions

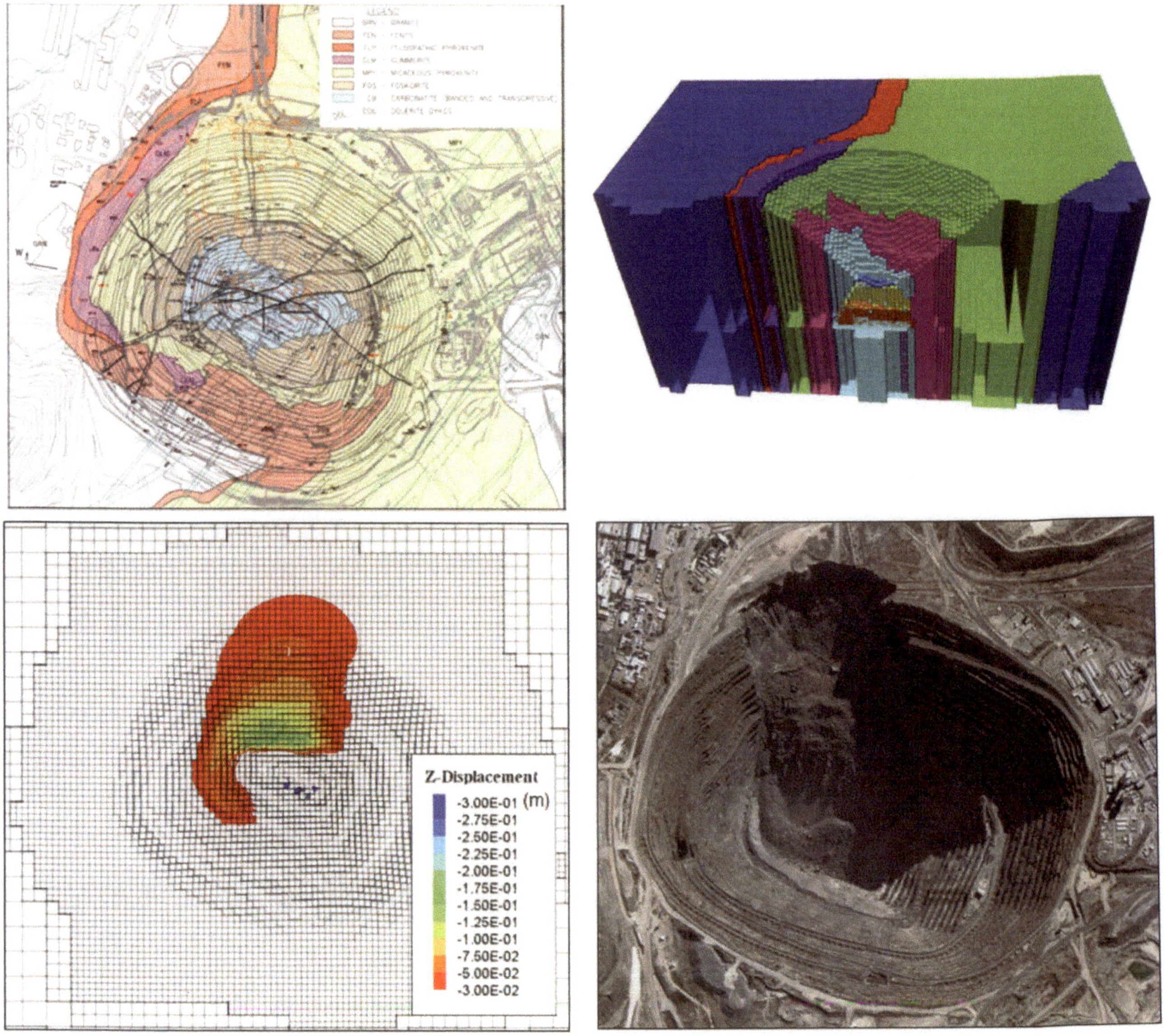

Fig. 7.19: Back-analysis of the Palabora pit slope failure caused by interactions with the block cave beneath the pit, showing the geology built into the model and modelled outline of the instability relative to that of the eventual collapse (modified from Woo *et al.* 2012).

for mechanisms that have shown evidence of dilation, but have not failed or collapsed. A state of equilibrium assessment requires considerable experience with the identified mechanism. For example, if slope movement and dilation have occurred with measurable shear and observed dilation (i.e. cracking, increased slope movement in radar or prisms, measured subsurface shear) beyond background movements, an assumption may be made that the FOS/SRF should be marginally above 1.0. In other words, if dilation beyond background rates, subsurface shear and tension cracking have been observed on a given slope, an SRF/FOS result of 1.4 would be considered unreasonable. The model's represented mechanism, inputs and overall calibration would be revisited.

In summary, achieving the behaviour, deformation magnitude and mechanism criteria is recommended as the primary step for model behaviour validation. Reconciliation of the state of equilibrium, based on model outputs, should then be reasonable and defined by the mechanism in question and by engineering judgement.

7.7.5 Using monitoring data to reconcile model predictions

A priority for the predictive effort is to identify if the model is stable for the given mining period based on the convergence state (i.e. stable/unstable or accelerating/decelerating trends). Then the deformation patterns should be explored and the histories of the velocities and displacements should be examined for the locations with the largest displacements. In the event of a decelerating velocity trend, the condition can be considered to be stable, as illustrated in Fig. 7.24. In addition to the history plots,

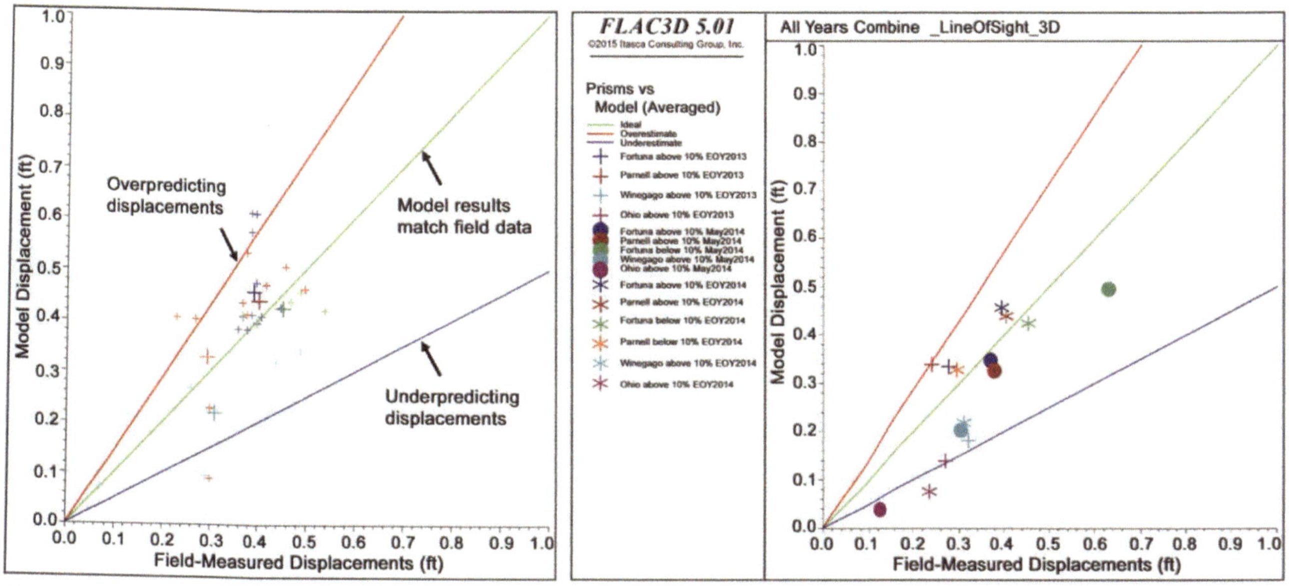

Fig. 7.20: Prism calibration chart showing (a) modelled *v.* prism measured field displacements, with individual and average prism values marked by small and large crosses, respectively, and (b) modelled *v.* measured displacements with respect to different time periods (different symbols indicate different time periods). Different colours of the crosses/symbols are for different geological units (source: Itasca). Red and blue lines in each plot shows the ±50% criteria lines (source: Itasca).

several sections should be scrutinised through the location of predicted deformation to explore details of the model's represented mechanisms. The predicted mechanisms should be compared to those demonstrated in calibration and considered in the context of known geology and rock mass conditions.

Predictive models provide an opportunity to run additional sensitivity cases to test different mining scenarios to support the decision-making process. One of the powerful tools for sensitivity analysis is to perform SRF simulations, which not only provide the margin of stability in a system, but also help define regions with possible stability issues. Flagging areas of concern can then ultimately guide monitoring targets; this is discussed further in the following section.

Once modelling projects have been completed and results are in, it is vital for the practitioner to keep reconciling model performance as mine designs are

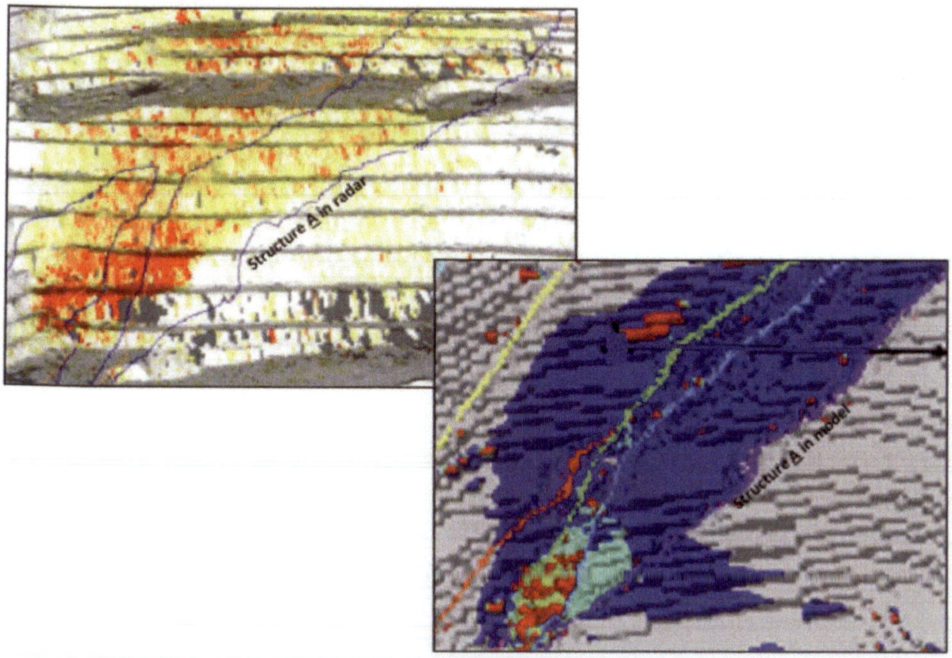

Fig. 7.21: Radar data compared to the model deformation patterns. Radar showing movement on a white-to-red scale. Model shows movement on a blue-to-red scale (source: Itasca).

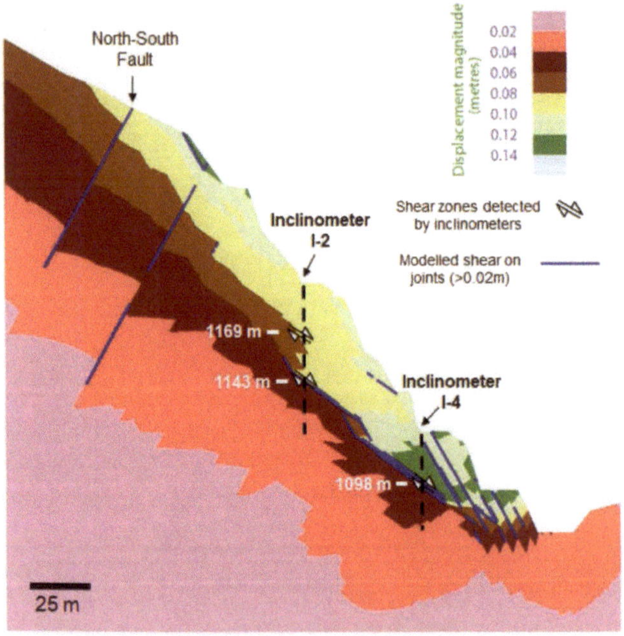

Fig. 7.22: Model calibration results from distinct-element analysis comparing location of modelled shear displacements with depths of shear zones identified in inclinometer data (source: E. Eberhardt).

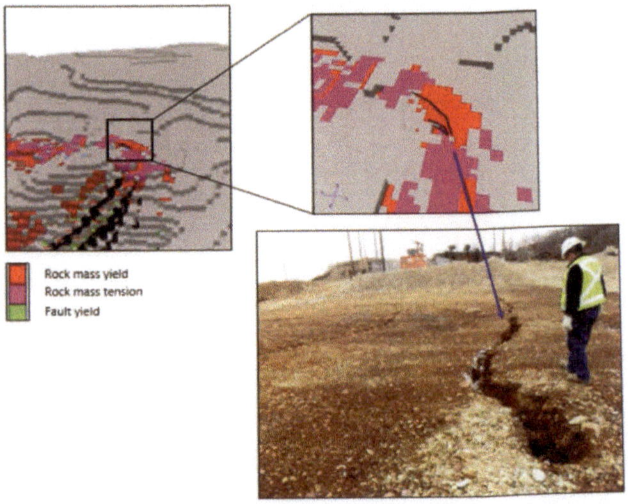

Fig. 7.23: Comparison of field-mapped tension cracks and model yield state plot (source: Itasca).

executed. Comparisons in this phase of work are not to the level of model calibration, but they provide verification that model predictions are reasonable. If major deviations from reconciliation to as-built performance are identified, model calibration may need to be revisited.

7.7.6 Using models to optimise instrument layout

One of the objectives of performance monitoring instrumentation is the assessment of design validation. The concept of design validation can easily be understood by considering performance of structures built using concrete. The design of concrete structures is based on the assumption that the concrete will have a certain strength, called the design strength. As the structure is being constructed, concrete samples are collected, cured and tested to make sure the concrete has achieved at least the design strength, thereby validating the design. In the same way, slope designs are based on a set of assumptions regarding the water pressures, location of key structures, rock strength and expected behaviour. Analyses are performed using numerical models to predict the range of expected slope behaviours. Since nearly all slopes are designed to be stable, the predictive behaviour often takes the form of displacements or strains (in numerical modelling parlance). Predicted displacements are particularly important where critical infrastructure and/or equipment (crushers, grinding mills, etc.) are located behind a pit crest. These features may be extremely sensitive to deformation, particularly if it is differential. In such cases, acceptable displacement ranges are provided and the slope is instrumented with prisms, inclinometers and sometimes InSAR to ensure the actual displacements are within expected ranges as the slope is excavated. Displacements above the expected range suggest the design is not valid, and signal the immediate need for

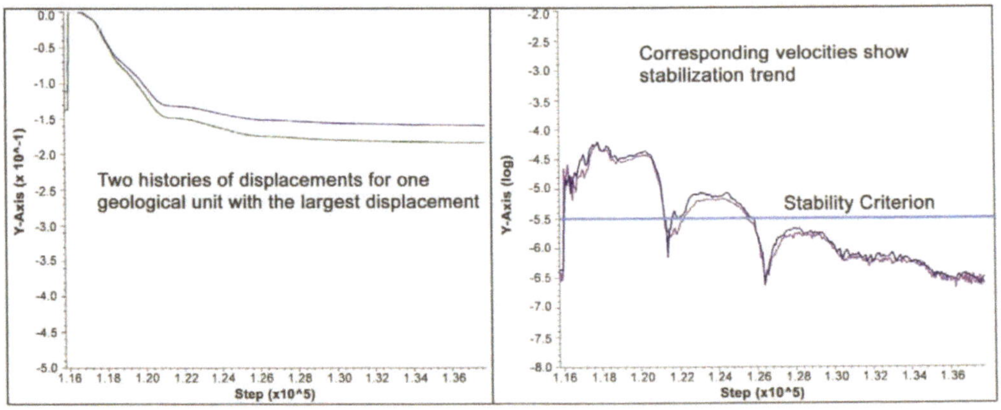

Fig. 7.24: Example of displacement and velocity histories for the location with the highest displacements (source: Itasca).

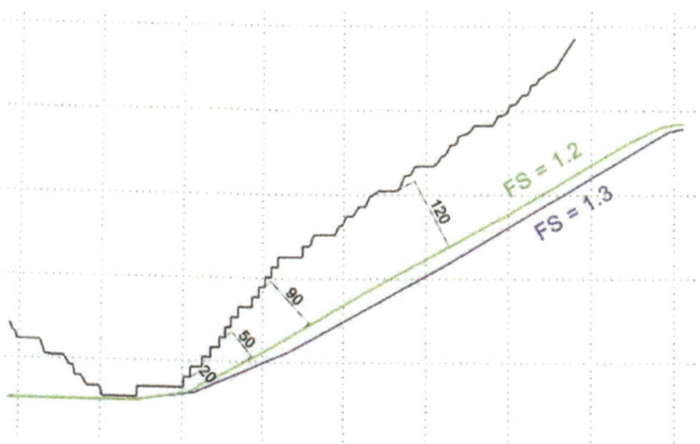

Fig. 7.25: Example slope showing target water levels. Measured water levels above the green line suggest a safety factor less than 1.2. Measured water levels between the green and blue line suggest a safety factor between 1.2 and 1.3 (source: Itasca).

further analysis and/or site characterisation. In other cases, involving brittle materials that are anticipated to weaken if they are excessively strained, instrumentation should be used to confirm that deformation remains below critical levels.

However, in practice, displacement or deformation criteria may be difficult to establish, as they depend on a thorough understanding of the failure mode and the potential consequences of any deformation. For example, a toppling slope that does not carry ramps or facilities near the crest can undergo several metres of annual movement without causing concern. In contrast, more brittle failures, such as daylighting wedges, can withstand very little displacement before failure occurs.

More commonly, instrumentation is used to validate key design assumptions, such as water levels or pressure distribution. Since most slopes are sensitive to water level or pressure, they need to be instrumented with piezometers to validate the design assumptions. Figure 7.25 provides an example of target water levels within a slope. Measured water levels above the targeted level of design tolerance, or criteria, indicate that the slope is in a condition of reduced safety margin.

In the cases mentioned above, numerical models can be used to optimise the instrumentation type and location by identifying key areas of the slope where design assumptions need to be validated. As mentioned previously, development of any instrumentation plan depends on a thorough understanding of the deformation mode and fundamental mechanisms. The use of numerical models to understand deformation modes and fundamental mechanisms is discussed in the next section.

7.7.7 Understanding fundamental mechanisms with numerical models

The fundamental mechanisms for slope deformation/failure are different for hard/competent rocks (R3–R6)

Table 7.2: Factors that influence the instability mechanisms in weak and strong rocks

Factor influencing the instability mechanisms	Weak rock (R0–R2)	Strong rock (R3–R6)
Cohesive strength of intact rock	Very important, as it is usually very low	Not a major issue, as it is usually relatively high
Strength of intact rock relative to strength of structure	Very important, as it is usually very low	Not a major issue, as it is usually relatively high
Pore pressure in the intact rock	Very important	Seldom an issue
Porosity and moisture content of intact rock	Very important	Seldom an issue
Grain size distribution of intact rock	Very important	Seldom an issue
Mineralogy of intact rock	Very important	Seldom an issue
Geological stress history	Very important	Seldom an issue
Presence of discontinuities	May be important in combination with weak intact rock	Very important, as these are usually the cause of failure

compared to weak rocks (≤R2). Slope failure in hard/strong rocks (R3–R6) is usually controlled by geological structure, while in weak rocks the rupture surface develops in the material matrix. In some cases, it may appear that there is little difference in the failure modes found in hard and weak rocks. The differences are attributed to various factors, summarised in Table 7.2, that control the instability mechanisms that ultimately lead to failure.

One of the advantages of numerical models is that the user does not specify the anticipated failure mode, although the failure mode is usually evident, based on features of geology model input. Rather, the model predicts the failure mode with the lowest safety factor. Examples of failure modes exhibited by numerical models are given below.

7.7.7.1 Failures involving geologic structural fabric

In slopes where stability is controlled by structure, sliding is likely to occur along pre-existing structures. These slides can be planar, when the mass slides on top of a single plane that strikes subparallel to the slope trend, or wedges, when the intersection of two or more discontinuities favours movement of the mass downslope, as shown in Fig. 7.26. It should be noted that, particularly in weak rocks, the structures on which sliding occurs do not have to be entirely daylighted in the slope, as is the case in non-daylighting wedges and step-path failures (see Fig. 7.27).

A mode of failure that is typically found in layered strong rocks with anisotropic strength is toppling, shown in Fig. 7.28. This mode of failure may also occur in weak, altered foliated rocks, such as those found in association with soft iron ores. A similar failure mode, flexural

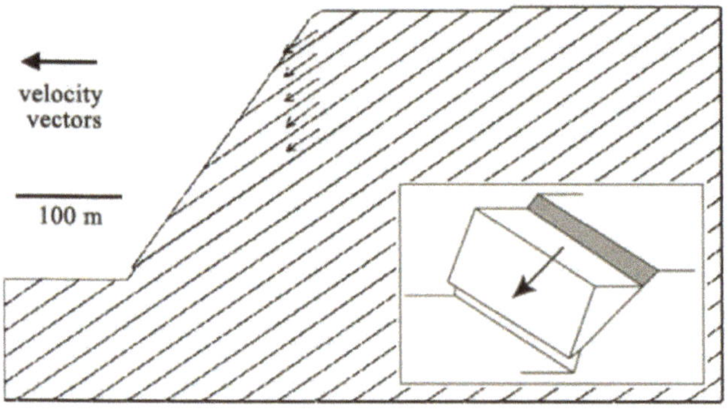

Fig. 7.26: Planar failure mode (see inset) modelled with rigid blocks using the distinct-element method (source: Lorig and Stead 2017).

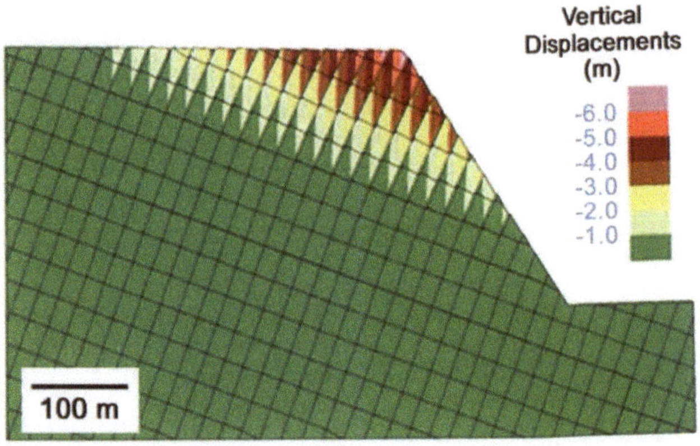

Fig. 7.28: Forward block toppling failure mode modelled with deformable blocks using the distinct-element method (source: Itasca).

toppling, occurs when there is one dominant closely spaced set of joints dipping steeply into the face, with insufficient cross-jointing to permit the free rotation of blocks. The columns bend out of the slope like cantilever beams. Figure 7.29 shows the results of an analysis with joints spaced at 20 m. The FOS is 1.3, being reduced as the joint spacing decreases. Because flexural toppling involves high stress gradients across any rock column, it is necessary to provide sufficiently fine zoning (i.e. smaller elements) to accurately represent the stress gradients due to bending.

7.7.7.2 Failures involving development of a rupture surface through the rock mass matrix

In weak rocks, slope instability typically involves a combination of failure both along the structure, if present, and through the intact matrix, or rock mass. The complicating factor in weak rocks is that the failure process may initiate in the rock mass before progressing to the geological structure. Consequently, instability modes of failure in weak rocks are not limited to simple sliding along geological structures. An example of this type of failure mode, involving the development of a rupture surface in the material matrix, is shown in Fig. 7.30 for the 1994 pit slope failure at the Twin Creeks mine in Nevada, USA.

7.7.7.3 Step-path failure developed along jointing and shear through the rock mass

In massive crystalline rock where adverse dipping discontinuities are not fully persistent, failure may occur through a combination of sliding along jointing and shearing of the rock mass (Eberhardt 2008). This progressive failure mechanism is seen to develop through a step-path failure mode; in the absence of an adversely dipping geological structure daylighting at the toe of the slope, it is accommodated through shearing of the rock mass at the toe, otherwise referred to as toe breakout (Fisher and Eberhardt 2007).

The development of internal shearing and toe breakout associated with the step-path failure mechanism limits the applicability of limit equilibrium analyses. Discontinuum-based numerical modelling, for example using the distinct-element method, is required to properly capture the internal shearing, large-strain deformations and complex active-passive rock mass interactions involved (Stead *et al.* 2006). Figure 7.31 shows the modelling of a step-path failure using the distinct-element

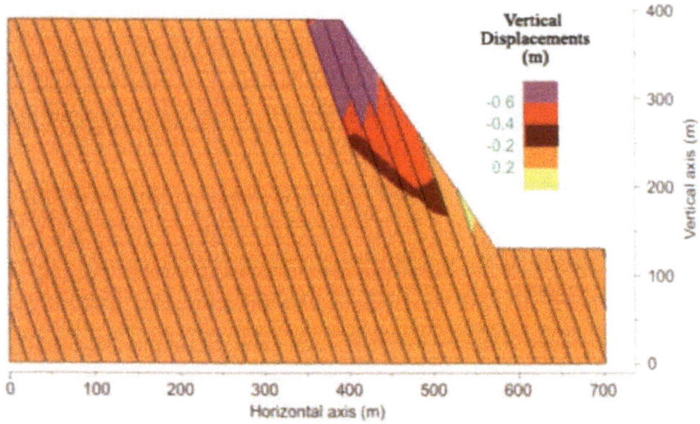

Fig. 7.27: Non-daylighting planar failure mode modelled with deformable blocks using the distinct-element method (source: Lorig and Stead 2017).

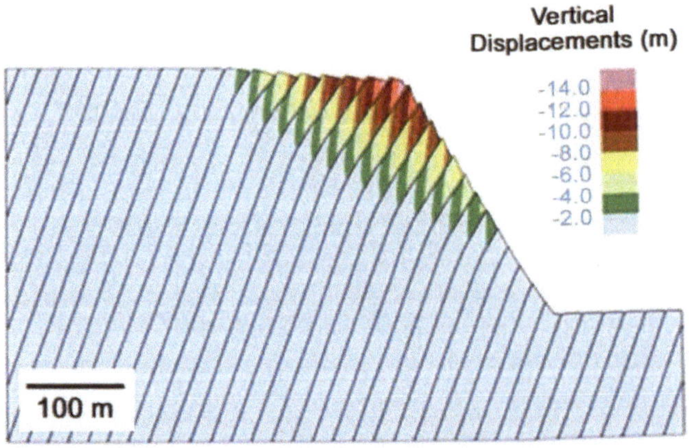

Fig. 7.29: Flexural toppling failure mode modelled with deformable blocks using the distinct-element method (source: Lorig and Stead 2017).

Fig. 7.30: Example of modes of failure observed in weak rocks, with finite-difference modelling of horizontal displacements for the Twin Creeks open pit slope failure. The mechanism observed in the 1994 slope failure could be closely reproduced by the model, increasing confidence in the predictive analysis and material properties used (source: Itasca).

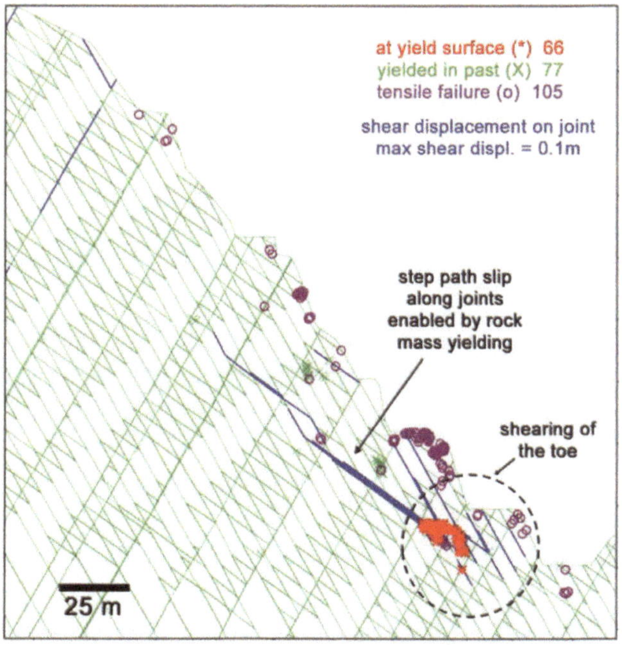

Fig. 7.31: Model results from a distinct-element analysis showing the development of a step-path failure and toe breakout caused by a combination of slip along the joint network and shearing of intact rock bridges (source: E. Eberhardt).

method. The results show the initiation and localisation of a shear surface near the toe of the slope where the limited persistence of the toe joints requires shear failure through the rock mass blocks for kinematic release and movement. This transitions into a step-path failure that develops along the slope parallel discontinuities, accompanied by small toppling movements along the fault sets dipping into the slope. The model showed good agreement with surface and shear displacements at depth agreeing with data derived from surface geodetic and inclinometer monitoring.

8

DATA UTILISATION AND COMMUNICATIONS

Phil de Graaf, Pat Jenks, Jeff Mattern, Mike Ness, Alex Pienaar, Nick Rose, Eric Schwarz, Robert Sharon, Ryan Turner, Fanie Wessels

8.1 Introduction

Communication of slope monitoring results is vital to the practical use of data derived from pit slope monitoring systems. Communication can take many forms, from verbal reporting on visual observations to routine reports of slope monitoring performance and to urgent advisories that call for the evacuation of specific areas in the pit. This chapter focuses on vital areas of data utilisation including:

1. the detection of anomalous changes in monitoring data and alarming in the event of hazardous conditions; and
2. the development of methods to communicate monitoring results to the workforce and other stakeholders, and the reporting of hazardous conditions and resulting actions.

A reliable monitoring system should record and identify anomalous slope movement. Although the most obvious purpose of a monitoring system is safety-related, slope deformation and performance monitoring also enhance the understanding of slope behaviour and assist in improving system design and implementation. For a monitoring system to be considered an effective early warning alarm, it needs to operate in real-time and should be linked to the mine's emergency communication system.

8.2 Observational mining

The observational approach, introduced in Chapter 2, is a common and practical method for dealing with unpredictable ground conditions during slope design and implementation.

The method is applicable where conditions upon which design must be based are not fully known, or where it is impractical to obtain sufficient design information before the beginning of mining. It is also a rational approach utilised to deal with unanticipated conditions encountered during mine development. Proposed by Dr Karl Terzaghi and introduced in Terzaghi and Peck (1948), the method (see also Peck 1969) embodies the following steps as adapted to mining:

1. exploration sufficient to establish at least the general nature, pattern and properties of the rock mass or soil material to be mined;
2. assessment of the most probable conditions and the most unfavourable conceivable deviations from these general conditions. Geology plays a major role in this assessment;
3. establishment of the design based on a working hypothesis of behaviour anticipated under the most probable conditions;
4. selection of quantities to be observed as mining proceeds and calculation of their anticipated values on the basis of the working hypothesis;
5. calculation of values of the same quantities under the most unfavourable conditions compatible with the available data concerning the subsurface conditions;
6. selection, in advance of a course of action or modification of the design, for foreseeable significant deviations of the observational findings from those predicted on the basis of the working hypothesis;
7. observed slope performance and evaluation of actual conditions using robust slope monitoring practices; and
8. modification of the design to suit the actual conditions and actual slope performance.

An iterative application of the observational method is often used to refine and optimise mine design as new data become available. Key to this approach are accurate and consistent monitoring data.

To obtain the greatest benefit from performance monitoring, it is necessary to have a good understanding of the possible range of outcomes and to have a clear plan of action for likely outcomes. Furthermore, it is vital to have a systematic approach to designing and managing the monitoring process, from the collection of data to actions resulting from interpretation of the data.

8.3 Mine plan stability evaluation

Performance monitoring results can provide insights on the local geology, geotechnical characteristics and influence of mining practices on the behaviour of the excavated slope.

Combining understanding of the failure mechanisms present at a mine site with slope performance monitoring results, a mine plan evaluation should provide insight as to where areas of stability concern exist in the plan. Such an evaluation can alert the operation to potential incipient stability problems.

A mine plan stability evaluation should:

- assess risk potential associated with limits of flexibility (e.g. single ramp system) should an instability occur that changes the mine plan;
- communicate stability concerns of the mine plan to geologists, surveyors, planners, mining engineers, operations supervisors and senior management;
- provide an opportunity to regularly review, improve and redirect the economics of the plan (calibrate mine optimisation);
- highlight hazard identification and management of risks (relating to ground conditions);
- emphasise the importance of monitoring, data collection and system maintenance;
- ensure adequate resources are available for system performance and maintenance;
- determine the extraction methods required to achieve maximum plan optimisation (access, tonnes, grade, profitability);
- review safety procedures related to instabilities; and
- agree to mining plans and schedules between all parties.

When evaluating a mine plan from a geotechnical point of view, the assessment of potential risk to personnel should be the focus. Limited access during development presents a higher level of risk to the operation than multi-access designs. Infrastructure placement should avoid problem areas or hotspots so that risk is minimised in the event that an instability occurs. Contingency planning to provide mine access for 'what if' cases should be presented and championed. Locations for critical infrastructure, including dewatering facilities, should be assessed for risk exposure and the results communicated. Planners and operators should be aware of the consequences should these features be impacted during mine development. The slope performance monitoring system design should reflect the risk potential of the mine plan.

8.4 Alarming instabilities and failures

8.4.1 Overview

The purpose of alarming is to determine when indications from monitoring instrumentation could result in a failure that will pose a material risk to the operation. Failure of an excavated slope is defined as when the rate of displacement is greater than the rate at which the failed mass can be mined safely and economically, or when movement produces unacceptable damage to a permanent facility (Zavodni 2001). Until a failure occurs, parts of a slope moving faster than the surrounding areas are often referred to as instabilities.

Alarms should be based on threshold values that exceed predetermined values and time periods. The determination of threshold values is improved from the application of relevant site-specific experience. Alarms are set so that the slope engineer, the workforce and management are aware of slope performance issues with minimal occurrence of false alarms.

Two broad approaches are typically utilised for improving the effectiveness of alarms. The first is to minimise information to the operator that is of no consequence. The second approach is to improve the visibility of the alarm and ensure clear expectations for response to the most important or critical alarms. Most practitioners rely on movement data as the primary indicator of relative slope performance; that is, excessive movement is the primary means of slope performance determination. Correspondingly, an alarm is set to alert the slope engineer to a significant change in the movement trend.

8.4.2 Types of alarms, alarm levels and triggers

The slope monitoring system can be set up to send an alarm, or message, when slope behaviour changes. Different alarm levels can be set, typically in order of increasing severity. Initial alerts are usually sent only to the geotechnical team to warn of a minor change in the slope monitoring trends, whereas the most serious alarm triggers execution of the operation's emergency preparedness and response plan (EPRP). Alarm thresholds can be based on various movement parameters. The most commonly used parameters include deformation displacement (either total displacement or incremental displacement), velocity or inverse velocity (acceleration). Other triggers can be integrated with deformation for alarming, such as seismic activity, blasting, strong storm events, surface water flow

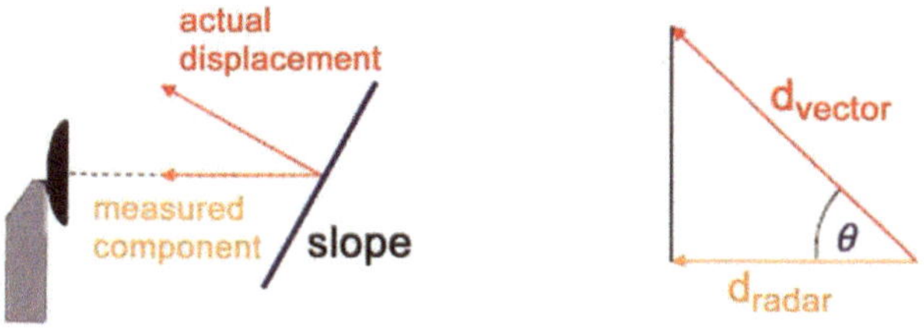

$$d_{radar} = d_{vector} \times \cos(\theta)$$

θ	cos(θ)
0°	1.00
15°	0.97
30°	0.87
45°	0.71
60°	0.50
75°	0.26
90°	0.00

Fig. 8.1: Effect of geometry on components of movement measured by LOS instruments (source: Pienaar and Joubert 2011).

and groundwater conditions (high pore-water pressures). The challenge is to determine which movement-related parameter(s) will be most effective for the operation and what threshold criteria to use.

An effective slope stability monitoring program will ensure that:

- additional geotechnical information about the slope behaviour is provided to engineers;
- safe operational practices are maintained; and
- advance warning of impending instability is given so that actions can be taken to minimise the impact of the slope failure (Call and Savely 1990).

8.4.3 Setting alarms and thresholds

The most often used and widely recognised approach to trend analysis for detecting anomalous slope movement is the time-displacement curve. Changes in movement rates that occur during the early stages of a developing instability are most recognisable to the slope engineer. Time-displacement information is typically used when there is very little change in movement. Total or incremental displacement can also be used in setting threshold levels. A shortcoming of total displacement time-series plots is the uncertainty of how much of the initial movement has been recorded. Challenges related to the installation and maintenance of monitoring stations near active operational areas often result in the loss of the initial displacement record (Zavodni 2001).

When setting slope movement thresholds, it is important to be aware of the limitations of the monitoring system, assumptions made, and details of the calculations. For example, is the velocity calculated as a line of best fit over a given period, or as the actual amount of displacement calculated by subtracting displacement at the start of the period from displacement at the end of the period? The latter method can be prone to anomalous results if spurious readings are inadvertently included in the calculation. In the case of line-of-sight (LOS) displacement, such as prism slope distance or slope radar measurements, the observed displacement, unless perfectly aligned with the slope deformation vector, is less than the actual slope displacement as illustrated in Fig. 8.1 (Pienaar and Joubert 2011). Section 4.4.3 includes relevant discussion of safety-critical monitoring alarm settings and the scaling factor.

Generally, the assessment of slope displacement data will focus on identifying changes in the displacement rate over time. For this reason, velocity is typically the preferred parameter for an alarm setting. The challenge involved with using velocity is illustrated in Fig. 8.2; the

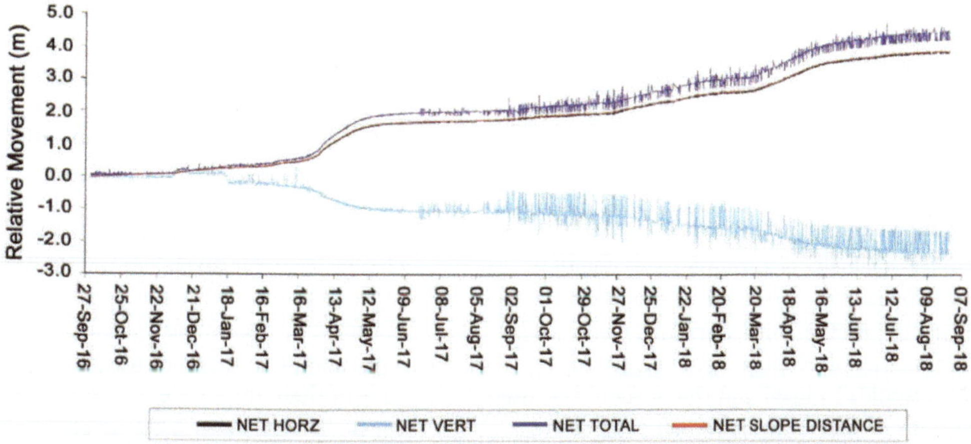

Fig. 8.2: Example of prism data demonstrating the smooth, more precise nature of EDM data (red line) with respect to net horizontal (northing and easting vector combined), elevation and total vector data (source: Teck Highland Valley Copper Partnership).

equilibrium of the slope is impacted only temporarily on a daily basis by events such as proximal blasting, a storm event or mine development. Figure 8.3 shows displacement and velocity time-series graphs, as measured by a prism monitoring system. The velocity graphs plot the moving average calculated over 24 h and three-day intervals for each prism in order to smooth the data and to clarify trends. In the displacement plot of Fig. 8.3a, several distinct periods of accelerated deformation can be identified. However, in the velocity graphs of Figs. 8.3b and c, these periods of increased movement are less obvious. In the 24 h velocity graph (Fig. 8.3b), the data noise obscures these distinct periods of increasing movement. However, if the velocity is calculated over a longer period (e.g. a three-day moving average in the case of Fig. 8.3c), the periods of increased slope deformation can easily be identified. A similar problem can be encountered in setting alarm thresholds. If the data are very noisy, the threshold could either result in many false alarms caused by data spikes or be set very high to avoid false alarms. These high, and not very useful, thresholds can miss changes in the slope behaviour, as reported in a change in trend of the velocity graph. Even if all the controls previously discussed are implemented, some level of random error will still occur. When setting alarm thresholds, this problem can be addressed through more averaging of the data by calculating velocity over a longer period.

Another technique used to smooth data from prisms is to use only LOS data from the electronic distance meter (EDM), which tend to be much smoother than total displacement due to the lack of angular error inherent in theodolite position systems, as indicated in the slope distance curve of Fig. 8.2. Practitioners are cautioned to be cognisant of the geometry limitation of LOS data as summarised above.

8.4.4 Inverse velocity method and event alarming

The inverse velocity method has become a popular added slope movement trend analysis tool for identifying accelerating slope behaviour and predicting time-to-failure. The method and its calculation are described in Sections 7.5.4 and 7.5.5. In many cases, the inverse velocity time-series plots as a straight line which can be used to predict the time of failure, as described and illustrated in Section 7.5.

However, inverse velocity is a complex variable that should be used with caution and requires experience in its use, especially in conjunction with alarming. Carlà *et al.* (2017b) recommend that the method not be used for setting alarm thresholds by itself, but that it should be used with other methods of data analysis. When progressive deformation is observed and alarm thresholds are reached, initial actions call for excluding personnel and equipment from affected areas. Once a clear progressive trend is seen, regular, repeated collapse forecasting can be applied using the inverse velocity method.

Inverse velocity time-to-failure predictions must be regarded as general estimates. The margin of error (i.e. the time difference between actual and predicted failure) can range from a few hours to several days; in some cases, a failure event may not occur. In other cases, predictions cannot be performed with adequate confidence. However, the operational pressure to re-enter areas where collapse has been forecast but has not occurred can be significant. Thus, the issue of determining when slope failure may occur and setting alarm thresholds should be managed with extreme caution, while recognising that inverse velocity functionality is programmed into most GB radar slope monitoring software and therefore is readily available, even to those without experience in its use (Carlà *et al.* 2017b).

In presenting the inverse velocity method for open pit slope monitoring and early warning, Rose and Hungr (2007) suggested the following general rules.

1. The method must not be applied in isolation; it must be accompanied by qualitative observations of slope behaviour, collection of data and ongoing analysis of the structural controls, rock mass conditions, stress and groundwater regime. Displacement monitoring is only one component of a complex process that comprises slope stability management.
2. The method should not be used for pit slopes expecting sudden brittle failure requiring little deformation, although the identification of such cases is at present a matter of judgement. Particular care should be exercised when dealing with relatively small failures in strong rock.
3. The monitoring data must be processed to remove the effects of instrument error and eliminate records distorted by local movements.
4. Failure forecasting relies on the identification of consistent trends. The possibility of trend changes, driven by observable or unknown factors, must always be kept in mind. Monitoring must be continued as long as possible before failure. The results must be constantly re-evaluated, and any established best-fit functions must be revised in view of the latest data (see Dick *et al.* 2015). If a stable trend exists, such as the initial trend shown in Fig. 8.4, predictions can be given and their reliability can be confirmed by continuing to evaluate the most recent data until the time of failure. If a trend change occurs (also shown in Fig. 8.4), the predictions must be promptly revised. It must be accepted that false alarms may occasionally occur.
5. The treatment of cyclic changes depends on the magnitude of their amplitude, relative to the distance from the horizontal axis on the inverse velocity plot. If the ratio between these two quantities is high, it may be necessary to assume that the low point of any given cycle may produce sudden rupture.

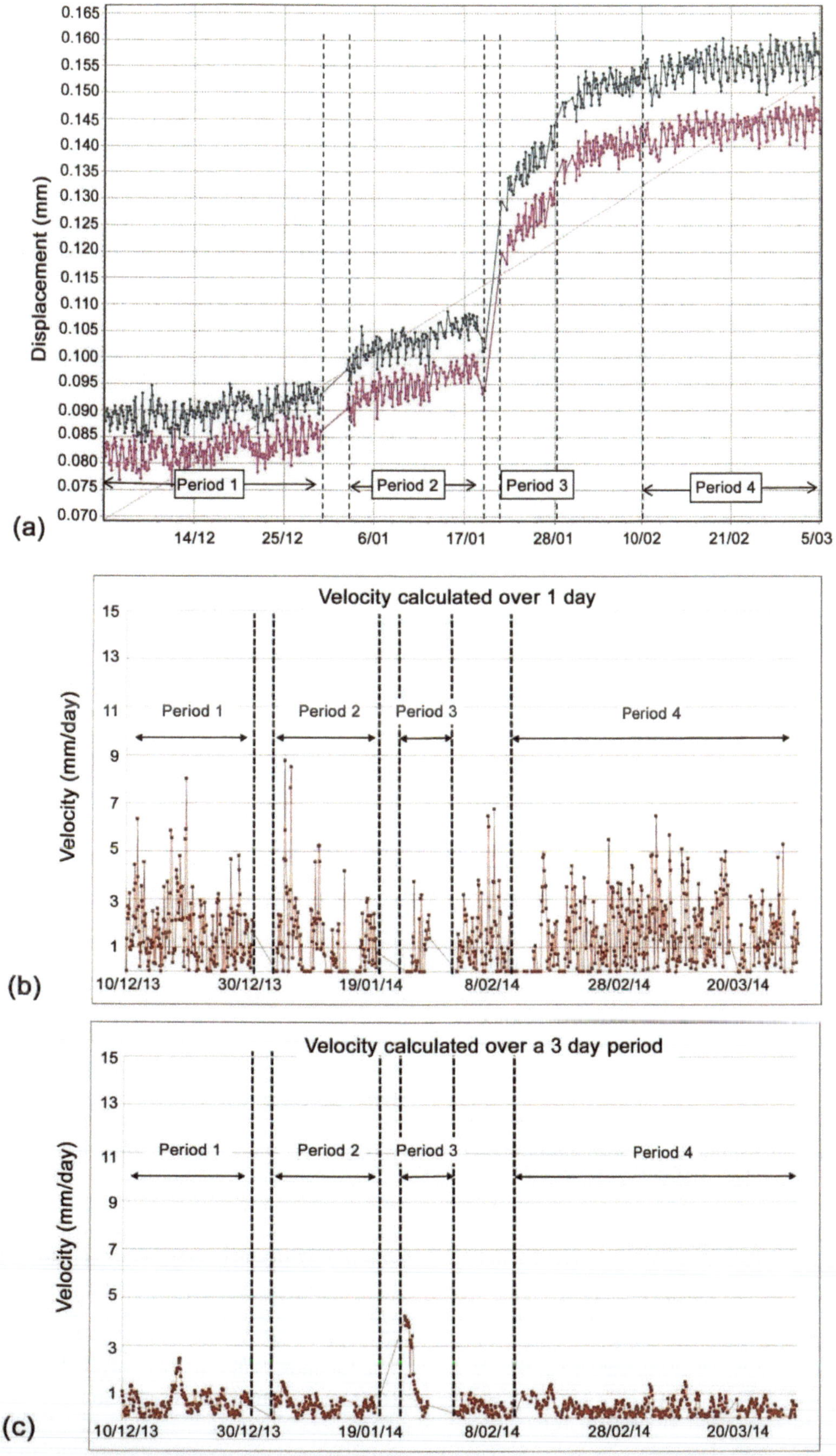

Fig. 8.3: (a) Displacement plot showing distinct periods of different slope behaviour and (b) 24 h and (c) three-day period velocity plots for the same data. The change in velocity is more clearly defined in the longer moving average period of plot (c).

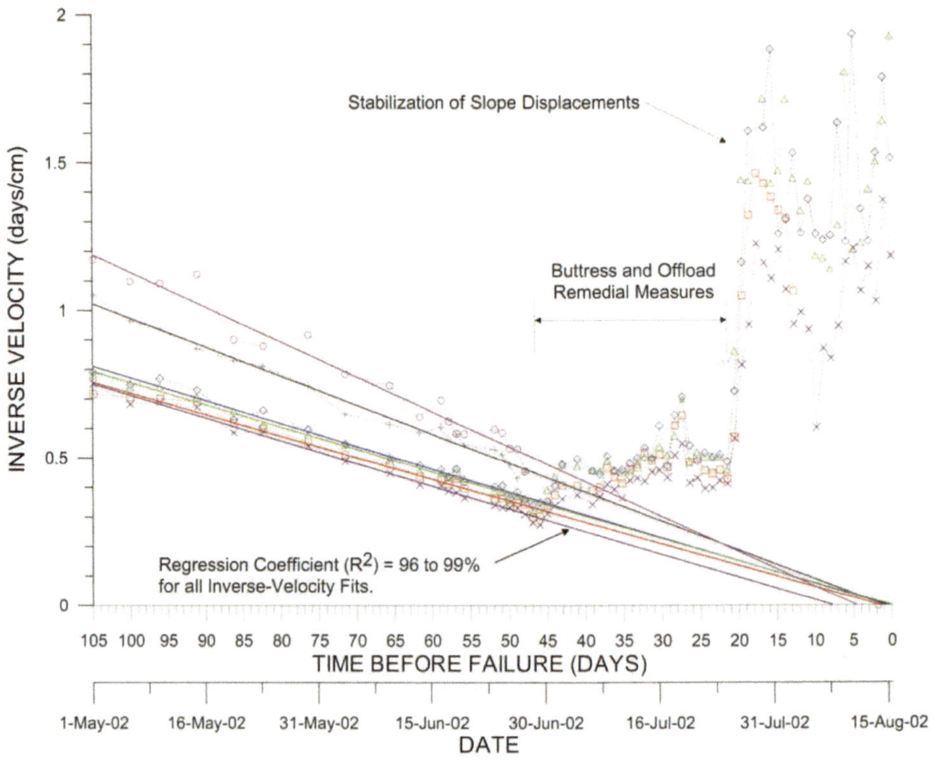

Fig. 8.4: Plot of inverse velocity v. time for six prisms, where the trend towards slope failure reversed and the slope returned to an equilibrium state in response to slope remediation efforts (source: after Rose and Hungr 2007; copyright 2006 with permission from Elsevier Ltd).

6. Data fitting using non-linear inverse velocity trend lines may provide a more accurate assessment of some longer-term trends but are more complex, which may limit practical use. Rose and Hungr (2007) recommend the use of linear fits, updated on an ongoing basis, to identify trend curvature or to signal the onset of trend changes. Since trend curvature may result from a particular manner of change of unknown driving factors, it does not seem advisable to look for a uniquely shaped mathematical function. Linear fits facilitate easy prediction of anticipated displacements, which can then be used for rapid confirmation of a consistent trend (or otherwise).

8.4.5 Cautions regarding alarm settings

Site-specific experience using reliable data from previous instability events provides an excellent opportunity to back-analyse monitoring data and interpret the onset of failure from displacement data. Using the velocity and time-to-failure at this point as a guide, a decision can be made as to how much warning will suit the operation. It is important to remember that this is only applicable if the expected failure mechanism (and contributing effects of structure, rock mass, water) is consistent with the back-analysed incident.

Often, there are no previous or relevant instabilities to back-analyse. In this case, the existing background velocity is used as an initial guide. As initial guidance, an alarm threshold should be 1.5–4 times higher than the background velocity. The quality of data, expected failure mechanisms and confidence in analysis processes also play a role in threshold determination.

The use of too high, and therefore not useful, alarm thresholds is often driven by an inappropriate combination of deformation and calculation interval parameters, especially in the case of slope monitoring radars. For alarms to be effective in the timely detection of changes in movement trends, calculation cycles for velocity determination must be very short (i.e. three hours or less). However, if the time window needed or used is too short, the result may be an increase in data noise that potentially produces more false alarms. To reduce the number of false alarms, users often arbitrarily increase the threshold level to very high, unrealistic levels. In this case, it is likely that monitoring alert/alarm objectives will not be achieved, and the alarm will be triggered too late. To keep the threshold levels as close as possible to background velocities and minimise potential for false alarms, the time window for velocity calculation cycles should be expanded. Caution must be taken not to make the time window too large, as this might mask a change in movement trend. There are no hard-and-fast rules in setting alarm thresholds; sound engineering judgement based on relevant experience is required. Dynamically testing new alarm thresholds by playing back historical data is good practice to determine the likely incidence of false alarms.

8.5 Practical aspects of alarm management and response

8.5.1 Trigger action response plan

A trigger action response plan (TARP) defines the predetermined responses for responsible individuals during the occurrence of an unwanted geotechnical event (the trigger). TARPs are usually tied to a mine site's geotechnical performance monitoring system. Monitoring inputs or triggers can vary significantly across a site and may include:

- inspections and observations;
- prism movement (velocity, slope distance, total vector distance);
- rockfalls;
- radar images indicating deformation and resulting radar alarms;
- lack of data or lack of data collection (equipment downtime);
- bad weather (precipitation, high winds, reduced visibility, rapid temperature changes);
- surface or groundwater inflows; and
- increases in groundwater levels (piezometers).

Other monitoring inputs can include changes in geologic conditions with respect to design, seismic events, lack of surface drainage control, or operational issues (e.g. poor excavation practice, poor highwall blasting).

8.5.2 Alert levels

TARPs rely on proper determination of monitoring alert levels. Alert levels are based on the site's experience, the type of instrumentation the alarm is based on, activity near the alarm area, and the potential hazard the alarm indicates. Equipment limitations must be taken into account with respect to accuracy, range, maximum and minimum velocities, and how limitations may impact TARPS.

Alert levels will depend on the site's risk tolerance, operational philosophy, site work culture, consequence history of previous unwanted events, and experience with hazard controls. The key to successful TARPs is associating hazards with actions, determining roles and responsibilities, and defining the actions required. Responses prescribed in a TARP are usually actions graduated from low levels to successively higher levels, as shown in Table 8.1.

TARPs ideally should reduce the tendency to normalise evidence and the temptation to explain away unpleasant signals which can be prevalent in high-risk cultures at some mining operations. TARPS usually conflict with the human tendency to resist change or, in some supervisors, to take on more responsibility. Complete understanding of an operation's ground control TARP should raise the site's risk awareness for unwanted geotechnical incidents or events. TARPs should be written clearly, be easy to understand and limit the amount of subjectivity and interpretation involved in decision-making.

An effective TARP must clearly describe the minimum set of actions required by site personnel in response to a change in mine conditions from normal. A change, or exceedance, in the rate of slope movement that is a departure from the nominal background rate is usually defined as movement or velocity that crosses a certain prescribed threshold, as shown in a typical prism time-series plot in Fig. 8.5.

TARPs form an integral part of an operation's risk management system, of which the slope monitoring program is a component. Establishment of TARPs and associated accountabilities for action can minimise impacts or consequences of ground movement; can assist in limiting equipment damage, loss or delays in the business plan; and can save lives. They are also developed to ensure operational safety (production-critical monitoring). Monitoring areas of instability, anticipated instability or rockfalls requires a clear set of instructions for key personnel so that hazardous conditions are managed properly. TARPs provide clear documentation to ensure responsible management.

Depending on the trigger and data type, data analysis is available from prism movement monitoring or GB radars that can trigger alarms which automatically notify key personnel when movement beyond a specified level is detected. Five TARP examples are included in Tables 8.2–8.5 and Fig. 8.6, each of which demonstrates a variety of attributes:

1. Table 8.2 represents a TARP that accounts for safety-critical monitoring when the GB radar breaks down or is unavailable. This TARP also presents a good example of how to implement pit evacuation procedures, since it triggers a standard operating procedure (SOP);
2. Table 8.3 is a qualitative TARP that provides trigger notifications, but without any predefined actions. These kinds of TARPS are brief and to the point but lack instructions for personnel. This can lead to confusion, so this type of TARP should be avoided;
3. Table 8.4 is an example of a detailed TARP for a saprolite slope susceptible to high pore pressures. The actions presented are insufficient to advise the workforce under emergency conditions. Actions must be specific and actionable;

Table 8.1: TARP example from the Goldstrike open pit mine in northern Nevada showing increasing displacement triggers and resulting actions by responsible parties

Condition	Trigger	Actions	Responsible person for implementing actions	Monitoring	Criteria for downgrading
Normal	Slopes in good condition with no hazards	N/A	N/A	Look for areas of cracking, slumping and bulging, and other areas of potential wall instability Assess need for placement of berms along slope toes and crests	N/A
Level 1 Low probability of safety impacts	Radar breakdown Prism monitoring breakdown	Data are collected manually/ conditions assessed visually until repairs completed	Geotechnical engineer	Visual and manual	Reliable monitoring connection re-established
	SKYBOX 5 PRISMS: monitored movement <1.2 inches per day; movement trends for greater than 1 day show no creep or acceleration **SKYBOX 7 PRISMS:** monitored movement <1.2 inches per day; movement trends for greater than 1 day show no creep or acceleration **SKYBOX 9 PRISMS:** monitored movement <2.0 inches per day; movement trends for greater than 1 day show no creep or acceleration **RADAR:** monitored movement <2.0 inches per day in bedrock and dumps, <4.0 inches per day in failed debris **VISUAL:** inspections and manual monitoring indicate minimal movement and no expansion or lengthening of cracks, rockfall or bulging	Continue with standard scheduled inspections	Geotechnical engineer Planning engineer Mine operations supervisors	Daily review of prism and radar data including velocity	No movement trends for over a week
Level 2 Moderate probability of safety impacts	**SKYBOX 5 PRISMS:** monitored movement >1.2 inches per day; movement trends for greater than 1 day show no creep or acceleration **SKYBOX 7 PRISMS:** monitored movement >1.2 inches per day; movement trends for greater than 1 day show no creep or acceleration **SKYBOX 9 PRISMS:** monitored movement >2.0 inches per day; movement trends for greater than 1 day show no creep or acceleration **RADAR:** monitored movement >2.0 inches per day in bedrock and dumps, >4.0 inches per day in failed debris **VISUAL:** inspections and manual monitoring indicate movement or expansion/lengthening of cracks, rockfall or bulging	Review data trends with chief geotechnical engineer and provide guidance to mine engineering and operations supervisors	Geotechnical engineer	Daily review of monitoring trends in areas of slope movement	At the discretion of the geotechnical engineer
	Rain event of 0.2 inches or greater in a 24 h period Snowmelt wets slopes and begins to channel down roadways and ramps	Increase frequency of visual inspections to at least twice per shift	Geotechnical engineer Planning engineer Mine operations supervisors	Twice-daily review of monitoring trends	At least 3 consecutive days of no movement AND infiltration of water into slopes has ceased
	Increase in rockfalls from normal conditions (rilling down slopes, isolated rockfalls, dust coming off the face)	Notify operations supervisors and geotechnical engineer	All	Continual observation from operators Twice-daily inspections by operations supervisors	At the discretion of the geotechnical engineer

Condition	Trigger	Actions	Responsible person for implementing actions	Monitoring	Criteria for downgrading
High risk of adverse safety impacts	**SKYBOX 5 PRISMS:** monitored movement >1.4 inches per day; movement trends for greater than 1 day show continued movement above threshold **SKYBOX 7 PRISMS:** monitored movement >1.4 inches per day; movement trends for greater than 1 day show continued movement above threshold **SKYBOX 9 PRISMS:** monitored movement >2.2 inches per day; movement trends for greater than 1 day show continued movement above threshold **RADAR:** confirmed movement >4.0 inches per day in bedrock and dumps **VISUAL:** inspections and manual monitoring indicate movement or expansion/lengthening of cracks, rockfall or bulging	Isolate affected area Review data trends and field observations Notify geotechnical design consultant(s) Provide guidance to mine engineering and operations supervisors	Chief geotechnical engineer or designee	At the discretion of the chief geotechnical engineer or designee	At the discretion of the chief geotechnical engineer Mine engineering superintendent Mine operations superintendent

Source: J. Mattern, Nevada Gold Mines.

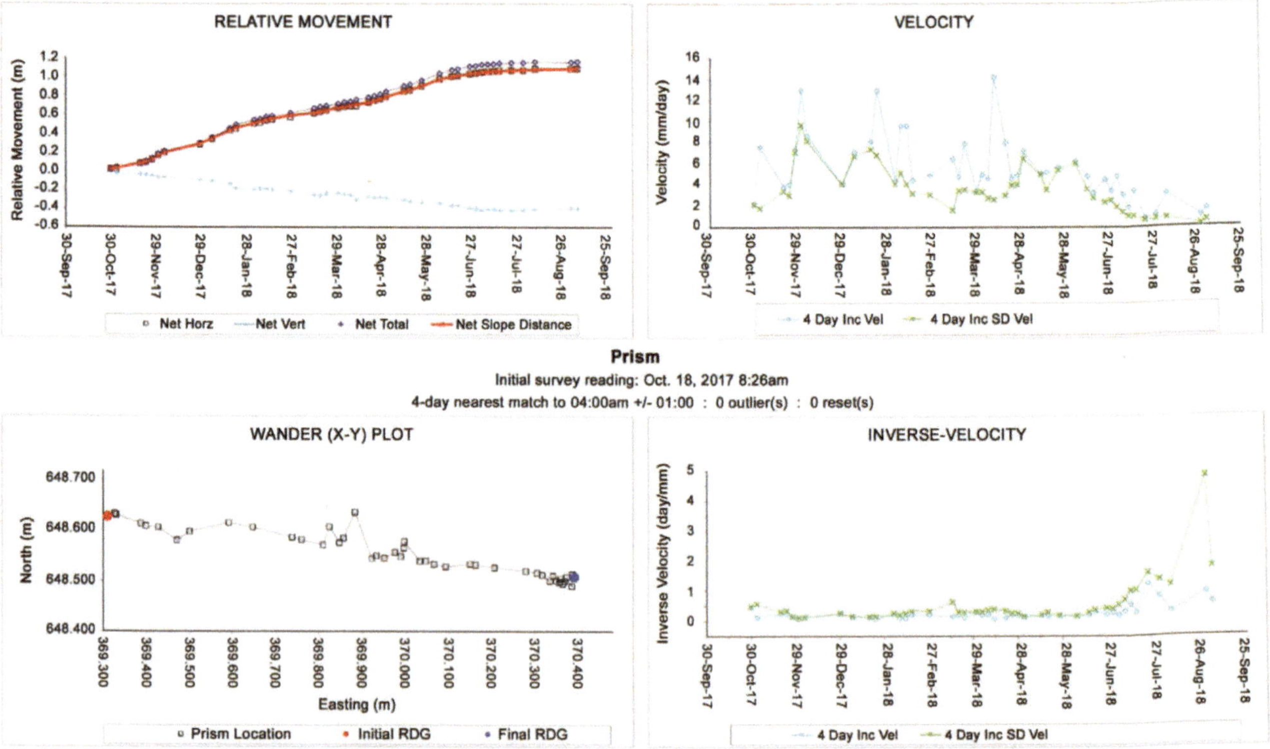

Fig. 8.5: Example of typical prism displacement, velocity, wander and inverse velocity plots with velocity thresholds indicated by dashed lines (source: Teck Highland Valley Copper Partnership).

Table 8.2: Example of a TARP that accounts for safety-critical monitoring radar that shuts down, or is unavailable. This is a good example of how to implement pit evacuation procedures since it triggers an SOP

Scenario/target	Movement rates #	Geotechnical action	Operations action
Radar being moved or serviced	No data	• Advise dispatch and open pit coordinator	• Consider applying highwall procedures • Other actions or limitations to mining as directed by the geotechnical section • Standoff the wall and no work beneath the area until completion of a job safety analysis (JSA) and sign-off by the geotechnical superintendent or senior geotechnical engineer
Radar not functioning (out of service, mechanical and/or software problem)	No data	• Advise dispatch and open pit coordinator	
Toe of South West Dyke Failure	Red Critical Alarm 50 mm/2 h	• Alert open pit coordinator, open pit manager/superintendent, mining manager, chief engineer, underground manager, senior production engineers, chief surveyor and chief geologist • Geotechnical inspection • Notify dispatch to implement pit evacuation procedure	• Implement pit evacuation procedure (SOP_OP_700_36 Pit Evacuation) from above and beneath the area
	Orange Geotechnical Alarm 25 mm/2 h	• Alert dispatch and open-pit coordinator • Geotechnical inspection	• Apply highwall procedures • No stopping or mining activities permitted in mobile demarcation zone beneath failure zone unless otherwise certified in writing by the geotechnical section in the demarcated zone • Demarcate no entry beneath the area as marked by demarcation tyres/blocks and cones unless otherwise certified in writing by the geotechnical section

Source: Porgera Open Pit, Barrick Niugini Ltd.

Table 8.3: Simplified TARP (qualitative) without predefined actions (i.e. only notifications). These TARPS can lead to confusion in the workforce when management is offsite or otherwise indisposed. The Immediate Actions under the Red Status are insufficient

Wall status level	**Criteria to determine wall status level**					**Action taken**	
	Radar warning (Phase E) By dispatcher (radar message)	**Prism movement rate** By geology (mm/24 h)	**Wall ravel** By senior mine shifter (observation)	**Rain event** By senior mine shifter (mm/24 h)	**Rapid melt of snowpack** By senior mine shifter (mm of snowpack)	Immediate action	Long-term action
Green	None	<50	Intermittent	<7.5	<15	Normal operations	Normal operations
Yellow	GeoTech slope alert or radar system warning	50–100	Moderate	7.5–25	15 to 300	Notify geology and mine operations leadership	Operate under restrictions determined by management
Red	Critical slope alert or radar system failure	>100	Significant	>25	>300	Evacuate affected area Notify geology and mine operations leadership	Determine and implement mitigation

4. Table 8.5 is an example of a TARP used where open pit slope stability significantly impacts underground mining activities. Acceleration trends monitored by the GB radar system are used to trigger both evacuation and re-entry procedures for the underground mining operation; and
5. Fig. 8.6 is an example of a detailed flowchart-style TARP specifically designed for assessing domain-specific GB radar-generated slope movement along with other hazard recognition. This style of TARP is generally less confusing and easier to present multiple responsibilities at the same threshold level.

Similar to the TARP shown in Table 8.2, thresholds to be considered for all TARPS should include what to do when specific instrumentation is out of service.

8.5.3 Trigger action response plan processes

When an alarm trigger occurs, the response of the workforce and implementation of the plan depends on employee 'buy-in' to the TARP and the belief that following the TARP will keep people safe. When writing a TARP, assignment of and input from responsible persons (RPs) are required to ensure its success. The process of involving RPs should be used for other critical mining documents such as the ground control management plan (GCMP) and the EPRP. Ownership of the TARP by the workforce and those supervising the workforce is paramount. Four key elements are vital to the TARP process.

1. *Status*: the communication of the TARP status is vital to ensuring RPs act in accordance with the TARP. Notifications of status can include email, cell phone text messages, audible and/or visual alerts over computer networks, sirens in strategic locations, and visual alarms such as flashing lights, among others. When TARP actions span more than a work shift, handover between shift supervisors becomes critical for a successful TARP: oncoming personnel must understand the status of any potential hazard. Communications to stakeholders, including regulators, third parties and slope design consultants, should also be considered.
2. *Coordination*: mine management, safety personnel and their consultants often assist in assessing hazard exposure, including the development of mitigation plans such as stand-off distances, buffer zones, barricading policies, evacuation and muster points, and in accounting for personnel. Input from the site's geotechnical engineer is essential to ensure safety during incidents or events.
3. *Sign-off and training*: sign-off of and training on TARP procedures should be regularly reviewed, updated and tracked. Records of personnel receiving TARP training, assessments of competency and any updates should be managed and maintained.
4. *Personnel requirements*: in the course of developing an implementation plan for the TARP, consideration must be given to the level of experience, oversight and availability of the RPs. Most sites have small geotechnical departments consisting of one or two people and often must rely on other departments, shifters or dispatchers to carry out TARP responsibilities. The response system in place must match the capabilities of human resources available to manage the system. Sensitivity to increasing risk potential is necessary, as operations mature and slopes become higher and/or steeper. As operations mature, effective geotechnical management functions often require additional geotechnical staff to manage the riskier and more complex system. This can

Table 8.4: A TARP that includes saprolite slopes susceptible to instability due to high pore pressure. Prisms and piezometers are used exclusively to monitor these slopes. Prism density is approximately 1 per 100 m^2 for the site using this TARP. Actions presented under Reporting Required are insufficient to adequately advise the workforce under emergency conditions. Actions must be specific and actionable

		Trigger events			
		Tension crack movement	**Prism movement**	**Radar deformation**	**Visual inspection**
Trigger Level 0	**Stable**	No tension cracks	No prism movement	No wall movement	No rockfalls observed
Trigger Level 1	**First indication of instability**	First tension crack(s) observed	Confirmed movement from two prisms in one location	Apparent wall movement indicated from one pixel	Minor rockfalls observed
Trigger Level 2	**Onset of movement**	First indication of movement across tension cracks (>5 mm)	Confirmed movement from three or more prisms in one location	Confirmed movement from two or more pixels	Bench cracking, minor wall deformation or additional rockfalls observed
Trigger Level 3	**Continuous movement**	Accelerated movement across tension cracks (>20 mm)	Acceleration of initial prism movement, with additional prisms showing movement	Acceleration of movement within initial movement area and expansion of pixels showing movement	Onset of bench-scale failures, separation of benches into blocks
Trigger Level 4	**Failure imminent**	Formation of additional tension cracks. Evidence that wall is coming apart	Rapid acceleration of prisms, progressing towards failure	Rapid expansion of area pixels showing movement	Active sloughing of benches, disintegration of wall becomes obvious

Any change in trigger level will require immediate inspection by the senior geotechnical engineer.
When any trigger level is reached it will remain in effect until otherwise determined by the senior geotechnical engineer.
For Trigger Levels 3–4, after rainfall of more than 25 mm in 24 h, all movement indicators will be reviewed by the senior geotechnical engineer within 24 h, including weekends and holidays.
For Trigger Level 2, all movement indicators will be reviewed on the next working day.

	Reactions to trigger events			
	Minimum frequency of pit inspections	**Minimum frequency of prism and VWP data review**	**Planned response**	**Required reporting**
Trigger Level 0	Weekly	Prisms monthly VWPs monthly	None	None
Trigger Level 1	Weekly	Prisms monthly VWPs monthly	Maintain existing control levels	Geotechnical engineer, mine superintendent and supervisors
Trigger Level 2	Mon/Wed/Fri	Prisms fortnightly VWPs fortnightly	Undertake risk assessment / place additional monitoring devices as required / review dewatering requirements	TL 1 reporting plus technical service superintendent / senior mine geologist
Trigger Level 3	Daily*	Prisms weekly VWPs weekly	Take corrective actions, including **barricade affected area** / inform affected personnel of ground conditions / redesign to reduce hazards / **engage consultants**** / review planning and practices	TL 2 reporting plus mine manager / mobile maintenance superintendent
Trigger Level 4	Twice Daily*	Prisms daily VWPs daily	**Isolate personnel and equipment** from unstable areas / define and create exclusion zone	TL 3 reporting plus general manager

* Pit inspections must continue through weekends and holidays.
** The site geotechnical consultant should be brought to the site for induction and an inspection of the affected pit area(s).

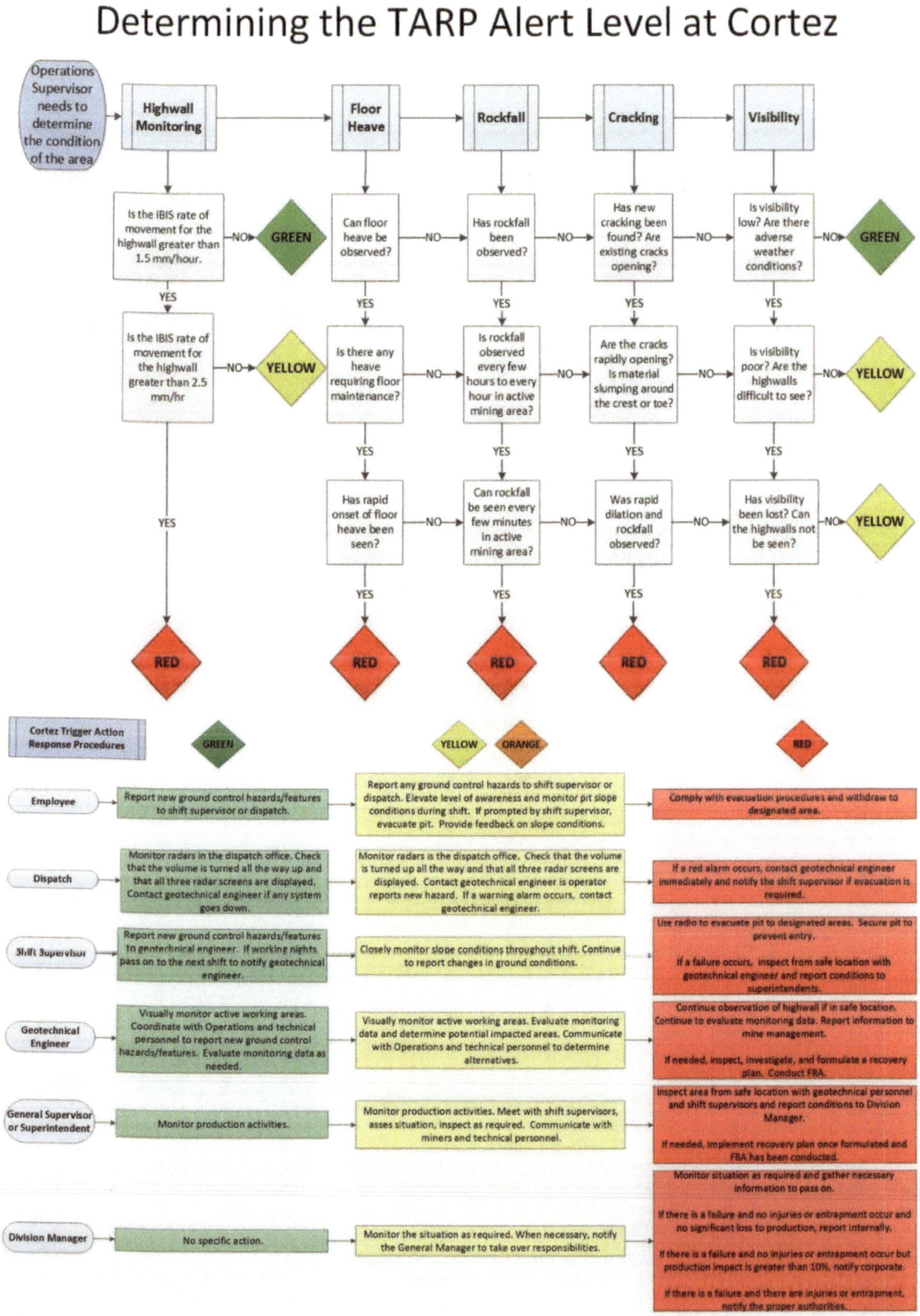

Fig. 8.6: A detailed TARP from a northern Nevada gold mine specific to radar slope movements on a specific wall and other hazard warning signs with specific predefined actions, structured as a flowchart. This type of TARP is generally less confusing and easier to present multiple responsibilities at the same threshold level (source: P. Jenks, Nevada Gold Mines).

Table 8.5: Example of a highly evolved TARP combining open pit with underground mine operations. Thresholds are based on time-to-failure as determined by the geotechnical engineer using acceleration trends monitored with the slope radar. Note the inclusion of a safety response, which is critical for TARP effectiveness to keep workers safe

Time	Response level	Triggers	Operations response	Technical response	Safety response	Management response
Very low probability of failure	1	**Monitoring** • Background movement associated with relaxation (<2.5 mm/day) • Regressive trends for 7 days below critical and instability thresholds in surface and underground instruments **Field observations** • Little to no rockfall activity • Dry conditions • Existing geologic structures (West Shear) showing little sign of deformation or dilation underground • Underground workings adjacent to highwall showing little signs of ground support issues	• Look for changing ground conditions • Report new ground control hazards or events to technical group	• Conduct routine inspections of 5BOP and 2BUG Download, analyse and report on monitoring data frequently • Conduct routine inverse velocity analyses using monitoring data	• Routine evacuation drills	• No specific action
Weeks/ months until failure	2	**Monitoring** • Deformation rates above background levels (12.5 mm/day) in West Highwall and breccia for more than 24 h • Inverse velocity trends indicate progressive deformation within weeks or months **Observations** • Infrequent rockfall activity from breccia and major structures in pit • Dry conditions • Existing geologic structures (West Shear) showing progressive acceleration for more than 48 h in underground • Underground workings adjacent to highwall having minor issues with ground support holding	• Look for changing ground conditions • Report new ground control hazards or events to technical group • Be aware of geotechnical areas of concern	• Conduct routine inspections of 5BOP and 2BUG • Download, analyse and report on monitoring data frequently • Conduct routine inverse velocity analyses using monitoring data	• Routine evacuation drills	• Monitor situation as necessary
Failure within 7 days	3	**Monitoring** • Deformation rates above instability thresholds (25 mm/day) in West Highwall, NW Corner and breccia for more than 24 h • Inverse velocity trends indicate failure within 7 days **Observations** • Rockfall at least once a day from breccia and/or major structures in pit • Heavy rainfall or runoff • Existing geologic structures (West Shear) showing progressive acceleration for more than 48 h in underground • Underground workings adjacent to highwall and major structures showing signs of cracking and ground support failure	• Look for changing ground conditions • Be ready to evacuate • Minimise radio chatter and make sure radio system is working correctly • Isolate and barricade failure area underground • Move equipment out of 5BOP or into locations away from portals when directed by management • Watch underground instruments for alarms that indicate evacuation	• Conduct daily inspections of 5BOP and 2BUG • Download, analyse and report on monitoring data daily • Conduct frequent inverse velocity analyses using monitoring data to update estimated failure time • Report updates to operations and management	• Review evacuation plan and muster points with all personnel • Ensure barricades/ barriers are available for use and are adequate • Review emergency response plan with shift supervisors and superintendents	• Monitor the situation as necessary and gather information to provide to corporate groups • Communicate with other sites potential need for resources/ manpower

Time	Response level	Triggers	Operations response	Technical response	Safety response	Management response
Failure within 24–48 h	4	**Monitoring** • Deformation rates above critical thresholds (50 mm/day) in West Highwall, NW Corner and breccia (radar will announce on Channel 11) • Inverse velocity trends indicate failure within 24 h **Observations** • Rockfall/ravelling at least every hour from breccia and/or major structures in pit • Heavy rainfall or runoff • Existing geologic structures (West Shear) showing progressive acceleration and dilation within hours • Rapid deterioration of underground workings and ground support adjacent to highwall and major structures	• Use stench and radio to evacuate the underground clear of 5BOP • Barricade access at entrance to 5BOP • Ensure all personnel are accounted for at muster point • Maintain spotter in observation shack	• Conduct daily inspections of 5BOP and 2BUG • Download, analyse and report on monitoring data daily • Conduct inverse velocity analyses using monitoring data to monitor change in estimated failure time • Provide updates to operations and management	• Notify Mine Safety and Health Administration (MSHA) of anticipated failure • Initiate the emergency response plan • Notify regional mine rescue team of situation	• Notify corporate groups and stakeholders of anticipated failure • Coordinate technical and operations group responses • Restrict mine site access only to essential personnel • Initiate incident command • Organise manpower and equipment as necessary for re-entry
<24 h to failure or unexpected failure	5	**Monitoring** • Inverse velocity trends indicate imminent failure based upon velocity trends or failure has occurred • Underground instruments indicate rapid deformation **Observations** • Series of failures or overall wall failure • Continuous rockfall/ravelling along major structures, at toe of breccia, above 4925 buttress • Existing geologic structures (West Shear) rapidly dilating or closing within hours • Slacking of ground support adjacent to highwalls and major structures	• Evacuate the underground clear of 5BOP • Barricade access at entrance to 5BOP • Ensure all personnel are accounted for at muster point • Maintain spotter in observation shack	• Inspect highwalls of 5BOP beyond barricades to watch for changing conditions • Download, analyse and report on monitoring data • Conduct inverse velocity analyses using monitoring data to monitor change in estimated failure time • Provide updates to operations and management • Re-establish monitoring and inspections post-failure	• Notify MSHA within 15 min of failure • Create incident command • Initiate the emergency response plan • Contact regional mine rescue team and outside services if necessary	• Notify corporate groups and stakeholders of situation • Re-entry or access to hazard zone only authorised through incident command • Restrict mine site access only to essential personnel • Organise manpower and equipment as necessary for re-entry

Source: R. Turner, Golden Sunlight mine, Montana, USA.

be counter-intuitive to most mining human resources (HR) departments, which often cut back staff as the mine matures or approaches mine closure. On the other hand, geotechnical personnel fatigue may result in the need for larger geotechnical departments to cover the site 24 h/day and 7 days/week, or to assign TARP duties to other departments.

8.5.4 The effect of instrumentation limitations on trigger action response plans

For TARPs to be effective, the capabilities and limits of instrumentation used to determine TARP levels must be considered. Since most TARPS utilise slope movement as the primary alarming criterion, the ability of instruments to measure movement should be accurate and precise enough to provide ample warning for operators to properly react to incipient slope movement. Importantly, however, the relationship between the slope distance measured by the instrument relative to total vector movement of the target needs to be clearly understood by the slope engineer.

Analysis of monitoring data should consider instrument limitations in order to avoid misinterpretation or a misunderstanding of information. This includes the time between readings (scan times for radar and laser scanners) when using velocity as the primary alarm. Typically, misinterpretation and lack of understanding of instrumentation limitations result in poor decisions being made – including false alarms or, more importantly, lack of action calling for access restrictions or evacuations when needed.

As mining companies increasingly employ slope monitoring systems using radar and laser scanners, diligence with respect to observing details of the dynamics of the slope and reacting to the failure mechanism can be compromised. Because radar covers the entire slope, operators and practitioners are lulled into a false sense of security as long as the radar does not generate any alarms. There are two potential traps which often occur together.

1. Displacements measured by radar are relative to the direction between the target and the slope, or LOS. As such, LOS movement may not represent the full component of the actual movement which potentially leads to the setup of thresholds that are too conservative (often done to minimise false alarms). This can lead to a false sense of comfort or safety with respect to the stability of the slope being monitored.
2. Coherence is a correlation measurement based on the range and amplitude between the current scan and the preceding scan, with values close to 1.0 indicating little difference in range and amplitude between the two scans. The coherence measurement is especially useful for establishing the time of slope collapse. Practitioners must understand the implications when the radar signal is out of coherence (values <1.0). It may signify the slope being monitored is moving faster than the radar's ability to measure the actual slope movement along the LOS.

Understanding what the true slope movement is can be enhanced by employing geodetic surveying using RTS and/or GNSS, and comparing total vector movement with LOS measurements at key targets. Otherwise, not understanding the LOS limitation of instrumentation and the loss of coherence with successive measurements of the radar means that the slope is moving faster than the radar's ability to track real movement. The loss of coherence and/or inadequate LOS measurements of the total vector movement will result in the interpretation of movement that is significantly less than what is actually occurring. Understanding how instrumentation is processing data with relation to true slope movements is critically important for the effective application of monitoring instruments that rely on LOS measurement only.

While radar and laser scanners are effective for slope monitoring, they cannot be the only tools used to manage geotechnical risk in a pit. Without a carefully considered strategy for the application of alarms for monitoring instruments, such as radar and laser scanners, some operations may be left exposed without appropriate documentation, processes or adequately trained site personnel. Such a strategy should include reinforcement of LOS measurements with RTS and/or GNSS to validate monitoring data.

Any deficiencies in training and/or documentation relating to alarm threshold justification should be recognised as significant business risks for any operation wishing to implement what should be seen as industry leading practice. Determining appropriate alarm thresholds is a key component of a site's geotechnical risk management strategy with regard to slope stability monitoring.

A thorough understanding of inherent geotechnical risk relating to a mine design is required when setting up a slope monitoring program. Additionally, the risk must be continually evaluated in the event that an unwanted geotechnical incident escalates. A process must be in place to communicate increasing risk to the appropriate level in the organisation at the right time, which is a fundamental objective of a TARP.

8.5.5 Connection to the emergency preparedness response plan

The most elevated TARP status usually calls for evacuation of large areas of the mine. These activities will likely require connections to the EPRP, which specifies procedures for safely managing sudden or unexpected situations across the mine site. The slope engineer should review the site's EPRP to ensure that ground control incidents are among the items covered. If ground control incidents are not covered by the EPRP, a planning session is recommended to include provisions for such emergencies.

A major benefit of providing guidance during a ground control-related emergency is the discovery of unrecognised hazardous conditions that could trigger an emergency situation, or be aggravated during an emergency situation. The planning process may bring to light deficiencies, such as a lack of resources (equipment, trained personnel, supplies) or items that can be rectified before an emergency occurs. In addition, an EPRP promotes safety awareness and demonstrates the organisation's commitment to the safety of the workforce.

Objectives of the ground control component of the site's EPRP include:

- preventing fatalities and injuries;
- mitigating potential for damage to mining equipment and infrastructure;
- protecting the environment and the community;
- expediting the resumption of normal operations;
- identifying evacuation routes, including alternates, and making these known to the workforce;
- keeping evacuation routes unobstructed;
- specifying safe locations for muster zones;
- assigning specific RPs for conducting head counts to ensure that everyone has left the danger zone;
- carrying out treatment of the injured and searching for the missing simultaneously with efforts to contain the emergency; and
- providing sources of medical aid or medical evacuation when normal facilities may be in the danger zone.

Experience has shown that the most common and sensible means of emergency evacuation of an open pit access route closed by an instability is by helicopter. Mine sites should have contractual arrangements with local aviation providers to ensure such services are available on demand in the event of an emergency. An item often not considered by an operation's EPRP is long-line rope evacuation gear. Distance and heights typical for mill building rescues may not be sufficient for excavated slope evacuations.

It is important that qualified members of several groups be asked to participate in the initial development of the EPRP and the supporting geotechnical TARP. Among these groups, the slope engineer and the health and safety department can provide valuable input and a means of wider workforce involvement. Appropriate mine management should also be included. Communication, training and periodic drills will ensure adequate performance in the event that the EPRP triggered by a geotechnical TARP must be carried out.

8.6 Reporting process

8.6.1 Reporting procedures

Before instruments have been installed, slope monitoring and reporting procedures should be prepared as part of the planning process of a slope monitoring strategy. These procedures should include:

- site characterisation, including failure mechanisms and assessment of risk exposure;
- a listing of data to be collected;
- equipment specifications, including servicing and calibration requirements;
- data collection methodologies to be utilised;
- processing and presentation procedures for data analysis and communication; and
- interpretation procedures, including alarm levels that tie into a TARP.

8.6.2 Data collection and management

Responsibility for the collection and management of data will have been determined during the planning stage and should preferably be under the direct control of the geotechnical engineer. Significant support from the operation's IT group will be required, as well as support from the department in charge of communications (wired and wireless).

Uniform development of hydrogeology and geotechnical databases using compatible software platforms for all systems should be used for the slope monitoring data system that is tied to the TARP(s). Technical personnel should discuss data management needs with their service providers to explore options to increase uniformity in the selection of instrumentation and telemetry, as well as data management. Proper data management and communication will enable:

- an improved ability to manage data such that changes are detected in a timely manner (this addresses challenges associated with ever-increasing quantities of data to analyse);
- improved performance, reliability and accuracy of instrumentation;
- improved systems to provide real-time monitoring capability;
- the development of integrated capabilities for interpretive contexts (e.g. adding daily rainfall records to prism time-series plots, piezometers, inclinometers, extensometers, etc., as needed);
- ensuring there are sufficient qualified staff on (or off) site for slope monitoring functions;
- improved TARPs, particularly TARP responses to inspections, selective mine sector closures based on observations, and reactions to and after strong storm events; and

- remote/automated monitoring and data acquisition systems to assist operations in making rapid and reliable decisions with more information.

8.6.3 Processing and presentation of data

8.6.3.1 Overview

An important role of the slope engineer includes the routine interpretation of movement trends. Trends are used to demonstrate how closely, or not, observed conditions align with predicted behaviour of excavated slopes.

Data processing and presentation monitoring should enable the slope engineer to make rapid and reliable assessments of information sufficient to protect people and assets. Ideally, data should be presented in a verifiable context of potential triggers that explain trend changes resulting from events such as:

- strong storm and rapid snow/ice melting;
- blasting and mining activity;
- changes in geology encountered during excavation such as lithology, alteration and faults; and
- changes in groundwater or surface water conditions resulting from installed drains, pumping of nearby wells, or changes in the surface water diversion system.

Automating the integration of databases from various performance monitoring instruments makes putting data in context much simpler and far faster than manually manipulating spreadsheets. The slope engineer is encouraged to contract with relevant service providers that specialise in the development of automated data integration systems.

8.6.3.2 Responsibility for data processing and interpretation

Responsibility for the processing and presentation of instrumentation data must be under the direct control of the site's geotechnical engineer. The site must have the resources necessary to process and report timely monitoring information results to protect people and assets. Skill-sets required of personnel assigned to data collection, processing and analysis are frequently underestimated, resulting in the accumulation of unprocessed data, failure to recognise an incipient failure event and failure to take appropriate action to mitigate the consequences of an incipient failure event. Experienced geotechnical engineers can be marginalised by having to support the function of a monitoring system and management of data rather than having time to assess and report results in a timely manner and to efficiently participate in executing corrective measures necessary to manage geotechnical risks.

8.6.3.3 Automated data analysis

Automated routines allow the continuous flow of data from the various instruments that make up the monitoring system to be more quickly compiled and analysed in context. This increases the likelihood that exceptions will be observed and causes will be understood in a timely manner, resulting in an improved ability to respond to potential instability developments and execute effective corrective measures.

Monitoring data generated from all instrumentation should be compiled in an office-based platform that can receive information, manipulate the data for analysis and generate graphical representations. These systems should be capable of generating automated messages and notifications used in communicating TARP effectiveness.

8.6.3.4 Use of maps when presenting data

When presenting data, it is important to utilise a plan map showing instrumentation locations and geotechnical domain boundaries. Useful background information can include data obtainable from interferometric synthetic aperture radar (InSAR) surveys, commonly used for the detection of suspect infrastructure movement across the mine site and beyond. InSAR is particularly useful for site-wide spatial deformation monitoring of large structures such as tailings embankments and heap-leach facilities, infrastructure located near large open pits, the ground surface above underground operations, and large landslide blocks.

Other mapped features can be added as needed, such as controlling faults, contacts and/or lithology and alteration. Results of measurements from GB and satellite radar, laser scanning and/or prism movement rate contours, and total vector arrows with plunge angles are useful when prisms, resolution cells/pixels or permanent scatterers are displaying significant movement. Placing radar pixels, or prism locations, on slope cross-sections can be useful in illustrating and explaining potential instability mechanisms to mine operators. Superimposing photographs with point data can be effective in communicating stability concerns in combination with contour maps of movement rates.

8.6.3.5 Reporting of results, conclusions and recommendations

After each dataset has been analysed and interpreted, results and conclusions must be communicated in the form of periodic monitoring reports submitted to stakeholders, particularly those personnel responsible for the implementation of any remedial actions indicated to be required. Management should be supplied with periodic summary reports from the monitoring program, even if no movement is detected. Reporting on at least a monthly basis is standard, in combination with daily status reports usually communicated through the chain-of-command to senior site management. Individual operations typically develop a reporting plan that suits their unique requirements. Operations may increase the frequency of detailed reports on ground control conditions to a weekly basis when challenging operating conditions persist.

8.7 Duty of care and external use of monitoring information

8.7.1 Overview

The external use of monitoring data and other relevant information borders on a wide range of key topics including, but not limited to, corporate external communications policy, corporate data retention policy and possibly regulatory requirements or regulatory compliance issues. Discussion and advice regarding legal and regulatory compliance topics are beyond the scope of these Guidelines. Instead, this discussion of external use of monitoring data is intended to present a widely accepted and practical view of a prudent engineering approach to the responsible management and communication of performance monitoring information. It is suggested that readers seek relevant corporate policy, legal and regulatory advice from qualified personnel charged with managing these issues in their own company.

8.7.2 Duty of care

As part of occupational health and safety responsibilities and obligations for employers and employees, each person has a duty to ensure that their action, or failure to act, does not harm others. Employers have a primary duty of care to ensure a safe workplace, so far as practicable. For example, Lang (1999) and the Department of Mines and Petroleum (2011) provide an overview of the duty of care expectations for the Western Australian mining environment.

In many countries, the common law of care has been legislated and applies to all employers at all levels, including corporations, and all workers, including supervisors and managers. Torlatch (1996) summarises the employer's duty of care to provide:

- a safe place to work;
- a safe system of work;
- competent and properly trained staff and supervisors; and
- fit-for-purpose plant and equipment.

In particular, professional engineers hold a critical safety role in engineering endeavours. In recognition of the Professional Code of Ethics, some jurisdictions require appropriately high standards of duty and conduct from engineers in discharging their duties to the public, their employers, their clients, their colleagues, their profession and themselves. High technical standards are extremely important, given the responsibility assumed by, and integral to, the undertakings of professional engineers (Marston 1996). In the active mining environment, slope performance monitoring is a critical safety function and an essential element in achieving demonstrated and documented duty of care. To this end, some jurisdictions have legislated the appointment of mining geotechnical engineers with prescribed accountabilities and obligations. Equally, many mining companies have detailed appointment letters for competent persons, outlining role requirements for key positions associated with the slope management plan (or GCMP) and TARP. The level of care required will vary according to the situation. The Department of Mines and Petroleum (2011) outlines the following principles to determine whether an action is reasonable:

- the standard of care will rise with the seriousness of the injury or harm that could result;
- the greater the likelihood of injury, the greater the care that should be taken to avoid it; and
- the easier it is to avoid the injury, the more reasonable it is to expect that appropriate measures will be taken to ensure it does not happen.

Efficient and effective systems and procedures can be engineered and prescribed in workflows and role descriptions. However, these can be nullified if individuals lack the authority, ethics and motivation to fulfil their obligations. Significant incidents (including fatalities) have occurred due to the negligence of individuals entrusted with the effective management of slope performance systems. In one such incident (despite a comprehensive array of monitoring devices, including multiple slope radars and prism monitoring systems), the following behavioural deficiencies were noted as contributing to a fatal outcome:

- inadequate understanding and training on the system requirements and limitations;
- omission to set any alarm thresholds on one slope radar and setting an excessively high threshold on the second radar;
- a poorly maintained prism monitoring system with no support system to attend to breakdowns; and
- no regular analysis of any monitoring data trends to support understanding of changing conditions.

The forensic investigation indicated that the onset of instability could have been predicted up to six months in advance, and a procedure for providing assurance that the system was being managed effectively could have prevented the fatality.

8.7.3 External use of monitoring information

In the course of mine development, the necessity may arise to provide geotechnical monitoring data, interpretations or other related work products to external (non-employee, offsite) stakeholders. Assuming that the geotechnical team at a mine site comprehensively protects personnel by controlling exposure to ground control hazards, matches mine plans and practices to geotechnical conditions encountered, and has clear documentation of those activities, any disclosure of information to external parties is an opportunity to demonstrate the skill and diligence of the geotechnical team. In principle, all managed data, reporting and geotechnical controls or procedures must be available for inspection for whatever reason, at any time. As mentioned previously, the geotechnical program must be of adequate quality, scope and transparency to maintain credibility with mine operations; otherwise, the objectives of performance monitoring cannot be achieved. By this point in the Guidelines, it should be perfectly clear that neither hiding nor altering data, nor intentional misinterpretation of data, should ever be permitted to occur; there is nothing to hide.

However, because uncertainty and engineering judgement are part and parcel of the interpretation of monitoring data, it is possible to generate different interpretations from one dataset. This implies that some interpretations made in real-time during mine development will be wrong. A measure of conservatism in the form of experience-based practical thinking is used to make reasonable recommendations from analysed data. For example, a shovel operator might be allowed to continue working below a slowly accelerating instability close to alarm thresholds during daylight shifts. However, the same condition at night, without good lighting or visual observation (actual heading and/or continuous radar imagery in real-time), might require the shovel operator to leave the face and work another heading until the light of day when safety to continue can be re-established. On the other hand, less conservatism might be used, as a business decision, to push the limits of long-term slope behaviour, as long as there is ongoing radar tracking to assure personnel safety. Each instability event (and therefore decision event) uses the combination of monitoring data, interpretation, weather conditions, geotechnical conditions, mining process, mine planning and other factors to form a completely unique decision process for each situation and/or stage. The collection of daily, weekly, monthly, and as-needed operational reports and recommendations creates an orderly history of the operation's geotechnical effort and diligence.

It is rarely appropriate to use simple discrete displacement trigger levels to open and close mining areas for safety purposes. Instead, the serious business of performance monitoring for risk mitigation requires continuous active problem-solving, supported by experience-based engineering judgement in combination with local slope behaviour experience gained over years of work at the mine. If something was to go wrong (e.g. instabilities such as sliding, rockfalls, etc.) and caused economic loss, failure to meet forecast, disruption or, far worse, injuries or fatalities, part of the investigation will focus on understanding whether ground control procedures were followed and whether those procedures were suitable or adequate. Logically, there will be a concentrated search for negligence as a factor in the incident. The position of strength is to be able to demonstrate diligence in all areas and to understand the true causes of a loss incident. From this viewpoint, consider the following experientially derived list of key points for management of a slope program that could allow indirect (or out of process) use of monitoring information.

- A standard instability document (slope failure report) should be available for every significant instability that has occurred, or is occurring, including descriptions and enough sequential monitoring data to understand the instability scale, type and behaviour. The life-of-mine collection of these reports forms a catalogue of slope stability management. This catalogue, or another log, should clearly show the history of notifications, recommendations and actions taken regarding the instability.
- Background stability monitoring data should be available as sequential maps or digital terrain model shapes produced with sufficient frequency to validate assertions of zero displacement stable ground during a specified time period.
- An accurate and reliable monitoring database with maps and other documentation showing instrument locations, calibrations and data for all displacement and piezometric instruments (measurements) should be readily available and have the ability to point to any particular instrument/measurement value location.
- An up-to-date reference map with current mine and surrounding area topography should be readily available for discussion with stakeholders. It should include relevant geological and geotechnical features, all instability locations, mine features (roads, pushbacks, major equipment), monitoring coverage areas and other geotechnical monitoring features of importance.
- Clear demonstrations of frequent operational communications should be abundantly located in key operation areas. These include up-to-date geotechnical hazard maps, daily and weekly mine plans with ground control annotations, relevant policy documents (e.g. posters) and a logbook showing at least monthly meetings communicating ground control topics to operating teams, with attendance lists.
- Evidence of the operation's plan to assure adequate monitoring equipment and personnel (i.e. GCMP) that especially includes five-year budgeting documents listing details of geotechnical monitoring instruments,

field work, maintenance and support for equipment and software, and manpower planning for geotechnical monitoring personnel. Many of these services are typically contracted to external specialists and instrumentation providers.

If these work products and working documents are standard practice and readily available, it is likely they will never be needed to demonstrate actual diligence, since incidents will likely have been prevented or mitigated in the first place.

In summary, the mine operator or owner has a substantial duty of care to ensure that obligations required by the mining licence, mining permits and regulators, with respect to safety and environmental stewardship, are executed to the best of the operator's abilities. With support from mine management, it is the responsibility of the individual in charge of the operation's geotechnical engineering program to escalate to management when the performance monitoring system is not sufficient to satisfy these obligations. To achieve this objective, the slope engineer and/or geotechnical engineer primarily responsible for the monitoring program must have the authority of a well-defined role to implement the operation's geotechnical risk management system, of which the monitoring program is a component. The importance of fully understanding the duty of care that is expected to be exercised by all responsible parties cannot be over-emphasised (Lang 1999).

9

MONITORING FOR SMALL OPERATIONS, TAILINGS FACILITIES, WASTE DUMPS AND OTHER INFRASTRUCTURE NEAR OPEN PITS

Geoff Beale, Henry Chiwaye, Paolo Farina, Mark Hawley, James Hogarth, M. Kim McCarter, Mike Ness, Marc Ruest, Robert Sharon, Michael Shelbourn, Peter Stacey, Ryan Turner, David Williams, Rocky Wu

9.1 Introduction

The motivation, design, implementation, data management and results reporting of performance monitoring for open pit slopes has been the primary focus of these Guidelines. Specific topics described to this point include:

- monitoring methods, including how various monitoring instrumentation works;
- application of various types of instrumentation, including their capabilities and limitations;
- practical aspects of monitoring system requirements, support and operational considerations;
- data acquisition, data transmission and management, with emphasis on automation;
- data analysis and utilisation, including the value of integrating data for validation, results interpretation and reliability; and
- results reporting, including the effective escalation of actions required in the event that monitoring data indicate an incipient failure condition.

An introduction to these topics, including historical perspective, is summarised in Chapter 2, with the details presented in Chapters 3 through 8. Even though topics in these Guidelines focus on open pit slopes, the monitoring methods in the physical mine setting actually have, in many cases, been developed and deployed to do more, including performance monitoring of mine infrastructure and underground mining operations. In fact, some of the instrumentation described in these Guidelines, such as time domain reflectometry (TDR) and inclinometers, was specifically designed for other purposes and subsequently found a useful purpose for monitoring the open pit mining environment.

The information presented thus far applies to mining operations of all sizes. Nevertheless, larger operations with proportionately more complex management and geotechnical risks will have differing requirements to efficiently manage their geotechnical programs. Large and small open pit operations often deal with adjacent underground mining operations; surface and underground mines require continuous and complementary performance monitoring. Geotechnical monitoring of site-wide infrastructure can take advantage of instrumentation types and integrated systems designed and implemented for monitoring the open pit, including the site's communications architecture. Topics in this chapter focus on smaller open pit operations that may experience risk management challenges that set them apart from larger operations. Specific attention is also given to the monitoring of large earth structures including waste rock dumps, stockpiles and tailings facilities. Two case histories are presented that provide examples of the details of the monitoring interaction between adjacent open pit and underground mining operations, and process plant facilities located near the highwall of a large open pit.

9.2 Small operations with limited budgets

9.2.1 Overview

Small open pit operations frequently involve either high-grade deposits of limited extent or initial surface mining to establish cash flow for an operation that will subsequently transition to underground mining. In both cases, the life-of-mine is likely to be short, and there is frequently a requirement for steep slopes to minimise stripping and increase return. As a consequence, mining could be by contractor, open pit engineering budgets are likely to be restricted and/or technical staffing with open pit experience could be limited. Even with these constraints, it is essential that there is a slope monitoring system to ensure the safety of workers and the recovery of the anticipated ore reserves.

The scope of a monitoring program for small operations should address the risk management issues associated with any other open pit, in terms of the following basic requirements:

- confirmation of the slope design model and detection of any unexpected adverse geological or other conditions that might require a change in the mine plan; and
- strategic surveillance monitoring to confirm the stability of the slopes and to detect the onset of any unexpected movements.

The influence of surface and/or groundwater could be an important component to address, for certain small operations. These components of the monitoring program can usually be undertaken by the mine staff or by a suitable consultant who is located nearby. Particularly in the case of movement monitoring, it is essential that technical personnel are available in proximity to the mine, at least for initial input to the monitoring system design and to provide support to address unexpected movements. The following sections summarise the basic requirements for the monitoring program for small open pits, including design model confirmation and slope movement monitoring.

9.2.2 Basic monitoring program

9.2.2.1 *Design model confirmation*

Confirmation of the slope design geotechnical model typically involves two aspects, namely:

- geological mapping of the lithologies and alteration, and the structural features (major structures and structural fabric) exposed as the pit walls are excavated; and
- achieving bench and inter-ramp slope configurations relative to the design recommendations.

The associated face mapping can be conducted by the mine geologist as part of the routine mapping program undertaken in the pit. However, it is important that the geologist performing the mapping has a basic understanding of the engineering geology, as well as the factors controlling the slope designs and the associated assumptions. Accordingly, it is essential that:

- the mine geologist has basic training in geotechnical data collection and evaluation; and
- the slope design report contains clear descriptions of the geotechnical model components upon which the slope designs are based, as well as the formulation of the recommended slope configurations and the associated implementation assumptions.

Frequently, these aspects are also included in a ground control management plan (GCMP) for the proposed open pit prepared by the design engineer/consultant. It is prudent to have regular reviews by a competent geotechnical engineer (typically the design consultant or a corporate review consultant). These reviews should be undertaken at least semi-annually during the initial years of mining and annually thereafter. Any deviations from the original design model and/or achieved slope configurations must be reviewed in detail with the design consultant, with a view to modifying the slope designs to ensure effective geotechnical risk management.

9.2.2.2 *Slope movement monitoring*

A slope movement monitoring program should be in place throughout the life of the open pit. Even where no movement is indicated, it is prudent to have a basic system to confirm the stability of the slopes. This surveillance monitoring system should be designed to be capable of easy and rapid expansion in the event that unexpected movement is detected. The basic components of a surveillance monitoring program for a small open pit should include:

- visual inspections of the pit, with specific emphasis on the slopes above the operating areas; and
- quantitative strategic or tactical monitoring of the slopes for any unexpected movement.

Both functions can generally be performed by mine staff after suitable training, as discussed below. However, it is prudent to have the results reviewed by a geotechnical engineer on a regular basis.

9.2.2.3 *Visual inspections*

The visual inspections should be performed on two levels, namely:

- daily reviews of the slopes above the operating areas by the geologist/engineer or technician, with more frequent observations by the mine supervisor; and
- weekly inspection of all pit crests, wide benches and ramps, including areas that are not currently in operation. These inspections should be made by the person primarily responsible for pit slope stability.

Mine operations supervisory staff should have basic training in the identification of potential instability at all scales, from rockfall to inter-ramp and overall slope movement. Operating personnel should have at least a basic introduction to slope stability, as part of their general training program. They should be encouraged to report any unusual conditions, including rockfall occurrences, to the mine supervision.

All observations by mine supervisors and operating personnel should be recorded in the daily log, which should be reviewed on a regular basis by the person responsible for slope stability. More detailed inspections by the technical staff should also be fully documented, preferably with photographs taken from specific locations.

9.2.2.4 *Strategic and tactical monitoring*

Monitoring of the pit walls can be undertaken by the mine survey team or a geotechnical technician. The most appropriate system normally consists of survey monitoring of prisms distributed around the pit crest and along ramps and wide accessible benches. A typical spacing for the prisms is 50–100 m horizontally and 50 m vertically, with successive rows offset. The actual surveying can be performed with a total station installed on first-order survey stations, typically concrete plinths, distributed around the pit crest. Normally, at least four base stations are required to provide adequate coverage of all pit slopes. The total stations can be installed manually. However, a more effective and accurate system typically involves robotic total station (RTS) units, which are either installed manually for each reading cycle or located permanently in protective shelters, as discussed in Section 5.2.

A minimum reading frequency for strategic, or surveillance, monitoring where normal or no movement is indicated is typically once per week for the slopes above operating areas and monthly for other sections of the pit slopes where risk of movement to the operation is low. However, the frequency for slopes above operating areas is often increased to daily or every second day, particularly where there are uncertainties in the slope designs and operational performance.

Data from the survey monitoring system can be reduced by software associated with the total station equipment. This software can typically be used to plot graphs of displacement against time for the individual prisms; these graphs then have to be evaluated manually. More sophisticated integrated monitoring software packages can also indicate anomalous movement rates beyond preset threshold levels (as specified in the trigger action response plan, TARP) and flag alarms accordingly. The slope designer should provide clear direction regarding areas of potential instability and the associated trigger levels of movement (again, as specified in the TARP), together with possible remedial measures, as part of the slope design report.

The system described above should be considered the basic minimum for any open pit. Where slopes have been designed with a relatively high-risk tolerance, a more rigorous monitoring system that provides increased operational safety through real-time monitoring should be considered. Such instrumentation may include radar or laser scanners, noting that either or both should not replace the basis geodetic slope monitoring system using reflective prisms (or GNSS). Such a system would normally involve telemetry of data from fixed total stations, with automated data reduction and alarming to the mine control and the geotechnical team, as described in Section 3.3. An example of surveillance monitoring coverage for a small open pit using ground-based (GB) radar is described in Section 4.4.4. In combination with the total station monitoring described above, the strengths and weaknesses of the radar placement for slope monitoring can be quantified. See Sections 4.4.3.2 and 8.4 for relevant discussion regarding pit geometry and radar measurements, and system alarming.

9.2.3 Potential system expansion

When designing the slope monitoring system, consideration should be given to the ability to rapidly enhance the program to ensure continued safe operating conditions in the pit if unexpected movement is detected. An initial step in the system upgrading would normally be the addition of one or more radar or laser scanning units, to monitor critical slopes above operating areas as well as slopes above or below access ramps.

In extreme cases, the basic movement monitoring system, whether prism surveying or radars/scanners, should be augmented by designated spotters to watch for rockfall and ravelling that may not be detected by the more sophisticated systems. The spotters should have radio communication to transmit warnings to the other personnel in the pit, and may also be required to give permission for operations or transit below critical areas.

When unexpected movement commences, it is important to involve an experienced geotechnical engineer in the monitoring and investigation program, particularly if qualified personnel are not available onsite.

9.3 Monitoring of tailings facilities

9.3.1 Overview

Traditionally, the monitoring of tailings facilities has included porewater pressures, the movement of survey monuments and, occasionally, inclinometer measurements. Emerging monitoring methods include laser scanning, thermal imaging, GB radar and site-wide satellite radar (InSAR) surveys. Each of these methods is briefly summarised in the context of tailings facilities; detailed treatments of the monitoring methods are provided in Chapters 3 and 4. This section describes the key triggers for tailings facility instability (i.e. seismicity, extreme rainfall, liquefaction), which are essential considerations for tailings facility performance monitoring. It also discusses the management and maintenance of tailings facility monitoring systems.

9.3.2 Performance monitoring

9.3.2.1 Porewater pressure monitoring

The stability of tailings embankments and their foundations depends primarily on groundwater phreatic levels and seepage being maintained at levels below design tolerances, and the susceptibility of the impounded tailings and foundation to contraction on loading and liquefaction. Porewater pressure monitoring, including associated estimates of seepage gradients, is a critical component in tracking compliance with design tolerances for most tailings facilities. Key considerations include validating design assumptions, guiding future development stages, warning of abnormal behaviour and confirming forecast post-operation (closure) trends. Standpipe (Casagrande-style) piezometers and vibrating wire piezometers (VWP), installed via borehole or direct placement, constitute the most widely used types of piezometers. Details on groundwater monitoring methods and instrumentation are presented in Section 3.8 and their application in Section 4.7.

Alternative piezometer monitoring methods employing pneumatic, hydraulic, strain gauge, fibre optic or TDR technologies are employed to a lesser degree at tailings facilities. Open wells or standpipe piezometers with long screened intervals can be used to indicate saturation (phreatic surface) levels in uniform, coarser-grained materials. However, critical drawbacks of such installations include water levels being indicative only of porewater pressures for hydrostatic conditions and fluid mixing (contamination) potential among zones intercepted by the well screens. Specialised field instrumentation can be installed to monitor negative (soil suction) pressures in finer-grained materials located above the phreatic surface; however, apart from closure cover performance assessment, negative pore pressures are generally discounted in geotechnical assessments of tailings facilities.

Piezometers are generally installed in the embankment fills and foundations of tailings facilities, but may also be located within the tailings mass to assess consolidation progress and underdrain (when installed) performance. For upstream-raised embankments, piezometric data are used to monitor construction porewater pressures and their subsequent dissipation. Nested piezometers are installed where perched phreatic surfaces are expected. Considerations for the type and arrangement of piezometers include availability, cost, reliability (adequate accuracy, precision, durability and expected service life), response time, precedence, user compatibility (ease of installation, operation and troubleshooting), limiting interference with construction, access and other nearby activities and, to an increasing extent, compatibility with automated data acquisition systems.

Piezometer data should be acquired, screened, processed, reviewed and reported upon at intervals appropriate to the associated risk of abnormal porewater pressure conditions in the structure or component being monitored, and with consideration of previously observed changes to and/or values of porewater pressure measurements. Changes from normal conditions, whether detected by visual inspections, observed in other types of instrumentation or resulting from climatic or seismic events, warrant increased porewater pressure monitoring activities and frequencies. Given the availability of low-cost, easy-to-use, wireless and battery-powered dataloggers for VWP and other electric piezometers, frequent, real-time data acquisition and analysis are recommended. Many failures of tailings dams occur rapidly (in minutes) and catastrophically. Their development can be missed completely by infrequent porewater pressure monitoring, leading to no warning of imminent failure. Even frequent, real-time monitoring of porewater pressures may be inadequate, particularly where liquefaction of susceptible tailings and/or foundation materials occurs.

9.3.2.2 Deformation monitoring

Deformation monitoring of tailings embankments traditionally includes periodic surveying of monuments placed on the embankment crest, along the toe and on any intermediate benches. Periodic inclinometer measurements may also be carried out in boreholes installed at the embankment crest, and along the toe and any intermediate benches. Such measurements do not lend themselves to automated, real-time monitoring and hence cannot provide early warning of a sudden, possibly catastrophic tailings embankment failure. RTS, for the continuous and automatic collection of measurements from prisms installed along the embankments, are increasingly used after recent occurrences of significant tailings dam failures. Only measurements from prisms and GNSS can provide true total vector data that can be used to reconstruct the total geometry of detected movements. Details on geodetic methods of deformation monitoring are presented in Section 3.3 and their application in Section 4.3.

9.3.2.3 Laser scan monitoring

Laser scanning of tailings facilities is typically an aerial surveying application. It measures the distance to a target by illuminating it with pulsed laser light then measuring the reflected pulses with a sensor attached to the laser

Fig. 9.1: LiDAR images of a 3D model of a tailings facility (modified from Chetty 2013).

source mounted on a UAV drone. Differences in laser return times and wavelengths can be used to make 3D representations of the target, such as those shown in Fig. 9.1. Laser scanning is becoming routinely used for monitoring the volume occupied by stored tailings and the extent of the decant pond. Details on monitoring using laser scanners and drones are presented in Section 3.4.3 and 3.5.2, respectively, and their application in Section 4.6.

9.3.2.4 *Thermal imaging*

Thermal sensors attached to a UAV drone can be used to estimate the moisture state and settled density of tailings. Calibration of the thermal sensors requires direct measurements of the moisture content and density of the tailings. Thermal imaging is typically used to monitor for the detection of seepage. Additional information on monitoring with UAV drones is presented in Sections 3.5.2 and 10.3.5.

9.3.2.5 *GB radar monitoring*

GB radar, applied to the monitoring of open pit slopes since the early 2000s, can also be used to measure deformation of tailings embankments. It has been used for this purpose since the failure of the Fundão tailings dam at Samarco in Brazil in 2015 (Carneiro *et al.* 2017; Clarkson *et al.* 2018) and is now in use at other tailings facilities. Figure 9.2 shows an example of a displacement map based on GB radar. In contrast to excavated rock slopes, tailings embankments are topographically different and may be vegetated, both of which contribute to noise in the radar response. Moreover, most tailings embankments do not experience the higher rate of movement measured by radar in open pit mines (e.g. mm/day to cm/day). To detect the more typical slow movements of tailings facilities (e.g. mm/month, mm/year) with GB radar, the noise induced by vegetation or changes in the electromagnetic properties of the dam surface (where coherence of the radar signal in the long term is maintained during the monitoring period) can be filtered out by selecting the appropriate time period over which to analyse the data. Because of the line-of-sight (LOS) nature of radar measurements, they need to be calibrated against survey measurements derived from prisms or GNSS sensors. Details on deformation monitoring using GB radar are presented in Section 3.4.2 and its application in Section 4.4.

9.3.2.6 *Satellite radar monitoring*

Satellite radar systems (InSAR) are commonly used by mining operations for site-wide deformation monitoring. InSAR is also ideally suited to deformation monitoring of large tailings facilities. Allowing for the LOS limitation and limited resolution of total vector movement when images for both orbital paths are utilised, stereo satellite images of the target allow 3D maps of the target region to be generated. Differences between successive maps over time allow deformations to be determined to a high degree of precision, as described in Section 3.5.3. InSAR measurements can be compared with available GB radar and other measurements, and need to be calibrated against geodetic survey measurements. Archived InSAR images can be analysed to estimate the pre-failure deformations of tailings embankments that subsequently failed. Utilisation of InSAR for deformation monitoring of tailings facilities has increased, following recent failure events. Examples include the Samarco Fundão tailings dam (Carneiro *et al.* 2017; De Athayde *et al.* 2017; Mura *et al.* 2018) and

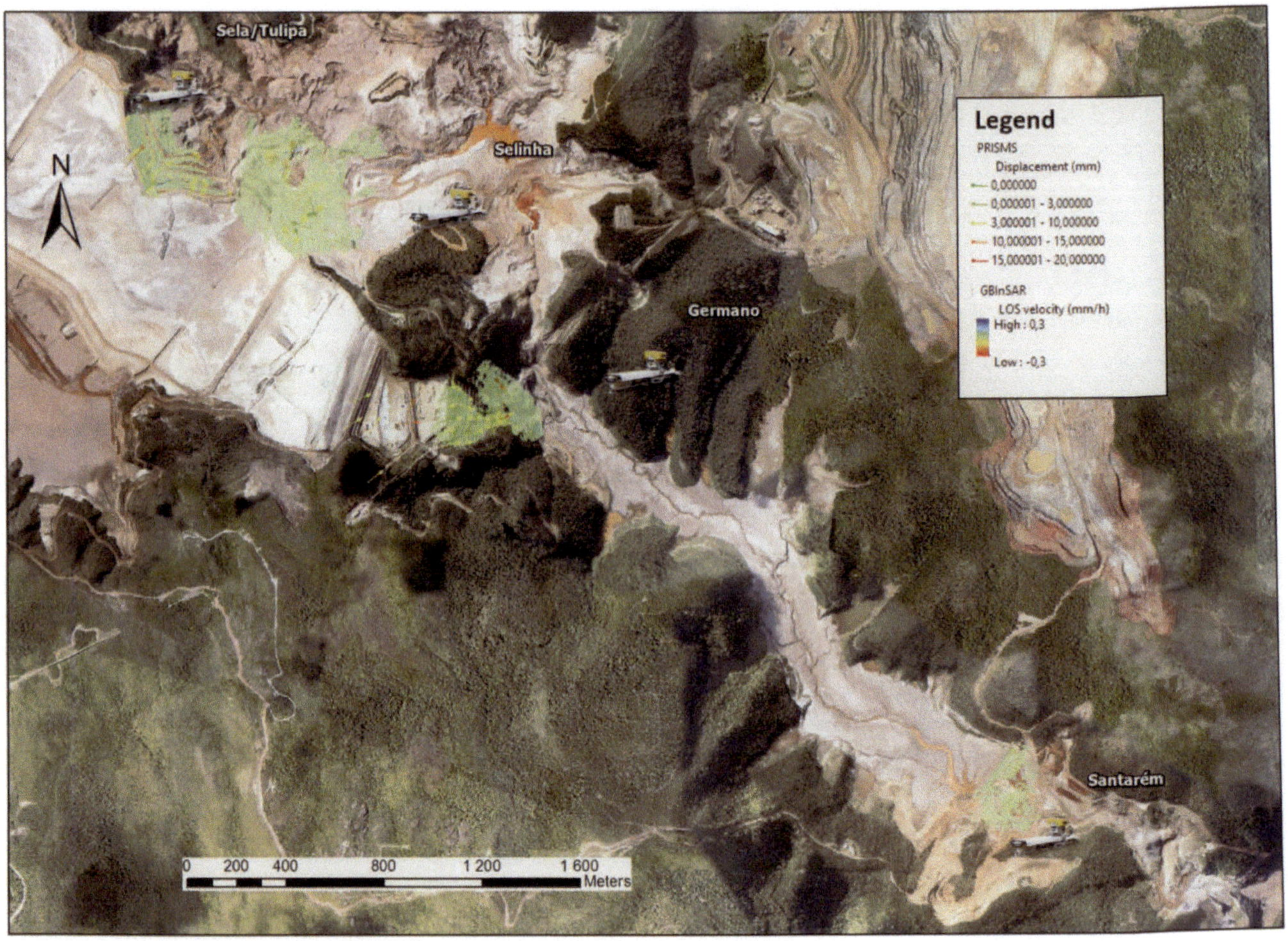

Fig. 9.2: Radar displacement map with prism vectors over the Sela/Tulipa, Selinha, Germano and Santarém tailing dams at the Samarco Complexo de Germano mine (modified from P. Farina and Carneiro *et al.* 2017).

Northern Tailings Storage Facility (NTSF) Embankment failure at Cadia Valley Operations (Jefferies *et al.* 2019). An example is shown in Fig. 9.3. InSAR measurements can be compared with available radar and survey measurements, and need to be calibrated against survey measurements. Details on site-wide deformation monitoring using InSAR are presented in Section 3.5.3, and its application in Section 4.5.

9.3.2.7 *Additional monitoring components*

Essential additional components of a tailings facility monitoring system include a full weather station located at the facility to provide meteorological data impacting the tailings water balance, and flow measurement at any drainage points and seeps from the facility. In addition, the solids concentration of the input tailings needs to be monitored, along with the rate of rise and deposition plan of the tailings, their settled dry density and occupied volume, and water inputs and outputs. Any changes to the nature of the tailings due to input ore/gangue changes need to be recorded. Relevant discussions on surface water monitoring and management, as well as meteorological monitoring, are presented in Section 3.7.

Stereo-optical satellite images of the target allow 3D maps to be generated, and differences between successive 3D maps over time allow 3D deformations to be determined. Satellite images can also be analysed to generate vegetation and wetness indices.

9.3.3 Seismicity

Seismicity refers to the frequency, intensity and spatial distribution of earthquakes. The characterisation of the earthquake source(s) and the associated ground motions that may affect a site or specific structure of interest, such as a tailings facility, are the main products of a seismic hazard assessment (SHA). An SHA is completed to support the initial design, operational risk assessment and/or closure planning of the facility. Earthquake monitoring systems for tailings facilities in known seismic regions usually comprise a network of accelerometers with automated triggering, data collection and event notification.

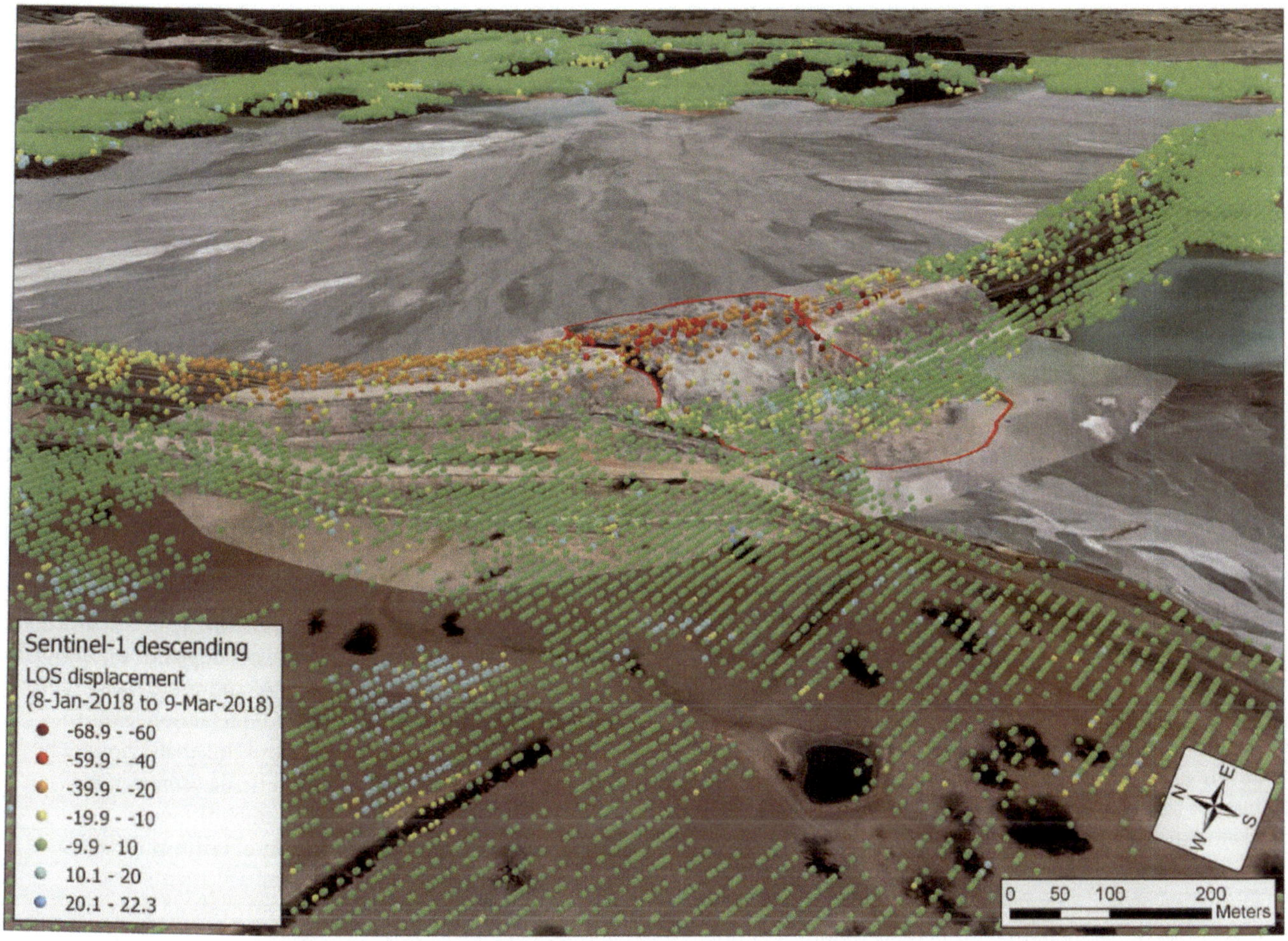

Fig. 9.3: A 3D view of the displacement map derived from InSAR data (Sentinel-1 descending orbits) covering three months before the March 2018 failure of the NTSF Embankment at Cadia Valley Operations, overlaying the post-failure aerial photo of the tailings dam (source: P. Farina).

The decision on whether to install an earthquake monitoring system for a tailings facility, and the complexity and capability of the accelerometer network, should consider various factors, including:

- the predicted seismicity for the site, including the confidence in, or quality of, the underlying SHA;
- the estimated potential for, and magnitude of, earthquake-induced deformation of the facility, including the foundation and critical appurtenant structures;
- the estimated consequences of large-scale earthquake-induced deformation of the facility, such as embankment slope failure, liquefaction flow or closure cover deformation;
- the potential for, and associated consequences of, secondary failure mechanisms, such as embankment settlement leading to loss of freeboard and overtopping, tearing of geosynthetic materials leading to internal erosion, or disruption of internal filter and core zones, leading to porewater pressure increase and slope instability;
- the ability to undertake a facility inspection in a timely manner after detection of significant ground motions (especially important for remote sites);
- the availability of key associated performance information, e.g. porewater pressure or embankment deformation readings recorded leading up to, during or soon after ground shaking;
- the ability to provide rapid data analysis and, as appropriate, calibrate existing design earthquake behaviour and/or plan and execute remedial work (relevant for remote and/or closed facilities as well as for active facilities);
- regulatory requirements and/or stakeholder expectations for earthquake monitoring;
- the existence of nearby seismic monitoring stations that could provide earthquake event data, such as a government agency or academic research networks;

- the applicability of earthquake ground motion and tailings facility performance data from the subject site to other operations in the portfolio, or for the industry in general; and
- the time intervals (exposure periods) of the facility under different scenarios. For example, in the operation and initial closure period the impoundment may be expected to hold a large decant pond and extensive, saturated tailings with a full-time site management presence, whereas the long-term closure setting may include a regraded embankment, dry cover system and well-drained tailings mass with no site staff.

There are uncertainties and risk potential inherent in the prediction of earthquake design ground motions, particularly for sites of low seismicity and short earthquake historical data, and corresponding tailings facility behaviour. Substantial embankment seismic design and impoundment water management are paramount, particularly in moderate- to high-seismicity regions. Earthquake monitoring of tailings facilities in such regions may provide the designer with an opportunity to compare predicted *v.* recorded responses, should significant (and thus low-probability) earthquake ground motions occur at the site and, if required and practicable, to recommend changes to the facility geometry, management or downstream conditions to improve earthquake safety. Real-time data from the earthquake monitoring system would also provide a reference for instigating emergency response actions, including immediate dam inspection and data collection for low-level earthquakes, and evacuation plans in the event of damaging earthquakes. In some jurisdictions, large tailings facilities may be legally required to establish seismic monitoring capability, as well as to develop tailings facility seismic designs and contribute to the regional earthquake database.

Where earthquake monitoring is required only to trigger a dam safety inspection for a minimum-threshold ground motion, a single, automated accelerometer installed at an appropriate location (often the tailings dam crest) may suffice. However, to provide more useful data for earthquake design verification, the minimum accelerometer array usually includes sensors at the tailings dam crest, toe and abutment, as well as a free-field sensor located outside the stress influence of the tailings facility. The recommended accelerometer locations, type of accelerometers and data management system will be specific for each individual tailings facility. The instrumentation array may require modification at stages of the facility's development and operation, and should be designed and installed with input from appropriate specialists.

Tailings facilities located in low-seismicity regions may not warrant site-based earthquake monitoring instrumentation. Few, if any, records would be expected during the operation and active closure phase, when the effects of earthquake-generated deformation may be of greater concern and resources (including rapid inspection, data analysis, materials and site staff) would be readily available. In the long term, such as during the passive closure phase and post-closure, appropriate seismic design criteria (e.g. consideration of the estimated maximum credible earthquake and/or a suitably low-likelihood probabilistic event) are advised, rather than reliance on earthquake monitoring.

9.3.4 Liquefaction

Liquefaction of susceptible tailings facilities and foundations can occur as a result of the short-term loss of shear strength due to the generation of excess porewater pressures and commensurate reduction in effective stress below that required to support the applied loading. Materials susceptible to liquefaction include loose, near-saturated, fine-grained sandy tailings. Liquefaction can be triggered by earthquake or static loading (e.g. embankment failure, embankment raising, tailings deposition), flooding events, nearby blasting, or ongoing and progressive creep deformations. The occurrence of liquefaction leads to strain-softening and potential flow-slide behaviour. The potential for liquefaction is of particular concern for upstream-raised tailings facility embankments.

The safe design and operation of tailings facilities require adequate downstream embankments or the establishment of a relatively dense, drained proximal tailings beach for successive upstream embankment raises. The loss of decant pond control, unanticipated changes to tailings characteristics, or other influencing factors can result in the development of contractive, loose, saturated to near-saturated tailings underlying the impoundment slope, and an increased risk of triggering collapse behaviour under applied static or dynamic loading. In some instances, the foundation soils underlying the tailings embankments may also be of concern for liquefaction risk and require appropriate removal or improvement as part of the tailings facility design and construction.

The risk of liquefaction potential reinforces the need for geotechnical parameter assessment and monitoring directed at the identification and evaluation of potentially liquefiable materials. Given the greater risk of tailings liquefaction and flow-slide behaviour associated with upstream-raised embankments, periodic assessments of liquefaction potential are warranted. These are most commonly performed by piezocone testing at an appropriate distribution and schedule for the development of the facility and the consequences of failure by liquefaction. However, given the often catastrophic (rapid and large-volume) nature of liquefaction failures and the difficulty in reliably predicting and monitoring liquefaction triggers, defensive design and operation practices are paramount. Particularly for upstream-raised facilities, defensive measures include restricted rate of rise for embankments and tailings, and the maintenance of small, centrally located decant ponds. Another defensive and proactive operating practice for monitoring a potential liquefaction trigger is to utilise an integrated and automated (ADAS) monitoring approach that could include surface and subsurface instrumentation as described above and in Chapters 3 and 4 of these Guidelines.

9.3.5 Monitoring system management and maintenance

Tailings facility monitoring should be automated, real-time and linked directly to TARPS and alarms for geotechnical stability and possible site evacuation. Monitoring systems should have in-built redundancy to allow calibration and to provide backup in the case of failure of some monitoring components of the system. The monitoring system should regularly be checked and maintained. Historical monitoring data should be recorded and maintained to provide performance data for the tailings facility, against which current and future performance should continuously be assessed.

Tailings embankment failures are often rapid, can be catastrophic where there is substantial release of tailings solids and water, and present risk to life downstream. For such worst-case scenarios, no currently available monitoring methodologies are capable of providing sufficient warning for the safe and timely evacuation of people at risk. In such cases, placing people at potential risk should be avoided.

The resolution of laser, thermal, GB radar and InSAR images will continue to improve, as will the methods and speed of interpretation. The frequency of satellite passes will also continue to increase, thus allowing their use for near real-time monitoring. These developments will improve the capability and speed of these remote methods of monitoring the deformation of tailings embankments.

9.4 Monitoring of waste dumps and stockpiles

9.4.1 Overview

The monitoring of waste dumps and stockpiles is treated in considerable detail in Chapter 12 of *Guidelines for Mine Waste Dumps and Stockpile Design* (Hawley and Cunning 2017). That chapter has been modified herein to eliminate redundancy. The reader is encouraged to refer to the original source for additional information.

Instability of waste dumps and stockpiles is usually preceded by warning signs, such as:

- increased rates of deformation;
- cracking and/or settlement of the platform;
- bulging of the face;
- bulging of the toe;
- bulging or heaving of the foundation in front of the toe (i.e. toe spreading);
- seepage on the face;
- increased pore pressures in the foundation or embankment; and
- increases in the ambient noise level within the embankment.

Many of these warning signs can be observed visually, and the importance of routine, periodic visual inspections to identify signs of developing instability early cannot be understated.

The purpose of waste dump and stockpile monitoring is to:

- maintain safe operating conditions;
- provide advance warning of developing instability so that mitigative measures can be implemented to lessen the impact on operations;
- provide information about the mechanism of instability;
- quantify displacements and movement rates; and
- establish and maintain a record of facility performance.

As noted by Clayton *et al.* (1995), instrumentation utilised to monitor waste dump and stockpile conditions should be as simple, reliable and robust as possible and capable of measuring the required parameters to an acceptable level of accuracy, with an acceptable speed of response to changes that might occur in the parameter being measured.

9.4.2 Visual inspections

Visual inspection is critical to ongoing safe operation of a waste dump or stockpile, and is the easiest, most common and most practical method of monitoring (BCMWRPRC 1991). Visual inspection is a key component of an effective waste dump and stockpile monitoring program. At many mines, operations staff, equipment operators, surveyors and other personnel who regularly visit the waste dumps and stockpiles are trained to recognise potential signs of developing instability, including:

- excessive or abnormal cracking;
- excessive crest deformation;
- excessive over-steepening of the crest;
- abnormal platform tilting;
- seepage breakout on the face;
- bulging of the face; and
- toe spreading.

Observations of any of these indicators should be brought to the attention of the geotechnical or mine engineering department and evaluated in the context of available monitoring data to help identify potential instabilities.

Inspections of the crest and active and inactive platforms should look for evidence of slumping, development of large, arcuate tension cracks, normal or obsequent scarps and/or tilting of the platform that might be indicative of a developing deep-seated instability. Inspections should also identify over-steepening of the crest beyond the normal angle of repose. Over-steepening can result from an accumulation of wet, fine-grained material at the crest and lead to rapid onset of sliver failures. Excessive over-steepening can be addressed by cutting down the crest and refilling the cut-down area with clean, coarse material. Tension cracks should be graded over to help prevent infiltration of runoff and maintain a smooth platform for optimum trafficability. Platforms should be graded away from the crest.

Inspections of the slope or face of the waste dump or stockpile should look for signs of bulging or seepage breakout. Bulging in the upper portion of the face is typically related to over-steepening caused by poor segregation, placement of fine-grained materials, excessive advance rate, or saturated material. Bulging at the toe can be an indication of developing instability along the interface between the embankment and the foundation, or shallow foundation failure. Bulging at the toe combined with heaving of the ground in front of the toe can be an indication of deep-seated instability. Both shallow foundation failure and deep-seated failure are often accompanied by cracking and settlement many tens of metres behind the dump crest, whereas shallow sliver failures are often accompanied by tension cracking closer to the crest. Where toe bulging is observed, a review of monitoring records should be carried out to help determine the cause. In many cases, minor bulging of the face does not lead to instability, and routine advancement of the crest and face will restore the normal angle of repose. However, if the bulging persists or expands, or is accompanied by heave beyond the toe, appropriate slope monitoring methods should be employed to confirm that movement rates are within acceptable limits and not increasing. If necessary, material placement should be discontinued or slowed until movement rates reduce to within acceptable limits.

Many waste dump and stockpile failures can be linked, at least in part, to foundation failure. As indicated above, visual indicators of foundation failure include heaving/deformation of the foundation in front of the toe. Adverse foundation conditions are typically managed by limiting dump height and/or advance rate or by removal of weak foundation materials in advance so the toe is founded on more competent materials. Any visual monitoring program should establish a systematic approach to the inspection and recording of observations. Where possible, routine visual inspections should be carried out by the same people. This continuity may help to identify subtle changes in the facility that may not be evident to someone who is less familiar with the facility. However, while continuity is important, periodic reviews by external geotechnical specialists should also be conducted to help ensure that inspections do not become too routine and to add the perspective of a different set of eyes.

9.4.3 Displacement monitoring systems

Monitoring of waste dumps and stockpiles by means other than visual observation occurs at most mines where these structures are subject to one or more of the following conditions:

- high face height (≥100 m);
- steep overall slope angles (≥25°);
- steep, adverse foundation slopes (≥15°);
- lack of topographical confinement;
- poor-quality foundation materials;
- poor-quality waste or stockpile materials;
- aggressive or unfavourable construction methodology;
- unfavourable piezometric or climatic conditions;
- high advancement/loading rates (≥50 m^2/d [crest advance in metres per day × face height] – see Chapter 3 in Hawley and Cunning 2017); and
- poor historical performance.

These attributes are important factors in the Waste Dump and Stockpile Stability Rating and Hazard Classification system described in Chapter 3 of Hawley and Cunning (2017). Any one factor, or combination of factors, can have a significant impact on stability.

The mechanics of the various modes of instability are well understood, with different modes of failure tending to manifest in different ways. While ongoing consolidation settlement routinely induces deformation and cracking in large waste dumps and stockpiles, the rate of deformation is generally linear or decreases over time. Failures, however, are typically preceded by increasing/accelerating rates of deformation. As indicated above, where a failure mechanism is deep-seated, evidence of deformations well behind the crest may be observed. Understanding the likely failure mechanism responsible for observed deformation patterns and rates is important when designing a monitoring program, including the type of monitoring system best suited to assessing that failure mechanism.

Several of the most common types of monitoring systems for waste dumps and stockpiles are listed below in general order of practical utilisation. Additional details are available in these Guidelines, in the sections shown in parentheses.

- visual inspection (Section 3.2.2);
- wireline extensometers (Sections 2.3.2, 3.2.4) – most common (manual and digital);
- prisms (Sections 3.3.2, 3.3.3) – especially for monitoring closed dumps;
- tell-tales (Section 3.2.3) – common to paint cracks;
- cross-crack monitoring (Sections 3.2.2, 3.2.3);
- GB radar (Section 3.4.2);
- satellite InSAR (Section 3.5.3);
- inclinometers (Sections 2.3.6.2, 3.6.3);
- GNSS (Section 3.3.4);
- laser scanners (Section 3.4.3);
- TDR (Section 3.6.2);
- laser distance measuring (LDM) system (Section 3.2.6);
- autonomous wirelessly networked (smart) sensors (Section 3.6.5); and
- acoustic emission monitoring (Section 9.4.3).

Specific considerations for waste dumps and stockpiles are discussed below.

Wireline extensometers. For waste dumps or stockpiles that are subject to a moderate or higher hazard of instability (i.e. a waste dump and stockpile hazard class [WHC] of III, IV or V; see Hawley and Cunning 2017, Chapter 3), the use of wireline extensometers, manually read and remotely telemetered, is common. However, additional instrumentation should be considered, to provide a record of total deformation. This could include periodic surveys of the facility using prisms or GNSS, satellite InSAR or laser scanning surveys (LiDAR).

GNSS. Like wireline extensometers, where remote GNSS is used to monitor waste dumps and stockpiles that are under construction, relocation and resetting is required as the crest and face advance. The amount of time between resets is dependent on the rate of advancement of the crest. Remote GNSS monitors located near the crest are at risk of loss due to sliver failures. Accordingly, remote GNSS systems may be better suited to the monitoring of inactive facilities where crest deformations and consolidation settlements related to advance rate are minimal and where latent deep-seated failure modes and settlement are more of a concern than shallow sliver failures.

GB radar. In the case of active waste dumps or stockpiles, deformations measured by radar may be masked by activities associated with construction, such as end dumping or dozer pushing along the crest, or sliding and roll-out of material dumped or pushed over the crest. However, there are some situations in which radar monitoring of active waste dumps or stockpiles may be useful. For example, where waste dumps or stockpiles are advancing over poor foundations, it may be possible to use a radar system to monitor for foundation deformations in front of the toe of the advancing embankment. In some cases, radar systems may be able to filter out normal operational background noise and movements associated with advancement of the face and crest, detect local bulging or anomalous deformations or hotspots on the advancing face, and possibly even detect over-steepening of the crest.

Laser scanning (LiDAR). The utilisation of laser scanners for real-time monitoring of waste dumps is increasing due to recent improvements in accuracy, image quality repeatability and data reliability. In particular, the use of these systems to conduct rapid topographical surveys of inaccessible slopes has increased in popularity. Details of laser scanner capabilities and limitations are described in Chapter 4. As with radar systems, laser scanners are well-suited for monitoring inactive dumps and stockpiles.

Acoustic emission monitoring. It has been observed that failures of waste dumps composed of coarse rock can be preceded by acoustic emissions for some time before failure (McCarter 1976). This noise is attributed to the fracturing of waste rock blocks caused by movement along the failure plane. By recording acoustic emissions using an array of at least three geophones installed on and around a waste dump or stockpile, it may be possible to determine the location of a specific acoustic event. It should be cautioned, however, that a source location solution based on three inputs is poorly constrained and will always be ambiguous. Acoustic emissions attenuate rapidly in unconsolidated fill materials, so it is important to place the sensor/geophone array as close as possible to the expected source of the emissions. This can be accomplished by installing the sensors in the foundation and/or within the embankment itself. Additionally, there may be sources of acoustic emission activity other than movement along the failure plane. Emissions within an unconsolidated pile of coarse material can result from rotation/shearing/fracture of the aggregate within the pile. Another limitation is that it is not possible to determine the magnitude of movement associated with the emissions. However, when used in conjunction with other instrumentation systems, such as inclinometers or TDRs, acoustic emission monitoring may help to develop a better picture of a failure surface. The presence of noise (audible/amplified by instrumentation) is perhaps the most reliable use of this monitoring method for early detection of impending failure.

InSAR. The use of satellite InSAR for broad-based mine site monitoring has gained popularity in recent years as the cost has come down. This system is of particular interest in the long-term monitoring of settlement and stability of old waste dumps.

Smart Markers. The use of Smart Markers to monitor internal deformation and settlement within waste dumps and stockpiles has not yet been trialled on a large scale, but it is expected that this monitoring system may warrant further consideration as improvements in the technology increase its utility.

9.4.4 Surface water and groundwater monitoring

9.4.4.1 *General considerations*

The amount and type of hydrological monitoring data required for a waste dump or major stockpile will depend on the site setting and the types of materials consigned to the facility. Hydrogeological monitoring can be minimal where facilities are located in arid climates and where the waste dump and stockpile materials are strong and not subject to degradation over time. In this case, monitoring may be limited to upgradient and downgradient groundwater chemistry for permit compliance. Conversely, an extensive array of instrumentation may be required for facilities located in wet or seasonally wet climates, where the waste dump or stockpile materials and/or foundations are weak, or where the topography is steep.

9.4.4.2 *Types of surface water and groundwater monitoring data*

The following types of surface water and groundwater monitoring may be required.

Climatological monitoring. Climatological monitoring typically forms part of the site-wide monitoring plan and is not usually specific to individual waste dump or stockpile locations. However, in wet areas and/or regions of steep topography, it may be necessary to operate a rainfall station close to the location of a particular facility. Additional details regarding meteorological monitoring are provided in Section 3.7.2.

Monitoring of upgradient conditions. Upgradient groundwater level and chemistry monitoring (and potentially surface water monitoring) will normally be required throughout the operating life of the facility and beyond (see also Sections 3.7 and 3.8). Monitoring requirements specific to the waste dump/stockpile operational stage are discussed below.

Monitoring within the facility. The following operational monitoring may be required for the waste dump/stockpile hydrogeology program:

- the nature and distribution of the materials being placed (e.g. material types, tonnages/volumes and spatial distribution);
- water content of the placed materials;
- the geochemical nature of the waste dump and stockpile materials, typically with a quarterly sampling program that involves acid-base accounting and static testing and aims to confirm the nature and variability of the materials;
- VWPs and/or soil moisture probes installed in the foundations or in the basal layer of the waste dump or stockpile;
- thermistors to assess freeze-back in the base of the facility or the outer (downgradient) toes;
- VWPs installed in the initial lifts;
- VWPs or standpipe piezometers installed from benches between lifts (the use of standpipe piezometers enables water quality sampling of the porewater chemistry);
- regular visual inspection and monitoring of any seepage zones (flow and chemistry); and
- regular visual inspection and monitoring of the surface water management system to identify any required maintenance or remediation. In some cases, the installation of surface water monitoring stations for key runoff channels may be beneficial for understanding the site water balance and/or potential downgradient changes in the surface and groundwater system.

Monitoring within the near-surface materials. If it is necessary to accurately assess recharge to the surface of the facility, the following types of monitoring may be required:

- VWPs and/or soil moisture probes installed in the near-surface materials and/or cover layer;
- suction lysimeters to allow porewater sampling of the near-surface materials and/or cover layer;
- mapping of snow banks, including measurements of the depth of accumulated snow and its water content;
- thermistors to assess the surface freeze–thaw layer and freeze-back; and
- visual inspection of any vegetation programs for the cover.

Monitoring of the foundations.: Hydrogeological monitoring requirements for the foundation may include:

- VWPs installed in the foundation to obtain pore pressure data before facility construction;
- using soil moisture probes and/or lysimeters installed during construction to monitor changes in moisture content; and
- piezometers installed in the toe area to monitor pore pressures to help confirm that foundation design criteria are being met.

Monitoring of downgradient conditions. Downgradient monitoring will usually include:

- groundwater levels and water quality monitoring in environmental wells installed within the underflow system in the toe area and downgradient of the facility (some type of downgradient groundwater monitoring is required for virtually all major waste dumps and stockpiles); and
- surface water monitoring stations for regular or continuous flow monitoring, sediment monitoring and water quality sampling (downstream surface water and sediment monitoring is usually required for all waste dumps and stockpiles where surface water has been diverted and/or for facilities that create any downgradient surface water flows). Where flows are continuous, frequent monitoring may be required or continuous recording stations may need to be established.

One or more downstream compliance points need to be established for surface water and groundwater. The optimum location of these compliance points will depend on the site setting and the overall layout of the mine site water management system. Ideally, the compliance points for the waste dump and stockpile facilities can be located downstream from the entire mine site so that they can benefit from mixing with other waters to optimise the chemistry. Corporate environmental staff and regulators often require trigger monitoring points for any runoff and seepage immediately below the waste dump or stockpile, so that the water quality performance of the facility can be assessed on an ongoing basis. Monitoring of water quality in the receiving environment may also be a permit requirement, particularly where discharge criteria are based on background values.

9.4.4.3 Pore pressure monitoring

Depending on potential failure modes, pore pressures in the facility or its foundation can be one of the most critical factors affecting stability. Where high pore pressures within the facility or its foundation pose a concern for stability, monitoring of pore pressure during construction and operations will be important, particularly where the materials within the facility or its foundation are fine-grained and subject to pore pressure generation due to loading. The potential for pore pressure generation and dissipation rates must be evaluated and the results incorporated into the analysis and design process.

Measuring groundwater levels and porewater pressures is typically accomplished using piezometers installed in the facility and/or its foundation. Foundation piezometers can be installed in boreholes drilled through the facility into the foundation, or in the foundation before construction. Some piezometer types can be installed on existing platforms before the start of construction of subsequent, overlying lifts.

Several piezometer types applicable to the monitoring of waste dumps and stockpiles are available. These include pneumatic, VW, strain gauge (i.e. electrical resistance), standpipe (slotted or porous media) and multi-port or multi-level piezometers. For more detailed descriptions and theory regarding piezometers and piezometric monitoring, including a discussion of their pros and cons, please refer to Beale and Read (2013).

9.4.5 Monitoring program, data assessment and response

9.4.5.1 General considerations

As with open pits, monitoring of waste dumps and stockpiles should be an integral component of the operation's geotechnical risk management process. The monitoring methods and requirements depend on the type and scale of the waste dumps and stockpiles, as well as the possible failure modes and overall hazard level of the facility.

9.4.5.2 Monitoring program

It is important to initiate an appropriate monitoring program as early as possible in the waste dump/stockpile development cycle in order to establish an initial baseline for assessing the performance of the structures. Without this reference baseline, it is difficult to evaluate the significance of monitoring results or to identify and interpret signs of developing instability.

The monitoring program should be developed in collaboration with geotechnical specialists experienced in the design, construction, operation and monitoring of waste dumps and stockpiles. It should be designed to recognise deformations that may be precursors to the possible instability modes that were identified during the investigation and design process. The required scope and intensity of the monitoring program will vary depending on the level of hazard, as summarised in Chapter 3 of Hawley and Cunning (2017). A flowchart outlining the basic steps involved in developing a waste dump or stockpile monitoring program is shown in Fig. 9.4. A summary of suggested initial monitoring, inspection and reporting guidelines for each waste dump hazard class (Hawley and Cunning 2017) is provided in Table 9.1. The guidelines are for stable dumps and should be considered a starting point. They should be modified to suit the

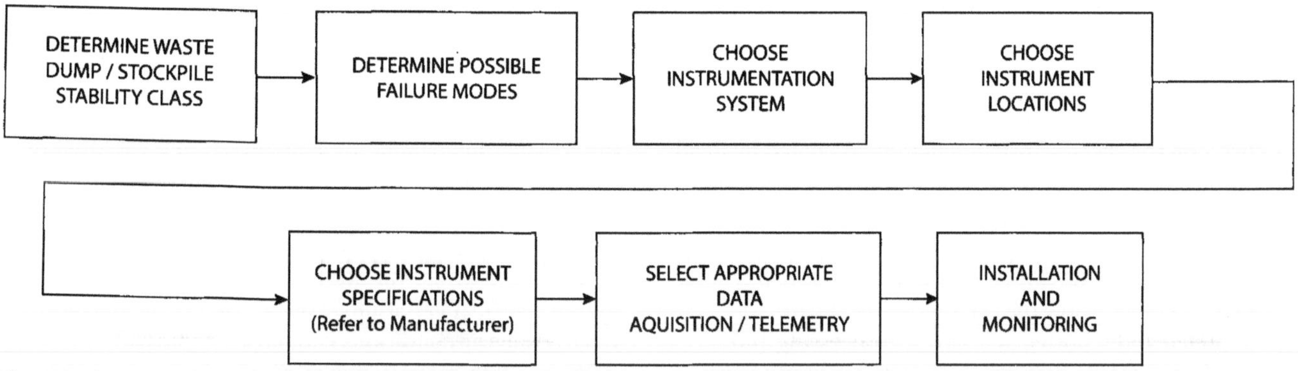

Fig. 9.4: Flowchart for development of a waste dump or stockpile monitoring program (modified from Hawley and Cunning 2017)

Table 9.1: Suggested initial monitoring and inspection guidelines

Waste dump and stockpile hazard class (WHC)	Waste dump and stockpile stability rating (WSR)	Monitoring methods	Inspection requirements[1,2]	Frequency[3]
I	80–100	• Visual	• Operations	• Daily visual inspection
			• Geotechnical	• Monthly visual inspection
II	60–80	• Visual Basic instrumentation: manual wireline extensometers, crack monitors, tell-tales, manually monitored prisms (monitoring frequency weekly–bimonthly) Piezometers as required	• Operations	• Visual inspection and monitoring at the start of each shift (every 12 h)
			• Geotechnical	• Weekly visual inspection; monthly data review
			• Independent reviewer	• Every 12–24 months
III	40–60	• Visual Basic instrumentation: manual wireline extensometers, crack monitors, tell-tales, manually monitored prisms (monitoring frequency daily–monthly) Redundancy (more than one system and/or multiple instruments) Piezometers as required	• Operations	• Visual inspection and monitoring every 6 h (twice per shift)
			• Geotechnical	• Daily visual inspection; weekly data review
			• Independent reviewer	• Annual
IV	20–40	• Visual Automatic, continuous (telemetered) wireline extensometers, LDM, RTS, radar Redundancy (multiple systems) Specialised instrumentation (tilt sensors, LDM, TDRs, inclinometers, acoustic emissions) as required Piezometers as required	• Operations	• Visual inspection and monitoring every 4 h
			• Geotechnical	• Visual inspections every 12 h; continuous (24/7) remote surveillance; biweekly data review
			• Independent reviewer	• Semi-annual
V	0–20	• Visual Automatic, continuous (telemetered) wireline extensometers, LDM, RTS, radar Redundancy (multiple systems) Specialised instrumentation (tilt sensors, LDM, TDRs, inclinometers, acoustic emissions) as required Piezometers as required	• Operations	• Visual inspection and monitoring every 2 h
			• Geotechnical	• Visual inspections every 8 h; continuous (24/7) remote surveillance; daily data review
			• Independent reviewer	• Quarterly

Notes:
1. All inspections should be summarised in a suitable format, circulated as appropriate and filed for future reference.
2. Geotechnical department staff and independent reviewer conducting inspections should be geotechnical specialists with appropriate experience in the design, construction and operation of waste dumps.
3. Frequency of inspections subject to local regulatory requirements.

Source: Hawley and Cunning (2017).

individual facility, as site-specific experience with the performance of the facility is gained.

9.4.5.3 *Data assessment and reporting*

Data processing requirements and presentation formats depend on the specific instrumentation. As noted by Read and Stacey (2009), monitoring programs have been known to fail because the data acquired were not used or interpreted properly, or were never used. The monitoring program must be developed with a clear sense of purpose, which will dictate how the accumulated data are interpreted and reported. The goal of processing and properly presenting data obtained from the monitoring system is to rapidly and reliably assess data, and to detect changes that require action.

Interpretation of data from the monitoring system typically involves looking for changes from the expected norms. For this reason, as discussed above, early implementation of a monitoring program is essential in establishing a baseline and determining the accuracy of components of the specific monitoring system that has been deployed. Baseline data may also be needed to determine nominal variations in the data due to diurnal and/or seasonal variations. For example, diurnal variations in atmospheric conditions can have a significant impact on EDM and RTS measurements, particularly where the differences in daytime and night-time temperatures are significant. Similarly, phreatic levels within a waste dump/stockpile or its foundation can fluctuate seasonally. These variations must be recognised and considered as part of the interpretation process. With respect to an EDM or RTS system, it may be possible to minimise the potential for diurnal effects by taking measurements at the same time each day. However, if instability develops and it becomes necessary to monitor prisms more often than daily, it will be necessary to filter out the atmospheric effects.

Reporting formats should be based on site requirements. Where monitoring systems are large or complex, it may be worth considering the use of an integrated geotechnical monitoring system developed specifically for mining applications. These systems are designed to allow real-time data acquisition and processing of multiple instrumentation systems, and typically include the ability to issue alerts and alarms. They also allow monitoring data to be stored in a central location that can be queried by authorised users from almost anywhere in the world. They help increase monitoring program reliability while reducing data acquisition and processing costs. Examples of integrated geotechnical monitoring systems that are currently available include Canary Systems, NavStar, Sixense Group, RST, Hexagon, Sensemetrics, M2M, SHMS and more. Some mines also develop site-specific integrated geotechnical monitoring data platforms that are customised to the suite of instruments in use at the mine, although the cost of developing such systems can be prohibitive.

9.4.5.4 Trigger Action Response Plan

A TARP should be established for the key monitoring method(s) used to assess the short-term performance and safety of the waste dump or stockpile. To be effective, the TARP should be based on changes in readings in excess of the monitoring method's normal error range or, in the case of piezometers, a predetermined piezometric pressure or head that may be associated with a specific design exceedance, factor of safety, or probability of failure. A generic example of a TARP for monitoring of a dump using wireline extensometers is given in Table 9.2.

The generic TARP does not provide specific movement thresholds as these will depend on many factors, including the monitoring method, the size and geometry of the facility, foundation conditions, material types, the method and speed of construction and, most importantly, the performance history of the facility. For waste dumps and stockpiles that have no performance history, suggested initial movement thresholds are provided in Table 9.3. Suggested thresholds depend on the potential hazard of instability as represented by the WHC. Lower thresholds are suggested for facilities with a higher hazard class. Note that the WHC does not specifically address the potential consequences of instability or risk. Hence, even lower thresholds may be appropriate for cases where the consequences of instability and risk are high. Note also that the suggested thresholds in Table 9.3 are intended for simple wireline monitoring systems (without intermediate supports). These guidelines are intended as a starting point only. They should be adapted and modified to suit site conditions as experience is gained.

9.5 Geotechnical monitoring of mine facilities

9.5.1 Overview

Performance monitoring beyond the immediate limits of the open pit slopes should include large critical infrastructure. Surveillance monitoring of mine facilities, on an as-needed basis, is well-served by the geodetic methods described in detail in Chapter 3. Instrumentation such as GNSS (Section 3.3.4) and precise levelling (Section 3.2.5) are widely used for purposes that include monitoring for subsidence and differential settlement. Successive chapters in these Guidelines have brought attention to the utilisation of data from multiple instruments in combination for data validation and reliability, for improving interpretive context and for interpretation of failure mechanisms. The remainder of this chapter presents examples of specific experience on how data from slope monitoring are combined with other facility monitoring in the vicinity of the open pit, where movement on slopes may impact adjacent ground or facilities. In some cases, data from alternative monitoring systems have been useful in the detection and mitigation of an unwanted geotechnical event, and for interpretation and delineation of a failure mechanism. Details of the utility of integrated performance monitoring for mine facilities are summarised in the following detailed case histories:

- Jwaneng mine, Botswana, which provides an example of a multi-instrumented operation designed and implemented for assessing a pit slope and adjacent critical infrastructure monitoring; and
- Golden Sunlight mine, Montana, USA, which provides an example of the interaction of an open pit slope and adjacent underground mine, and operational risk management.

9.5.2 Case history 1: Jwaneng mine

9.5.2.1 Overview

This case history provides a classic example of using instrumentation for monitoring critical mine infrastructure located in the vicinity of the open pit. The

Table 9.2: Example TARP for a waste dump monitored using wireline extensometers

Alert	Green	Yellow	Orange	Red
Platform condition	No new tension cracks; no recent movement on old tension cracks	Existing cracks opening and new tension cracks developing close to the crest	Cracking extends well behind the crest; open cracks; obvious differential settlement and/or development of steps or grabens on the platform	Rapidly coalescing, arcuate cracks that extend well behind the crest; deep, open tension cracks and/or grabens; platform not trafficable
Crest, face and toe condition	Normal toe, face and crest geometry; no toe spreading or heaving, face bulging, crest over-steepening or excessive ravelling	Minor changes in dump geometry (minor toe spreading/heaving, face bulging, crest over-steepening); minor face and crest ravelling	Pronounced toe spreading/heaving and/or face bulging; substantial over-steepening of crest; small sliver failures; active face and crest ravelling/breakback	Extensive toe spreading/heaving and/or face bulging; active face and crest slumping/ravelling, frequent sliver failures; active instability
Deformation rate	<X mm/h; extensometer monitoring frequency every A h; dump advance rate <D m/d	Rate of movement >X mm/h and <Y mm/h, rate increasing; extensometer monitoring frequency every B h; dump advance rate <E m/d	Rate of movement >Y mm/h and <Z mm/h, rate increasing; extensometer monitoring frequency every C h	Rate of movement >Z mm/h, rate increasing
Responses				
Operations control	Normal operational controls	Contact mine geotechnical engineer; notify all affected personnel of yellow alarm level	Contact mine geotechnical engineer; notify all affected personnel of orange alarm level	Contact emergency response committee and mine geotechnical engineer; notify all affected personnel of red alarm level
Mine workers	Continue with normal duties	Become familiar with location and potential change in waste dump condition during shift; report significant changes in conditions to shift supervisor	Elevate level of awareness; visually observe and monitor waste dump and platform conditions; report significant changes in conditions to supervisor	Comply with emergency evacuation procedures and remain outside waste dump and potential runout area
Geotechnical engineer	Routine inspections and monitoring	Visually inspect area; determine frequency of inspections, monitoring and need to reduce loading rate; notify management of any recommended changes; communicate with mine workers and advise on the location, nature and expected conditions associated with a potential failure	Evaluate the monitoring data and provide recommendation for TARP level advance; determine frequency of inspections, increased monitoring and mitigative measures; notify management of any recommended actions	Inspect, investigate and formulate recovery plan (risk assessment required); report findings to management
Shift supervisor	Continue with normal operations	Monitor conditions throughout shift; report any noticeable change in conditions to the geotechnical engineer; consult with geotechnical engineer on need to reduce loading rate; report any change of conditions or change in TARP level to next shift	Close dump, idle trucks or redirect to alternative dumps; communicate with workforce that orange level has been reached; closely monitor conditions throughout shift; report any noticeable change in conditions to geotechnical engineer; report any change of conditions or change in TARP level to next shift	Communicate with workforce that a red level has been reached and withdraw personnel and equipment to a safe location; secure to prevent entry; inspect area from outside the instability zone and report to mine superintendent/mine manager; implement recovery plan once formulated (risk assessment required); inspect area from outside the instability zone and report to mine manager; implement recovery plan once formulated (risk assessment required)
Mine superintendent	Monitor production	Monitor production; communicate with geotechnical engineer	Liaise with shift supervisor; assess situation and inspect as required; communicate with geotechnical engineer; notify mine manager of situation as appropriate	Inspect area from outside the instability zone and report to mine manager; implement recovery plan once formulated (risk assessment required)
Mine manager	Normal operations	Monitor situation as required	Prepare to evacuate affected areas; monitor situation as required	Evacuate affected areas; approve recovery plan; notify corporate, mine inspectors and emergency services; monitor situation as required

Source: Hawley and Cunning (2017).

Table 9.3: Suggested simple wireline movement thresholds for waste dumps and stockpiles with no performance history

Waste dump and stockpile hazard class (WHC)	Alert level							
	Normal		Caution		Warning		Close	
	Monitoring frequency (h)	Movement threshold (cm/h; *cm/d*)	Monitoring frequency (h)	Movement threshold (cm/h; *cm/d*)	Monitoring frequency (h)	Movement threshold (cm/h; *cm/d*)	Monitoring frequency (h)	Movement threshold (cm/h; *cm/d*)
I	Daily visual monitoring only		12	3.0 *(72)*	6	4.0 *(96)*	4	5.0 *(120)*
II	12	1.5 *(36)*	6	2.0 *(48)*	4	3.0 *(72)*	2	4.0 *(96)*
III	6	1.0 *(24)*	4	1.5 *(36)*	2	2.0 *(48)*	2	3.0 *(72)*
IV	4	0.75 *(18)*	2	1.0 *(24)*	2	1.5 *(36)*	1	2.0 *(48)*
V	2	0.5 *(12)*	2	0.75 *(18)*	2	1.0 *(24)*	1	1.5 *(36)*

Notes:
1. Movement rates should be above the threshold on two consecutive readings before advancing to a higher alert level and below the threshold on four consecutive readings before dropping to a lower alert level.
2. Closed dumps should not be reopened until the rate has dropped to the Caution level for at least four consecutive readings.
3. Thresholds are for simple wireline extensometer monitoring (manual or automatic/telemetered systems) without intermediate supports and with the wireline positioned in the middle of the portion of the crest that is being advanced.
4. Thresholds are intended as a starting point for facilities that have no performance history and should be adapted and modified to suit site conditions and as experience is gained.

Source: Hawley and Cunning (2017).

example describes how practitioners at Debswana Diamond Co.'s Jwaneng mine, located in south central Botswana, developed a comprehensive system for monitoring the potential impact of open pit mining on a critical component of the ore processing facility.

The Cut 8 pushback at Jwaneng was designed to deepen the pit to 640 m below the original ground surface. The pit crest of the pushback is ~180 m from the edge of the scrubbing plant, which is part of the critical main treatment plant. Numerical analysis was conducted using the distinct element code UDEC to assess how mining might impact the plant foundation. The simulations provided estimates of the surface tensile strain and the limiting tensile strain, which were interpreted using damage criteria (Burland *et al.* 2001). Results from the horizontal strain and the limiting tensile strain criteria indicated that at a distance of ~180 m from the pit crest, the horizontal strain is at 0.13%, implying that minor damage would be expected to plant infrastructure (Lorig and Alvarez 2012). This translates to the formation of cracks without any damage to the integrity of the structures.

To validate the results of the numerical analysis, the operation designed an integrated deformation monitoring system to achieve the following:

- provide deformation measurements in the slope that can be used to validate and calibrate the numerical models used for large-scale slope stability;
- provide notification of deep-seated deformations within the slope that might indicate large-scale instability;
- provide surface deformation measurements between the pit crest and processing plant; and
- provide upfront warning of areas in the plant that are deforming in response to mining.

System components are classified according to utilisation for monitoring slopes, surface subsidence, infrastructure and subsurface deformation. Figure 9.5 illustrates the system classification. The following section describes how these systems were used in the program.

9.5.2.2 *Monitoring system components*

The 11 different monitoring systems in use at Jwaneng can be conveniently divided into three categories: remote sensing, geotechnical and geodetic monitoring. Table 9.4 shows the full list. Geodetic systems are managed by the mine surveyors while the rest are directly managed by the geotechnical team. The following sections describe each system and how it was used in the integrated monitoring of the mine facilities.

9.5.2.3 *Geodetic monitoring system*

The geodetic monitoring system used at Jwaneng includes prism monitoring using RTS, GNSS, Continuous Operating Reference Stations (CORS) and laser scanning, as well as precise levelling. Jwaneng has a surveyor

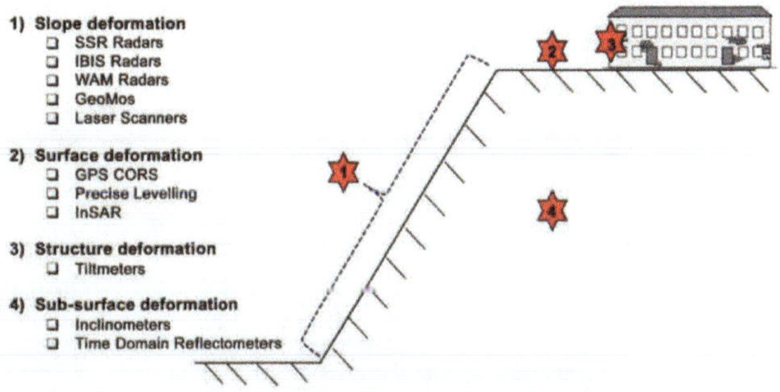

Fig. 9.5: Summary of the deformation monitoring system at Jwaneng, categorised according to monitoring purpose (source: Debswana, Jwaneng mine).

Table 9.4: List of the monitoring systems currently in use at Jwaneng

Instrument	Category
GB radar (RAR)	Remote sensing
GB radar (SAR)	Remote sensing
Satellite radar (InSAR)	Remote sensing
Work area monitor (WAM)	Remote sensing
Tiltmeter	Geotechnical
Inclinometer	Geotechnical
Time domain reflectometer (TDR)	Geotechnical
Prism monitoring	Geodetic
Laser scanner	Geodetic
GNSS-CORS	Geodetic
Precise levelling	Geodetic

Source: Debswana, Jwaneng mine.

dedicated to slope monitoring, who is responsible for the setup and operation of the geodetic monitoring systems. GNSS-CORS is a network of ground-based CORS using GNSS receivers.

Survey control. The most critical component of any survey network is survey control. The Jwaneng survey control network has evolved with mine development. Over time, a reduction in the number of control beacons occurred as they were mined out by pit expansion, including the Cut 8 pushback. The remaining control network was problematic as most of the remaining survey control beacons were located on the west side of the pit, resulting in poor survey geometry.

As a result, a project to improve survey control for monitoring and other general survey requirements was undertaken before mining Cut 8. Twenty-four additional survey beacons were installed around the mine. The beacons were constructed on piled ground, drilling through a 20 m-thick Kalahari Sand layer until reaching hard rock, as shown in Fig. 9.6. This was necessary to ensure that the control network was not affected by ground settlement in the soft Kalahari Sand. The network was surveyed using GPS static surveys, and all data were post-processed. The vertical control was fixed by differential levelling. Upon completion of the expansion of the survey control, each monitoring station had at least four reference stations with significantly improved geometry.

Prism monitoring. Prism monitoring of the pit walls is achieved with two total stations that provide total pit coverage. The monitoring prisms are installed by drilling at least 2 m below ground surface and inserting a steel

Fig. 9.6: Construction cycle of a survey beacon at Jwaneng (source: Debswana, Jwaneng mine).

Fig. 9.7: A typical prism setup at Jwaneng and (right) a total station in the shelter (source: Debswana, Jwaneng mine).

pipe. Targets are generally installed at a spacing of 50 m horizontally and three benches vertically, providing 40–50 m spacing. The total stations are solar-powered. A meteo-sensor housed in a Stevenson screen shelter is connected at each station for atmospheric correction. The instruments are housed in a shelter (Fig. 9.7) for protection from dust, water and other harsh conditions. The shelter has a blast protection cage with flaps which are closed during blasting to protect the front glass panels and the instrument from fly rock.

The operating software is installed in a computer located in the office away from the open pit. The open pit is divided into prioritised zones organised into target groups. The monitoring cycles are set according to priority hierarchy. High-priority groups include the reference group (control stations) and special groups. The reference group is measured before and at the end of every cycle measurement. Special groups are targets located in areas suspected to be unstable as well as in areas declared unstable, which are measured more frequently. The monitoring plan is then sent to the total stations in the pit as commands through a radio connected to the office computer.

The radio at the office sends commands to all radios in the pit through the repeater radio which has LOS to all radios, as shown in Fig. 9.8. The radios at the total stations then send measurements back to the computer in the office, on which the operating software is running. From the office computer, the data go onto a server on the mine local area network (LAN) through a network switch. All data and operating software licences reside on the server. Software on the office computer carries out data processing and calculates the movement by comparing the new data with the null measurements for each measured target. Data visualisation and analysis are carried out using the software analyser which runs on the office computer. Geotechnical and survey personnel who need to visualise the data do so by remotely logging in to the office computer.

Due to the number of systems using wireless communication in the pit, such as the drill navigation system, the truck-mounted fatigue monitoring system, the dispatch system and others, radio interference in the pit is challenging. Interference at times impacts the ability of the total stations to record measurements. At the time of writing, an initiative to eliminate interference across Jwaneng is underway, through the installation of a single network for all communication devices in the pit. Once this has been implemented, the total stations will use the new pit-wide radio network for all communication.

Loss of monitoring targets on open pit slopes commonly occurs as a consequence of berm and ramp reconstruction. Dusty pit conditions degrade the quality of measurements and result in increased data noise. To deal with dust, all prisms showing low visibility (low target measurement success rate) are cleaned.

GNSS-CORS. In November 2012, Jwaneng commissioned a GNSS-CORS network with the aim of monitoring ground movement around the main treatment plant. This network consists of six reference stations which collect raw data continuously, with provision for processing. They are all solar-powered with battery backup. GNSS data processing software is installed on the same computer as software for processing prism data from the

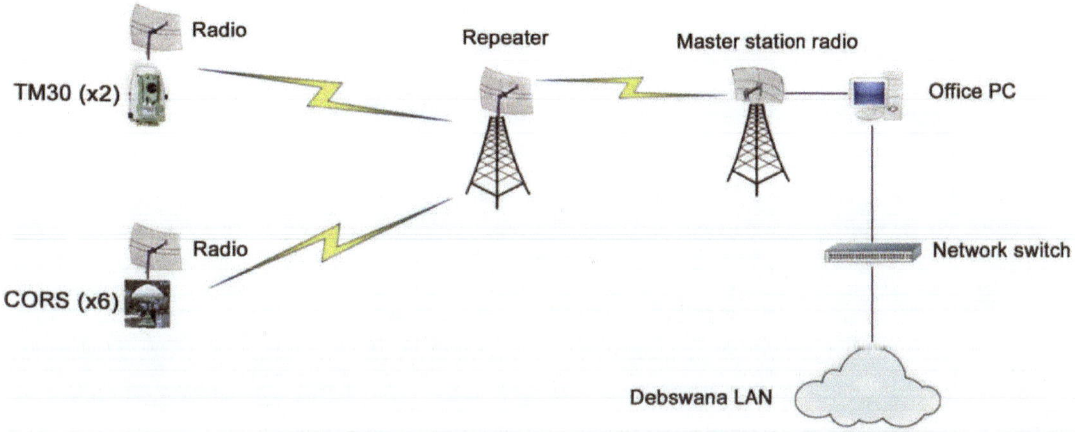

Fig. 9.8: Communication network setup for the prism monitoring system (source: Debswana, Jwaneng mine).

Fig. 9.9: CORS installation at the Jwaneng mine (source: Debswana, Jwaneng mine).

total stations. Raw data from the GNSS-CORS stations are sent to the office computer through a radio network used by the total station radio network (Fig. 9.8). Data from the reference station are sent to the master radio connected to the office computer through the repeater radio. Post-processing of raw data from the stations is done with GNSS operating software according to the positioning products every 10 min, 1 h, 6 h, 12 h and 24 h. The computed position data are uploaded into the same software used to operate data generated by the total stations. An example of a CORS installation is shown in Fig. 9.9.

Laser scanning. Laser scanning is utilised to monitor selected areas, including a meshed highwall in the open pit, shown in Fig. 9.10. The laser scanner utilised at Jwaneng is a high-speed and high-resolution 3D laser long-range scanner that uses the time-of-flight (TOF) method. Inside the scanner is a TOF laser range finder, which calculates the distance to the target by timing the round-trip time of a pulse of light (Laserscanning Europe GmbH 2015).

Six targets have been set specifically for scanner orientation. These targets are fixed, are of known coordinates and are measured before scanning of any area of interest. The measurements are used to re-orient the scanner as well as to calculate an atmospheric range correction factor, which is then applied to the measurements. These targets are well-distributed around the pit for favourable geometry. The laser scanner and the computer which controls the scanner and processes the data are housed at the scanner station in the pit. Communication from the scanner station to the server is by LAN. All data are stored in the server and available for analysis on demand.

The main challenge is dust, which can reduce the target's reflectivity. Hence, some targets may be missed in measurement correction calculations and can otherwise impact the quality of measurements.

Fig. 9.10: The laser scanner and the Cut 7 meshed highwall (source: Debswana, Jwaneng mine).

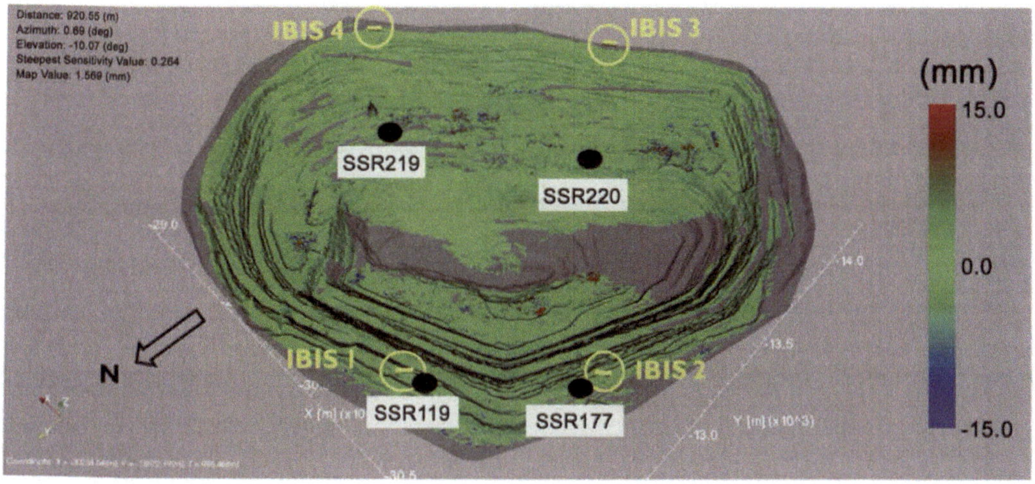

Fig. 9.11: Locations and monitoring coverage for the GB radars in the Jwaneng open pit (source: Debswana, Jwaneng mine).

Precise levelling. Differential levelling refers to any method of using a level to directly measure the difference in elevation between two or more points. Precise levelling is an accurate method of differential levelling, which uses highly accurate levels and a more rigorous observing procedure. It aims to achieve high orders of accuracy in levelling (World Hydrological Cycle Observation System 2004). The main aim of precise levelling is to monitor subsidence around the process plant area.

A total of 32 survey benchmarks are installed around the plant area. These benchmarks are monitored through precise levelling on a quarterly basis. A baseline levelling run was established, and subsequent levelling runs are conducted and compared to the baseline levelling run. The precise levelling job is currently outsourced to a survey consulting company which carries out the levelling and hands over the data to the Jwaneng mine surveyors. After data quality checks, the Jwaneng surveyors then pass the data to the geotechnical team for analysis and interpretation.

9.5.2.4 Remote sensing instrumentation

Jwaneng utilises four remote sensing systems, namely ground-based RAR and SAR radars (GB radar) and satellite radar (InSAR). The GB radars are used exclusively for monitoring the pit walls while InSAR covers the entire mine lease area. The Jwaneng Geotechnical Section is responsible for open pit slope monitoring; all slopes are monitored by one or more radar systems. Locations of the GB radars are shown in Fig. 9.11.

Data communications architecture. The communications network for the RAR radars at Jwaneng mine is illustrated in Fig. 9.12. Monitoring is linked to the mine LAN via a

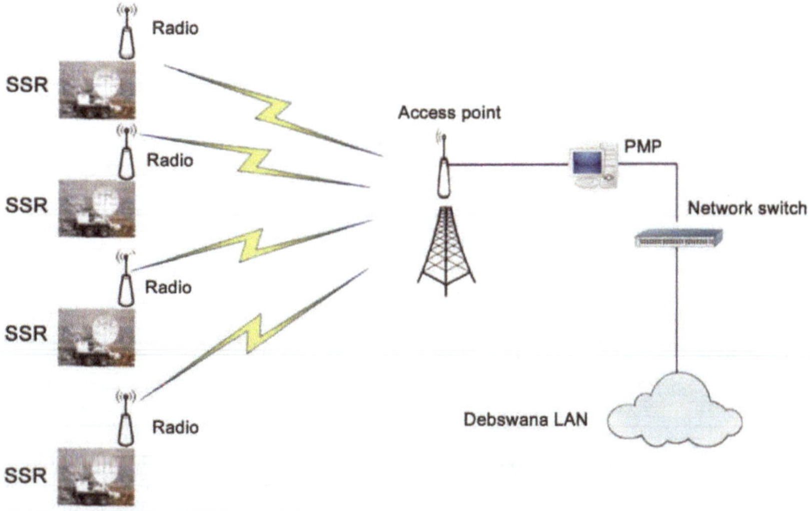

Fig. 9.12: Communication network setup for the four RAR radars at Jwaneng mine. The PMP is the computer on which real-time monitoring takes place (source: Debswana, Jwaneng mine).

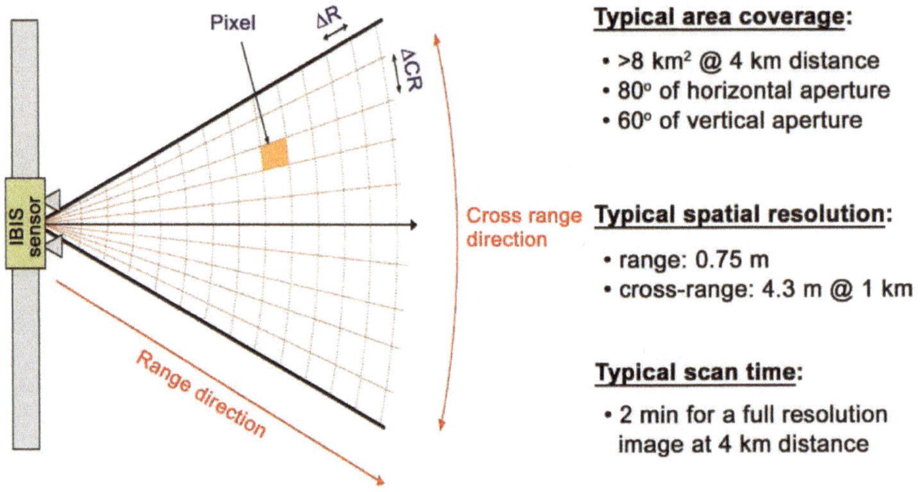

Fig. 9.13: Typical area coverage, spatial resolution and scan time for the SAR radars (source: IDS IBIS training manual).

network switch which enables geotechnical staff to remotely access the radar systems. Real-time monitoring is facilitated from a computer, called the primary monitoring point (PMP), located in the central control room.

GB radar (RAR and SAR). At a distance of 1.8 km, the width of the open pit created a challenge for monitoring with the RAR. This range to target resulted in pixel sizes of ~16 m × 16 m, which meant that the RAR units in service could no longer effectively monitor bench-scale stability. The long range also increased the influence of atmospheric effects on data accuracy. To address these issues, two RAR units were moved into the pit to reduce the range, thereby reducing the pixel sizes and the atmospheric artefacts. The pixel size for the relocated in-pit RAR units reduced to ~5 m × 5 m, making the system suitable for safety-critical monitoring (Bellett and Noon 2013). System health and data quality were checked at the end of each 8 h shift using a standard checklist developed by the manufacturer. Once enough space was available in the pit, the other two RAR units were moved into the pit as well.

Four SAR units were utilised by the operation for long-range monitoring in positions indicated in Fig. 9.11. At 1.8 km, the range resolution (Fig. 9.13) was 0.5 m and the cross-range resolution was ~7.7 m. The higher resolution available from the SAR units enabled the detection of bench-scale instabilities.

The SAR monitoring software allows the output from all four radars to be viewed on a single screen. The purpose of the SAR system is to enable long-term monitoring from permanent positions that provide communication to mine dispatch via radio, as shown in Fig. 9.14. As in the case of the RAR system, a network switch is used to isolate the SAR units from the mine LAN. However, all monitoring data are backed up to the network server. The SAR units are powered from the mains line

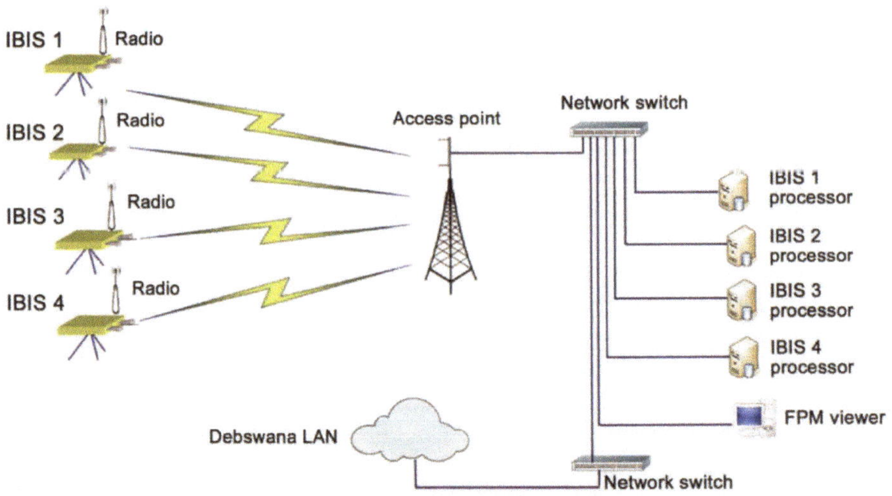

Fig. 9.14: Communication network setup for the four SAR radars at Jwaneng. The FPM is the computer on which real-time monitoring takes place (source: Debwana, Jwaneng mine).

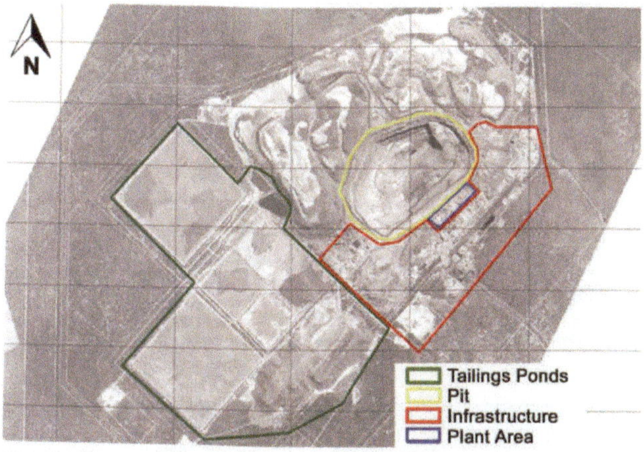

Fig. 9.15: Jwaneng's area of interest for site-wide satellite radar monitoring (source: Debswana, Jwaneng mine).

with standby battery packs that can power the system for ~6 h in the event of a power failure. Most of the technical challenges with the SAR system are related to the software or communication network. Due to the small number of moving parts, mechanical breakdowns rarely occur.

In addition to the SAR and RAR systems, two work area monitor radars (WAM) were deployed. The WAM system is a small truck-mounted RAR that is used for safety-critical monitoring in high-risk, active mine development areas. These systems have been designed specifically for ease of use with the output being a local area alarm to warn workers of slope movement in their vicinity (Noon *et al.* 2011). The WAM system is utilised as an easily deployed and mobile short-term, short-range safety tool.

Satellite radar for site-wide monitoring. Jwaneng utilises satellite radar (InSAR) for periodic site-wide data collection. Data from InSAR are utilised by the operation in both orbital passes, with a ground resolution of ~1 m. Processing of satellite radar data is carried out by an external consulting company. The area of interest, including the open pit, tailings facility, process plant and other infrastructure, is shown in Fig. 9.15. Interpreted reports and displacement data files are generated by the service provider on a quarterly basis.

9.5.2.5 Geotechnical sensors

The geotechnical sensors utilised at Jwaneng to monitor the pit and adjacent plant area include tiltmeters, inclinometers, piezometers and TDR cables. The locations of these sensors are shown in Fig. 9.16.

Data are continuously collected from the geotechnical sensors by using dataloggers. Communication between the dataloggers and a computer in the mine office is facilitated by GSM/GPRS modems. The gateway between the GPRS mobile network and the monitoring stations is established via a private access point name (APN) provided by a local cellular service provider. This enables the monitoring

Fig. 9.16: Aerial image of the Jwaneng open pit and process plant area showing the locations of tiltmeters (green), inclinometers (red) and TDRs (blue). Note the concentration of instruments in the plant area (source: Debswana, Jwaneng mine).

stations to remotely exchange data with a server located in the mine office. A schematic of the basic communication layout is shown in Fig. 9.17. Data acquisition software is used for receiving data from the dataloggers and for sending instructions to the dataloggers from the computer. This setup enables the user to view the data in real-time.

The entire setup is powered by a 12 V valve-regulated sealed lead acid (gel) battery which in turn is charged by a 20 W solar panel. Periodically, the solar panels need to be cleaned to avoid accumulation of dust which affects the battery charging capability. Otherwise, the field setup is virtually maintenance-free.

Tiltmeters. Eight uniaxial tiltmeters are installed on critical infrastructure in and around the process plant to monitor any tilt induced on buildings in response to mining activity. The tiltmeter type contains a micro electromechanical system (MEMS) sensor which offers

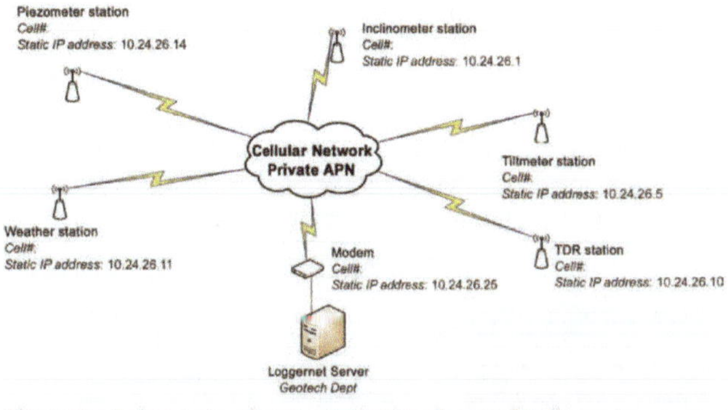

Fig. 9.17: Schematic of communication layout for the geotechnical sensors (source: Debswana, Jwaneng mine).

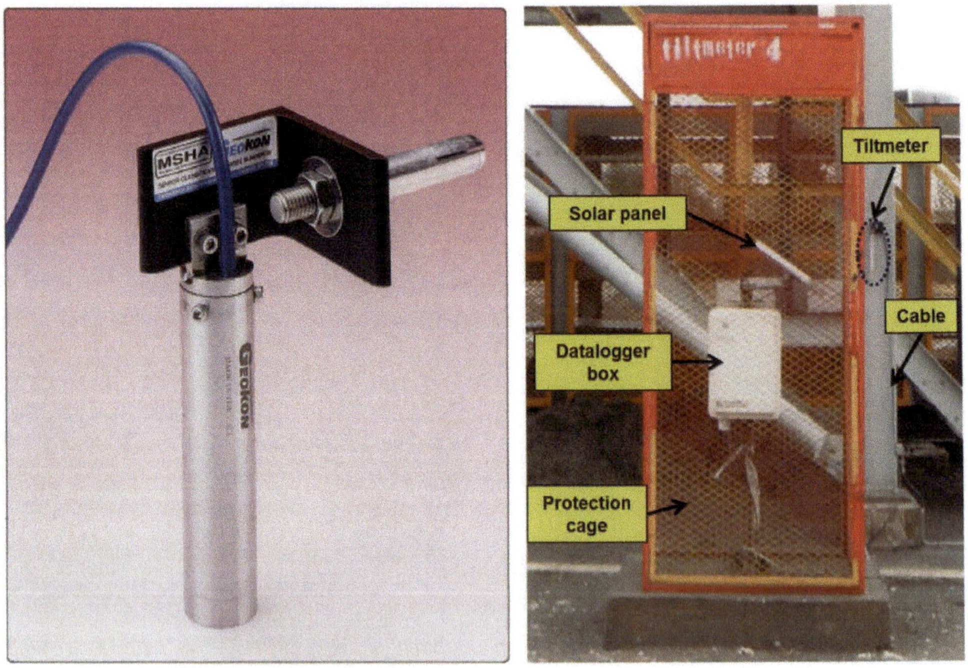

Fig. 9.18: (Left) Uniaxial MEMS tiltmeter with mounting bracket assembly. (Right) A typical installation at Jwaneng, showing the tiltmeter and datalogger box (source: Mhutsiwa *et al.* 2013).

high sensitivity and accuracy (Geokon 2014). Figure 9.18 shows a close-up of the uniaxial tiltmeter type used and a typical tiltmeter setup at the mine.

Tiltmeters are attached to structures by means of an adjustable bracket, allowing for simple and rapid installation. The data are presented as a plot of tilt (°) *v.* time. A TARP has been developed for the tiltmeters, based on the recommendations of the Australian Mine Subsidence Board (Anon 1975) and (Hartman 1992). The alarm trigger levels are shown in Table 9.5. The green threshold is basically the background noise level, so any reading exceeding this level indicates a possibility of real deformation of the structure. Orange alarms trigger detailed investigations by the geotechnical team, and red alarms result in the engagement of the engineering department to carry out analyses of the integrity of the structure. Tiltmeter data are typically analysed on a monthly basis and included in the monthly infrastructure stability report.

Inclinometers and piezometers. Eight fixed in-place inclinometers (IPI) are installed in the plant area in locations shown in Fig. 9.16. The IPIs are installed in 100 m deep, 135 mm diameter boreholes to accommodate the cables from all tilt sensors. The IPI consists of a string of MEMS tilt sensors mounted on lengths (3 m, in the case of Jwaneng) of stainless steel tubing which are linked by universal joints (Geokon 2015). The string of sensors is installed inside the casing, with all sensor cables passing to the surface where they are connected to dataloggers. The IPIs are read automatically, providing continuous monitoring.

Installing an inclinometer in a 100 m deep hole proved to be challenging. In some cases, grout would seep into the casing through the joints. Investigation with a borehole camera during installation showed that casing separation had taken place, resulting in the grout seeping through (Fig. 9.19). This required a modification to the installation method; so, instead of grouting, the inclinometer boreholes were packed with 6.7 mm concrete pebbles. VWPs were also installed in the inclinometer holes by tying them to the outside of the inclinometer casing then grouting them in place.

Table 9.5: Alarm threshold levels for the Jwaneng tiltmeters

Alarm	Threshold (mm/m)	Threshold (°)
Green	2.0	0.1
Orange	7.0	0.4
Red	10.0	0.6

Source: Debswana, Jwaneng mine.

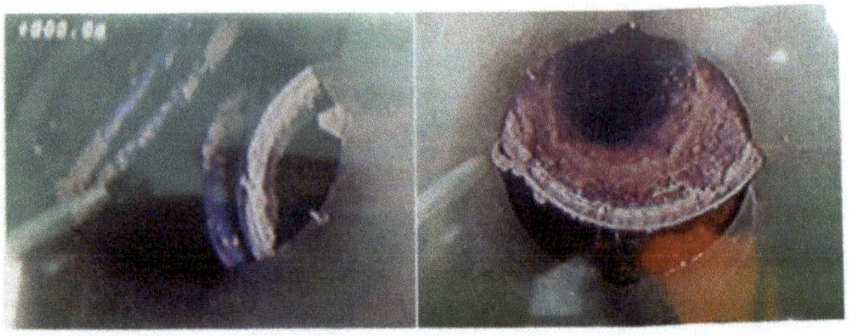

Fig. 9.19: Borehole camera images showing inclinometer casing separation in some of the holes (source: Mhutsiwa *et al.* 2013).

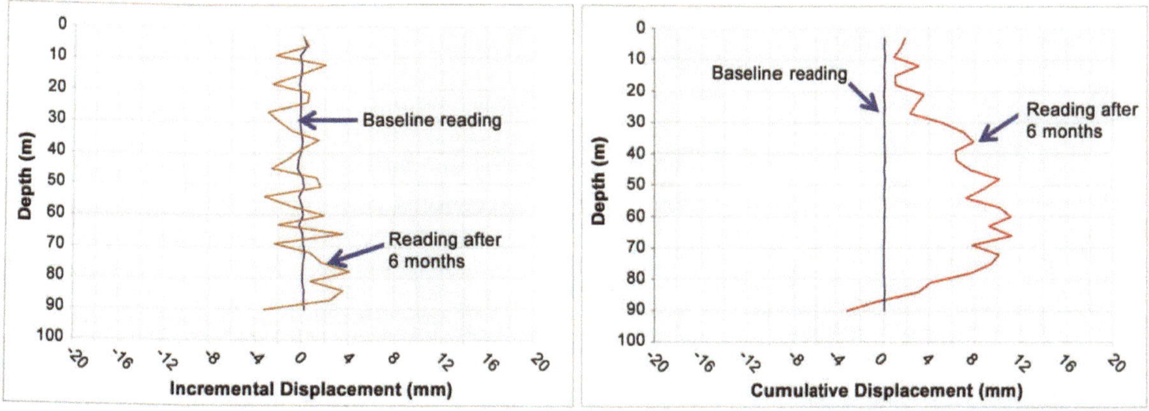

Fig. 9.20: Comparison of inclinometer data using incremental (left side) and cumulative (right side) displacement profiles. The noise is much easier to visualise and interpret in the incremental plots than in the cumulative plots (source: Debswana, Jwaneng mine).

Like the tiltmeters, the IPIs are connected to dataloggers and communicate with the mine office computer via GPRS modems, as shown in Fig. 9.17. Data display and analysis are similar to that done with the tiltmeters. Initially, the inclinometers were configured to display cumulative displacement profiles. The problem with this approach was that, at times, it created the impression that there was significant ground movement when in reality it was just noise from the sensors. To avoid these problems, the inclinometer data display was changed so that incremental displacement was plotted as well. Using this approach, noise is easily distinguishable from real ground movement, as illustrated in Fig. 9.20. Inclinometer data are typically analysed on a monthly basis and included in the process plant stability report. If any trends are picked up, the frequency of data analysis is increased.

TDR and piezometers. Six coaxial cables for TDR monitoring were installed around the pit perimeter at Jwaneng, as indicated in Fig. 9.16. TDR cable lengths range from 225 m to 400 m. Cables were installed with VWPs then fully grouted in the boreholes, similar to the application for inclinometers. The installation of TDRs was relatively easy and straightforward, especially given the use of NQ/NX diameter holes.

The TDRs at Jwaneng are connected to the GPRS network used by the tiltmeters and inclinometers, as described above. Compared to inclinometers, TDRs are much cheaper and easier to construct and can be installed to greater depths (Beck and Kane 1996). A limitation of TDR compared to inclinometers is that TDR only gives an indication of where shear displacement has taken place; it does not give a quantitative or directional measure of the displacement. Furthermore, the coaxial cables used for TDR are only suitable for monitoring localised shearing deformations (i.e. narrow shear zones) (Dowding *et al.* 2003). At Jwaneng, most of the rock mass movement is along bedding planes and as such is highly localised, thus making coaxial TDRs suitable for monitoring the slopes.

9.5.2.6 *Integrated slope monitoring data management and visualisation*

The increased availability and volume of real-time monitoring data bring new challenges in managing the data flow and converting it into information upon which informed engineering judgements can be made. To aid in achieving this, Jwaneng has contracted a geotechnical instrumentation system integrator to support real-time data management and performance monitoring. The system includes a GIS web-based platform which can process any type of data from sensors, dataloggers and files, with real-time acquisition and analysis, and advanced alarm functions (Soldata 2015). This is meant to provide an integrated solution that manages all data from geotechnical instrumentation and can also be used as a centralised information system for ground control, including blasting data, visual inspections and reports.

9.5.3 Case history 2: Golden Sunlight mine

9.5.3.1 *Overview*

Barrick's Golden Sunlight mine is an underground gold mining operation that is located directly behind and below the west highwall of the Mineral Hill open pit, a former surface operation that was developed intermittently from 1983 to 2015. The most recent phase of surface mining in the Mineral Hill pit, shown in Fig. 9.21, was discontinued in April 2015 due to hazardous rockfall conditions and indications of progressive slope instability. The west highwall instability encompasses the entire west slope (466 m in height by 580 m maximum width), with an estimated mass of 42 Mt.

9.5.3.2 *Surface mining and the west highwall*

Surface mining began in 1983 in the Mineral Hill open pit, and production at its height produced over 300 000 ounces of gold per year. After six phases of surface mining and push backs, the pit achieved a depth of 550 m and a diameter of 920 m. Bench heights range from 8 m to 30 m,

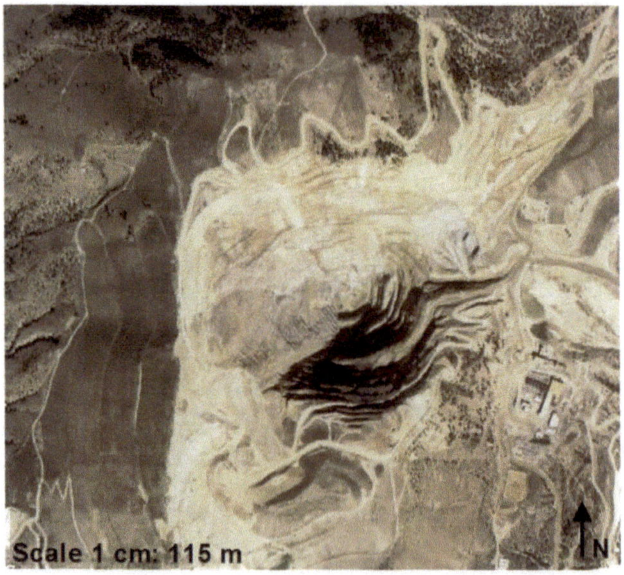

Fig. 9.21: Flyover image of the Golden Sunlight mine and the Mineral Hill open pit from 2014. Note the distortion and colour change over much of the west quadrant of the open pit; north is at the top of the photo (source: R. Turner and R. Wu).

and average bench face and inter-ramp angles are 65° and 44°, respectively.

Experience from mine development indicated that elevated groundwater pore pressure and low-frequency vibrations from blasting and mining contributed to acceleration of the west highwall. As a result, the mine installed horizontal drains to reduce pore pressures and refined blasting procedures, including free-faced shots and pre-splitting, to control vibration energy. Rockfall proved to be an ongoing hazard due to localised bench failures, leading to the installation of ring-nets and energy barriers in 2014 to reduce the exposure of mine personnel working on the pit floor. Slope monitoring during surface mining was conducted with a combination of RTS, GB radar, shape-acceleration arrays (inclinometers), TDR cables and wireline extensometers until surface mining ceased in April 2015.

Instability on the west highwall in the Mineral Hill open pit occurs due to displacement along persistent structures, rather than from failure of the rock mass. Fault zones that cross the west highwall act as release planes for the west highwall instability together with contacts between the breccia pipe and the Belt Supergroup. The Lone Eagle fault zone, Gray fault, Fenner fault and West Shear fault are the primary release features that bound the instability zone to the north, south and at depth, as shown in Fig. 9.22.

The existence of a basal structure called the West Shear fault zone was confirmed with a 2012–2013 drilling campaign. It can be observed underground due to the visible offset of excavations that cross it. The breccia, siltstone, sandstone and shale encountered at the mine typically have rock mass rating (Bieniawski 1976) qualities ranging from 'Fair' to 'Good' (RMR 41–80). Uniaxial compressive strength values range from 53 MPa to 98 MPa. The correlation of instability events with blasting and precipitation indicates that displacement is triggered

Fig. 9.22: The west highwall of the Mineral Hill open pit as it appeared in 2013 with major faults and lithologies. The West Shear fault zone is the basal release structure for the west highwall instability and is only observable in the underground mine (see Section 9.5.3.3) (source: R. Turner and R. Wu).

Fig. 9.23: (Left) Remote mucking is used in all stopes and is particularly important for stopes that daylight into the pit. (Right) Development mining is completed with jumbos and bolters (source: R. Turner and R. Wu).

by decreases in frictional resistance along discontinuity surfaces due to low-frequency vibrations and increased water pressures.

9.5.3.3 Underground mining

Phase 2B of underground mine development (2BUG) began in May 2015. It uses the long hole open stoping method within and below the instability zone of the west highwall. Stopes located in the hangingwall of the West Shear fault zone were daylighted to the ground surface of the highwall by design (see Fig. 9.23), which allowed for the visual observation of stope blasting from the crest of the open pit. Development mining is conducted using drill and blast methods with jumbos, bolters and remote muckers, where typical drift dimensions are 5 m in height by 5 m in width. Drifts and stopes have been mined in the hangingwall and footwall of the West Shear fault zone. Since February 2017, all mine development has been in the footwall, as visualised in Fig. 9.24.

Typical ground support consists of 1.8 m long, 39 mm split-set bolts in the ribs and back. Where spans increase or ground conditions decrease, inflatable rockbolts are used. Shotcrete is reserved for drifts that cross actively displacing faults and is used where thick clay zones cannot be supported. It was found that shotcrete is more useful as a tell-tale of movement along the faults than as primary ground support, because it tends to degrade after 4 mm of displacement. Above the 4550 elevation, drifts that cross into the hangingwall of the West Shear zone have a typical life-span of 6 months until incipient localised failure occurs, despite the addition of extra ground support.

Underground mine development has been adaptive to the hazard posed by the west highwall instability, and it is normal practice to modify designs and schedules based upon proximity to the active hangingwall of the instability. From January 2015 to June 2017, the west highwall experienced up to 15.9 m of displacement, as indicated by RTS monitoring of prisms in Fig. 9.25. In the event that the highwall fails, runout analyses indicate potential for debris runout to impact the portals to the mine. This potential condition dictates that the mine operations require stringent TARP procedures for evacuation and that secondary egresses be in place to prevent entrapment. To respond to potential risk exposure, a third portal was

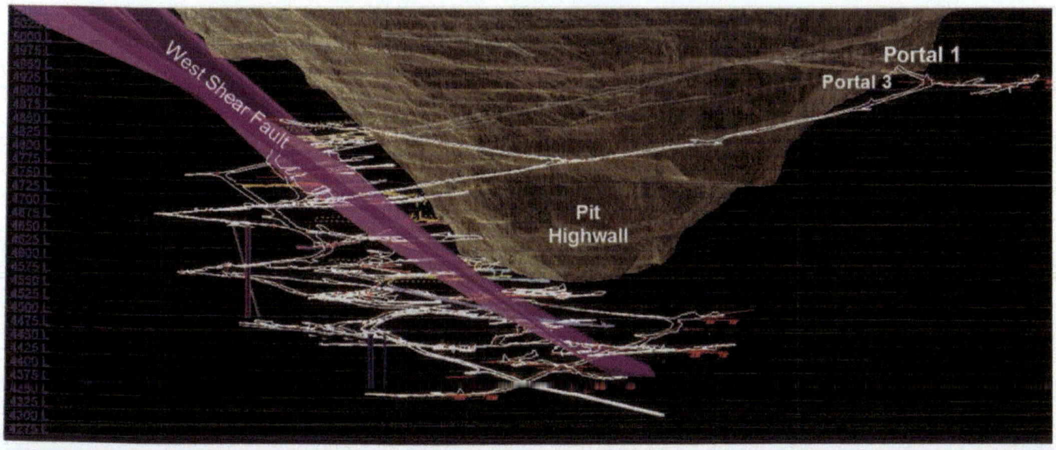

Fig. 9.24: Looking north in this visualisation of the underground mine development above and below the West Shear zone (shown in magenta), and relative to the Mineral Hill open pit shape (source: R. Turner and R. Wu).

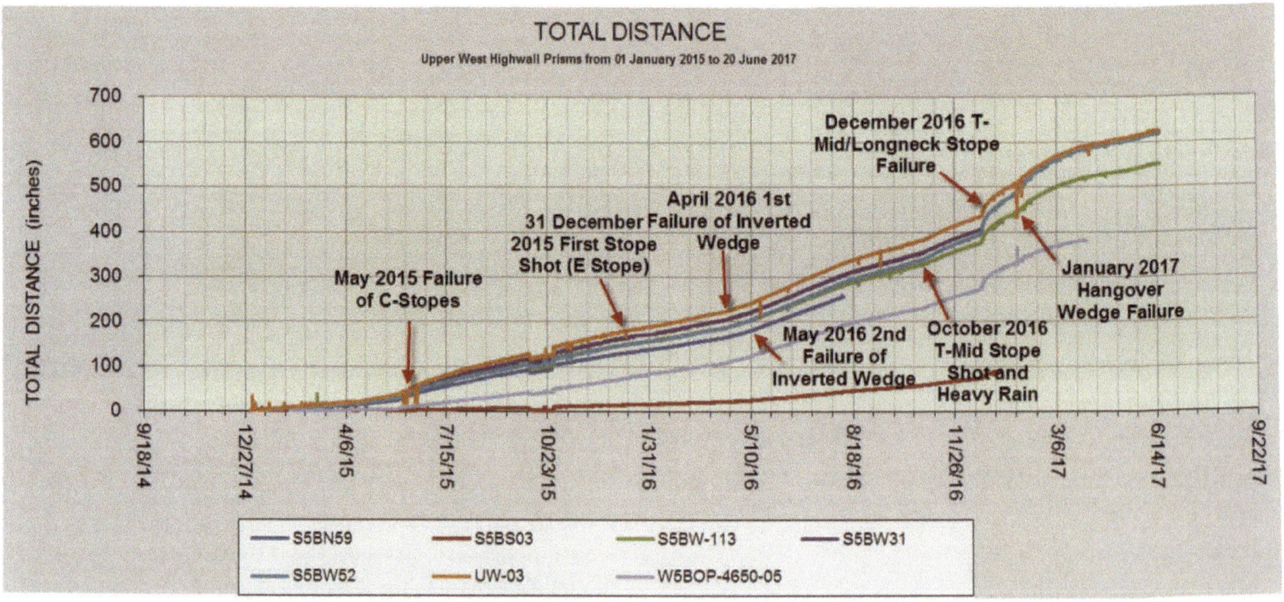

Fig. 9.25: Total vector (3D) prism time-displacement plot for the west highwall from 1 January 2015 to 20 June 2017; over 15.9 m of displacement has occurred in the upper west highwall since the beginning of 2015. Individual events of stope blasting, localised slope failures and strong storm events that triggered the highwall to accelerate are indicated (source: R. Turner and R. Wu).

constructed above the estimated runout zone in 2016 to provide an additional air intake and escape route.

9.5.3.4 *West highwall instability*

A review of engineering reports from the past 20 years indicates that instability and failures in the west highwall of the Mineral Hill pit are recurring events that coincide with active mining. The preceding phase 4, 5 and 5B surface mining stages all experienced localised and multi-bench scale instabilities in the upper west highwall and north-west corner of the pit. As the earlier mining phases reached the Mineral Hill breccia (very strong rock mass), active instabilities appear to have regressed. However, for the most recent phase of surface mining, upon reaching the Mineral Hill breccia, instability progressed rather than regressed.

Based on radar and prism data, as well as from observations of the highwall following production pattern shots, it was apparent that blasting was a trigger for progressive acceleration of the entire west highwall. Note the acceleration trends in Fig. 9.25 that occurred in April and May 2016, associated with a stope blast and failure to

Fig. 9.26: During a stope shot initiated from the 675D level, the inverted wedge pictured here failed. The wedge had moved enough to activate critical alarms on the radar before the blast. This event occurred in two stages from late April to early May 2016, as indicated in Fig. 9.25 (source: R. Turner and R. Wu).

the ground surface shown in Fig. 9.26. Working with the explosives supplier, the operation developed the following blasting techniques to minimise the vibrations directed towards the highwall during stope blasts:

- pre-split blasting against the final highwall using air-decked holes to decouple the highwall from vibration energy from trim blasts;
- buffer patterns and production patterns required at least one free face;
- production patterns were limited to a maximum charge weight and tonnes of rock blasted;
- signature holes were used at each elevation to determine the most ideal vibration frequencies;
- points of initiation for each shot had to be placed where it would pull towards the free face; and
- each shot was designed, simulated, reviewed and signed-off by Barrick and the explosives supplier, filmed and analysed.

After implementing new risk management procedures (described below) and blasting techniques, full production mining resumed in 5BOP.

9.5.3.5 Risk management and risk mitigation

Safe management of a mining operation with the amount of slope movement and rockfall hazard potential at Golden Sunlight requires a GCMP that effectively uses communications, operational procedures, hazard training and engineering controls. Mine personnel are updated daily on highwall conditions via the TARP, daily underground mine geotechnical inspections and daily reports. Stopes that are shot in the hangingwall of the instability cause the slope to temporarily accelerate and behave as an apparent progressive failure, followed by rapid deceleration and a return to normal creep movement. Stope shot re-entry procedures (see Section 9.5.3.10) require monitoring to verify that the slope is in a regressive state before underground mining is allowed to resume. Ground control training focuses on hazard recognition, and miners are engaged underground to address individual problems. Drawdown of pore pressures in the highwall and mine dewatering is facilitated by a drainage drift at the bottom of the underground mine that discharges water via a well to the surface. Drifts that cross major structures will typically deform slowly due to slip along the fault and will eventually collapse due to rock mass failure at the underground opening. Designing these drifts requires dedicated ground control designs based on numerical and probabilistic analyses, monitoring devices that are accessible to miners, and scheduling based upon the limited stand-up time. Stopes are designed using empirical and deterministic methods, and their influence on highwall stability is assessed with numerical modelling. Development and production within the instability zone of the west highwall continue to trigger accelerations of the slope, and ongoing mining is dependent upon real-time monitoring in the surface and underground, predictive analyses, geotechnical designs and procedures for working under hazardous conditions.

Since 2012, the following controls and procedures have been utilised to reduce the risk from the west highwall:

- daily geotechnical inspections;
- on-call 24/7 geotechnical support;
- redundant monitoring for data validation (radar, prism, GPS, etc.);
- use of inverse velocity analyses when progressive conditions exist;
- highwall illumination using stadium lights;
- trained highwall spotters (when progressive conditions exist);
- blasting requires everyone to clear the Mineral Hill open pit;
- blasting in hangingwall requires multiple checks;
- progressive highwall movement conditions require that mining is halted until regressive conditions are confirmed by monitoring and inspection;
- monthly inspections and review by consultants and corporate geotechnical engineers; and
- geotechnical analysis (kinematic, limit equilibrium and numerical).

9.5.3.6 Monitoring and instrumentation

The implementation of surface and underground monitoring systems allows the Golden Sunlight mine to safely operate in the Mineral Hill open pit. Remote total stations, slope monitoring radar, GPS, multi-point borehole extensometers (MPBX), Rock-Its wireline subsidence monitors, sloughmeters, wireline extensometers and instrumented split-sets are used to monitor the west highwall and the hangingwall of the West Shear zone. These instruments are shown in Fig. 9.27. Initially, 100 prisms were installed on the bench faces of the west highwall before the cessation of surface mining in April 2015 and were shot four times every hour by an RTS. Utilising a real-time, fully automated data monitoring platform together with a radio communication system, alarms triggered by the acceleration of individual prisms would be transmitted over the operations radio channel and technical personnel would be contacted to check the alarms.

In the course of underground mine development, rockfall and creeping slope movements have taken a toll on the prisms in the west highwall and it has become too hazardous to replace them. Consequently, the operation relies more upon the slope monitoring radar system and GPS units for monitoring. The GB radar system is located on the east highwall of the pit and scans the west highwall every four minutes. Individual instability regions are

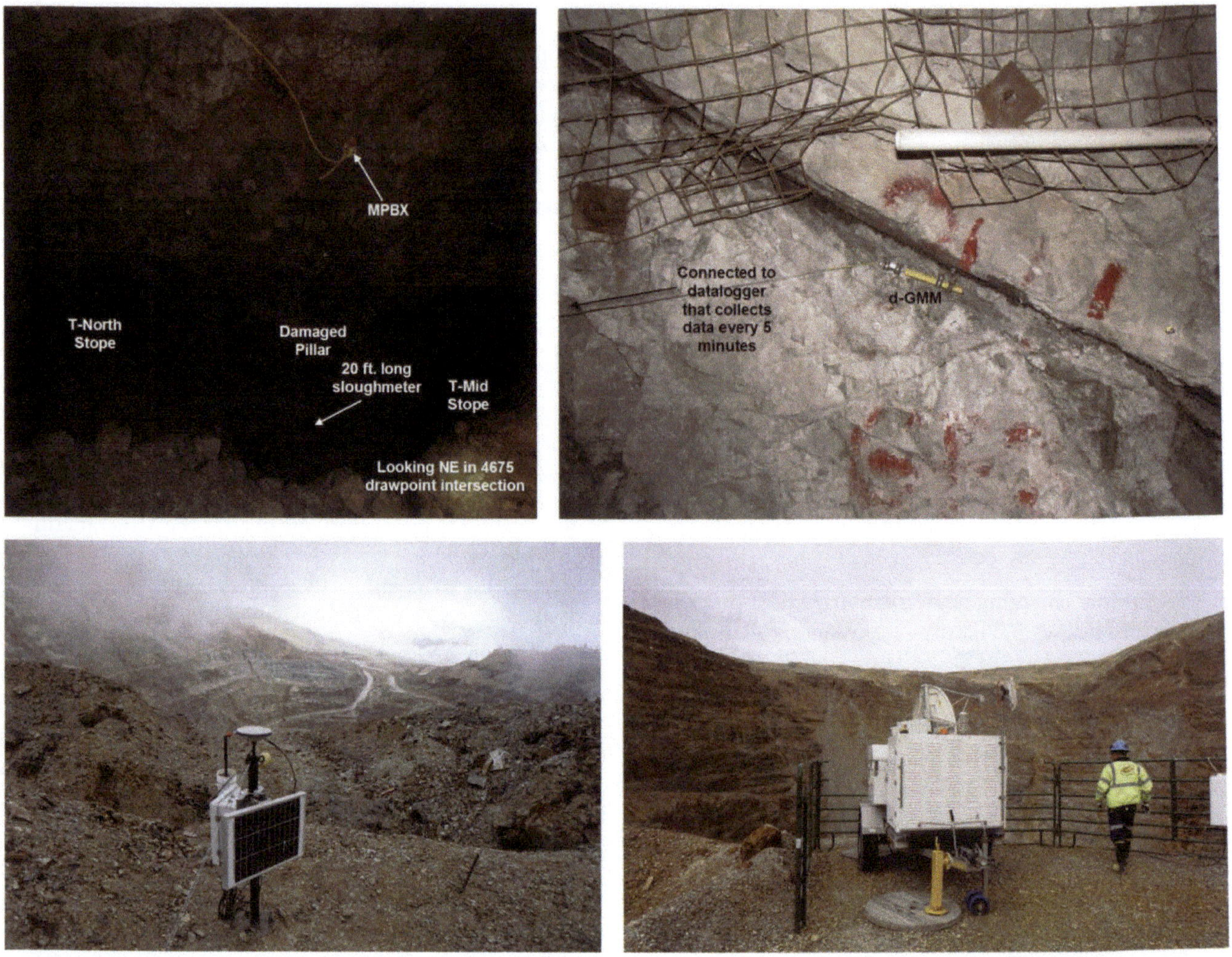

Fig. 9.27: The instruments used to monitor the west highwall on the surface and underground include MPBX and ground movement monitors (top left and top right), GPS units (bottom left) and GB radar (bottom right) (source: R. Turner and R. Wu).

broken out on the radar (shown in Fig. 9.28) so that when critical alarms become active, it is easier to find the location of excessive movement. Breaking out the regions also allows custom alarm settings to be created. These are based upon back-analyses of past failures, the proximity to underground operations, the size of the instability zone and the spacing between points captured by the radar. The radar system plays a crucial role in determining when the west highwall is in a progressive or regressive state of movement and, when the mine is evacuated, the operations and technical groups are able to base re-entry plans on the updated trends. An example is illustrated in Fig. 9.29.

No monitoring system is perfect. Both the total stations and radar systems are susceptible to rapid swings in temperature and humidity, which causes the refractivity index, a measure of the density of the air, to shift rapidly as well. The radar and total stations utilise weather stations and algorithms to correct for changes in refractivity. However, refractivity changes can cause false alarms when apparent slope distance velocities and magnitudes change. Examining refractivity in the time-series can be useful for interpreting whether the radar system is reliably tracking real movement or apparent movement.

9.5.3.7 Integration of open pit and underground mine instrumentation

Supplementing the surface instrumentation, underground instrumentation is deployed in drifts that intersect the West Shear fault zone and in critical facilities and intersections. MPBX and instrumented split-sets are used to detect subsidence in intersections and shearing in pillars adjacent to stopes. Wireline extensometers (Rock-Its) and sloughmeters are instruments that allow mine workers to check displacement in the field without having to download and interpret the data, which gives them the prerogative to barricade the area if a hazard exists. Underground instrumentation located in the hangingwall of the West Shear fault zone typically confirms trends observed from the radar and prism systems, while instruments located in the footwall zone confirm that

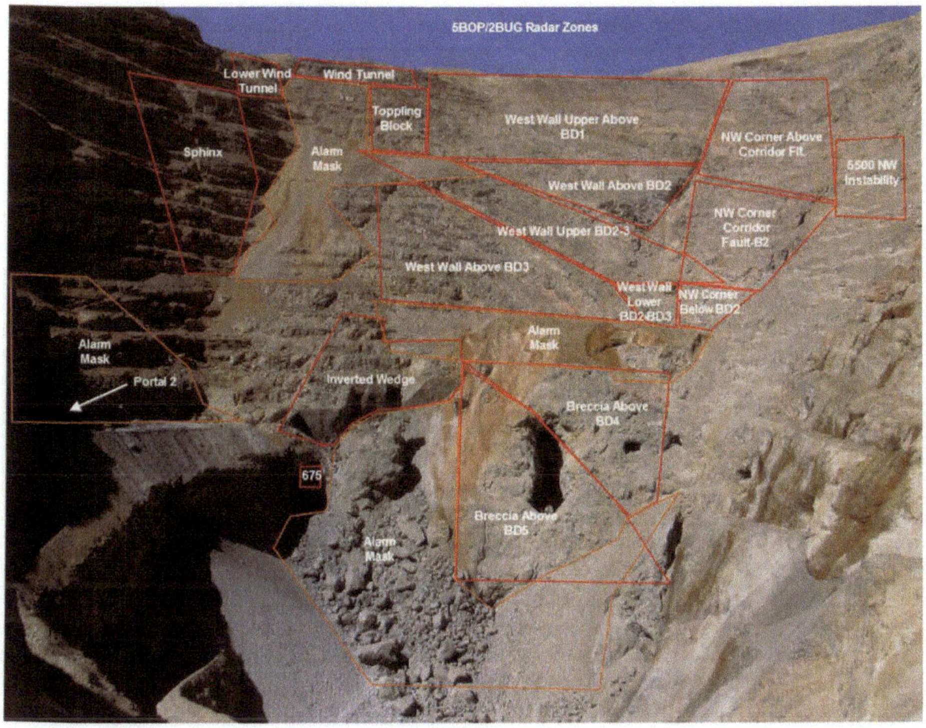

Fig. 9.28: Looking west in the Mineral Hill open pit at the various high threat zones covered by the slope stability radar. Each zone has individually calculated geotechnical and critical alarm settings based upon back-analysed failure events, proximity to underground operations, the instability mechanism and/or the size of instability in the zone (source: R. Turner and R. Wu).

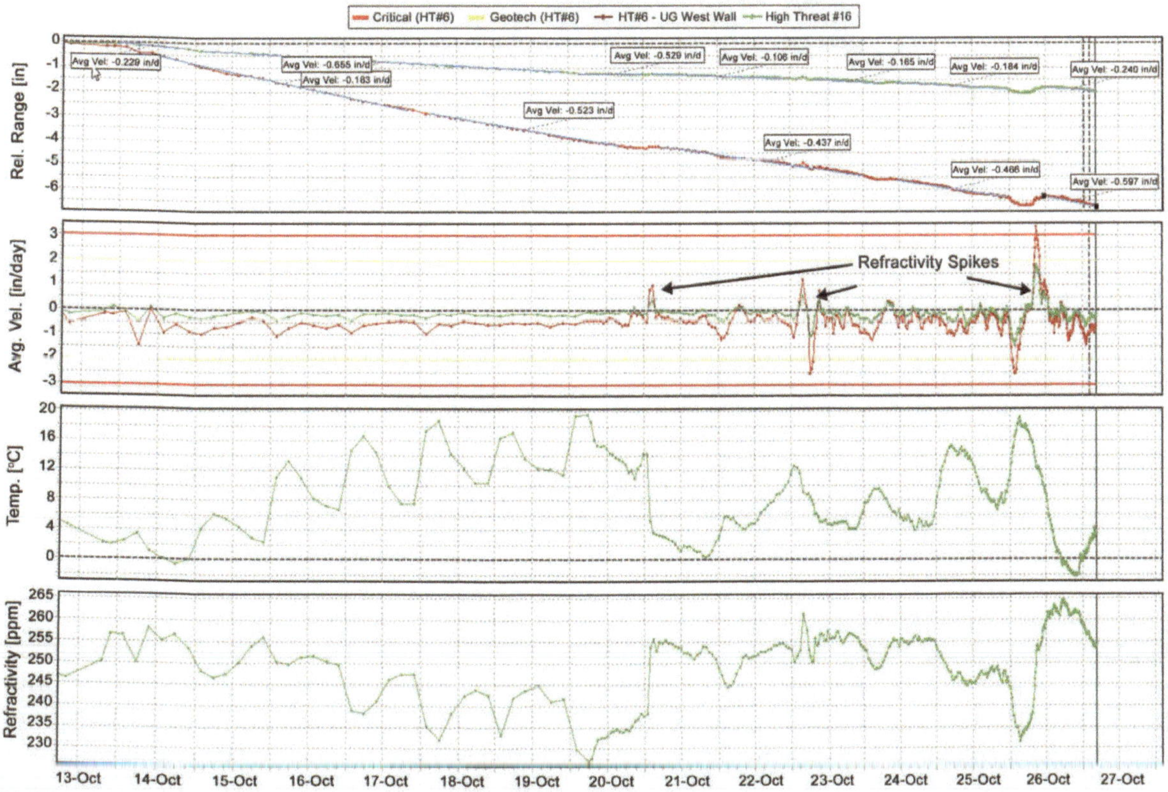

Fig. 9.29: Radar time-series trends from 13–27 October 2016 were influenced by swings in temperature and refractivity, shown in the lower two graphs. Refractivity spikes occur when the air density changes rapidly due to swings in temperature and humidity, such as when a thunderstorm approaches the mine. Rapid changes in air density cause energy from the radar to deflect, despite the use of atmospheric correction; false alarms can then occur (source: R. Turner and R. Wu).

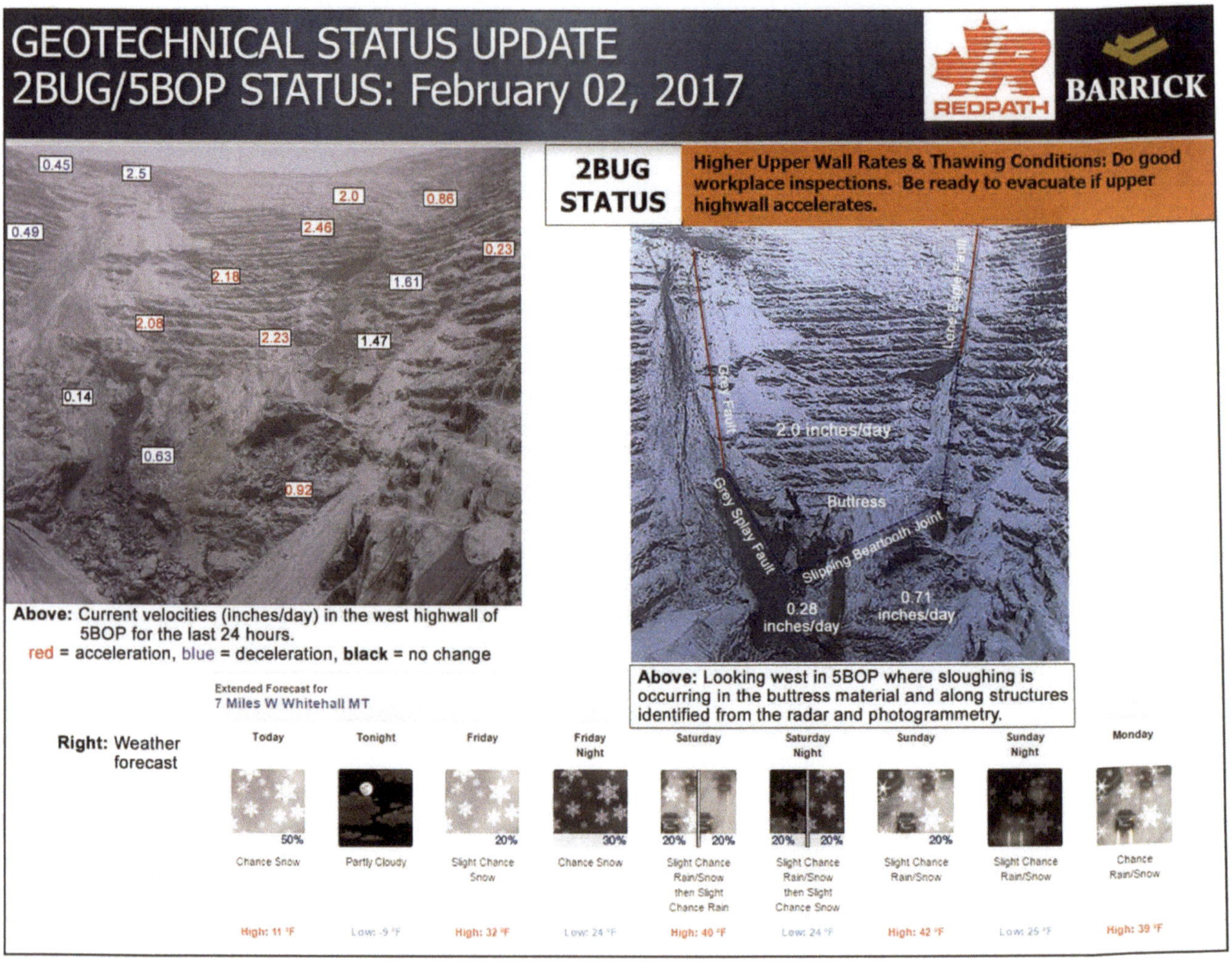

Fig. 9.30: Example of a Daily TARP status report displayed to mine personnel in the lineout office. This document briefs both shifts on changes in highwall condition (source: R. Turner and R. Wu).

only localised movement in the underground excavation is occurring.

Combined, the underground mine instrumentation, the total station and radar system play a crucial role in determining when the west highwall is in progressive or regressive states of movement. When the mine is evacuated, the operations and technical groups are able to base re-entry plans on the monitoring trends.

9.5.3.8 *Communications to mine operators*

Although implementing and maintaining the various monitoring systems is important, it is more crucial to the operation that clear and easily understood interpretations of the monitoring data are provided to all technical and non-technical mine personnel. Daily monitoring reports summarise total station and radar data, and communicate notes and photographs from that day's geotechnical inspection. This report is condensed and issued as a presentation called the Daily TARP update, which is displayed on a large HD monitor in the lineout room for all mine personnel to view during their shift change. A monthly hazard map displays all of the locations in the mine that are of geotechnical concern; this is also displayed in the lineout room. Ground support reviews, numerical modelling results and post-event summaries are submitted in memoranda to key stakeholders. The most important form of communication is the engagement of mine personnel in the field, where engineers and geologists interact with the mine workers and point out hazards in the underground and on the surface, as illustrated in Fig. 9.30.

9.5.3.9 *Trigger Action Response Plan*

From 2012 to 2017, several generations of TARPs evolved for surface and underground operations at the Golden Sunlight mine, based on development experience. A version of the TARP utilised in June 2017 is described in Section 8.5 and shown in Table 8.5. Trigger alarm levels for the TARP include a combination of monitoring trends, inverse velocity projections, weather conditions, physical conditions

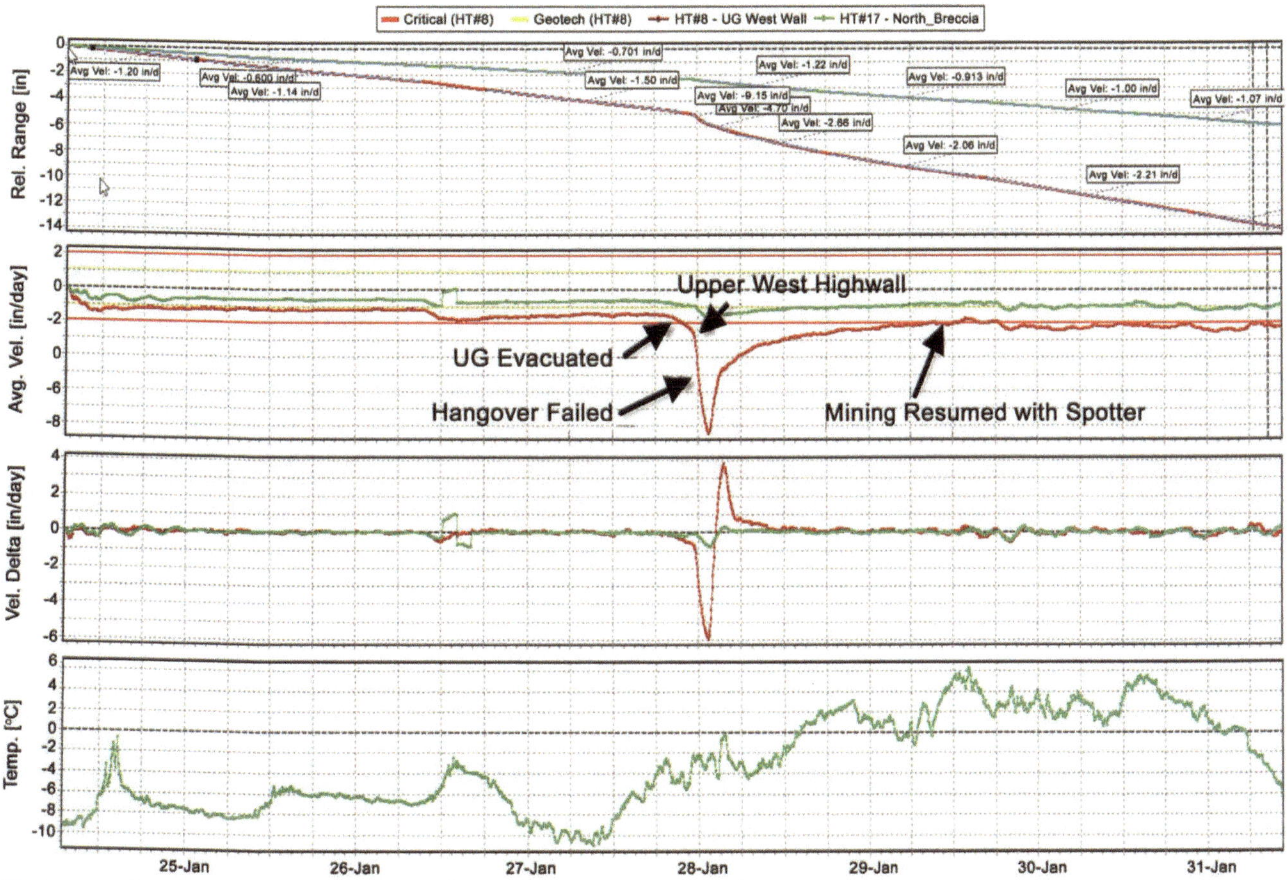

Fig. 9.31: Radar time-series of the west highwall from 24–31 January 2017. When the Hangover Zone above the T-Mid stope failed, the event caused the upper west highwall to accelerate up to 216 mm/d. From the time of the failure event on 28 January until mining resumed on 29 January (~36 h), the highwall moved 115 mm. Mining resumed when the velocity in the upper west highwall dropped below the critical rate (50 mm/d), with restrictions in place (i.e. mining to occur during daylight hours only with a spotter in place) (source: R. Turner and R. Wu).

observed on the surface and underground, and conditions adjacent to the West Shear zone underground. Initial versions of the TARP used trigger alarm levels set to specific velocity thresholds for the west highwall of the Mineral Hill pit. However, this proved ineffective when a failure or hazard event occurred below the threshold. After attempting to update the velocity thresholds each time conditions changed, it became apparent that it made more sense to fit the TARP to progressive or regressive conditions of slope movement (i.e. by setting triggers upon sustained acceleration and deceleration). The TARP is effective for determining the role of mine personnel when highwall conditions are hazardous, and the response is more immediate when the triggers are allowed to evolve with experience.

9.5.3.10 Re-entry plan

Following a sustained period of acceleration of highwall movement, surface and underground operations are halted and the pit is evacuated to beyond the projected point of failure runout. The pit is barricaded to prevent re-entry, and the operations group goes on standby while the technical group collects and interprets monitoring data. Re-entry into the Mineral Hill pit is not considered until sustained deceleration is confirmed by two or more monitoring systems, normally the radar system and the prism network. Deformation trends and the temperature for 24–31 January 2017, measured by the slope stability radar, are shown in Fig. 9.31. A failure/hazard trigger event in an area called the Hangover Zone was experienced on 28 January. After the failure of the Hangover Zone, a re-entry standard operating procedure was implemented, as shown in Fig. 9.32.

9.5.3.11 Lessons learned

Multiple lessons were learned while managing the hazards posed by the west highwall to the surface and underground operations at the Golden Sunlight mine, including:

- use best practice blasting techniques when dealing with hard rock to minimise vibration energy that can otherwise induce accelerated highwall movement;
- budget for core drilling, rock mechanics testing and geological modelling for pre-feasibility investigations that will lead to optimising mine design and reducing risk exposure;

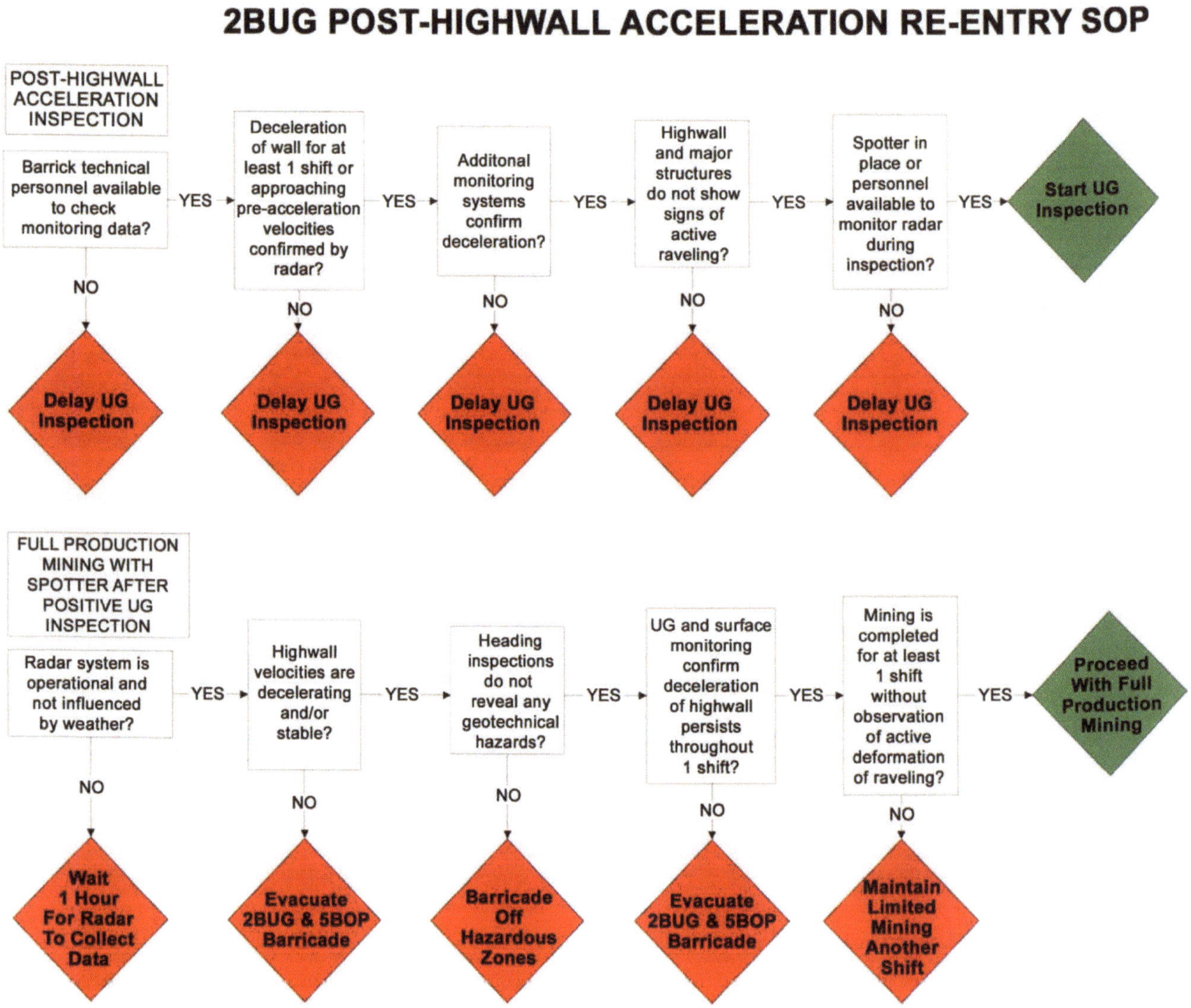

Fig. 9.32: Following the failure of the Hangover Zone in January 2017, a re-entry standard operating procedure was implemented that focused on deceleration rather than fixed velocity thresholds (source: R. Turner and R. Wu).

- complete 3D numerical modelling that is appropriate for assessing potential structurally controlled failure mechanisms (continuum- and discontinuum-based codes);
- thorough mapping and modelling were key to predicting the location of faults and potential release structures (i.e. recognise the value of an accurate structural model);
- budget for a reliable and validation/calibration-oriented monitoring system (e.g. prisms and slope monitoring radar) through all phases of mining that can be used for validating performance monitoring data and for ensuring total vector movement is well-understood;
- utilise monitoring equipment to its fullest extent and work with manufacturers and service providers to continually improve technology and obtain reliable data, not just equipment availability;
- utilise all data for interpretive context (including integrating results of surface and underground mine instrumentation) to assist in predicting ground behaviour, making better decisions when movement exceeded alarm thresholds, interpreting movement mechanisms and recognising apparent data errors;
- organise data and documents for effective communications of factual data and real alarm events;
- train mine personnel to think like geotechnical engineers. This is well worth the time invested –

utilising the trained eyes of the workforce exposed to hazard potential daily is an effective and safe way to execute mine development;

- involving operations personnel with risk management decisions and giving them the power to make the call when the situation is determined to be too hazardous to proceed was essential in getting their buy-in and utilising their extensive experience;
- the use of consulting firms can provide invaluable input based upon practical experience; and
- remember that the role of a geotechnical engineer is to protect every person who enters the open pit and underground mine.

10

MONITORING SYSTEM GUIDANCE, INNOVATION AND FUTURE APPLICATIONS

Isabel Barton, Erik Eberhardt, Ben Haugen, Moe Momayez, Emre Onsel, Brad Ross, Robert Sharon, Doug Stead, Chad Williams

10.1 Introduction

As emphasised throughout these Guidelines, performance monitoring of open pit slopes is essential for ensuring safe and economic open pit development and is a critical part of a mine operation's geotechnical risk management program. This includes assessing site-specific needs required to design, implement and maintain the slope monitoring system, to produce and manage reliable data and to report and act confidently on results in real-time. Emphasis is also placed on the human resources required to manage the monitoring system, including the appropriate skill-sets and continuing training requirements. It is important to understand monitoring program design requirements, including the essential capabilities and limitations of instrumentation that comprise the monitoring system, as discussed in Section 4.2. Opportunities for improvement are evident from the continuing advances in instrument capability, data communications, speed and integration of data that come from improvements in computing, and the ability to acquire data using remote methods that will continue to be required of mine operators/owners.

Innovation in the development of instrumentation and remote monitoring, investigation and problem-solving will be driven by mine operators who seek to automate their operations and remove people from potentially hazardous conditions. Motivation for increasing automation of mining operations is driven by safety performance, economy (budgets and staffing) and risk management.

Because technology is changing so quickly, geotechnical professionals could benefit from a standard approach to designing monitoring systems to assist in developing an effective slope performance monitoring program. A risk-based slope monitoring guidance process is presented in Section 10.2, building upon components described in Chapters 1 through 9 of these Guidelines. Following the monitoring guidance process, several innovative performance monitoring applications are described. These involve experience-based approaches with instrumentation currently in common use. In Section 10.3, basic approaches to performance monitoring are reviewed, including investigative and predictive monitoring, with emphasis on unique problem-solving situations that assist the slope engineer in the assessment of failure mechanisms and modification of the slope design based on local development experience.

As mining operations evolve towards increasing automation, it is anticipated that geotechnical performance monitoring systems will also become increasingly automated. Data management, interpretation and results reporting from instrumentation will become increasingly automated. A primary goal of such a performance monitoring system will be to manage critical procedural controls in real-time that will maximise geomechanics risk management in the automated mining environment. A sample of innovative geotechnical monitoring methods under development, some of them still visionary, is presented in Sections 10.4 and 10.5, including:

- utilisation of microsensors (nano-sensors) for mine environmental and deformation monitoring;
- imaging spectroscopy using remote mapping methods;
- autonomous mapping, inspection and analysis using fully autonomous UAVs; and
- virtual reality for open pit visualisation, inspection and monitoring.

Each of these methods will advance improvements in data storage/management and interpretation. It is noteworthy that the current priority for developing some of these innovative visualisation and monitoring tools is for underground mining applications. In time, however, the capability of such instrumentation will evolve towards open pit applications, driven by mine safety, to overcome access restrictions and to improve on existing methods for advancing geotechnical risk management.

Although the applications described in Section 10.4 are still in the development phase, including ongoing basic research, the early work described here indicates promising results that can contribute significantly to the future of geotechnical engineering for the mining industry. At the time of writing, a priority of the next phase of the Large Open Pit project (LOP III) is focused in part on research funding directed at innovative performance monitoring that will support the open pit mining activities envisioned for the future.

10.2 Risk-based slope monitoring framework

10.2.1 Overview

Geotechnical performance monitoring of open pit slopes is considered to be a requirement for keeping people safe in modern surface mines. The risk-based slope monitoring framework (SMF) process is introduced in this section as a tool to assist performance monitoring system design. The method is used to rate the geotechnical conditions, as well as the design and geotechnical performance, to determine a slope stability rating that addresses the likelihood of an unwanted geotechnical event occurring in each defined sector of an open pit slope. The slope stability rating is then combined with the consequence rating from a mining operation's risk assessment process to determine an overall qualitative risk rating for each sector of the mine, using a risk matrix.

Once the risk rating is known, the monitoring framework from the SMF process is used to help develop a mining operation's monitoring system, as well as to indicate the required frequency of monitoring, inspections and reporting. Although the monitoring guidance is not prescriptive as to the types of instrumentation to be used, a suggested list of monitoring instruments and their attributes for meeting the guidance is included.

In late 2018, Newmont Mining (now Newmont Goldcorp Mining) approached the Geotechnical Center of Excellence (GCE) at the University of Arizona to create a process that could be used to assist geotechnical engineers in the selection of appropriate geotechnical monitoring instrumentation that will meet the needs of any open pit mining operation. The objective of the project was to develop a fit-for-purpose process for determining an effective monitoring strategy for open pit mines that meets the following goals:

- achieves risk management objectives;
- keeps miners safe;
- protects assets and the environment; and
- maximises safe ore recovery.

The process required the following characteristics:

- repeatable and reliable;
- adequately addresses the residual geotechnical risks of the mine design and development;
- simple and efficient;
- easy to communicate to all levels of the business;
- able to adapt to, and evolve with, changes in a dynamic mining operation; and
- able to adapt to changes in slope monitoring and mine development technology.

The project team includes members from the GCE, Piteau Associates and Newmont. By March 2019, the project team tested the SMF process developed at a Newmont mine located near Elko, Nevada, USA. Once the testing was completed, the monitoring framework tool was applied at three additional operations in the Carlin Trend in Nevada. Two of the operations were active and two provided access to underground operations. By early April 2019, 14 experts had reviewed the process and made improvements.

The SMF process at an operation should be updated on a regular basis during mine development, as hazards change and as monitoring methods evolve. It is also expected that individual mines will re-evaluate their monitoring system requirements as mining, geologic and geotechnical understanding improve.

The SMF process is founded on the relevant experience of mining geotechnical practitioners and is detailed in the literature. Some of the key references that influenced this work include those by Hamman (2009), de Graaf and Wessels (2016), Hawley and Cunning (2017), Rose *et al.* (2018) and Hoek and Brown (2019).

10.2.2 Slope monitoring framework process overview

The SMF process comprises five steps:

1. divide the open pit into logical slope monitoring sectors, based on geologic and geotechnical characteristics, wall performance and mine design, geometry and sequencing;
2. rate the likelihood of failure for each sector using the slope stability rating system (SSRS);
3. rate the consequence of failure for each sector using the company-specific method for assessing geotechnical risk potential;

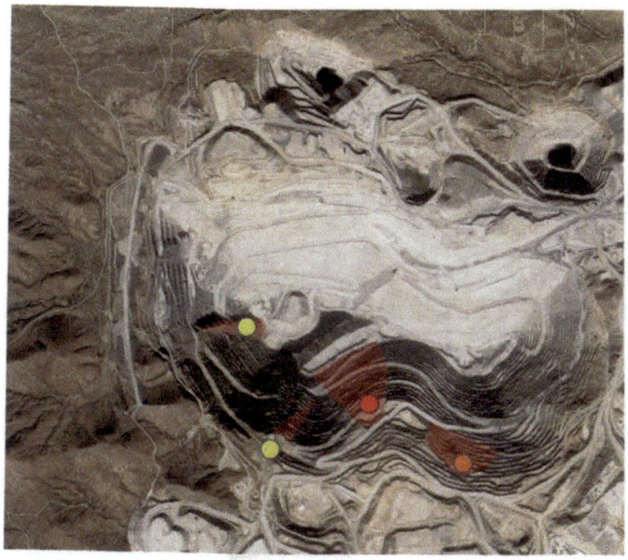

Fig. 10.1: Example of known hazards and risk rankings for a hypothetical large open pit (source: C. Williams).

4. define the monitoring guidance based on the overall risk rating; and
5. develop a geotechnical monitoring plan based on the monitoring guidance matrix.

The starting point for applying the SMF process is to gather the known geotechnical data for the mine and build a simplified geotechnical monitoring sectors map. This map will include the known geotechnical hazards, as shown in Fig. 10.1, as well as the sectors ranked for potential hazards shown in Fig. 10.2. Note that these illustrations represent a hypothetical example of a large open pit mine and are not based on real data. The colour of each dot on the map corresponds to the level of geotechnical risk associated with that location or domain: green indicates low risk, yellow indicates moderate risk, orange indicates high risk and red indicates critical risk. The known hazards map (Fig. 10.1) identifies locations where failures have already occurred (red shapes) and risk potential is interpreted. The second map shown in Fig. 10.2 (potential hazards) is divided into simple geotechnical monitoring sectors based on the following:

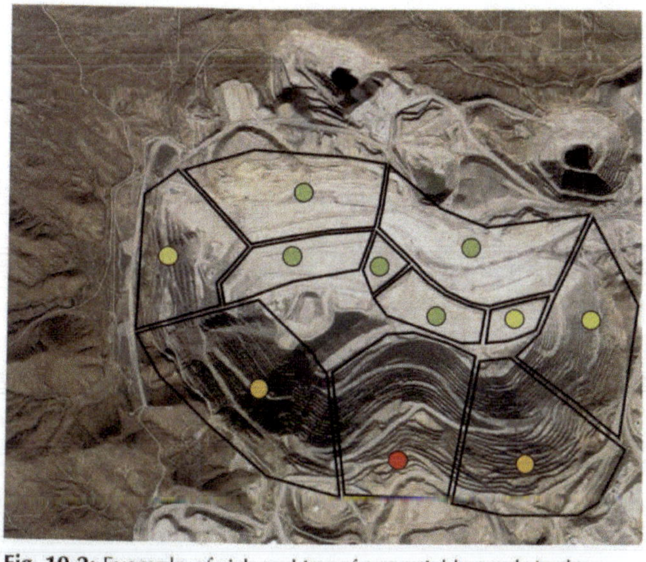

Fig. 10.2: Example of risk ranking of potential hazards in the established monitoring sectors for a hypothetical large open pit (source: C. Williams).

- *geotechnical considerations*: large structures (faults), geologic contacts or major changes in the rock mass characteristics, areas where pore pressure may be critical, wall performance; and
- *mine design considerations*: slope geometry and subsequent line-of-sight (LOS) limitations, mine development sequencing, mine access and major infrastructure.

10.2.3 Determination of likelihood classification

The second component of the SMF process is the SSRS. This tool combines the geotechnical conditions index (GCI) with the design, performance and implementation index (DPI) to determine a likelihood classification. The design of this component is adapted from the waste dump and stockpile stability rating and hazard classification system described in Hawley and Cunning (2017). The third component is the consequence evaluation that determines levels of risk, based on specific company standards as defined by its risk classification system. In the final component, the likelihood classification is combined with the consequence classification to determine a final risk classification for each monitoring sector, which is also identified with the company's risk classification system. From here, the monitoring plan for each sector can be determined. The structure of the SMF process is illustrated in Fig. 10.3.

The SSRS is a tool that can be used to estimate the qualitative likelihood of potential slope failures for individual monitoring sectors in an open pit mine. However, this method is not intended to address individual rockfall events (see Section 5.1.3.2). The SSRS is organised so that multiple sectors can be rated on one spreadsheet page. This ensures consistent ratings between sectors and reduces the time to complete the SSRS, since some ratings can be consistent for several sectors of the open pit. For this process, the slope engineer uses two hazard rating components: geotechnical conditions index and design performance index.

Geotechnical conditions index. The GCI is a measure of the physical conditions present in each monitoring domain of the open pit. Seven rock mass and geologic structure

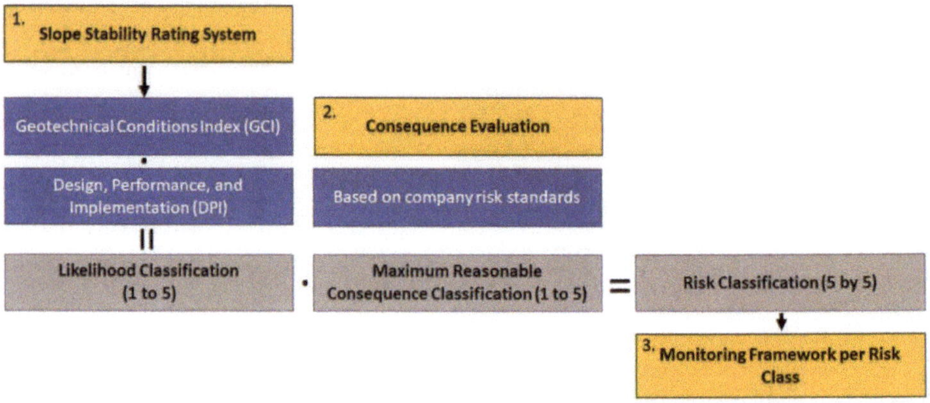

Fig. 10.3: The SMF process used to define the risk per monitoring sector. In this example, a risk classification based on five categories of likelihood and five categories of consequence is used. This scheme can change, based on each company's risk classification system (source: C. Williams).

conditions and six ground and surface water conditions are rated each on a scale of 0–10, with 0 representing a very poor condition and 10 representing a very good condition. Table 10.1 lists the geotechnical conditions that are included in the GCI. For example, the range of factors that define the intact rock strength used in the GCI is shown in Table 10.2. This table was developed from Hawley and Cunning (2017) and is consistent with ISRM classification standards. When using the GCI, only the resulting rating would need to be determined for each sector in the open pit. It is noted that some conditions may not be applicable to a specific open pit or sector, in which case 'not applicable' (N/A) can be assigned. The total score for all conditions that are rated are summed and divided by the total maximum scores of the rated conditions to determine a percentage. That percentage is then scaled to a value between 0 and 50 to be used in the slope stability rating.

Design performance index. The DPI is a measure of the factors that the mining operation understands and can control, including how well the conditions are understood, how challenges/conditions are addressed in the design, how well the mine plan is being implemented and how well the slopes are performing. Five categories for water controls, slope design and slope performance are rated on a scale of 0–10. Low numbers represent poor conditions applied to mine design, implementation and performance. Table 10.3 lists the controls, design and performance parameters that are considered in the DPI component of the SMF process. For example, the range of factors that define controlled blasting used in the DPI is shown in Table 10.4. When using the DPI, only the resulting rating would need to be determined for each sector in the open pit. Again, some design and performance parameters may not be applicable to a specific open pit or sector, in which case N/A can be assigned. The total score for all conditions that are rated are summed and divided by the total maximum scores of the rated conditions, resulting in a percentage. That percentage is then scaled to a value between 0 and 50 to be used in the slope stability rating.

The GCI and DPI scores are then used to calculate a likelihood classification for each pit sector that has been

Table 10.1: Factors included in the GCI

1. Rock mass and structure conditions	2. Surface and ground water conditions
1.1. Intact rock strength	2.1. Precipitation
1.2. Rock mass strength	2.2. Surface water
1.3. Rock mass damage	2.3. Phreatic surface relative to pit design
1.4. Rock mass deterioration	2.4. Perched groundwater systems
1.5. Rock mass fabric/structure/strength	2.5. Ability to drain rock mass and structure
1.6. In situ stress conditions	2.6. Recharge potential
1.7. Geologic complexity	

Table 10.2: Example of GCI index options and ranges that apply to the measure of intact rock strength

1.1 Intact rock strength		
Description	**Factors**	**Rating**
Very poor conditions	Very weak rock R0–R1 UCS <5 MPa	0
Poor conditions	Weak rock R2 UCS 5–25 MPa	2.5
Moderate conditions	Medium strength rock R3 UCS 25–50 MPa	5.0
Favourable conditions	Strong rock R4 UCS 50–100 MPa	7.5
Very favourable conditions	Very strong rock R5–R6 UCS >100 MPa	10.0

Table 10.3: Factors considered in the DPI

1. Surface and ground water controls, monitoring and reconciliation	2. Slope design	3. Slope performance and operational controls
1.1. Surface water controls	2.1. Slope height	3.1. Controlled blasting
1.2. Depressurisation measures	2.2. Slope shape (confinement)	3.2. Sinking rate
1.3. Pore pressure management	2.3. Design parameter reliability	3.3. Wall performance
1.4. Hydrologic study	2.4. Stability analysis	3.4. Phase of excavation
1.5. Pore pressure targets and reconciliation	2.5. Design flexibility	3.5. Design reconciliation and verification

Table 10.4: Example of DPI options and range for controlled blasting

3.1 Controlled blasting		
Description	**Factors**	**Rating**
Very poor design/ performance	• Production blast • No free face	0
Poor design/ performance	• Trim blast • Free face	2.5
Moderate design/ performance	• Pre-split/trim blast • Good vibration control • Good frequency control	5.0
Favourable design/ performance	• Pre-split/trim blast • Free faced • Good vibration control • Good frequency control	7.5
Very favourable design/ performance	• No blasting • Mechanical excavation	10.0

rated. Table 10.5 is an example of the results of a likelihood classification assessment from two monitoring sectors of a hypothetical open pit. Sector 1 has a GCI rating of 20 and a DPI rating of 13, and the resulting slope stability rating is 33. Sector 2 has a GCI of 28 and a DPI of 21, and the resulting slope stability rating is 49.

The results are next plotted on the SSRS matrix (see Fig. 10.4), which indicates the likelihood of slope instability. Sector 1 from the hypothetical example (rating of 33) plots in the orange zone, indicating a high likelihood of having a geotechnical event. Sector 2 (rating of 49) plots in the yellow zone, indicating a moderate likelihood of having a geotechnical event.

The SSRS converts the slope stability rating to a numeric likelihood rating, as shown in Tables 10.6 and 10.7. This likelihood rating will be combined with a consequence classification and applied to the mining company's risk matrix, to generate an overall risk rating. It is the overall risk rating that is used to build the geotechnical monitoring plan on a monitoring sector basis. For the hypothetical example, the slope stability rating of 33 results in a likelihood rating of 4, indicating a 'high' likelihood of a geotechnical event.

10.2.4 Evaluation of consequence

The risk-based monitoring guidance, as with most risk-based systems, utilises likelihood and consequence. The intent of the consequence rating is to be specific to each company's definition of risk tolerance. Therefore, the company's consequence classifications should be utilised for rating the potential consequence determined by the SMF process for each monitoring sector of the open pit.

As with most mining company risk-based systems, several factors should be considered if a geotechnical event occurs within any sector when assigning consequence ratings. For each of these factors, a consequence rating is assigned that ranges from 1 (lowest consequence) to 5 (highest consequence) for each sector. Typical consequence factors include, but are not limited to, the potential impacts that can reasonably occur in the following areas:

- *health and safety*: potential for multiple fatalities, a single fatality, or serious injury;
- *cash flow*: potential impact to income resulting from the loss of equipment, ramps, reserves and production, as well as early closure costs; potential lost income for the duration of impacted infrastructure, reserves and production;
- *infrastructure*: potential impact from the loss of portals for underground mines, power systems, dewatering systems (wells, galleries, etc.) and buildings;
- *environmental*: potential impact to surface and ground water, air quality and wildlife; and

Table 10.5: Example SSRS for a hypothetical open pit

	Sectors									
	1	2	3	4	5	6	7	8	9	10
GCI	20	28	-	-	-	-	-	-	-	-
DPI	13	21	-	-	-	-	-	-	-	-
Slope stability rating	33	49	-	-	-	-	-	-	-	-

Table 10.6: Description of SSRS likelihood classes and conversion of the slope stability rating into a numeric likelihood rating

Likelihood for 5 by 5 risk matrix	Score	Likelihood class	Description
5	<20	Very High	The event will most likely occur The event occurs daily
4	20–40	High	The event is expected to occur The event occurs weekly/monthly
3	40–60	Moderate	The event will occur under some circumstances The event occurs annually
2	60–80	Low	The event has occurred elsewhere The event occurs every 10 years
1	80–100	Very Low	The event may occur in exceptional circumstance

Table 10.7: SSRS – example of a likelihood rating

	Sectors									
	1	2	3	4	5	6	7	8	9	10
Slope stability rating	33	49	-	-	-	-	-	-	-	-
Likelihood rating	4	3	-	-	-	-	-	-	-	-

- *community and reputation*: potential impact to communities, economies and perceptions of the company (licence to operate).

Each factor is assigned a consequence rating for each sector. In this example, the rating levels are 1 – Minor, 2 – Moderate, 3 – Serious, 4 – Major, and 5 – Severe. Once the consequence factors are assigned to each sector, the SMF process will automatically assign the maximum rating for each sector, based on the highest rated level for that sector. Therefore, if health and safety is rated 4 (major level consequence) for a sector and the other four factors are rated 3 (serious level consequence), the process will apply a consequence rating of 4 for the sector.

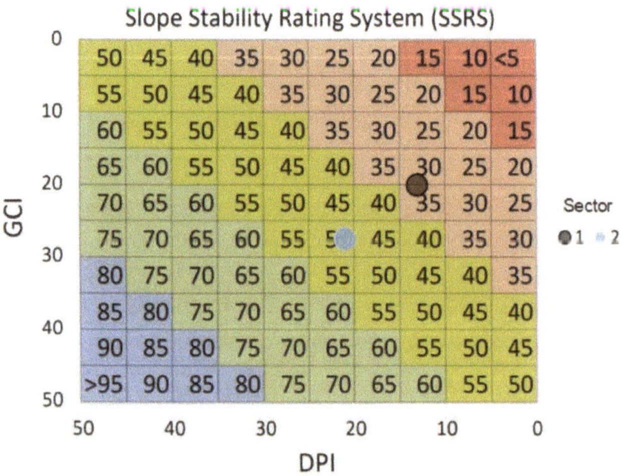

Fig. 10.4: The SSRS matrix. Sector 1 from the hypothetical example plots in the orange zone (high potential range), as indicated by the dark dot. Sector 2 from the hypothetical example plots in the yellow zone (moderate potential range), as indicated by the blue dot (source: C. Williams).

10.2.5 Risk rating

The SMF process plots the overall risk rating using a 5 by 5 matrix, as shown in the example in Fig. 10.5. For this example, there are four risk rating categories – low, moderate, high and critical. Therefore, if a sector has a moderate (level 3) likelihood and moderate (level 3) consequence, then the risk rating will be high, as indicated by the black dot (Fig. 10.5).

10.2.6 Monitoring system design framework

The monitoring system design guidance for each open pit sector is based on the risk rating determined by the SMF process for that sector. As the risk rating increases, so does the need for additional and redundant monitoring instrumentation, alarms, inspections and reporting. The suggested monitoring framework provided by the SMF process is summarised in Table 10.8. It is based on the following:

- design of a monitoring system is based on an understanding of the potential failure mechanism, and

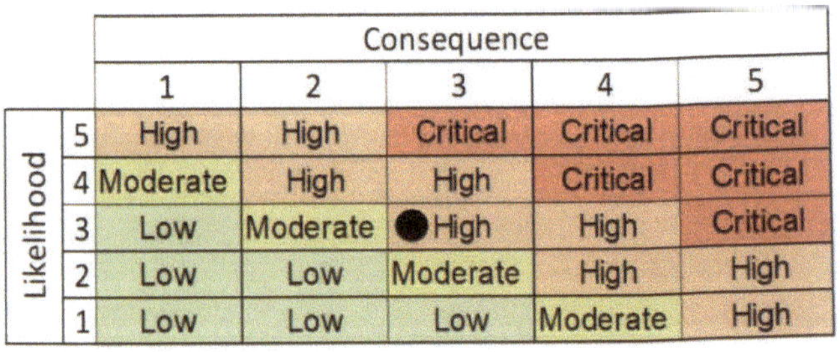

Fig. 10.5: The risk matrix applied to this hypothetical example. The actual configuration of the risk matrix will be in accordance with the scheme used by the mining company applying the SMF process (source: C. Williams).

Table 10.8: Monitoring suggestions matrix

	Monitoring suggestions			
	Low risk	**Moderate risk**	**High risk**	**Critical risk**
	• Data validated and reviewed by engineer weekly/monthly • Data should be validated across multiple instruments	• Maintain non-real-time monitoring systems that appropriately address the mechanism • Data validated and reviewed by engineer daily • Data should be validated across multiple instruments • Consider real-time monitoring system with appropriate alarms	• Maintain non-real-time monitoring systems that appropriately address the mechanism • Data validated and reviewed by engineer each shift • Data should be validated across multiple instruments • Consider one or more real-time monitoring systems with appropriate alarms	• Maintain non-real-time monitoring systems that appropriately address the mechanism • Data validated and reviewed by engineer each shift • Data should be validated across multiple instruments • Consider multiple real-time monitoring systems with appropriate alarms (at least one area wide coverage system)
	Monitoring frequency			
	Low risk	**Moderate risk**	**High risk**	**Critical risk**
Alarming	As needed	Checked daily	Per shift; consider real-time alarming	Per shift to real-time alarming
Visual Inspections	Weekly	Daily	Per shift	Per shift to constant (with spotter)
Reporting	Monthly	Weekly/monthly	Daily/weekly	Daily

used to assist interpretation of the mechanism, track movements and alert personnel in the event of an incipient failure;

- monitoring instruments that take measurements and transmit data in real-time are appropriate for alarming in moderate to critical risk rated sectors;
- monitoring instruments that are not real-time are often necessary for identifying and understanding failure mechanisms, as well as assessing the nature of longer-term trends;
- monitoring instruments such as radar, laser scanners and photogrammetry can monitor large areas and provide an overall understanding of the movement of a failure mass;
- geodetic and LOS monitoring instruments can provide critical information to validate data from other instrumentation;
- the required frequency of measurements increases as the risk increases; and
- visual inspections assist the interpretation, validation and reporting on monitoring data. They are an important component of the risk management process that ensures the monitoring system is working as designed.

The monitoring system design framework matrix (Table 10.8) and monitoring instrumentation selection that follows it are not intended to be prescriptive for developing a slope monitoring system. However, the SMF process does offer a standardised approach to designing a performance monitoring system through consideration of various components of a risk-based assessment of potential instability. For example, the process provides suggestions for deciding when real-time instrumentation is needed and the types of instrumentation that should be considered to ensure data validation is satisfied. The slope engineer should apply experience and knowledge of the mine's geotechnical conditions and the types of instruments that are available to build, maintain and evolve the monitoring system. Ultimately, the monitoring plan should address:

- known and potential hazards (understanding the potential failure mechanisms);
- all potential failure modes;
- the evolution of the mine plan; and
- the ability to cross-validate data to ensure reliability.

10.2.7 Monitoring instrumentation selection

The SMF process offers a suggested framework to select slope performance monitoring instrumentation that complements the monitoring guidance generated from the previous discussion, as summarised in Table 10.9. As previously described, monitoring instrument selection should be tempered by an assessment of local conditions,

Table 10.9: Monitoring system guidance list

Equipment types	Frequency	Coverage	Data type
Manual crack meters	Non-real-time	Point	Displacement across a crack
Wireline extensometers	Real-time	Linear	Displacement across cracks
Robotic total station (RTS)	Real-time	Points	LOS displacement, 3D vector
Global positioning systems (GPS/GNSS)	Real-time	Points	Displacement, 3D vector
Ground-based (GB) radar	Real-time	Broad surface area – front view	LOS displacement
Satellite-based InSAR	Non-real-time	Broad surface area – top view	LOS displacement, 2D vectors in some cases
Ground-based LiDAR or photogrammetry	Non-real-time	Broad surface area – front view	Change detection for 3D displacements
Drone-based LiDAR or photogrammetry	Non-real-time	Broad surface area – top/front view	Change detection for 3D displacements
Time domain reflectometry (TDR)	Real-time	Down hole line	Location of damage to a cable
Manual inclinometers	Non-real-time	Down hole line	Down hole displacements, 2D vectors
In-place inclinometers	Real-time	Down hole points	Deformation at a point in the borehole
Shape accelerometer array (SAA)	Real-time	Down hole line	3D displacement along a line
Vibrating wire piezometers (VWP)	Real-time	Point	Pore pressure
Microseismic	Real-time	Volume under surface	Location and magnitude of microseismic events
Documented field inspections	Non-real-time	Area	Written, photos, samples and index tests

performance experience and local instrumentation availability, including the support network.

As monitoring technology is changing rapidly, serious effort is required to understand the capabilities and limitations of current monitoring equipment to ensure that the monitoring system is, and remains, effective and reliable. The skill-sets of geotechnical professionals, technicians and other mine support personnel required to maintain an effective and reliable performance monitoring system must evolve in step with the inevitable technology changes. This is particularly important to recognise and act on, as mining operations evolve towards automation.

10.3 Innovative applications of investigative and predictive slope monitoring

10.3.1 Overview

The use of slope performance monitoring, to ensure safe operations and successful execution of the mine plan, requires a baseline understanding of the slope behaviour relative to its geological and geotechnical characteristics and of the predictive indicators that can be acted upon in terms of early warning alarm thresholds and for alert planning. The two purposes of performance monitoring, modified from Moore *et al.* (1991), are:

1. *investigative monitoring* to provide an understanding of the slope behaviour over time and typical responses to external events (e.g. precipitation, blasting, etc.). This is also commonly referred to as surveillance monitoring; and
2. *predictive monitoring* to provide a warning of a change in behaviour, enabling the possibility of limiting damage or intervening to prevent hazardous sliding. This is also commonly referred to as safety-critical monitoring.

For both purposes, issues of uncertainty relating to the geologic conditions, slope kinematics and evolving slope behaviour attributed to mine development can persist and challenge the problem definition and resolution. The increasing depth to which open pits are extending and the complexity introduced where operations transition to underground mining represent significant challenges in the management of risk imposed by the initiation and development of deep-seated pit slope movements.

To address these challenges, efforts are being directed towards innovative applications of monitoring technologies capable of improving the detection, characterisation and understanding of complex slope deformations, and improving early warning and alert planning. Of special interest is the spatial and temporal data that can be provided by remote sensing technologies, such as change detection using laser scanners (LiDAR), photogrammetry and GB and satellite radar. Examples of investigative and predictive monitoring are presented in the following sections, with some thoughts about future developments.

10.3.2 Investigative monitoring: rock bridge and damage characterisation

The assumed rock bridge content and discontinuity connectivity within an open pit slope can have a major influence on slope design (Moffitt *et al.* 2007; Dight and Baczynski 2009; Elmo *et al.* 2011; Havaej and Stead 2016). As the rock bridge percentage increases, the connectivity of discontinuities decreases and the factor of safety increases. Rock bridge content and discontinuity connectivity are initially controlled by the structural geology, but can change through blast-induced and gravitational/stress-induced slope damage. Both will reduce the rock bridge content by generating fractures that may propagate through intact rock between discontinuities. Near the ground surface, this may result in increased rockfall and superficial failures. At depth, this may increase the potential for toe-breakout of non-daylighting pit slope failures or, as demonstrated by Beale *et al.* (2013), may lead to a redistribution of pore water pressures at large distances behind the slope face due to changes in the connectivity of fracture permeability.

Although connectivity and rock bridge content are often treated in conventional slope analyses as a 2D feature, in practice they will have 3D characteristics. Rock bridges may exist within the plane of the discontinuity influencing connectivity along the dip and strike, thereby controlling the discontinuity strength and the potential failure volume. Rock bridges may also exist between offset discontinuities within the same set or between different discontinuity sets, resulting in a complex 3D step-path failure surface geometry. Jennings (1970) recognised the importance of considering step-paths along dip and strike. It is clear that directional rock bridge content and connectivity play an important role in slope kinematics and performance; however, the 3D complexity creates considerable challenges in reliably characterising key geometrical properties such as discontinuity spacing and persistence, especially when using conventional mapping techniques.

Tuckey and Stead (2016) carried out a detailed investigation utilising remote sensing methods for mapping discontinuity persistence, rock bridges and damage in rock slopes using examples from three open pit mines. It included the use of digital photogrammetry and LiDAR, which were shown to be excellent tools for digital discontinuity window-based trace mapping. To manage key sources of bias and error, varied slope face orientations were mapped and analysed. Based on these analyses, Tuckey and Stead adopted the P_{21} fracture intensity measure (Zhang and Einstein 2000) and supplemented it with measures of rock bridge intensity (R_{21}) and blast damage intensity (B_{21}) (Fig. 10.6). Although these represent 2D measures, they play an important role as input and validation data for 3D discrete fracture network (DFN) simulations where, in a typical workflow, 3D DFNs are iteratively refined by comparing equivalent trace intensity values from simulated synthetic window maps against the measured trace intensity from the remote sensing mapping data (Elmo *et al.* 2014).

Tuckey and Stead (2016) further used photogrammetry models to characterise adversely dipping discontinuity

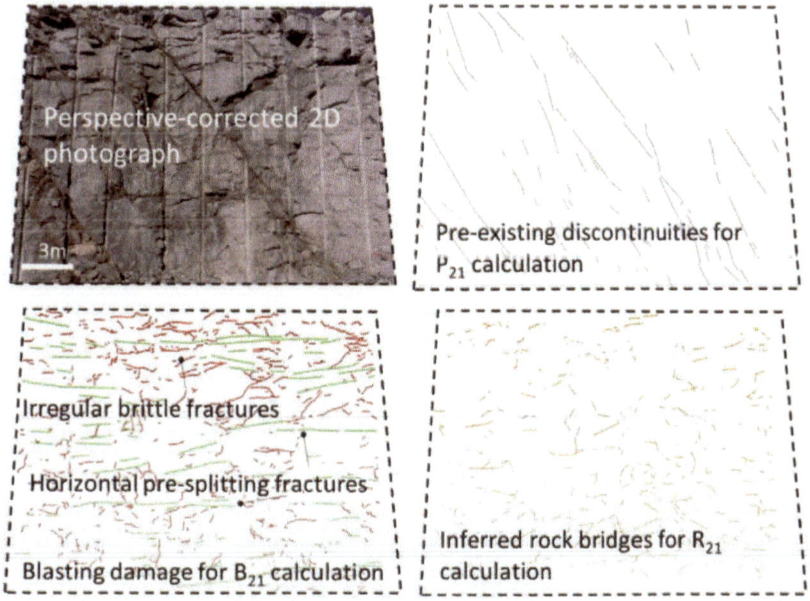

$$\text{Fracture intensity } P_{21} = \frac{\sum \text{discontinuity trace lengths}}{\text{window surface area}}$$

$$\text{Rock bridge intensity } R_{21} = \frac{\sum \text{intact rock bridge trace lengths}}{\text{window surface area}}$$

$$\text{Blast damage intensity } B_{21} = \frac{\sum \text{blast induced fracture trace lengths}}{\text{window surface area}}$$

Fig. 10.6: Example of digital trace window mapping stages undertaken using photogrammetry at an operating mine site, including the fracture, rock bridge and blast damage intensities (source: Tuckey and Stead 2016).

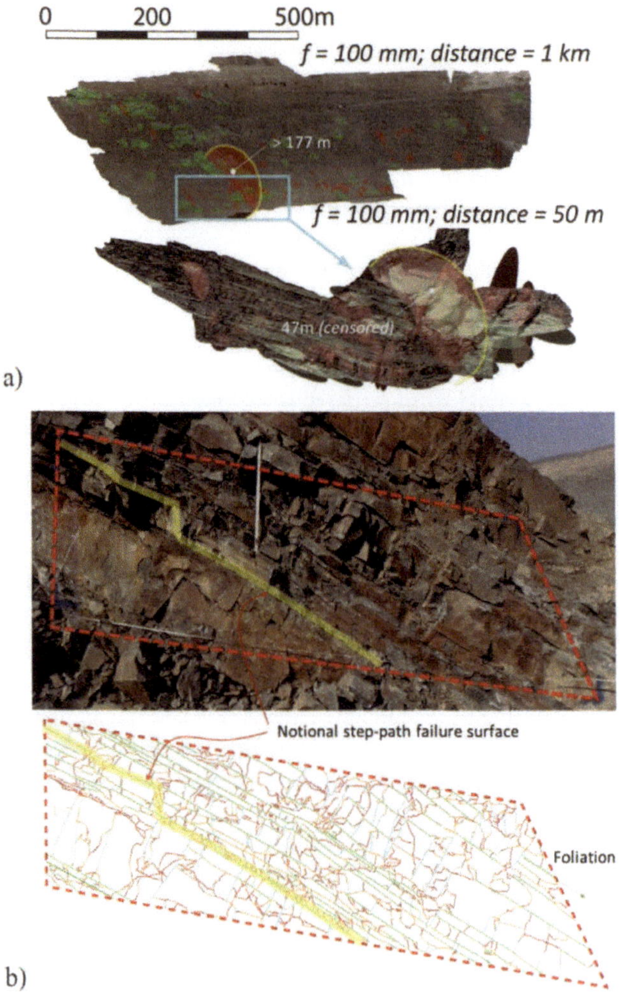

Fig. 10.7: (a) Photogrammetric model built from two overlapping photogrammetry surveys at a mine site showing the potential importance of mapping scale on the censoring of first-order structures such as faults. (b) Example of detailed digital window mapping of fractures and damage. Note 1 m rule in photo for scale (modified from Tuckey and Stead 2016).

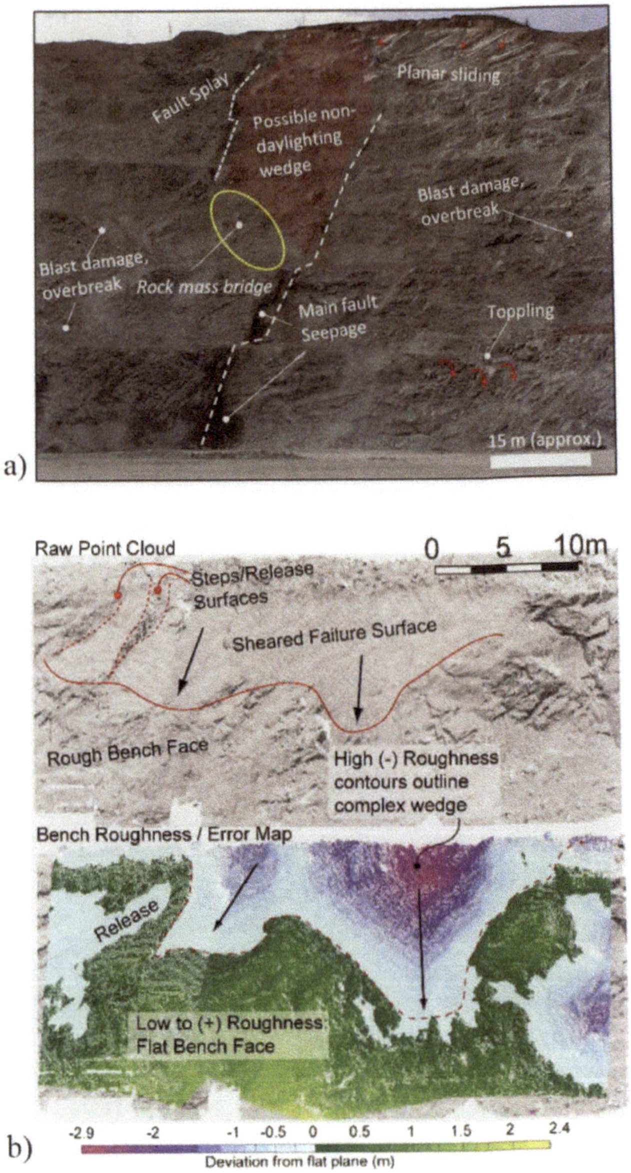

Fig. 10.8: (a) Major structural features and slope failure mechanisms mapped across an open pit slope. (b) Corresponding LiDAR imagery and mapping of complex bench-scale instability (modified from Tuckey and Stead 2016).

planes with step-path geometries along dip and strike (Fig. 10.7). The importance of photogrammetric mapping at the appropriate scale can be clearly shown: large first-order structures, such as faults, can be easily censored by an incorrect choice in focal length and/or imaging distance (Fig. 10.7a). Similarly, small-scale damage is best mapped by close-range digital trace mapping. The use of digital window mapping of fracture traces allowed detailed characterisations of fracture, rock bridge and blast damage intensity at this mine (Fig. 10.7b). GB LiDAR surveys can also be used to characterise bench-scale instability and damage (gravitational and blast-induced) on a local and mine-wide basis (Fig. 10.8). Tuckey (2012), Lee (2011) and Lupogo (2017) present developments in the applications of remote sensing to the characterisation of kinematic-gravitational and blast-induced damage to mine slopes. Important aspects of damage mapping that should be considered are the delineation of intact rock fractures (blast and gravitational/stress induced), fracture dilation and growth, changes in fracture connectivity, and damage-induced slope face roughness or morphology.

It is suggested that future research on rock bridge and damage characterisation in pit slopes should focus on the development of a fully integrated approach combining LiDAR and photogrammetry models with thermal and hyperspectral imaging (see Section 10.4.3). This will allow slope kinematics and damage to be investigated and the importance of variations in rock alteration and seepage to be more clearly understood. The structural controls on kinematics and rock mass damage should be explored in more detail. Ongoing work emphasises correlation between the data obtained from remote sensing imagery (with borehole and geophysical data) and the observed slope displacements using radar, LiDAR and GNSS. Change detection approaches applied to remote sensing LiDAR and photogrammetry should be used to characterise damage and thereby aid in the interpretation of spatial and temporal variations in pit slope deformation in magnitude and displacement vector direction.

10.3.3 Investigative and predictive monitoring: block cave–open pit interactions

Open pit mining operations increasingly involve the design and development of deeper and larger open pits, often with plans to transition into an underground mass mining operation. These larger, more complicated mine designs represent significant challenges to managing the risk imposed by the activation and development of deep-seated pit slope movements. Detecting the location, spatial boundaries and temporal characteristics of a potential pit slope failure are critical requirements in assessing risk, defining early warning alarm thresholds and/or devising mitigation options. Considerable effort has been expended on developing methodologies to better characterise, understand and predict potential pit slope failures through the integrated use of rock mass characterisation, monitoring and advanced numerical modelling.

As described in Section 3.5.3, satellite radar (InSAR) images can be used to track movement on the scale of centimetres to millimetres for a surface area resolution of several square metres. This ability to detect movement changes with significant resolution provides a means to monitor mining-induced deformation that develops across an irregular surface topography. Synthetic aperture radar (SAR) acquisition footprints that cover a broad area of the Earth's surface allow a bird's-eye view and full coverage of a mine site, including the pit slopes and surrounding area behind the slope crests where critical infrastructure might be located, as illustrated in Fig. 10.9.

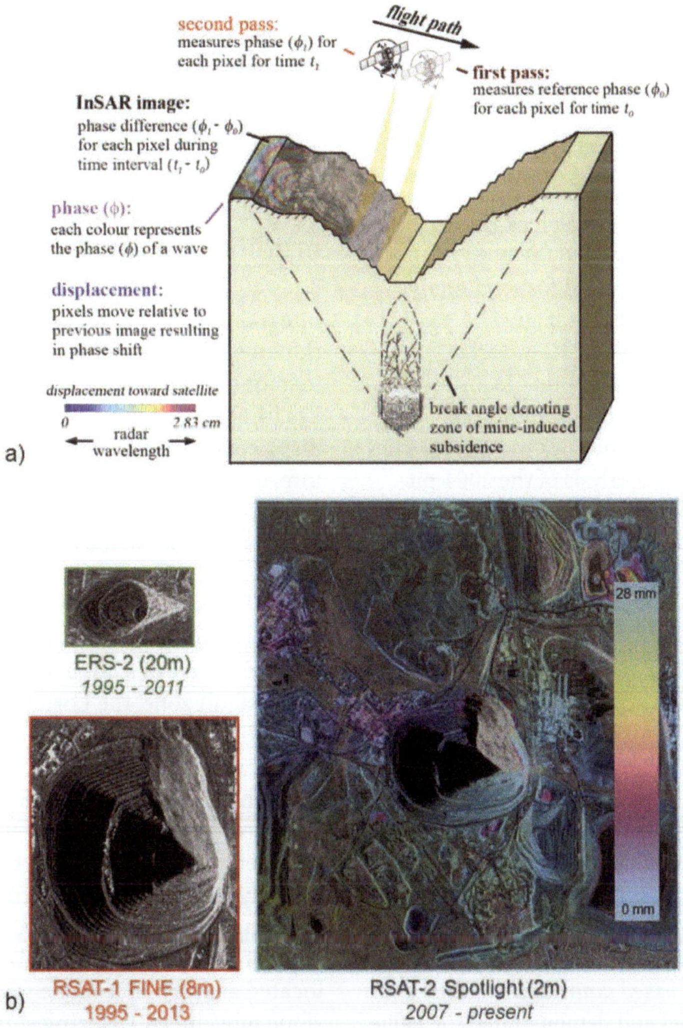

Fig. 10.9: (a) Principles of satellite radar interferometry (source: Rabus *et al.* 2009). (b) Improvements in SAR resolution over time and example of a bird's-eye view of mining-induced deformations for Palabora between 10 April and 15 July 2009 (source: MDA).

The capabilities of InSAR were investigated by Rabus *et al.* (2009) in a detailed study involving the targeted capture of SAR data for several open pit operations following the launch of the Canadian Space Agency's RADARSAT-2. The study included mine sites that covered a diverse range of latitudes, from Palabora in South Africa in the southern hemisphere (23°59'S) to Ekati in the Canadian Arctic (64°44'N). The satellite's 'spotlight' imaging mode was used, which provided a spatial resolution of 1 m × 3 m, as well as left and right look directions along the satellite's orbital path relative to the location of the open pit and orientation of the pit walls. Before processing the SAR data, several InSAR processing algorithm enhancements were tailored to the special challenges posed by the open pit environment (Rabus *et al.* 2009). These included the treatment of (i) steep slopes leading into the pit, which can be several hundred metres deep; (ii) dense local atmospheric effects in the pit, causing significant atmospheric error due to seasonal and weather-related variations of water vapour; and (iii) strong motion gradients and/or possible height errors between the high-resolution digital elevation models (DEM) for the mine site and the highly dynamic surfaces present in the form of moving pit slopes.

Woo *et al.* (2012) used these advanced InSAR processing algorithms to analyse the mining-induced ground deformations and their interactions with the open pit slopes at Palabora. Discussing the 2004 pit slope failure triggered shortly after transitioning to a block caving operation, Moss *et al.* (2006) emphasised the need for reliable, well-constrained and calibrated forward-analyses of caving-induced surface deformations. Figure 10.10 shows the results from a detailed FLAC3D model of Palabora developed by Woo *et al.* (2012) that incorporates the mine geology, cave geometries for different time intervals, *in situ* stress conditions and rock mass properties for each geological unit. A back-analysis of the 2004 pit slope failure was used to initially calibrate the material properties and validate the model.

The best fit set of input properties was then used for modelling of the expected caving-induced subsidence for the period 2009–2010 (Fig. 10.11). This predicted that the cave interactions would have the greatest influence on the west pit wall, with minor movement of the north wall. The FLAC3D models were then compared to the RADARSAT-2 InSAR data for the same period as they became available (Fig. 10.11). The data showed similar displacement magnitudes and patterns, with ongoing activity to the east of the 2004 north-west wall failure (primarily near the crest) and to the west extending across half of the west wall. Figure 10.12 compares the displacement-time trends for a point located near the main underground access shaft.

The close fit provided increased confidence in the predictive 3D numerical model and demonstrated the value of InSAR as a means to not only calibrate and validate numerical models, but also to identify and map spatial movements across a large open pit and beyond the pit limits. This ability is important for protecting critical mine infrastructure. Together, an advanced information product was delivered that integrated geology and geotechnical data, 3D numerical modelling and InSAR, to assist the mine's decision-makers in managing the risk resulting from the complex interactions between the open pit slopes and block cave mining beneath them (Woo *et al.* 2012).

Several constellations of satellite radars, described in Section 3.5.3, produce publicly available images that cover most of the Earth's surface. In the 1990s, the revisit frequency of InSAR imagery was in the range of 24–35 days. At the time of writing, InSAR images are available almost daily if a user is interested in capturing data from all of the satellite radars in service. The strength of InSAR as a method to monitor ground movement with precision lies in its ability to provide broad surface area, top-view coverage that cannot be done with ground-based instrumentation. InSAR is ideally suited to monitoring large critical infrastructure, such as tailings facilities or mine shaft headframes, and provides monitoring coverage over areas that are not easily accessible or are unsafe to access. As shown by the preceding example and other examples in these Guidelines, data from InSAR are useful for data validation, numerical model calibration, interpretation of failure modes, identifying the limits of movement and detecting new movement.

10.3.4 Integrating investigative and predictive monitoring: 3D slope kinematics

Since its introduction, GB radar has quickly become an integral monitoring tool for managing unstable open pit slopes, enabling quick identification of the size, extent and temporal behaviour of a developing failure. However, one key limitation is that GB radar can only provide LOS displacement measurements, meaning that important 3D information on the displacement kinematics of the slope is not provided. This served as motivation for a novel experiment conducted by the University of British Columbia, IDS and Teck's Highland Valley Copper mine. In the experiment, two SAR systems were deployed simultaneously to collect continuous LOS displacement data in 'stereo' of a large, moving open pit slope bisected by a large fault zone, the Lornex Fault (Fig. 10.13). The Lornex Fault dips ~75° into the slope and is interpreted as playing a critical role in controlling the kinematics of the upper slope movement (Rose and Scholz 2009).

Two SAR-type GB-radar units were positioned as shown in Fig. 10.14a. The first unit (Site A) was positioned roughly parallel to the general direction of slope movement, ~800 m and 1500 m away from the toe and crest of the slope, respectively. The second unit (Site B) was located at an oblique angle to the projected direction of slope movement, ensuring sufficient spatial overlap with the first radar unit (Fig. 10.14b). The distance of the monitored points for the second setup ranged between

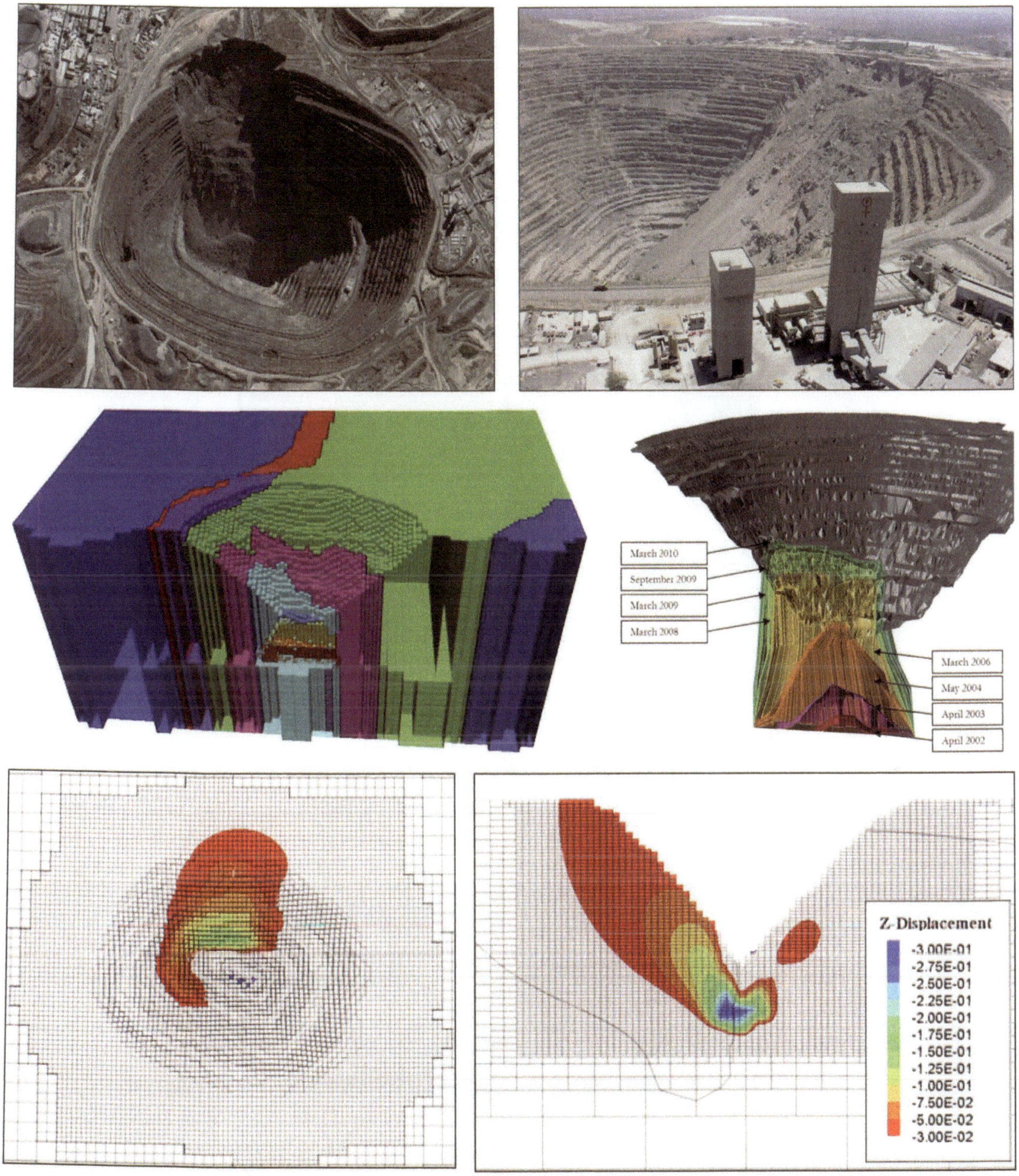

Fig. 10.10: FLAC3D model of Palabora, showing the detailed geology and cave geometries for different time intervals built into the model and back-analyses reproducing a 2004 pit slope failure triggered shortly after transitioning to a block caving operation (modified from Woo *et al.* 2012; photo of failure after Ngidi and Boshoff 2007).

1200 m and 2500 m. Both monitoring sites were founded on stable rock outside the range of any fly rock generated from production blasts in the pit (Fig. 10.14c).

The data obtained showed absolute movement in different distinct areas of the pit wall. The different perspectives provided by the two radar locations also captured different components of the slope displacements (Fig. 10.15). The comparison of the two displacement patterns highlights the importance of understanding the slope deformation mechanisms relative to the radar

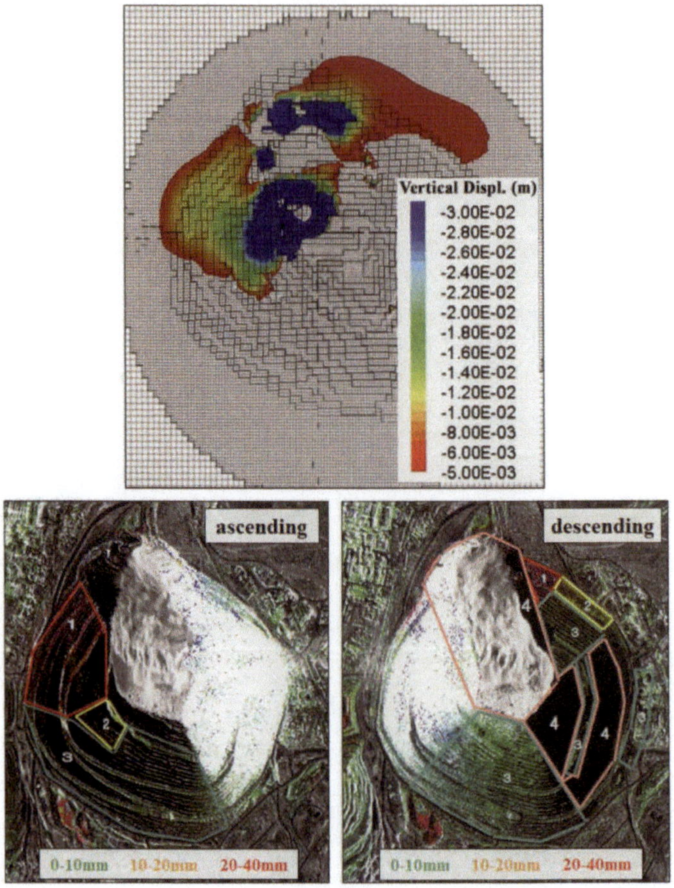

Fig. 10.11: Forward-modelling of expected caving-induced vertical displacements for Palabora, based on a calibrated FLAC3D model, with comparison to RADARSAT-2 data for the same period (modified from Woo *et al.* 2012).

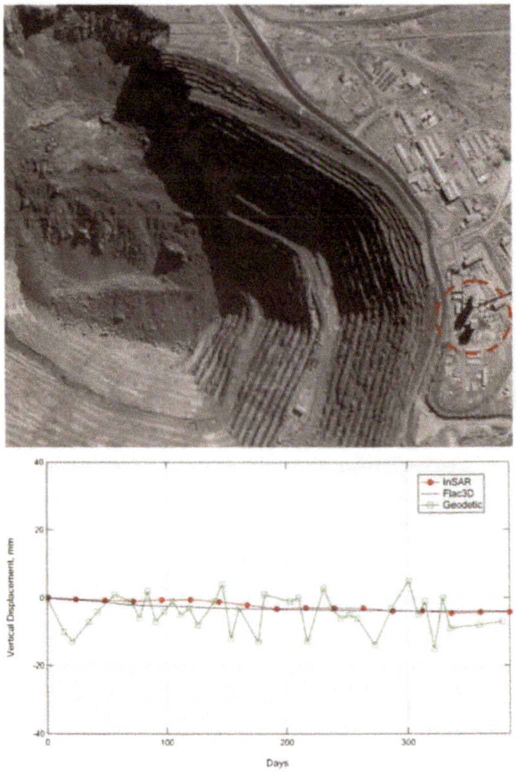

Fig. 10.12: Comparison of calibrated FLAC3D model, InSAR measured displacements and geodetic prism data between 2009 and 2010, for a point located near the main access shaft (modified from Woo *et al.* 2012).

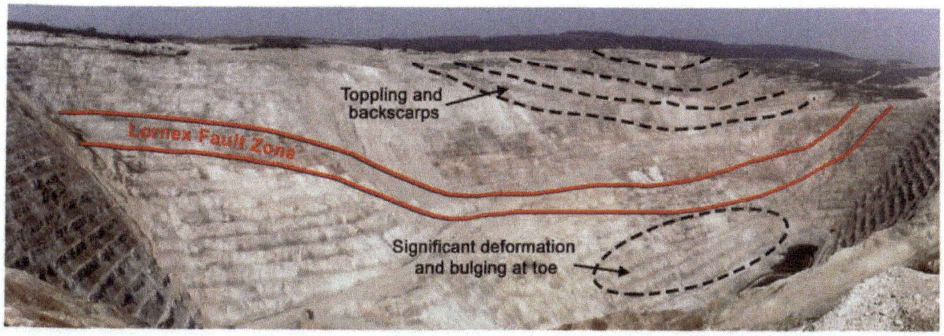

Fig. 10.13: West wall of the Lornex pit at Teck's Highland Valley Copper mine, showing the trace of the Lornex Fault zone and the locations of the graben-like backscarps and toe bulging that have developed with pit slope movements (modified from Severin *et al.* 2014).

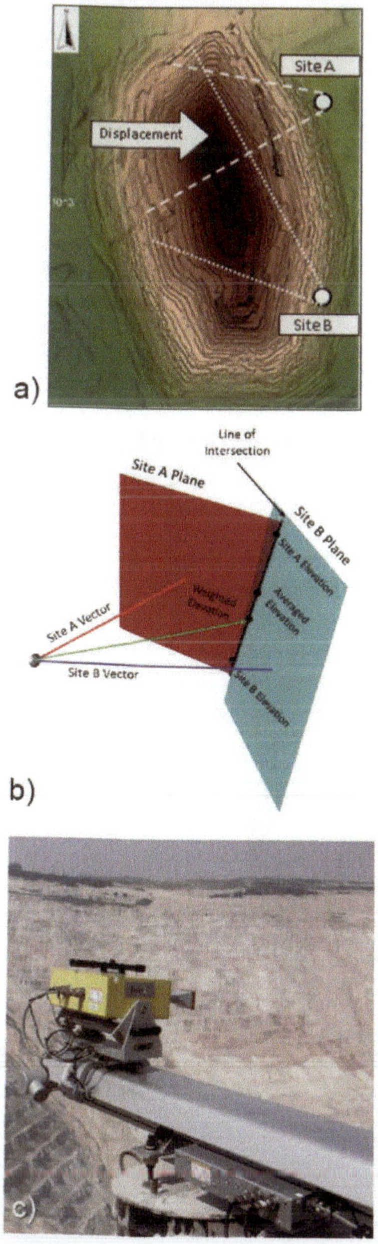

Fig. 10.14: Experiment setup, including (a) location of radar monitoring systems, (b) overlap relative to LOS vectors and (c) completed installation (modified from Severin *et al.* 2014).

placement before setting early warning alarm thresholds. The combined displacements obtained by Severin *et al.* (2014) led to the patterns observed in Fig. 10.16. Displacement values ranged between 0 mm/d and 307 mm/d, and as much as 750 mm of cumulative displacement over the four-day experiment period. Because the two sets of vectors have been combined, displacements do not need to be referenced as being towards or away from the LOS radar units; rather, they indicate absolute values.

By plotting the displacement vectors, several zones of movement within the slope were observed which could be compared to the controlling influence of the Lornex Fault (Fig. 10.13) that crosses the slope. In the upper third of the slope, the vectors suggest translational sliding with down-dropping block movements in response to the space created by the translational sliding (see Fig. 10.16 and interpretive section in Fig. 10.17). In the western half of the middle third of the slope, the displacement vectors suggest toppling and shearing of the rock mass accommodated by dilation and loss of confinement in response to bulging of the toe in the lower third of the slope below the Lornex Fault. As the pit wall curves towards the south, this transition to sliding occurs along the same fault, affecting the upper third of the slope. In the lower third of the slope,

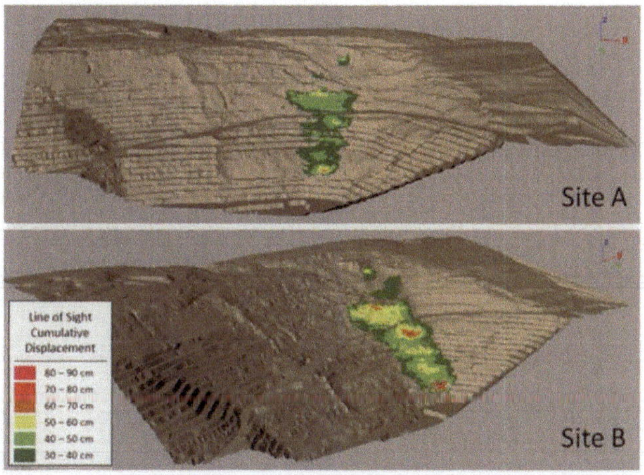

Fig. 10.15: Radar measured LOS displacement patterns from two different setup locations, Site A and Site B as shown in Fig. 10.14a (source: Severin *et al.* 2014).

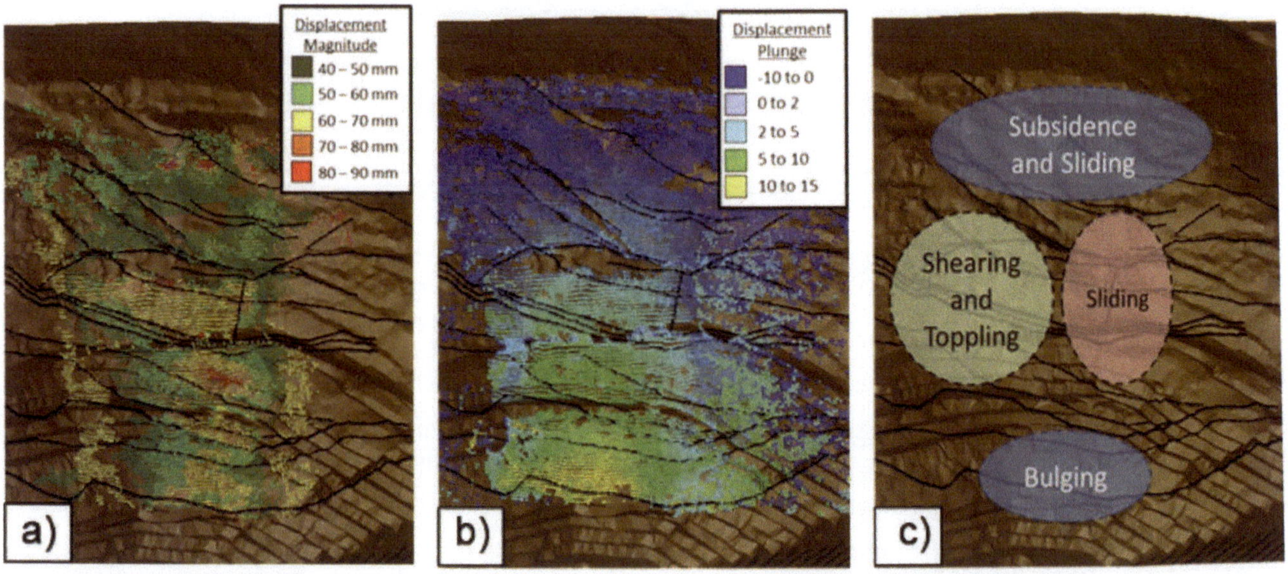

Fig. 10.16: Results of the 3D radar experiment, showing (a) displacement vector magnitudes, (b) vector plunges and (c) interpreted kinematics (modified from Severin *et al.* 2014).

the vectors indicate that behaviour is dominated by bulging of the weak, tectonically damaged rock at the toe of the slope below the Lornex Fault.

Severin *et al.*'s (2014) 3D radar experiment and corresponding displacement vector map demonstrated that localised zones of movement related to geological structures and changes in pit wall slope orientation could be resolved. It took full advantage of the complete spatial pit wall coverage provided by GB radar monitoring. This understanding can potentially add significant value to a mining project when assessing early warning alarm thresholds and risks presented by a large open pit slope, or where consideration is being given to a pit wall pushback to deepen the pit. In such cases, the influence of the controlling nature of major faults and smaller structures on the pit wall movements can be investigated and better understood with respect to pit geometry, slope deformation kinematics, evolving failure mechanisms and ultimately geotechnical risk.

10.4 Emerging technologies

10.4.1 Overview

A major transformation currently underway in the mining industry is the evolution towards automated applications that encompass all mining processes. Technology exists to support autonomous mining for some applications, including remote loaders in underground and open pit operations, blast hole drilling and automated hoisting

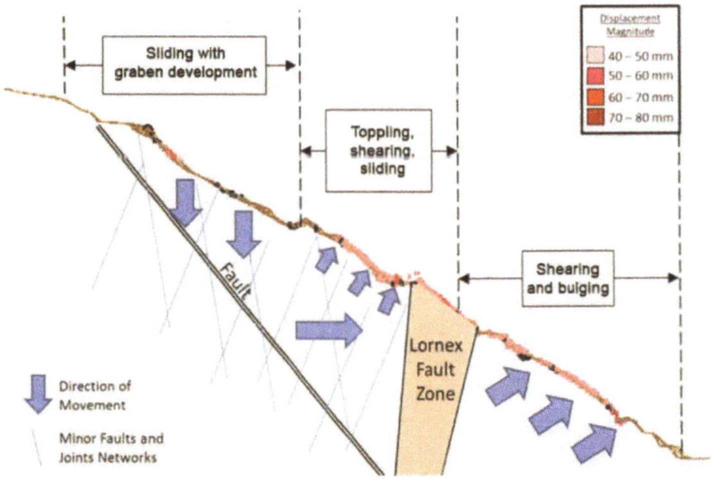

Fig. 10.17: Interpretation of pit slope kinematics for the north-west wall of the Lornex pit derived from 3D radar displacement vectors and the mapped geology (modified from Severin *et al.* 2014).

operations. Some elements of technology remain to be conceived before fully automated operations can be realised.

An example of the value of fully automated mining operations is the mine maintenance function, where goals include the development of full awareness of operational status and real-time integration of machine status and task management. Some mining companies are pursuing the vision that all major equipment will utilise the sensors necessary to provide real-time critical equipment performance and integrity information. For example, the equipment data collected and evaluated would include vibration, temperature and pressure, and the machine performance data would include throughput, quality, yield and/or flow rate. Models based on multivariate regression or neural networks would use historical data on failures to predict the occurrence of the next failure, based on changing equipment and process parameters. Replacement parts would be ordered when required by monitoring results rather than as currently prescribed by the original equipment manufacturer. The process would allow maintenance to be performed when required, thus reducing cost and unplanned shutdowns. Critical goals of the full automation process would be to significantly reduce cost and to make transparent the safety and operational risk related to maintenance, which will resonate with other mining disciplines.

It is important for geotechnical practitioners to recognise that technologies will continue to evolve, and therefore they should not to be restrained by the limitations of currently available technology. A fully autonomous mine will operate with fewer people entering the mine on a regular basis. In such an operation, the need for performance monitoring may change, especially in locations where no personnel ever enter the area of interest. Conventional thinking about instrumenting and performance monitoring will have to change.

In a completely automated mining environment, it is anticipated that geotechnical performance monitoring systems will also be automated and that data will be compiled and managed at a location removed from the operation. Management, interpretation and results reporting of data from instrumentation will also be automated. Data that are managed in an integrated system environment will make it possible for the geotechnical practitioner to interpret results and report exceptions in real-time. Importantly, a primary goal of such a performance monitoring system will be to manage critical procedural controls in real-time in order to maximise the benefit of geomechanics risk management in an automated mining environment.

Slope performance monitoring instrumentation that is already suitable for remote and inaccessible areas include GB and satellite-based radar systems, as well as drones capable of performing functions such as laser scanning, photogrammetry, imaging spectroscopy and ground-penetrating radar. A future application may include a SAR system adapted and mounted on a drone, and used for deformation monitoring of large areas such as tailings embankments and waste rock dumps, complemented by virtual reality for visualisation of the data (discussed in Section 10.5). Mining environments already exist where access is limited due to safety and weather-related restrictions, challenging topography, community unrest and vandalism, to name a few reasons. At the time of writing, in some open pit operations where autonomous haul trucks operate, access to the open pit is not permitted while such equipment is in operation. It is not a significant stretch of the imagination to foresee that human access to mines in the future should assume similarly limited access.

10.4.2 Nano-sensor platform for mine environment monitoring

10.4.2.1 Motivation

The University of Arizona is developing an economical, scalable, ubiquitous modular nano-sensor platform for monitoring vital statistics of individual miners, ambient air quality, ground deformation and characterisation, and geolocation of people and equipment relative to areas of unstable ground. This technology, called SMART Suite (System for Managing Advanced Response Technology), will provide a much-needed lifeline network for evacuation in emergency situations. The integrated sensing devices and network for transmitting data and communication signals present an opportunity to monitor worker health and working conditions in a mining operation remotely, continuously and in real-time.

In recent years, an impressive array of sensors and wireless network technologies for monitoring nearly all aspects of the mining process have become available. Equipment health monitoring systems dominate this area, and the number of companies manufacturing personal protective equipment and wearable technologies for monitoring worker fatigue continues to grow. In the wireless connectivity space for mines, a limited number of well-established manufacturers occupy nearly 100% of the market. Providing sensing capability is not the primary focus for these companies. However, using their application programming interface, data can be sent from a sensor over the network.

There are many companies that instrument everything from rollers on a conveyor belt to determining wear in an ultra-class haul truck and make the data available to clients, either by storing them locally or sending the data over their proprietary mesh network. One of the main challenges with this process is that vendor-specific networks do not communicate with each other, making it difficult to access and interrogate sensors in real-time from one platform to another.

The worker health monitoring market has become active in the past few years. In addition to startups such as 'Smart Cap' (fatigue monitoring), several equipment manufacturers provide wearable technologies to measure bio signals such as heart rate, blood pressure, blood oxygen, skin and galvanic temperatures, core body temperature and more. The data collected by commercially available equipment are typically stored in a logger for later analysis. At the time of writing, bio signal monitoring lacks capability for remote data communications.

10.4.2.2 Concept formulation and system design

The SMART Suite's ground-breaking technology platform delivers state-of-the-art, up-to-the-second monitoring of worker vital signs and the mine environment, providing a lifeline network for evacuation and emergency response, air quality, ground stability detection, geolocation tracking of personnel and equipment, and communication. It is a ubiquitous modular sensor platform that can be scaled to meet the demands of data collection, data analytics and mine-wide monitoring in surface and underground operations. The platform itself consists of two major components: the wireless Nano Access Points (NAP) platform and the wireless Nano Personal (NP). The wireless capability of the two devices allows strategic placement of many sensor types for maximising comprehensive data acquisition on demand in real-time.

The wireless NAP platform is powered by a nano-sensor technology that is modular and scalable, with options for up to 256 different types of sensors per unit (internal and external sensors), including bio, physical and chemical sensors. A network of tens to hundreds of NAP units can be quickly deployed throughout the mine in minutes to hours in any environment. Physical sensors such as accelerometer, gyroscope, temperature, humidity and volatile organic compound detectors are vital in monitoring working conditions remotely, continuously and in real-time.

Currently, commercially available off-the-shelf sensors are built into the NAP units. The platform can interface with any type of analogue, digital and nano-sensors. These include but are not limited to GNSS, pressure, 3-axis accelerometer, gyroscopes, proximity, compass, touch sensing, magnetometer, anti-tamper, pH, hydrogen ions, biometric, temperature, ultrasonic sensing, inductive, humidity, air flow speed, current, capacitance, gas, non-dispersive infrared and ambient light sensors.

Each NAP unit could (in developing applications) also be equipped with geolocation sensors, as well as a two-way radio. Sensors can be programmed to monitor changes continuously, intermittently or as needed. Because the sensors and gateways are relatively inexpensive and self-powered, redundancy can be built into the network to improve coverage (spatial resolution) and prevent blind spots in case of accidents and/or sensor/gateway malfunction. Bluetooth, Wi-Fi and VHF protocols are employed for communication.

10.4.2.3 SMART Suite characteristics and operation

The characteristics of the SMART Suite platform include:

- small and low-cost ubiquitous modular sensor platform;
- up to 256 unique bio signal, physical and chemical sensors;
- emergency flashlight using LEDs;
- directional arrows pointing towards an escape route or exit;
- open architecture hardware and software;
- big data analytics-ready; and
- kits ready to deploy to any end-user application.

Characteristics of the data management platform include:

- NAP units with data streaming capabilities (units with data streaming and communications capabilities are in development);
- data from stationary (air and rock mass) and mobile (biometric) units stored in the NAP units and in the cloud; and
- use of OSIsoft's PI System (application software for real-time data management) to permanently store the data in a PI Historian. The PI System can also be used for querying, analysing and visualising data; the data can also be accessed from other platforms and/or third-party software for analysis and visualisation.

Figure 10.18 shows the structure of the SMART Suite system at the time of writing. Currently, up to 256 micro-electromechanical systems (MEMS) can be installed on one organic substrate ~2 cm across. The system's highly adaptable open hardware architecture allows any existing third-party sensor to be integrated into the SMART Suite system. NAP units can be operated in continuous or intermittent modes, or interrogated as needed. The latter two cases provide the most economical option, allowing the devices to operate for a much longer period on a single battery charge.

The system operates using low-power components, and it is expected that future versions will feature long-lasting safe, non-flammable batteries. It requires minimal maintenance (other than charging batteries). Sensor deployment will be optimised adaptively based on cost and quantity required, in response to the changing mine environment. Components are designed to military specifications and are expected to withstand the harsh environments encountered in mining operations. Analyses of system performance and added-value for mine safety will motivate advancing plans to improve the platform for mining industry application.

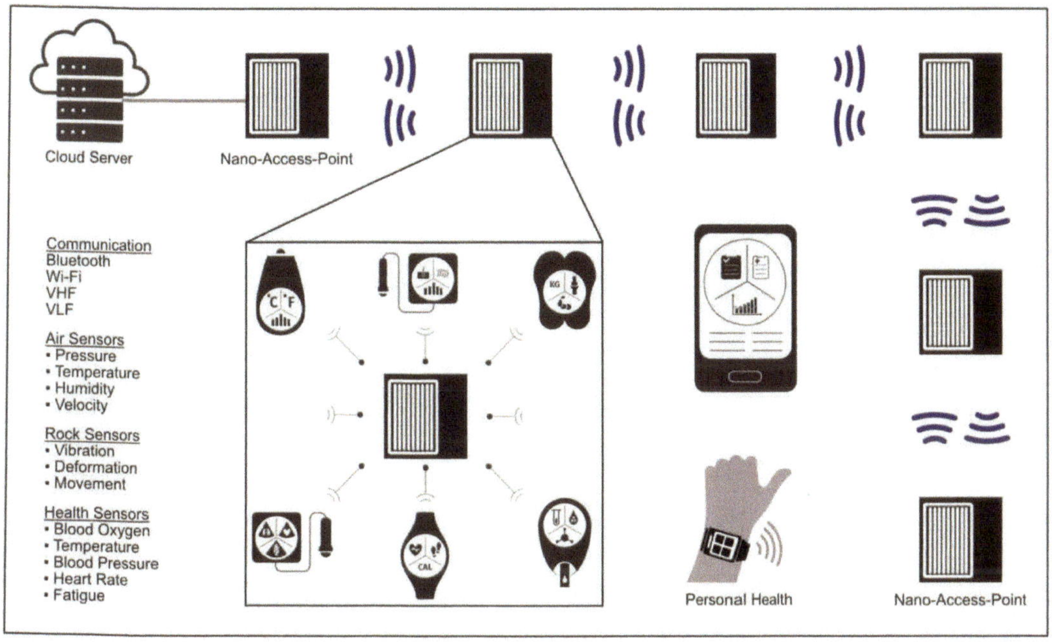

Fig. 10.18: SMART Suite system structure: the nano-sensor point that contains up to 256 MEMS measures ~2 cm across (source: M. Momayez).

10.4.2.4 Proof-of-concept demonstration plan

A planned proof-of-concept demonstration project will be housed at the University of Arizona's San Xavier Underground Laboratory (SX mine). The facility shown in Fig. 10.19 is being used as a teaching and research site. SX mine is a former silver-bearing lead and zinc mine comprising a skarn-type deposit. The facility has three different levels, including a surface level adit, and 30 m and 45 m deep levels (the 100 ft and 150 ft levels, respectively), both of which are serviced by the primary shaft (Fig. 10.20). As an underground laboratory, SX mine provides a unique research environment allowing short- and long-term studies to be conducted without disrupting production. It is an ideal location to install instruments to collect long-term deformation and other mine environment data.

To run the SMART Suite proof-of-concept demonstration project, the SX mine training facility has a control room (see Fig. 10.19) equipped with a server to run the PI system, large monitors and a small electronics workshop. It is connected to the training facility with two single and multi-mode fibre optic cables, providing the opportunity to stream data and communication signals at gigabit rates. Figure 10.21 shows the provisioned location of the NAP units in the adit and on the 100 ft and 150 ft levels.

Fig. 10.19: Aerial view of SX mine and facilities (source: M. Momayez).

10.4.3 Delineating zones of mineral-related geotechnical weakness through imaging spectroscopy

10.4.3.1 Motivation

The stability of open pit highwalls depends on many factors, most of them already treated in these Guidelines.

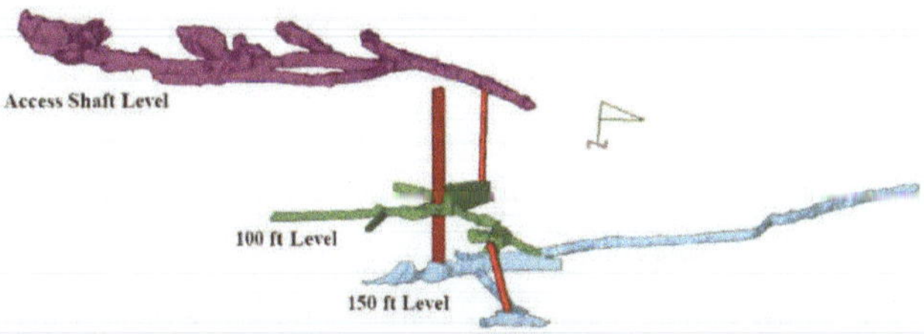

Fig. 10.20: 3D model of SX mine showing the layout of the different levels (source: M. Momayez).

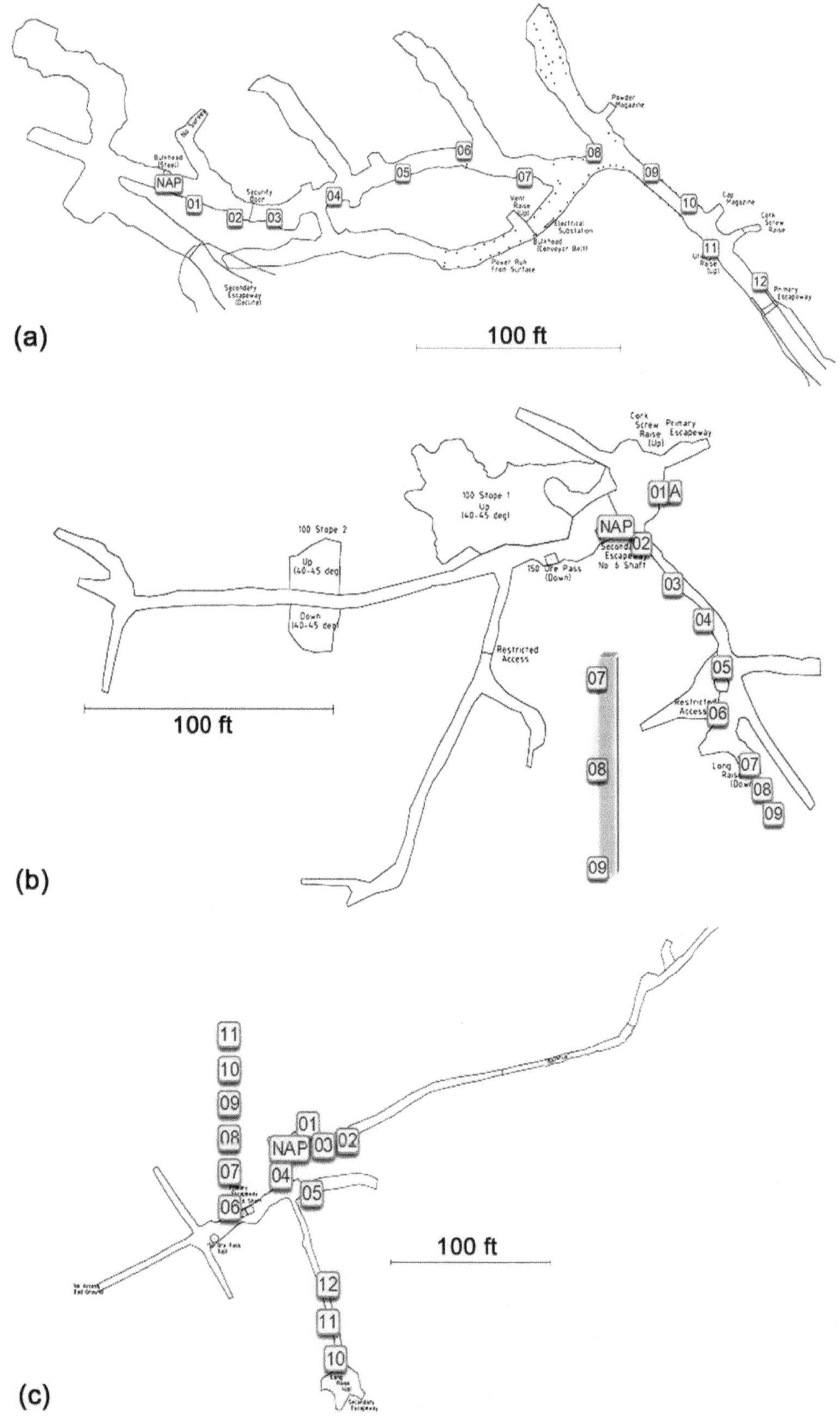

Fig. 10.21: Location of the environmental nodes for the SMART Suite proof-of-concept demonstration project for (a) the surface adit level; (b) the 100 ft level, with nodes 7–9 installed in the long raise connecting the 100 ft and 150 ft levels; and (c) the 150 ft level, with nodes 6–11 installed in the primary shaft to monitor shaft deformations (source: M. Momayez).

One emerging technology not discussed in detail is the recent development and use of full waveform hyperspectral LiDAR imaging to identify mineralogy-related zones of weakness in a highwall that may contribute to the initiation of slope movement and failure. This technique, new to mining, has the potential to contribute to faster and better identification of geotechnically weak zones, to help focus monitoring efforts on failure-prone areas and to contribute to a better understanding of the relationship between mineralogy and rock mechanics.

The structural sources of rock mass weakness – joints, faults and other discontinuities – have proven to be identifiable using remote sensing instruments like LiDAR. Cell mapping has long been a routine technique for discontinuity measurement, and techniques such as LiDAR are slowly automating it and making discontinuity mapping possible from increasingly large distances and for inaccessible areas.

Identification of mineralogical sources of weakness pose a greater challenge to remote mapping, for several reasons. Unlike faults, minerals are not easily visible, let alone identifiable, at large distances. The principal mineralogical sources of geotechnical weakness are phyllosilicate minerals (i.e. clays). The main challenge with identifying such mineralogical sources of weakness is that most of them cannot be distinguished in the field. Mapping the mineralogy of highwalls is a challenge at the best of times – exposures are often inaccessible, sometimes unsafe, and frequently contain minerals that are not identifiable by field methods due to their fine grain size and indistinct physical characteristics. Clays and other fine-grained phyllosilicates are among the most challenging minerals and usually cannot be identified in the field. Reliable characterisation of highwall mineralogy requires samples to be taken from the field and brought to the laboratory, where they undergo extensive sample preparation, X-ray diffractometry, infrared spectrometry and other techniques to identify their mineral content. This laboratory-based characterisation can be expensive and time-consuming, and the characteristics of the rock mass still have to be interpolated between sampled points.

Recent technological developments with LiDAR used in a remote sensing capacity incorporate hyperspectral time-of-flight sensors that measure the intensity of electromagnetic energy in the infrared-wavelength spectrum. Since some of the incoming energy is selectively absorbed by the vibration of certain mineralogical chemical bonds (e.g. those in clays), low reflective intensity in the emitted spectrum can be used to identify the structure and chemistry of the mineralogy by matching absorption features to chemical bands as illustrated in Fig. 10.22. Thus, infrared radiation is diagnostic of the mineralogical components of the rock and can be used to identify lithological variations (Stead *et al.* 2019). Clark (1999) provides an extensive review of the infrared spectroscopy of rocks and minerals.

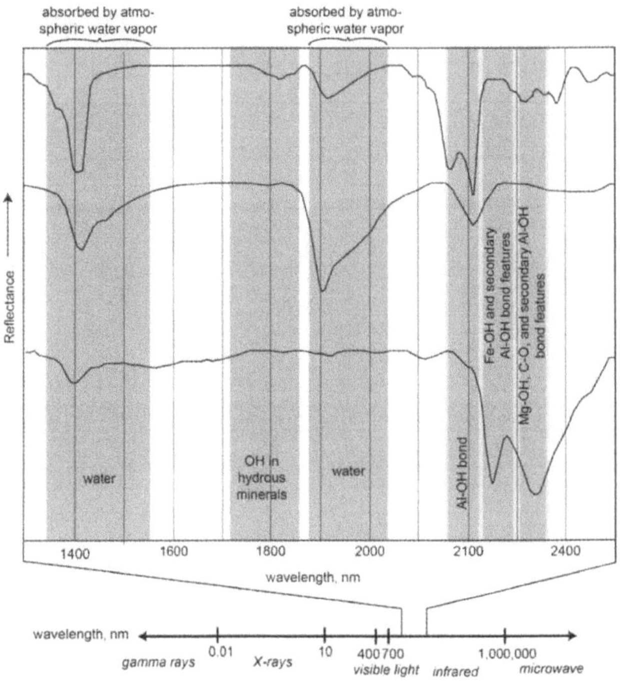

Fig. 10.22: Example of a short-wave infrared spectrum, showing the relationship of absorption features to chemical bond types (source: Barton *et al.* 2019).

The sensitivity of infrared spectroscopy to clays and other phyllosilicates, a common mineralogical source of rock mass weakness, makes this a potentially useful technique for geotechnical engineering applications. This can help to focus proactive monitoring and mitigation efforts where they will be most needed.

10.4.3.2 Imaging spectroscopy

Imaging spectroscopy has become a routine technique in other industries, but it is relatively new to mining where it has the potential to remotely map mineral types across the surface of an open pit slope. Imaging spectroscopy is a type of infrared spectroscopy, but is conducted with some distance between the scanner and the scanned object (Fig. 10.23). Variants include multi-spectral for spectroscopy over a small number of select bands, or hyperspectral for spectroscopy over most or all of the specified wavelength range. As a remote sensing tool, it produces an image consisting of spatial pixels in two dimensions. Each pixel has an associated infrared spectrum, representing the reflectance profile of that particular pixel. The dataset produced by a scan is thus a hypercube with a total of $2 + n$ dimensions, where n is the spectral resolution or number of bands in which the spectrometer measures. The mineralogy present in each pixel can be identified by comparing its spectrum to mineral reference spectrums by any of several methods.

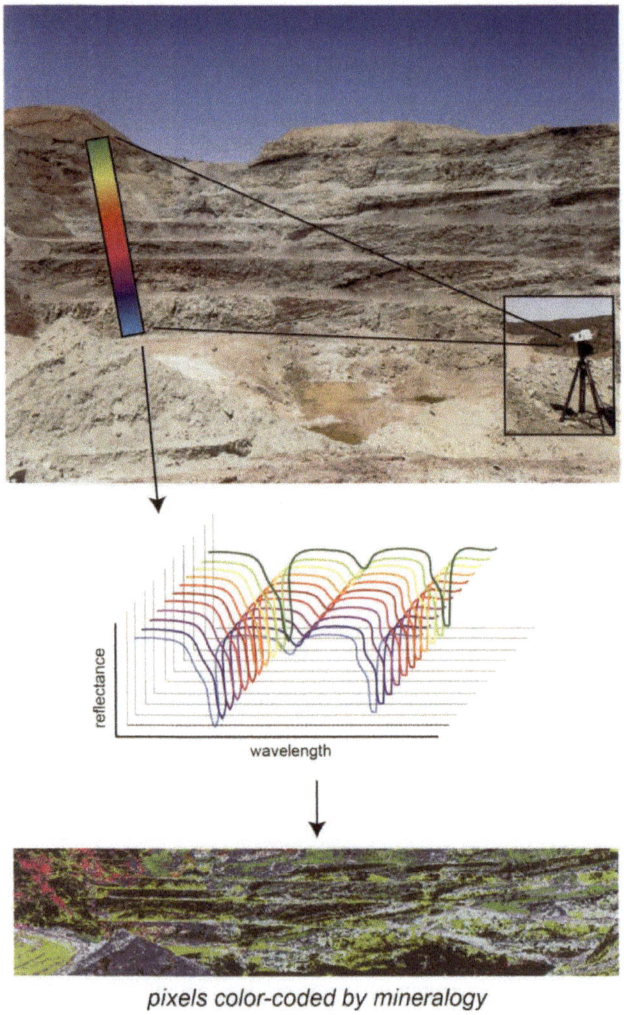

Fig. 10.23: Schematic illustration of imaging spectroscopy of a mine highwall using a tripod-based system. A tripod placed at a suitable location performs a line scan of the slope, measuring a spectrum at each pixel in the line, which moves from the designated start position through the specified arc. Each spectrum is then used to identify the mineralogy present in each pixel, which can be colour-coded by mineralogy (source: I. Barton).

Imaging spectroscopy can be delivered from several different remote platforms. Satellite-based spectrometers, such as Landsat, produce images with individual pixels that can be metres or tens of metres across. More useful to mining are spectrometers mounted on UAVs or drones, which can fly within 10 m of the target and produce images with pixel sizes of a few centimetres (see Section 10.4.4.2 for UAV flight limitations). Tripod-based imaging spectroscopy (as shown in Fig. 10.23) is particularly useful for vertical or steeply dipping surfaces; in such cases, the spatial resolution depends on the distance between the scanner and the surface. For sloped surfaces such as mine highwalls, this typically results in pixel size and spatial resolution varying within the image, as not all points on the scanned highwall are equidistant from the spectrometer.

Although widely used for decades for large-scale materials characterisation in industries such as agriculture and forestry, imaging spectroscopy has only recently been investigated for applications in mining. A major reason for the delay is that mining environments are far more complex than farmlands or forests, containing a much greater variety of spectrally active species. In addition, the fact that minerals occur as a fine heterogeneity in rocks means that spectral mixing is inevitable. These factors, plus the large size of hyperspectral datasets (a single 15 min scan commonly generates >20 GB of data), make the processing and interpretation of imaging spectroscopy data a particular challenge for mining.

Newly developed spectrometers with ultra-high spectral and spatial resolution, and advances in computing power, have made imaging spectroscopy a potentially viable technique for mineralogical characterisation of rock masses around mine sites. A pilot study conducted in June 2018 by the University of Arizona, Freeport-McMoran Inc., Headwall Photonics, Hexagon Mining and Lisbon Valley Mining showed promising results for rapid onsite characterisation of geotechnically problematic minerals in mine highwalls.

10.4.3.3 Imaging spectroscopy workflow for geotechnical slope characterisation

A hyperspectral scan may be carried out with the spectrometer mounted on a rotating tripod (Fig. 10.23) or a drone (Fig. 10.24). In the former case, the tripod is set up at a safe and convenient location and the scanning equipment mounted. A standard of calibrated reflectivity, such as a Spectralon tarp, is scanned first to provide a white reference. Then the scanner measures the intensity of light emitted from the scanned face in a straight vertical line that rotates through an arc of specified angle. Data are previewed on an attached computer and transmitted to a high-capacity storage drive.

For a drone-based scan, the spectrometers and a data storage drive are mounted on a UAV with a pre-programmed flight path. Figure 10.24 shows an example of a drone-based spectrometer starting on a highwall scan at a large open pit mine. During the first few minutes of the flight, the drone rotates and makes figure-eights to calibrate its GPS and magnetic sensors. When ready to fly, it first hovers over a calibrated reflectance standard, usually a Spectralon tarp spread out on the ground near the base. The scan area is usually divided into a back-and-forth pattern with several parallel legs, as shown in Fig. 10.24. The scan lines on adjacent legs overlap by roughly 20% on each side due to fluctuations in the line of ground being scanned as the drone pitches, rolls and yaws. The scan proceeds until the area is complete or (more

Fig. 10.24: Drone-mounted hyperspectral scanner shortly after take-off at an open pit mine, preparing for a highwall scan. The image on right shows a schematic illustration of the drone-based imaging spectroscopy system. Drone-mounted sensors pointing to nadir perform a line scan of the specified area, divided into parallel legs arranged so that the scan lines on each leg overlap by roughly 20% on each side, ensuring complete coverage (source: Headwall Photonics).

commonly) the drone depletes its battery and has to return to base. During the scan, data are continuously transmitted to the storage drive, whose contents are downloaded upon return to base. The data then have to be ortho-rectified, which removes the image distortions caused by roll, pitch and yaw, and the data collected on separate legs of the flight have to be stitched together, before data reduction.

At this stage, the data consist of readings of the intensity of light emitted or reflected by a surface at different infrared wavelengths. This is useful, but it has to be corrected for variability in the intensity of incoming sunlight, which is unevenly distributed across wavelength bands, and for non-uniformity in the sensitivity of the scanner. Using the measured intensity of light emitted from the reflectance standard, the scan data associated with each pixel are converted from intensities to reflectance spectrums.

Sunlight coming through the Earth's atmosphere is strongly absorbed at particular wavelengths, especially those associated with the O-H bond vibration around roughly 1400 nm and 1900 nm. Very little light in those wavelengths reaches the Earth's surface, so the signal-to-noise ratio is very low in that spectral band. These wavelengths are masked or excluded from further data reduction processes to avoid the problems associated with the low signal-to-noise ratio.

The reflectance spectrum associated with each pixel must be compared to a library of mineral reference spectrums in order to be identified. Pre-made libraries are available from the US Geological Survey and from the Jet Propulsion Laboratory, and their use is extensively described by Kokaly *et al.* (2017). These spectral libraries are adequate for most applications, but the pilot study of imaging spectroscopy for mining applications found that they tend to miss certain mineral types (carbonates and sulphates in particular) and yield higher rates of false positives than site-specific mineral spectral libraries created from samples at the scanned site (Barton *et al.* 2019). A variety of matching techniques are available to facilitate matching between pixel spectrums in the scanned image and spectrums in the reference library,

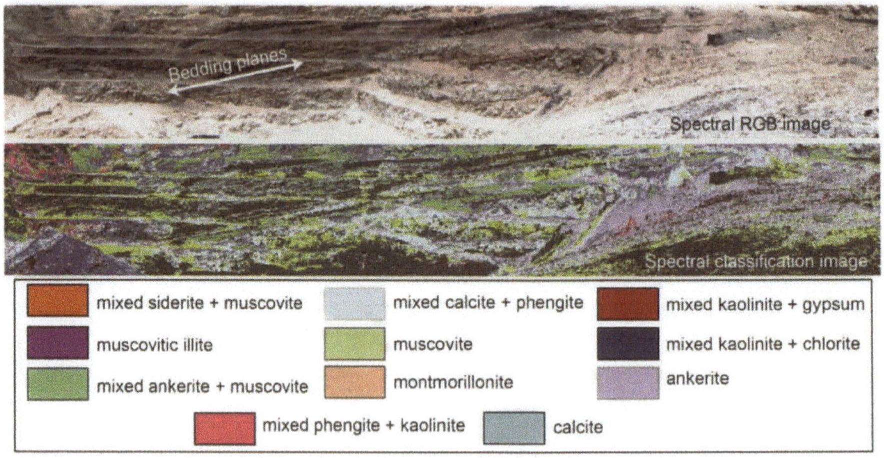

Fig. 10.25: Map of highwall mineralogy at an open pit mine, created by tripod-based imaging spectroscopy (source: Barton *et al.* 2019).

including spectral angle mapping, orthogonal subspace projection and principal component analysis (Manolakis *et al.* 2016).

Each pixel that matches a spectrum in the mineral reference library, to within a specified tolerance, can be colour-coded or otherwise attributed with the spectral match. The result is a pseudo-colour image showing mineralogy, as illustrated in Fig. 10.25 for a small section of a mine highwall. In addition to showing the distribution of spectrally active minerals in the rock, the hyperspectral image clearly indicates the locations and orientations of some of the structural features, in this case bedding planes.

Once the pseudo-colour composite image has been created, it can be used for identifying zones of mineralogical weakness, such as the particularly clay-rich sections of Fig. 10.25. However, this is not necessarily the most useful final form of the hyperspectral data. Although convenient for visualisation, versions of the hyperspectral mineral data that have been geo-referenced or otherwise unified with the corresponding spatial data could be imported into block models or geotechnical models (Fig. 10.26).

10.4.3.4 System development: current and continuing work

The adaptation of imaging spectroscopy to geotechnics and pit slope engineering is in its early stages, with little focused work at mining operations. Research investigating imaging spectroscopy (Barton *et al.* 2019) and the use of hyperspectral cameras mounted with terrestrial LiDAR (Stead *et al.* 2019) have successfully established the viability of the technique for detecting mineralogical zones of weakness in open pits and serve as a proof-of-concept for imaging spectroscopy in the mining environment.

However, these studies have also identified several key development areas necessary to better adapt imaging spectroscopy to mining and, in particular, to geotechnical applications. The first requirement is to streamline and automate as much of the data reduction as possible. The current, highly manual process of converting the

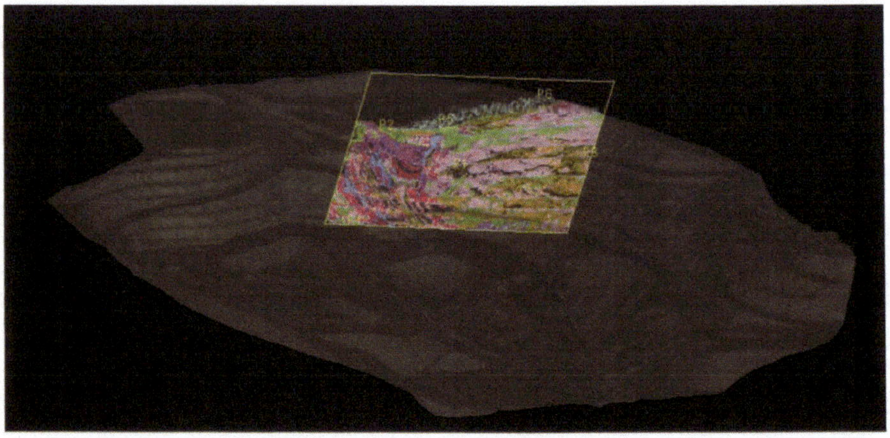

Fig. 10.26: Example of a classified spectral image draped over a 3D pit topography surface, created using 3D-Reshaper and a DEM supplied by the mine (source: J. Lyons-Baral, Hexagon Mining).

measured infrared light intensity data to a map of mineral distribution takes a significant amount of time and requires too much human input to be viable for mining operations where rapid turnaround and minimal human input are necessary. This development area can be broken down into two parts:

- an automated computer program that imports the measured intensities, locates and imports the corresponding white reference, and converts the intensities to reflectance spectrums; and
- an algorithm that automatically compares the spectrum of each pixel in the image to the spectrums in a mineral reference library.

Another development area is the reference library itself. For best results, the reference library must be derived from samples taken at the site being scanned or from an area with the same geology and alteration patterns (e.g. a different part of the same mine). As for the available characterisation methods (e.g. MLA, QEMSCAN), a site-specific mineral spectral reference library can be built over time to provide a complete catalogue of onsite spectrums. This library could be available in the data processing software for use with any scan.

The final development area is interfacing with modelling programs. This requires hyperspectral data to be geo-referenced so that they match a specific area of the rock mass and have known spatial coordinates. This can be done by co-mounting GNSS and LiDAR along with the hyperspectral systems or by post-scan draping of the image over a DEM of the scan area (Fig. 10.26).

Once the spectral mineralogy of the highwall is known, high-clay areas or other zones with particularly weak minerals can be selected for intensive stability monitoring and/or geotechnical testing. This will help focus the limited resources available for geotechnical monitoring in the areas of most need, helping the mining industry make better use of its technology, time and effort. Although imaging spectroscopy applications for mining are still in the development phase, the work described here indicates promising results that can contribute to the future of geotechnical engineering.

10.4.4 Autonomous mapping, inspection and analysis with unmanned aerial vehicles

10.4.4.1 Motivation

The use of UAVs in open pit environments has increased dramatically. Reductions in price and increases in the reliability and ease of use of commercial off-the-shelf (COTS) solutions have been strong motivators for mines to incorporate UAVs into their operations. Most mines' UAV programs are focused on the collection of photographic imagery and topographic data for survey and mapping purposes, but there are many other uses for UAVs that are relevant to slope monitoring, including hazard identification and tracking, blast performance evaluation and rapid-response imaging and analysis.

While some mines have the resources and personnel to pursue more advanced applications of UAV data collection in their slope management programs, many do not. The depth and breadth of most open pit UAV data collection programs are limited to the original intent of COTS solutions: topography and ortho-imagery. The importance and value of these data types are immense, but there is unrealised potential for UAVs in open pit mining to increase safety, improve efficiency and reduce workloads for slope engineers.

A UAV system has the potential to autonomously collect multiple types of 3D data and identify slope hazards and failures. To realise this potential, research is focusing on enabling UAV systems to automatically and safely avoid obstacles, persons and flight/property boundaries without input from an operator. A vision is for these UAV systems to make intelligent decisions, such as whether to return to base in case of bad weather or equipment malfunction, or determine the most critical hazard area to perform mapping and analysis work. The definition of truly autonomous systems includes the ability to perform complex tasks and achieve specific goals safely and efficiently, without operator inputs. At the time of writing, UAVs designed for open pit environments do not have these capabilities. The development of underground-specific UAV systems, however, has brought the industry closer to achieving the lofty goal of autonomy and may hold the key to the future of UAVs in open pit slope monitoring.

10.4.4.2 The link between underground and open pit

Most UAV systems developed for open pit and surface mapping applications rely on pre-determined flight plans to operate, and cannot function when their GPS and radio frequency (RF) control signals are interrupted. Among other things, electrical storms, power lines, cell phone towers, bridges and the passing of large equipment can interrupt GNSS and RF signals, and result in mission failure or the collection of incomplete or inaccurate data. Given how critical it is to obtain accurate and timely slope monitoring data, the reliability of existing UAV systems for this application is questionable. Moreover, to avoid obstacles and maintain GNSS and RF signals, most UAV flight plans require altitudes of 50 m or more above ground surface, limiting the quality of the imagery and 3D slope data they can collect. Yet the restrictions of signal interruption, flight altitude, object avoidance and safety in underground mining operations have driven a new set of UAV technologies that may dramatically change how UAVs are used in the open pit.

Underground mining operations have begun to implement mapping, inspection and safety programs that rely heavily on the use of UAVs. As expected, the challenges of using UAVs underground differ greatly from those of open pit environments: there is no GNSS signal, RF communication between the operator and drone is easily interrupted, and the spaces in which the drone can navigate are restricted on all sides (e.g. simply gaining altitude will not avoid a collision). For these reasons, most underground UAV work has been limited to direct pilot control. Automated flight plan and data collection are rarely used because of the risk of crashing the UAV and of striking persons or equipment.

In recent years, however, many groups have aimed to address the challenges of flying UAVs underground. Their primary goal has been to limit or completely remove the restrictions on flight path, signal interruption, operator intervention and collision risks from the equation. The net result has been a set of automation and intelligent systems tools that enable truly autonomous flight, the automated achievement of specific tasks and the rapid and safe collection of 3D and imagery data in close quarters and in otherwise inaccessible spaces.

10.4.4.3 Autonomy and its benefits in slope management

Autonomous UAVs have the potential to reduce risk and improve efficiency in open pit slope monitoring in several ways. First, they eliminate the need to maintain the LOS: the drone can react to a changing environment, avoid obstacles and generally perform its tasks unaccompanied by an operator and without intervention. Second, the drone operator, typically a surveyor or slope engineer, can perform other tasks while the drone flies its missions and may not be required to install ground control more than once during the first mission. Third, low-altitude flights that maximise the quality, quantity and accuracy of the data collected can be achieved without the threat of mission failure due to GNSS or RF signal loss. Fourth, and finally, open pit mines can be made safer with this technology by eliminating the need for personnel and equipment to enter dangerous or risk-prone terrain, while enabling entirely new data streams that will help monitor and characterise slope conditions.

The primary benefit of autonomous systems is that they enable work to be performed quickly and efficiently without putting people in harm's way. Allowing personnel to remain a safe distance from slope hazards is of particular benefit in the mining industry, particularly in slope monitoring, because it eliminates a crucial factor from the risk equation: exposure. Complete UAV autonomy frees valuable time for engineering staff, allowing them to collect other data and perform other necessary slope management tasks while the drone does its work. While some may question how safe it is to release fully autonomous UAVs to perform slope inspections or collect survey data, it could be argued that an autonomous drone is safer than any other data collection method, primarily because it does not rely on external data streams or operator inputs to ensure it maintains a safe distance from equipment, personnel and obstacles in its path.

LOS is always a concern when operating UAVs in the open pit environment. Many open pits have exclusion zones where UAV operators cannot enter, or have physically inaccessible areas that may otherwise serve as ideal launch and viewing positions for drone flights. Access restriction limits the operator's efficiency when collecting the required data. If the drone must be flown out of the operator's view, the operator must find a new observation position. Unfortunately, in some cases, there are no safe positions where operators can maintain full LOS at all times. Where brief LOS interruptions are acceptable (i.e. the UAV is only out of sight for a few seconds), it is possible that the drone may lose RF connection to the operator's control unit and begin to hover in place or return to the launch location automatically – effectively ending the data collection.

Operators must ensure they have appropriate ground control beneath their flight path. Without proper ground control, the images and 3D data they produce may be skewed, stretched or translated. Placing ground control is time-consuming and, like the operator's position, is subject to access restrictions. Moreover, because of the ever-changing nature of the open pit environment, new ground control must be installed often to facilitate repeat surveys. UAV-based slope monitoring tasks are thus inherently time-consuming and may in fact be less efficient than traditional monitoring and measurement methods (e.g. RTS, GNSS, GB-LiDAR or radar, wireline extensometers, inclinometers, etc.).

Another challenge in the open pit environment is maintaining a GNSS signal. Many COTS drones require a continuous GNSS signal in order to fly. In some cases, the operator cannot manually fly the drone without a GNSS signal – losing it results either in mission failure or flight delays. Because open pits have steep walls and usually have large metallic bodies moving about (e.g. haul trucks), the GNSS signal inside a pit can be spotty. GNSS signal reliability can be improved by increasing the flight altitude, but this limits the quality and accuracy of data acquired for slope monitoring purposes.

With regard to safety, autonomous UAVs have the potential to completely supplant the more dangerous tasks in slope management, including highwall structure mapping, manual installation of monitoring equipment, accessing steep slopes, and entering actively moving terrain for forensic and investigative purposes. That is not to say there are no risks associated with unattended autonomous systems flying above personnel, equipment and infrastructure; rather, those risks may be smaller than the active risks the slope engineers are required to take to collect the data needed to perform their jobs. Autonomous UAVs will enable the collection of data that currently cannot be obtained without entering dangerous or risk-prone areas (e.g. displacement measurements on an actively deforming slope). Continuously advancing technologies and safety controls will reduce the inherent risks with autonomous drone systems, resulting in a net increase in mine safety.

10.4.4.4 The future of UAVs in open pit slope monitoring

In the open pit mining environment, drones are now widely accepted tools for collecting topographic data and imagery. The data are undeniably useful for general survey and operations management tasks, but fall short of the UAV's true potential for slope management. The technology being developed for autonomous systems in underground mining will likely serve as a stepping stone for the development of new open pit-specific UAV systems that can be much more useful for slope monitoring and could completely change the way slopes are managed in open pits. It is no wonder that open pit operations adopted UAV applications first. In the same sense, though, the somewhat simple operational constraints of the open pit have led the technology to plateau a bit. There have been some improvements on the hardware end, such as increased battery life and flight time, but there have been no dramatic shifts in the type or quality of data that COTS drones collect (especially for imagery and topography). The challenges faced by drone operators in underground mines have pushed the industry to develop cutting-edge solutions that will, hopefully, be leveraged to develop new open pit UAV technologies.

An example of a state-of-the-art autonomous underground drone system is shown in Fig. 10.27. This subterranean-specific UAV combines 3D scanning LiDAR, infrared structure core imagery, 4K video and laser and infrared range finders to safely navigate tight spaces without assistance from GNSS and RF signal inputs from an operator. It is being developed by robotics and autonomous systems experts in collaboration with underground mining experts, with the aim of creating a system that can safely achieve an objective underground (e.g. recognise trapped miners in an emergency or identify poor-quality roof support) at the push of a button and without human intervention during the flight. The applications of this type of autonomy are many, including search and rescue, safety inspections, excavation mapping, hazard identification, general survey and change detection.

The autonomous system shown in Fig. 10.27 is designed for underground mine use, but because 'plug and play' sensors are used (e.g. a COTS LiDAR unit) any type of sensor that is small enough to fit on the platform could be used. It is envisioned that such a system could be adapted for open pit slope monitoring. Because this system uses deep-learning and artificial intelligence-enabled software to control its operation, it can be adapted for almost any environment. Systems like this have the potential to fly complex missions and automatically achieve goals like identifying tension cracks, measuring slope movement at any scale, or recording the position and condition of ground-based instruments. They could even be used to perform general inspections that include close-up imagery and video of slopes, and the collection and rapid processing of 3D point cloud data for measuring slope geometry and changes therein. All of this could be performed with less need for time-consuming placement of ground control, from a single safe distance away from a targeted slope and other hazards without concerns over the LOS limitation.

Fig. 10.27: The subterranean fully autonomous drone system being developed by AutoModality and RESPEC for NIOSH. This UAV has an onboard computer and specialised software that integrates data from multiple COTS sensors to fly 100% autonomously underground and to achieve a specific objective without a GNSS or RF signal. This type of system could be adapted for open pit slope monitoring applications (source: B. Haugen).

Undoubtedly, the adaptation of fully autonomous underground UAV systems for open pit slope monitoring will require additional research and development. The trajectory and pace of UAV technology indicate that the question is not whether, but when, this transition will take place. When it does, autonomous UAV systems stand to reduce the workload of the slope engineer, while safely and efficiently providing much-needed data that are more accurate and reliable than those currently available. The flexibility and adaptability of intelligent UAV systems may enable the collection of new and varied data, and could fundamentally change how open pit slopes are monitored and managed.

10.5 Mixed reality for visualisation in open pit monitoring

10.5.1 Overview

As the quantity and diversity of mapping and monitoring data collected for open pit mines continue to increase, there is an increasing need to ensure optimal data storage and interpretation methods. The data obtained for a typical open pit mining project amass over time, increasing from the conceptual through to the pre-feasibility and feasibility stages, into detailed design and construction for operations, and ending in mine closure. The volume and diversity of open pit mine big data require the application of state-of-the-art visual analytics to allow communication between stakeholders, mine operating staff, mining engineers and geotechnical engineers.

The efficient and clear presentation of mining and geotechnical data is essential to allow optimal selection of instrumentation and spatio-temporal interpretation of recorded mine instrumentation data throughout the mine life-cycle. In recent years there has been broad recognition of the importance of presenting open pit mine data in 3D formats with the use of commercial mine design software, in addition to geotechnical and remote sensing data platforms. Only recently, however, have techniques of virtual, augmented and mixed reality seen major advances in their application in computer graphics. These advances, along with increased computing hardware performance capabilities, provide immense potential in surface mining data visualisation and for improving data collection and slope monitoring.

10.5.2 The virtual reality continuum

Virtual reality is a well-established area of computer graphics research. Milgram and Kishino (1984) described what they termed the reality–virtuality continuum, which ranges from the real-world environment to the virtual environment (Fig. 10.28). The real environment here relates to the open pit mine: it may range in scale from a single borehole core drilled during exploration, to a geotechnical instrumentation hole (i.e. borehole inclinometer/piezometer) and through to the entire open pit mine property. It may even include the presence of an underground mine beneath the open pit.

A virtual reality (VR) environment consists of virtual objects, such as computer-generated graphic simulations of an entire open pit or specific inter-ramp slopes or benches. In a mixed reality (MR) environment, the real world and

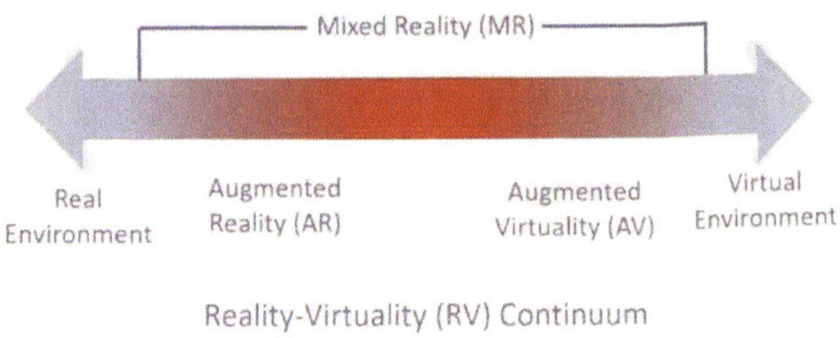

Fig. 10.28: The reality–virtuality continuum (modified from Milgram *et al.* 1994).

virtual world objects are presented together. This can be in the form of an augmented reality (AR) environment, where computer-generated virtual objects are added to a real mine environment, or an augmented virtuality (AV) environment, where real world objects, such as video or textural data, are added to a computer-generated virtual environment (Milgram *et al.* 1994).

The VR environment can be through a computer monitor or fully immersive. In the latter case, it is possible for the engineer to be fully immersed within the open pit, allowing virtual examination of a rock mass or individual mine benches. The engineer has the ability to virtually inspect sections of a pit constructed using remote sensing data such as photogrammetry or LiDAR. Multiple datasets can be draped on the virtual models to allow inspection of areas showing significant GB radar displacements, for example, and to conduct structural geological measurements on the virtual models. The power to interrogate multiple datasets in a virtual environment has enormous potential for improving understanding of open pit slope deformation.

Figure 10.29 provides examples from Simon Fraser University's MR visualisation laboratory applied to data collection using a Microsoft HoloLens, including a virtual drillhole core and a virtual discontinuity surface 10–15 cm in length. The virtual core box imagery was acquired using Structure from Motion (SfM) photogrammetry and a simple cell phone (Onsel *et al.* 2018, 2019). The use of MR/VR techniques has major potential for improving quality

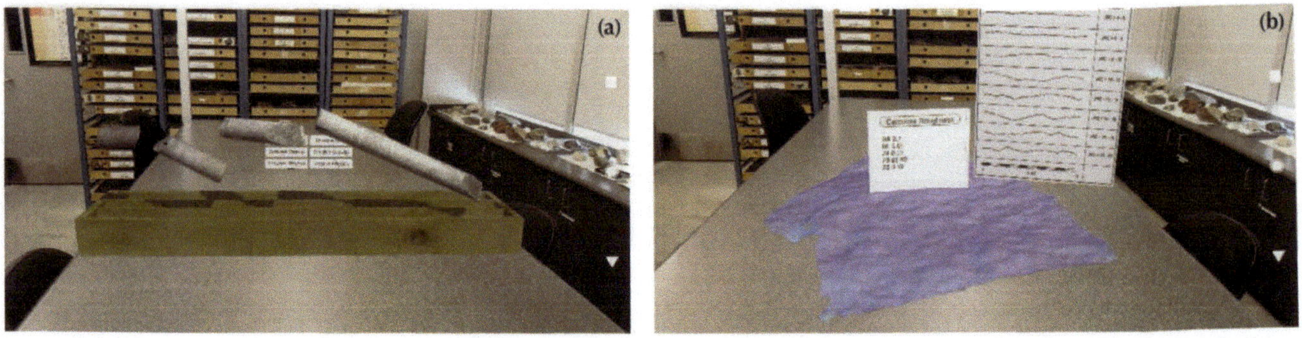

Fig. 10.29: Application of VR/MR using a Microsoft HoloLens. (a) Virtual core box facility acquired using SfM photogrammetry from cell phone images. (b) Virtual joint profiler showing a holographic image of a laser-scanned joint with automatic directional calculation of roughness coefficients and comparison with a joint roughness coefficient chart (source: D. Stead and E. Onsel).

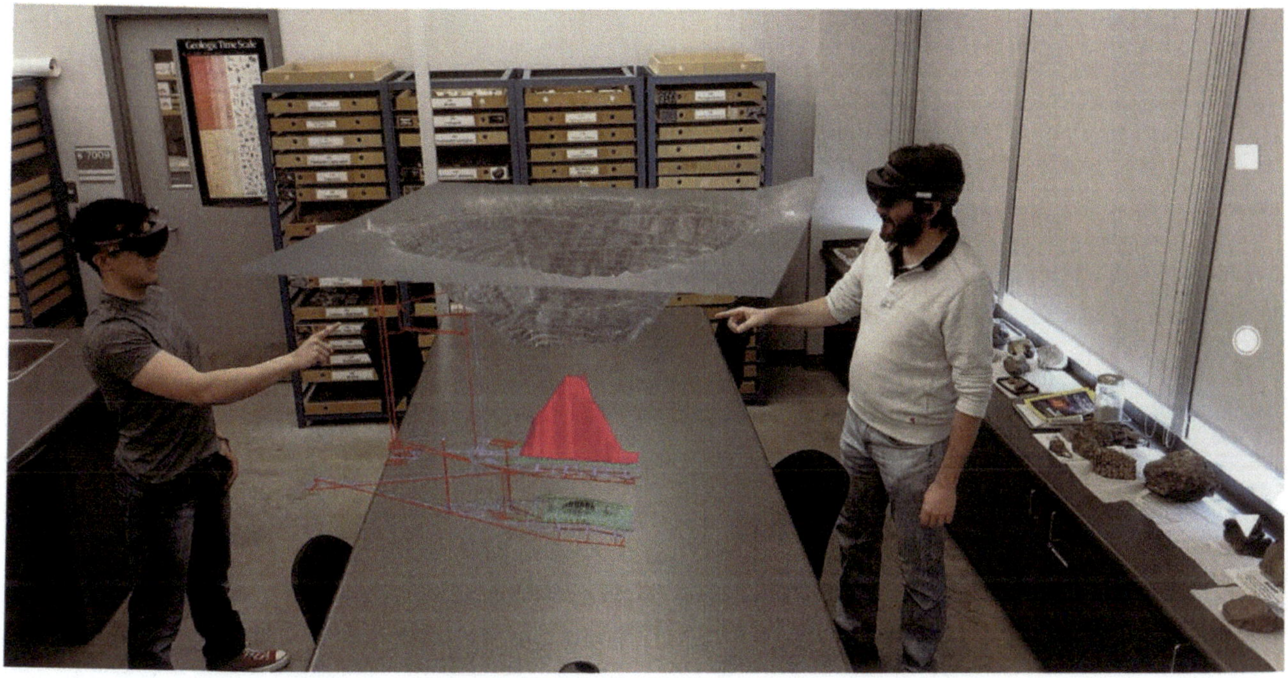

Fig. 10.30: Use of Microsoft HoloLens to examine underground–open pit interaction (source: D. Stead and E. Onsel).

control in core logging. Core logging data are often acquired by junior engineers and geologists. Construction of a virtual core shed allows the senior engineer to query data on selected remote stored core boxes in an office environment. Major benefits include improved training, communication and data interpretation with subsequent benefits for slope analysis, instrumentation and pit design. Figure 10.29b shows holographic visualisation of a laser-scanned joint surface. A virtual core profiler allows real-time calculation of directional roughness coefficients, according to Tatone and Grasselli (2010), and direct comparison with Barton's joint roughness coefficient chart. Joint roughness imaged in the virtual core box can similarly be measured holographically.

Several researchers and companies have used VR/MR methods to visualise open pits and rock faces holographically for varied applications. In VR applications, the user is immersed within the virtual environment. Powerful training opportunities exist; for example, the use of 360° VR rooms where a group can view an environment and be trained in a highly interactive and visual manner. At the University of Pretoria's mining engineering facility, Maritz and de Beer (2018) demonstrated the use of VR in training applications for underground mine rock support. The advent of technologies such as the Microsoft HoloLens has provided access to MR viewing where the user visualises objects within a geo-referenced environment such as an office or boardroom. To date, the use of MR techniques in mining engineering has included the production of holographic open pits on boardroom tables to allow communication between mine engineers in a mine production and truck haulage capacity. New innovative software platforms for MR/VR mine visualisation that combine 3D terrain and survey data, engineering designs and real-time computer graphics, enable experts (and community members for outreach programs) to walk around and interact with a virtual version of a project, from high-level overviews to true-life 'boots on the ground' viewpoints and from any point in time (Lato 2019). Li *et al.* (2019) provide an example of using MR/VR and holographic images to communicate the life-stages of an Alberta oil sands mine, emphasising the environmental reclamation of the site.

The ability to visualise and communicate holographically the progression of block caving and open pit–underground interaction is illustrated in Fig. 10.30. The open pit holograph in this case was constructed from photogrammetry of the open pit and is visualised using the Microsoft HoloLens. The potential use of MR in the mine design stages has facilitated the development of holographic visualisation of 3D numerical modelling data. The ability to export simulation results from numerical models already exists (Rocscience 2019). Figure 10.31 shows an example of a holographic visualisation of underground–open pit interaction based on 3D numerical modelling. The potential for holographic comparison of geomechanical model simulation data with multiple 3D datasets, including geology, rock mass and monitoring, is clear.

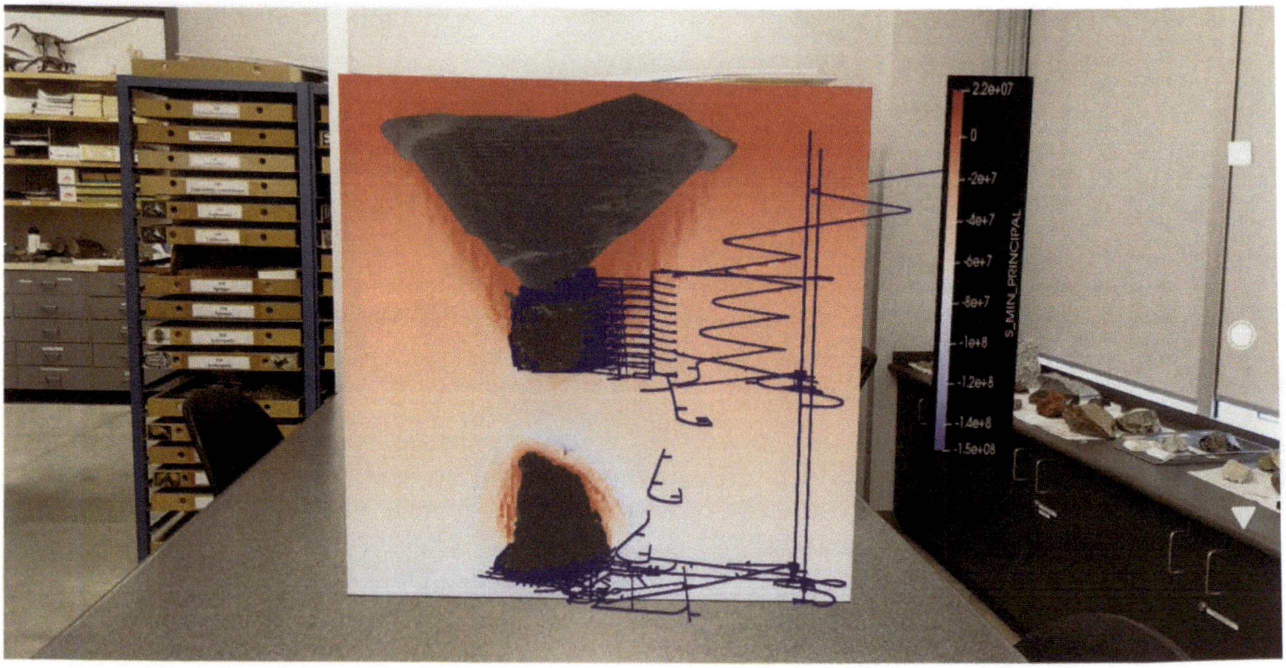

Fig. 10.31: Visualisation of underground–open pit numerical model simulation data using ABAQUS (source: D. Stead, E. Onsel; data courtesy of Beck Engineering).

10.5.3 Applications of mixed reality in pit slope mapping

Recent progress has been made in the use of holographic methods for mapping rock faces. Mapping using the Microsoft HoloLens can be undertaken directly and indirectly (Mysiorek *et al.* 2019a; Onsel *et al.* 2019). Direct mapping of rock slopes can be done using the Microsoft HoloLens which images the outcrop, producing a triangular mesh from which a holographic model can be derived. Mapping of geological units can be undertaken at the face by drawing lines on the outcrop virtual model and storing the data attributes digitally. Figure 10.32a shows a section of a rock face imaged using the Microsoft HoloLens, and Fig. 10.32b shows how the HoloLens can

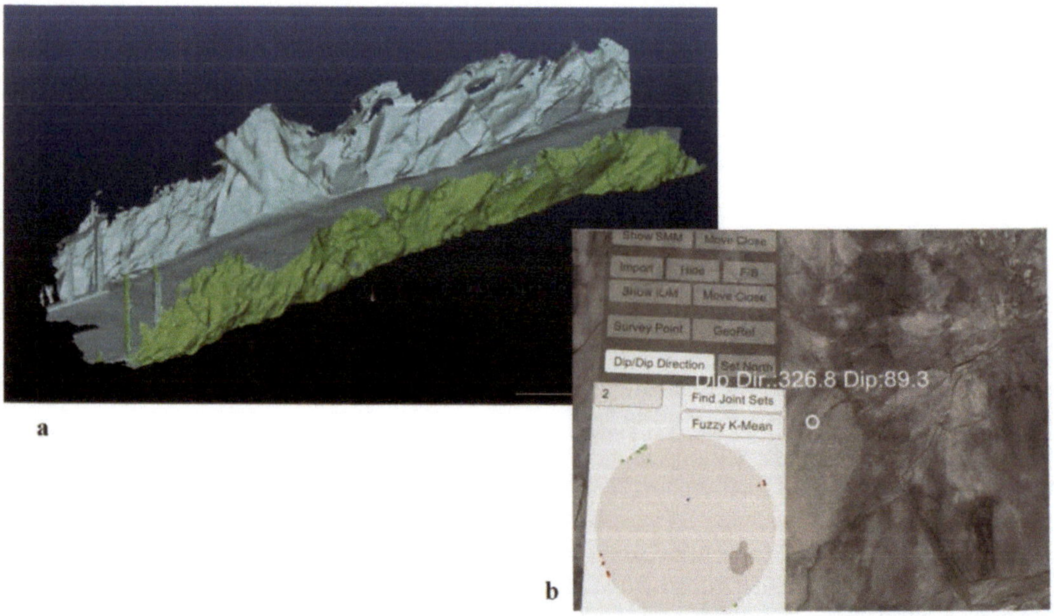

Fig. 10.32: Imaging of a rock face using Microsoft HoloLens, showing (a) section of a rock face visualised using the HoloLens and (b) direct measurement of the joint dip, and dip direction, using the HoloLens (source: D. Stead, E. Onsel and SRK Vancouver).

measure the dip, and dip direction, of joints by directing the engineer's view on the desired structure. The joints measured with the HoloLens are plotted directly on a stereo-net, where discontinuity sets can be identified. Remote mapping of open pit slopes, due to safety and limited access to bench slopes, is often undertaken using office-based or indirect techniques including photogrammetry or LiDAR, supplemented wherever possible by conventional field mapping and core logging. Indirect mapping from holographic virtual models of benches and pit slopes is now possible, and its use is anticipated to increase in the future. AR holographic rock slope mapping allows improved communication and interpretation of data. It is possible to undertake holographic discontinuity surveys, including line and window techniques. The ability to use MR models, augmenting virtual models of the rock slope with relevant structural or geological datasets while mapping, is a major advantage. Holographic mapping also allows improved communication and quality control by facilitating expert review by engineers and geologists of different experience and backgrounds.

10.5.4 Applications of mixed reality in open pit slope performance monitoring

MR/VR methods have significant potential for improving the presentation, interpretation, validation and communication of mine geotechnical models (e.g. geology, structure, rock mass and hydrogeology) and geotechnical domains. A summary of potential applications is presented in Table 10.10. The spatial and temporal visualisation of bench performance, considering rockfall hazards and blasting, is easily accomplished. Mysiorek *et al.* (2019b) show how rockfall can be seeded on holographic 3D models of rock slopes where the hazards and the performance of mitigation measures can be assessed and communicated effectively. Inter-ramp and overall slope design and performance can be investigated by incorporating geomechanical model results and draping on slope geometry data, as shown in Fig. 10.32. Slope monitoring is an area which can benefit substantially from the use of MR techniques. Data from slope displacement monitoring (GNSS, extensometers, tiltmeters, laser scanning, photogrammetry, GB radar and InSAR) can be shown in a range of formats, including simple images of graphs augmenting virtual pit slope models, time-varying virtual 3D contour plots, or simply draped on virtual open pit slope models. Subsurface monitoring data (borehole inclinometers, TDR, extensometers, piezometers and microseismicity) can similarly be portrayed as time-varying MR models. A major future development in open pit slope visualisation will be the storage of multi-mine datasets in a holographic database. Such a database could include real-world information, such as photographs and videos, in addition to virtual models. This development would allow the full reality–virtuality continuum to be considered in open pit slope design at all mine stages. As mining operations increasingly move towards remote and non-access methods, the need to optimise the storage, retrieval and visualisation of big data will become paramount. It is suggested that holographic visualisation, as outlined above, will be a major asset to future mining endeavours.

Table 10.10: Areas of potential application of VR/MR technology in open pit mining

Mine activity	Specific VR/MR applications
Economics	Virtual grade distribution and mineral economics models
Slope planning	Mine plan visualisation
Geological model	Direct – HoloLens outcrop mapping and core logging Indirect – holographic mapping and geophysical data visualisation
Structural model	HoloLens structural mapping, virtual core logging and virtual televiewer logs
Rock mass characterisation	Virtual geotechnical core logging, line and window mapping of discontinuities, and virtual DFN models
Hydrogeology model	Visualisation of piezometric installations, groundwater dewatering and depressurisation
Slope design	Slope design model and input data visualisation, virtual models of limit equilibrium and numerical modelling results
Slope performance (bench, inter-ramp and overall slope)	Virtual spatio-temporal bench/berm geometry models showing changes in pit slope geometry; holographic rockfall simulation, hazard assessment and mitigation; holographic slope failure visualisation
Blasting	Virtual models showing spatial blast performance variation and blast damage domains
Surface monitoring	Spatio-temporal VR/MR models of displacement results as virtual graphs or overlain on virtual pit slopes (e.g. GPS, extensometers, GB radar, satellite InSAR, LiDAR, photogrammetry)
Subsurface monitoring	Standalone virtual models of subsurface monitoring graphical results or incorporated within virtual pit slope models (e.g. inclinometers, TDR, extensometers, piezometers, microseismicity)
Uncertainty and risk	Holographic communication of data uncertainty, spatial geostatistical data and probabilistic slope analysis
Geo-environmental	VR models showing mine tailings, waste rock piles, mine closure and site reclamation

Glossary

Abbreviations

ADAS: automated data acquisition system
AOI: area of interest
API: application programming interface
ATS: see RTS
DEM: digital elevation model, pertaining to ground surface topography
DFN: discrete fracture network
DTM: digital terrain model
EDM: electronic distance meter
GB: ground-based; also defined in Keywords and conventions
GNSS: global navigation satellite system
GPS: global positioning system
LOS: line-of-sight, used interchangeably with slope distance
RTS: robotic total station, used interchangeably with ATS (automated total station)
TDR: time domain reflectometry
TLS: terrestrial laser scanner
UAV: unmanned aerial vehicle
VWP: vibrating wire piezometer
WHC: waste dump and stockpile hazard class

Keywords and conventions

Case history: historical summary of an event or sequence of events, used to make a point or provide a relevant experience
Case study: a term often used interchangeably with case history, but meant as a summary of an investigation
Deformation monitoring: used interchangeably with deformation survey
Drainholes: used interchangeably with horizontal drains
Emergency Preparedness and Response Plan: abbreviated as EPRP, used interchangeably with Emergency Response Plan (ERP)
Georeference: refers to relating a local measurement or data to a system of geographic coordinates
Geotechnical model: used interchangeably with geomechanical model
Ground based: abbreviated as GB, for example in reference to GB radar, GB InSAR, GB LiDAR, and used interchangeably with terrestrial
Ground-based radar: abbreviated to GB radar, and used interchangeably with terrestrial radar, slope monitoring radar and slope radar
Ground control incident: a broad term, generally understood by mine management and safety personnel, that includes all types of ground movement, including rockfall events, slope instability/failure, or fall of ground in an underground mining operation
Highwall: used interchangeably with excavated slope and pit slope
InSAR: space-borne interferometric synthetic aperture radar, used interchangeably with satellite InSAR and satellite radar, technically the same as SAR
LiDAR: light detection and ranging, used interchangeably with laser scanning
Meta-stable: slope region that has experienced greater than normal movement in the recent past but may currently be regarded as marginally stable; a zone to watch
Mine operator: the mine owner or managing company
Open pit: used interchangeably with pit, with a preference for open pit; the term open cast is often used for coal mines
Performance monitoring: a term often applied to tactical, campaign or safety critical monitoring, but is also applied to investigative, strategic, surveillance or systematic monitoring
RaDAR: radio detection and ranging, commonly simplified to radar
Real-time monitoring: defined as data that become available shortly after a measurement is made, used interchangeably with near real-time monitoring
Risk-based: a design approach applied with considerable relevant development experience, founded on leading industry practice, standards of design criteria and a thorough understanding of risk potential
Slope: used interchangeably with excavated slope
Slope distance: used interchangeably with line-of-sight (LOS)
Strategic monitoring: used interchangeably with surveillance monitoring or investigative monitoring
Tactical monitoring: used interchangeably with predictive monitoring or safety critical monitoring
Terrestrial photogrammetry: used interchangeably with photogrammetry
Theodolite: a precision optical instrument having a telescopic sight used for measuring angles between designated visible points in the horizontal and vertical planes; used interchangeably with total station
Total station: used interchangeably with robotic total station (RTS) or automated total station (ATS); see also theodolite
UAVs: unmanned aerial vehicles, used interchangeably with drones and UAV drones

References

Afeni TB (2011) An approach to eradicate the effects of atmospheric variations on total station distance measurement in a surface mine environment, the impact of taking measurement through a glass medium. Research Project submitted to the Faculty of Engineering and the Built Environment. University of the Witwatersrand, South Africa.

Afeni TB, Cawood F (2010) Do the properties of glass matter when taking total station distance measurements through an observation window? In *Proceedings of the International Society for Mine Surveying XIV International Congress*. 20–24 September, Sun City, South Africa. p. 167.

Afeni TB, Cawood FT (2013) Slope monitoring using total station: what are the challenges and how should these be mitigated? *South African Journal of Geomatics* **2**(1), 41–53.

Aitken HGJ (1976a) *Syntony and Spark: The Origins of Radio*. Princeton University Press, Princeton, NJ.

Aitken HGJ (1976b) *The Continuous Wave: Technology and American Radio, 1900–1932*. Princeton University Press, Princeton, NJ.

Aki K, Richards PG (2002) *Quantitative Seismology*. University Science Books, Mill Valley, CA.

Anon (1975) *Subsidence Engineers' Handbook*. National Coal Board, Mining Department, London.

Armstrong J, Rose ND (2009) Mine operation and management of progressive slope deformation on the south wall of the Barrick Goldstrike Betze-Post Open Pit. In *Proceedings of the International Symposium on Rock Slope Stability in Open Pit and Civil Engineering*. 9–11 November, Santiago. (Ed. J Read) University de los Andes, Santiago.

Audigé E, Fagan C, Beth M (2017) How to perform real-time monitoring when your reference network is moving? Innovative solution applied to Britomart Station on the CRL project in Auckland. In *Proceedings of the 16th Australasian Tunnelling Conference*. 30 October–1 November, Sydney. pp. 198–207.

AusSpec (2008) *Spectral Interpretation Field Manual*. GMEX 3rd edn. AusSpec International, Arrowtown, New Zealand.

Bannister A, Raymond S, Baker R (1998) *Surveying*. 7th edn. Addison-Wesley-Longman, Pearson Education, Reading, MA.

Barton IF, Gabriel M, Barton MD (2019) Pilot study of hyperspectral remote sensing for large-scale mineral mapping in active mining environments. *Proceedings of Copper* **2019**, 18–21. [August, Vancouver, Canada.]

BCMWRPRC (1991) *Mined Rock and Overburden Piles: Investigation and Design Manual (Interim Guidelines)*. British Columbia Mine Waste Rock Pile Research Committee, Vancouver.

Beale G, Read J (2013) *Guidelines for Evaluating Water in Pit Slope Stability*. CSIRO Publishing, Melbourne.

Beale G, Milmo P, Raynor M, Price M, Donzé F (2013) Preparing a conceptual hydrogeological model. In *Guidelines for Evaluating Water in Pit Slope Stability*. (Eds G Beale, J Read) pp. 153–214. CSIRO Publishing, Melbourne.

Beck TJ, Kane WF (1996) Current and potential uses of time domain reflectometry for geotechnical monitoring. In *Proceedings of the 47th Highway Geology Symposium*. Cody, WY. Wyoming Department of Transportation.

Bellett P, Noon D (2013) *Real Aperture Radar for Safety-Critical Slope Monitoring*. GroundProbe, Green Paper 101, Brisbane.

Bertacchini E, Capra A, Castagnetti C, Corsini A (2011) Atmospheric corrections for topographic monitoring systems in landslides. In *Proceedings of FIG Working Week: Bridging the Gap between Cultures*. 18–22 May, Marrakech, Morocco.

Betz J (2016) *Engineering Satellite-Based Navigation and Timing: Global Navigation Satellite Systems, Signals, and Receivers*. IEEE Press.

Bieniawski ZT (1976) Rock mass classification in rock engineering. In *Exploration for Rock Engineering* (Ed. ZT Bieniawski) **1**, pp. 97–106. Balkema, Cape Town.

Brown N, Kaloustian S, Roeckle M (2007) Monitoring of open pit mines using combined GNSS satellite receivers and robotic total stations. In *Proceedings of the International Symposium on Rock Slope Stability in Open Pit Mining and Civil Engineering*. 12–14 September, Perth. (Ed. Y Potvin) pp. 417–429. Australian Centre for Geomechanics, Perth.

Burland JB, Standing JR, Jardine FM (2001) Assessment methods used in design. In *Building Response to Tunnelling: Case Studies from Construction of the Jubilee Line Extension*. (Eds JB Burland, JR Standing, FM Jardine) pp. 23–43. Thomas Telford Publishing, London.

Bye A (2009) *Going Deeper: Deeper Open Pits*. International Mining, Hertfordshire, UK.

Cabrejo A, Harries N (2012) *Effective Slope Monitoring for Open Cut Coal Mines*. Project C17023. Australian Coal Industry Research Program, Brisbane.

Cahill J, Lee M (2006) Ground control at Leinster Nickel operations. In *Proceedings of the International Symposium of Rock Slopes in Open Pit Mining and Civil Engineering*. 3–6 April, Cape Town. pp. 321–334. South African Institute of Mining and Metallurgy, Johannesburg.

Call RD (1982) Monitoring pit slope behavior. In *Proceedings of the 3rd International Conference on Stability in Surface Mining*. 1–3 June 1981, Vancouver, Canada. (Ed. CO Brawner), **3**. Society of Mining Engineers of the American Institute of Mining, Metallurgical, and Petroleum Engineers, New York.

Call RD, Savely JP (1990) Open pit rock mechanics. In *Surface Mining*. 2nd edn. (Ed. BA Kennedy) pp. 860–882. Society for Mining, Metallurgy and Exploration, CO.

Call RD, Ryan TM, Barkely RC (1993) Geotechnical analyses for open pit mining in areas of large-scale slope instability. In *Proceedings of the International Congress on Mine Design*. 23–26 August, Kingston, ON, Canada. (Ed. JF Archibald) pp. 45–56. A.A. Balkema, Rotterdam.

CANMET (1977) *Pit Slope Manual*. Energy Mines and Resources Canada, Canada Centre for Mineral and Energy Technology, Ottawa.

Carlà T, Intrieri E, Farina P, Casagli N (2017a) A new method to identify impending failure in rock slopes. *International Journal of Rock Mechanics and Mining Sciences* **93**, 76–81. doi:10.1016/j.ijrmms.2017.01.015

Carlà T, Farina P, Intrieri E, Botsialas K, Casagli N (2017b) On the monitoring and early-warning of brittle slope failures in hard rock masses: examples from an open pit mine. *Engineering Geology* **228**(13), 71–81. doi:10.1016/j.enggeo.2017.08.007

Carneiro SR, Costa Ferreira F, Castilho BDM, Farina P (2017) Programa de monitoramento de deformações integrado para barragem de rejeitos: o caso do Complexo de Germano. In *Proceedings of Comitê Brasileiro de Barragens, XXXI – Seminário Nacional de Grandes Barragens (SNGB).* 15–18 May, Belo Horizonte, Brazil.

Carter M, Bentley SP (1985) The geometry of slip surfaces beneath landslides: predictions from surface measurements. *Canadian Geotechnical Journal* **22**, 234–238. doi:10.1139/t85-031

Casagli N, Pini G (1993) Analisi cinematica della stabilità in versanti naturali e fronti di scavo in roccia. In *Proceedings of 3° Convegno Nazionale dei Giovani Ricercatori in Geologia Applicata.* 28–30 October, Potenza, Italy.

Cawood FT, Stacey TR (2006) Survey and geotechnical slope monitoring considerations. *Journal of the South African Institute of Mining and Metallurgy* **106**(7), 495–501.

Chetty P (2013) Monitoring of mine tailings using satellite and lidar data. In *Proceedings of the South African Smart Grid Initiative (SASGI).* 22–24 July, Kempton Park, South Africa.

Chrzanowski A, Wilkins R (2006) Accuracy evaluation of geodetic monitoring of deformations in large open pit mines. In *Proceedings of the 3rd IAG Symposium on Geodesy for Geotechnical and Structural Engineering and 12th FIG Symposium on Deformation and Monitoring.* 22–24 May, Baden, Austria.

Clark R (1999) Spectroscopy of rocks and minerals and principles of spectroscopy. In *Remote Sensing for the Earth Sciences: Manual of Remote Sensing.* 3rd edn. (Eds AN Rencz, RA Ryerson) pp. 3–58. Wiley, Hoboken, NJ.

Clarkson L, Carneiro S, Castilho BDM, Williams D, Ruest M, Llano M, Noon D (2018) State-of-the-art monitoring techniques for Samarco tailings dams. In *Proceedings of the Slope Stability Conference.* 10–13 April, Seville, Spain.

Clayton CRI, Matthews MC, Simons NE (1995) *Site Investigation.* 2nd edn. Department of Civil Engineering, University of Surrey, UK. <www.geotechnique.info>.

Crook D (1995) PPM correction variables. *The Ontario Land Surveyor (Institute of Survey Technology of Ontario)*, Oshawa, Summer (online). <https://www.krcmar.ca/resource-articles/1995_Summer_PPM%20Correction%20Variables_1.pdf>.

Cruden DM (1986) The geometry of slip surfaces beneath landslides: predictions from surface measurements. *Canadian Geotechnical Journal* **23**, 94[discussion]. doi:10.1139/t86-012

Curry S, Sawyer R (1991) Getting the most out of least squares. *California Surveyor* **95**, 22–29.

Darcy H (1856) *Les fontaines publiques de la ville de Dijon.* Victor Dalmont, Paris.

Davis RE, Anderson JM, Foote FS, Mikhail EM (1981) *Surveying: Theory and Practice.* 6th edn. McGraw-Hill, New York.

Day AP, Seery JM (2007) Monitoring of a large wall failure at Tom Price iron ore mine. In *Proceedings of the International Symposium on Rock Slope Stability in Open Pit Mining and Civil Engineering.* 12–14 September, Perth. (Ed. Y Potvin) pp. 333–340. Australian Centre for Geomechanics, Perth.

De Athayde C, Sánchez C, Carneiro S, Castilho BDM (2017) O uso da tecnologia insar (interferometria de radar de abertura sintética) para o monitoramento da estabilidade de barragens. In *Proceedings of Comitê Brasileiro de Barragens, XXXI – Seminário Nacional de Grandes Barragens (SNGB).* 15–18 May, Belo Horizonte, Brazil.

de Graaf PJH, Wessels SDN (2013) Slope monitoring and data visualisation state-of-the-art: advancing to Rio Tinto Iron Ore's Mine of the Future™. In *Proceedings of the International Symposium on Slope Stability in Open Pit Mining and Civil Engineering.* 25–27 September, Brisbane. (Ed. PM Dight) pp. 803–814. Australian Centre for Geomechanics, Perth.

de Graaf P, Wessels S (2015) Rio Tinto Iron Ore's Pilbara Geotechnical Management System: a framework for managing geotechnical risks across multiple operations with multiple pits. In *Proceedings of the International Symposium on Rock Slope Stability in Open Pit Mining and Civil Engineering.* 12–14 October, Cape Town. South African Institute of Mining and Metallurgy, Johannesburg.

de Graaf PJH, Wessels SDN (2016) A framework for managing geotechnical risk across multiple operations. *Journal of the Southern African Institute of Mining and Metallurgy* **116**(5), 367–377. doi:10.17159/2411-9717/2016/v116n5a2

Debswana Diamond Mining Co. (2015) *About Us.* Debswana Diamond Mining Co., Orapa, Botswana. <http://www.debswana.com/About-Us/Pages/default.aspx>.

Denny M (2007) *Blip, Ping, and Buzz: Making Sense of Radar and Sonar.* Johns Hopkins University Press, Boston.

Dept of Mines and Petroleum (2011) *General Duty of Care in Western Australian Mines – Guideline.* 2nd edn. Department of Mines and Petroleum, Western Australia.

Dick GJ, Eberhardt E, Cabrejo-Liévano AG, Stead D, Rose ND (2014) Development of an early-warning time-of-failure analysis methodology for open-pit mine slopes utilizing ground-based slope stability radar monitoring data. *Canadian Geotechnical Journal* **52**(4), 515–529. doi:10.1139/cgj-2014-0028

Dight P, Baczynski NRC (2009) Rock bridges and their influence on slope stability. In *Proceedings of the International Symposium on Rock Slope Stability in Open Pit and Civil Engineering.* 9–11 November, Santiago, Chile. (Ed. J Read) University de los Andes, Santiago.

Domenico PA, Schwartz FW (1998) *Physical and Chemical Hydrogeology.* 2nd edn. John Wiley & Sons, New York.

Dowding CH, Dussud ML, Kane WF, O'Connor KM (2003) Monitoring deformation in rock and soil with TDR sensor cables. Part 1: concept and case history. *Geotechnical News* **21**(2), 51–59.

Doyle JB, Reese JD (2011) Slope monitoring and back analysis of east fault failure, Bingham Canyon Mine, Utah. In *Proceedings of the International Symposium on Rock Slope Stability in Open Pit Mining and Civil Engineering.* 18–21 September, Vancouver. Paper 269. Canadian Rock Mechanics Association.

Dunnicliff J (1993) *Geotechnical Instrumentation for Monitoring and Field Performance.* John Wiley & Sons, New York.

Dunnicliff J, Powderham AJ (2001) Recommendations for procurement of geotechnical instruments and field instrumentation services. *Geotechnical News* **19**(3), 30–35.

Eberhardt E (2008) The role of advanced numerical methods and geotechnical field measurements in understanding complex deep-seated rock slope failure mechanisms. *Canadian Geotechnical Journal* **45**(4), 484–510. doi:10.1139/T07-116

Eberhardt E, Stead D (2011) Geotechnical instrumentation. In *SME Mining Engineering Handbook.* 3rd edn. (Ed. P Darling) pp. 551–571. Society for Mining, Metallurgy and Exploration, CO.

Eberhardt E, Spillmann T, Maurer H, Willenberg H, Loew S, Stead D (2004) The Randa Rockslide Laboratory: establishing brittle and ductile instability mechanisms using numerical modelling and microseismicity. In *Proceedings of the 9th International Symposium on Landslides.* 28 June–2 July, Rio de Janeiro. (Eds W Lacerda, M Ehrlich, SAB Fontoura, ASF Sayao) pp. 481–487. A.A. Balkema, Leiden, Netherlands.

Elmo D, Clayton C, Rogers S, Beddoes R, Greer S (2011) Numerical simulations of potential rock bridge failure within a naturally fractured rock mass. In *Proceedings of the International Symposium on Rock Slope Stability in Open Pit Mining and Civil Engineering.* 18–21 September, Vancouver. Paper 272. Canadian Rock Mechanics Association.

Elmo D, Liu Y, Rogers S (2014) Principles of discrete fracture network modelling for geotechnical applications. In *Proceedings of the 1st International Conference on Discrete Fracture Network Engineering.* 19–22 October, Vancouver. Paper 238. Gibson Group Management, Richmond, BC.

Faraday M (1839–1855) *Experimental Researches in Electricity.* Vols 1, 2, 3. Cambridge University Press, Cambridge.

Farina P, Casagli N, Ferretti A (2007) Radar-interpretation of InSAR measurements for landslide investigations in civil protection practices. In *Proceedings of the 1st North American Landslide Conference.* 3–8 June, Vail, CO. pp. 272–283. American Society of Civil Engineers, Boston, VA.

Fenelli GB, Picarelli L (1990) The pore pressure field built-up in a rapidly eroded soil mass. *Canadian Geotechnical Journal* **27**(3), 387–392. doi:10.1139/t90-049

Feng Q (2012) Practical application of 3D laser scanning techniques to underground projects. Technical Report No. 114. Rock Engineering Research Foundation, Stockholm.

Ferretti A, Fumagalli A, Novali F, Prati C, Rocca F, Rucci A (2011) A new algorithm for processing interferometric data-stacks: SqueeSAR. *IEEE Transactions on Geoscience and Remote Sensing* **49**(9), 3460–3470. doi:10.1109/TGRS.2011.2124465

Ferretti A, Colombo D, Fumagalli A, Novali F, Rucci A (2015) InSAR data for monitoring land subsidence: time to think big. *Proceedings of the International Association of Hydrological Sciences* **372**, 331–334. doi:10.5194/piahs-372-331-2015

Fisher BR, Eberhardt E (2007) Numerical modeling and shear strength estimates of bi-planar dip slope failures. In *Proceedings of the 11th Congress of the International Society for Rock Mechanics.* 9–13 July, Lisbon. (Eds L Ribeiro e Sousa, C Olalla, N Grossmann) pp. 633–636. Taylor & Francis, London.

Freeze RA, Cherry JA (1979) *Groundwater.* Prentice Hall, NJ.

Fukuzono T (1985) A new method for predicting the failure time of a slope. In *Proceedings of the 4th International Conference and Field Workshop on Landslides.* 23–31 August, Tokyo. pp. 145–150. Japan Landslide Society, Tokyo.

Geokon (2014) *MEMS Tiltmeter.* Geokon, Lebanon, NH.

Geokon (2015) *MEMS In Place Inclinometer Systems.* Geokon, Lebanon, NH.

Gigli G, Casagli N (2011) Semi-automatic extraction of rock mass structural data from high resolution LIDAR point clouds. *International Journal of Rock Mechanics and Mining Sciences* **48**(2), 187–198. doi:10.1016/j.ijrmms.2010.11.009

Gill S (2013) Management of critical alarms: connecting the dots. *Process Safety Progress* **32**(1), 66–71. doi:10.1002/prs.11549

Ginting A, Stawski M, Widiadi R (2011) Geotechnical risk management and mitigation at Grasberg Open Pit, PT Freeport Indonesia. In *Proceedings of the International Symposium on Rock Slope Stability in Open Pit Mining and Civil Engineering.* 18–21 September, Vancouver. Paper 122. Canadian Rock Mechanics Association.

Girard JM, McHugh E (2000) Detecting problems with mine slope stability. In *Proceedings of the 31st Annual Institute on Mining Health, Safety and Research.* 27–30 August, Roanoke, VA.

Goodman RE, Bray JW (1976) Toppling of rock slopes. In *Proceedings of the Special Conference on Rock Engineering for Foundations and Slopes* **2**. pp. 201–234. American Society of Civil Engineers, Boulder, CO.

Hamman ECF (2009) Qualitative geotechnical hazard and risk assessment. In *Proceedings of the International Symposium on Rock Slope Stability in Open Pit and Civil Engineering.* 9–11 November, Santiago. (Ed. J Read) University de los Andes, Santiago.

Harries N, Noon D, Rowley K (2006) Case studies of slope stability radar used in open cut mines. In *Proceedings, Stability of Rock Slopes in Open Pit Mining and Civil Engineering Situations.* 3–6 April, Cape Town. pp. 335–342. South African Institute of Mining and Metallurgy, Johannesburg.

Hartman HL (1992) *SME Mining Engineering Handbook.* Society for Mining, Metallurgy and Exploration, Littleton, CO.

Havaej M, Stead D (2016) Investigating the role of kinematics and damage in the failure of rock slopes. *Computers and Geotechnics* **78**, 181–193. doi:10.1016/j.compgeo.2016.05.014

Hawley M, Cunning J (2017) *Guidelines for Mine Waste and Stockpile Design.* CSIRO Publishing, Melbourne.

Hawley M, Marisett S, Beale G, Stacey P (2009) Performance assessment and monitoring. In *Guidelines for Open Pit Slope Design.* (Eds J Read, P Stacey) pp. 327–380. CSIRO Publishing, Melbourne.

Henning W (2011) *Survey Guidelines for Real Time GNSS Networks.* National Geodetic Survey, Silver Spring, MD.

Hoek E (2007) *Practical Rock Engineering.* Rocscience, Toronto.

Hoek E, Bray J (1974) *Rock Slope Engineering.* Institute of Mining and Metallurgy, London.

Hoek E, Bray J (1981) *Rock Slope Engineering.* 3rd edn. Institute of Mining and Metallurgy, London.

Hoek E, Brown ET (2019) The Hoek–Brown failure criterion and GSI – 2018 edition. *Journal of Rock Mechanics and Geotechnical Engineering* **11**(3), 445–463. doi:10.1016/j.jrmge.2018.08.001

Hoek E, Rippere KH, Stacey PF (2001) Large-scale slope designs: a review of the state of the art. In *Proceedings of Slope Stability in Surface Mining.* 26–28 February, Denver. (Eds WA Hustrulid, MK McCarter, DJA van Zyl) pp. 3–10. Society for Mining, Metallurgy and Exploration, CO.

Hogarth J, Hawley M, Beale G (2017) Instrumentation and monitoring. In *Guidelines for Mine Waste Dump and Stockpile Design.* (Eds M Hawley, J Cunning) pp. 207–231. CSIRO Publishing, Melbourne.

Hudson JA, Harrison JP (1997) *Engineering Rock Mechanics.* Pergamon, Netherlands.

Hutchison BJ, Chambers J, Macqueen G, Salas E, Huang N (2015) The use of rockfall statistics in managing small rockfalls in two Australian open pit mines. In *Proceedings of the International Symposium on Rock Slope Stability in Open Pit Mining and Civil Engineering.* 12–14 October, Cape Town. South African Institute of Mining and Metallurgy, Johannesburg.

Hvorslev MJ (1951) *Time Lag and Soil Permeability in Ground-Water Observations.* US Army Corps of Engineers, Bulletin No. 36, Waterways Experiment Station, Vicksburg, MS.

IDS (2014) *IBIS-FM: Interferometric Mine Slope Radar.* IDS, Pisa, Italy.

Jacob CE (1940) On the flow of water in an elastic artesian aquifer. *Transactions - American Geophysical Union* **22**, 574–586.

Jefferies M, Morgenstern NR, Van Zyl D, Wates J (2019) *Report on NTSF Embankment Failure: Cadia Valley Operations for Ashurst Australia.* H356804-00000-22A-230-0001. Newcrest Mining, Melbourne.

Jennings JE (1970) A mathematical theory for the calculation of the stability of slopes in open cast mines. In *Proceedings of the Symposium on the Theoretical Background to the Planning of Open Pit Mines.* 29 August–4 September, Johannesburg. (Ed. PWJ van Rensburg) pp. 87–102. South African Institute of Mining and Metallurgy, Johannesburg/AA Balkema, Amsterdam.

Jones E, Andrews P, Holley S (2011) The Wallaby Mine: maintaining pit wall stability for continued underground mining. In *Proceedings of the International Symposium on Rock Slope Stability in Open Pit Mining and Civil Engineering.* 18–21 September, Vancouver. Paper 188. Canadian Rock Mechanics Association.

Jooste MA, Cawood FT (2006) Survey slope stability monitoring: lessons from Venetia Diamond Mine. In *Proceedings of the International Symposium on Stability of Rock Slopes in Open Pit Mining and Civil Engineering.* 3–6 April, Cape Town. pp. 361–374. South African Institute of Mining and Metallurgy, Johannesburg.

Kayesa G (2006) Prediction of slope failure at Letlhakane Mine with the Geomos slope monitoring system. In *Proceedings of the International Symposium on Stability of Rock Slopes in Open Pit Mining and Civil Engineering.* 3–6 April, Cape Town. pp. 605–622. South African Institute of Mining and Metallurgy, Johannesburg.

Kemeny J (2003) The time-dependent reduction of sliding cohesion due to rock bridges along discontinuities: a fracture mechanics approach. *Rock Mechanics and Rock Engineering* **36**(1), 27–38. doi:10.1007/s00603-002-0032-2

Kennedy B, Niermeyer KE (1970) Slope monitoring systems used in the prediction of a major slope failure at the Chuquicamata Mine, Chile. In *Proceedings of the Symposium on the Theoretical Background to the Planning of Open Pit Mines.* 29 August–4 September, Johannesburg. (Ed. PWJ van Rensburg) pp. 215–225. South African Institute of Mining and Metallurgy, Johannesburg/AA Balkema, Amsterdam.

Kilburn C (2012) Precursory deformation and fracture before brittle rock failure and potential application to volcanic unrest. *Journal of Geophysical Research* **117**, B02211. doi:10.1029/2011JB008703

Klappstein B, Bond G, Maston W (2014) Implementation of real-time geotechnical monitoring at an open pit mountain coal mine in western Canada. *Mining Revue* **20**(3), 2–14.

Kokaly R, Clark R, Swayze G, Livo K, Hoefen T, Pearson N, Wise R, Benzel W, Lowers H, Driscoll R, Klein A (2017) *USGS Spectral Library Version 7.* USGS Geological Survey, Data Series 1035, Denver.

Kostak B, Rybar J (1993) Alarming limits in slope deformation monitoring. In *Landslides: Proceedings of the Seventh International Conference and Field Workshop* (Eds Novasad & Wagner). A.A. Balkema, Rotterdam.

Kruseman GP, de Ridder NA (1994) *Analysis and Evaluation of Pumping Test Data.* 2nd edn. International Institute for Land Reclamation and Improvement, Wageningen, Netherlands.

Lang AM (1999) Geotechnical mining regulations in Western Australia. In *Rock Support and Reinforcement Practice in Mining.* (Eds AG Thompson, E Villaescusa, CR Windsor) AA Balkema, Rotterdam.

Langley RB (1990) Why is the GPS signal so complex? *GPS World* **1**(3), 56–59.

Langley RB (1991a) The GPS receiver: an introduction. *GPS World* **2**(1), 50–53.

Langley RB (1991b) The mathematics of GPS. *GPS World* **2**(7), 45–50.

Langley RB (1998) A primer on GPS antennas. *GPS World* **9**(7), 73–77.

Laserscanning Europe GmbH (2015) *RIEGL VZ-4000.* Laserscanning Europe GmbH, Magdeburg, Germany. <http://www.laserscanning-europe.com/en/glossary/riegl-vz-4000>.

Lato M (2019) Communicating complex engineering projects using holographic visualization. *LiDAR Magazine* **9**(1).

Lauer H, Brown H (1920) *Radio Engineering Principles.* McGraw-Hill, New York.

Lee SG (2011) Characterizing highwall slopes at the Line Creek Mine, British Columbia, using terrestrial photogrammetry. Masters of Science thesis. Simon Fraser University, Canada.

Leica Geosystems (2004) *Reporter.* No. 50. Leica Geosystems, St Gallen, Switzerland.

Li I, Lato M, Magnusson G, Roca J, Reid E (2019) Augmented reality and applied earth science: a new tool for site characterization. In *Proceedings of the 26th Symposium of the Vancouver Geotechnical Society.* 31 May, Vancouver. Vancouver Geotechnical Society, Vancouver.

Little MJ (2006) Slope monitoring strategy at Prust open pit operation. In *Proceedings of the International Symposium on Stability of Rock Slopes in Open Pit Mining and Civil Engineering.* 3–6 April, Cape Town. pp. 211–230. South African Institute of Mining and Metallurgy, Johannesburg.

Lohman SW (1972) *Definitions of Selected Ground-Water Terms: Revisions and Conceptual Refinements.* Water Supply Paper 1988. US Geological Survey, Washington, DC.

Lomberg C, Ream I, O'Rourke R, Read J (2013) Monitoring and design reconciliation. In *Guidelines for Evaluating Water in Pit Slope Stability.* (Eds G Beale, J Read) pp. 363–389. CSIRO Publishing, Melbourne.

Lorig L, Alvarez C (2012) *Subsidence Analysis for Jwaneng Cut 8: 2012 Update.* Consulting Report. Itasca S.A., Santiago, Chile.

Lorig L, Stead D (2017) Numerical analysis in rock slopes. In *Rock Slope Engineering.* 5th edn. (Ed. DC Wyllie) pp. 319–349. CRC Press, Boca Raton, FL.

Luo X, Salvoni M, Dight P (2018) Microseismic events for slope stability analysis: a case study at an open pit mine. *Journal of the Southern African Institute of Mining and Metallurgy* **118**, 205–210. doi:10.17159/2411-9717/2018/v118n3a2

Lupogo K (2017) Characterization of blast damage in rock slopes: an integrated field-numerical modeling approach. PhD thesis. Simon Fraser University, Canada.

Lynch RA, Malovichko DA (2006) Seismology and slope stability in open pit mines. In *Proceedings of the International Symposium on Stability of Rock Slopes in Open Pit Mining and Civil Engineering.* 3–6 April, Cape Town. pp. 375–390. South African Institute of Mining and Metallurgy, Johannesburg.

Lynch RA, Wuite R, Smith BS, Cichowicz A (2005) Microseismic monitoring of open pit slopes. In *Proceedings of the 6th International Symposium on Rockbursts and Seismicity in Mines.* 9–11 March, Perth. (Eds Y Potvin, M Hudyma) pp. 581–592. Australian Centre for Geomechanics, Perth.

Macdonald BC, Sharon R, Muir S, Iannacone JP, Falorni G, Michaud JS (2016) InSAR derived synoptic ground surface deformation map of the Porgera Mine, Papua New Guinea. In *Proceedings of the First Asia Pacific Slope Stability in Mining Conference.* 6–8 September, Brisbane. (Ed. PM Dight) pp. 673–683. Australian Centre for Geomechanics, Perth.

Machan G, Bennet VG (2008) *Use of Inclinometers for Geotechnical Instrumentation on Transportation Projects.* Transport Research Board, Washington, DC.

Macqueen GK, Salas EI, Hutchison BJ (2013) Application of radar monitoring at Savage River Mine, Tasmania. In *Proceedings of the International Symposium on Rock Slope Stability in Open Pit Mining and Civil Engineering.* 25–27 September, Brisbane. (Ed. PM Dight) pp. 1011–1020. Australian Centre for Geomechanics, Perth.

Manolakis D, Lockwood R, Cooley T (2016) *Hyperspectral Imaging Remote Sensing: Physics, Sensors, and Algorithms.* Cambridge University Press, Cambridge, UK.

Mao J, Nindl D (2009) *Surveying Reflectors: Characteristics and Influences.* White Paper. Leica Geosystems, St Gallen, Switzerland.

Maritz JA, de Beer K (2018) Using games to teach support design. In *Proceedings, 10th Asian Rock Mechanics Symposium,* Singapore. ISRM-ARMS10-2018-079.

Marr WE (2000) Advances in geotechnical measurements. In *Judgment and Innovation: The Heritage and Future of the Geotechnical Engineering Profession.* (Eds F Silva, E Kavazanjian). Geotechnical Special Publication **111**, pp. 9–12. American Society of Civil Engineers, Reston, VA.

Marr WA (2007) Why monitor geotechnical performance? In *Proceedings of the 7th International Symposium on Field Measurements in Geomechanics.* 24–27 September, Boston. (Ed. J DiMaggio) pp. 76–103. American Society of Civil Engineers, Reston, VA.

Marston DL (1996) *Law for Professional Engineers.* McGraw-Hill Ryerson, Toronto.

Martin CD, Stacey P (2018) *Guidelines for Open Pit Slope Design in Weak Rocks.* CSIRO Publishing, Melbourne.

Matheson GD (1983) *Rock Stability Assessment in Preliminary Site Investigations: Graphical Methods.* Report 1039. Transport and Road Research Laboratory, UK.

Maxwell JC (1873) *A Treatise on Electricity and Magnetism.* Vols 1, 2. Cambridge University Press, Cambridge, UK.

McCarter MK (1976) Monitoring stability of high waste dumps. In *Proceedings of the SME-AIME Fall Meeting and Exhibit.* 1–3 September, Denver.

McHugh EL, Long DG, Sabine C (2004) Applications of ground-based radar to mine slope monitoring. In *Proceedings of the ASPRS Annual Conference.* 23–28 May, Denver. American Society for Photogrammetry and Remote Sensing – Imaging and Geospatial Society, Bethesda, MD.

McKenna GT (1995) Grouted-in installation of piezometers in boreholes. *Canadian Geotechnical Journal* **32**(2), 355–363. doi:10.1139/t95-035

Mendecki AJ (1997) *Seismic Monitoring in Mines.* Springer, Berlin.

Mertie JB (1947) The delineation of parallel folds. *Geological Society of America Bulletin* **58**, 779–802. doi:10.1130/0016-7606(1947)58[779:DOPFAM]2.0.CO;2

Mhutsiwa O, Mongwa T, Mafa B (2013) *Installation of Geotech Monitoring Instruments at Jwaneng Mine 2013.* Installation Report. Aridus Hydro Geotechnik, Gaborone, Botswana.

Mikkelsen PE (2000) *Grouting-In Piezometers.* Slope Indicator.

Mikkelsen PE (2002) Cement-bentonite grout backfill for borehole instruments. *Geotechnical News* **20**, 38–42.

Milgram P, Kishino F (1984) A taxonomy of mixed reality visual displays. *IEICE Transactions on Information and Systems* **77**(12), 1321–1329.

Milgram P, Takemura H, Utsumi A, Kishino F (1994) Augmented reality: a class of displays on the reality–virtuality continuum. In *Proceedings of Telemanipulator and Telepresence Technologies.* 31 October–1 November, Boston. (Ed. H Das) pp. 282–292. SPIE 2351, Bellingham, WA.

Mine Subsidence Engineering Consultants (2015) *Mine Subsidence Damage to Building Structures.* Mine Subsidence Engineering Consultants Pty Ltd., Sydney.

Misra P, Enge P (2012) *Global Positioning System: Signals, Measurements, and Performance.* Ganga Jamuna Press, Lincoln, MA.

Moffitt K, Rogers S, Beddoes R, Greer S (2007) Analysis of slope stability in strong, fractured rock at the Diavik Diamond Mine, NWT. In *Rock Mechanics: Meeting Society's Challenges and Demands.* (Eds E Eberhardt, D Stead, T Morrison) **2**, pp. 1245–1250. Taylor & Francis, UK.

Moore DP, Imrie AS, Baker DG (1991) Rockslide risk reduction using monitoring. In *Proceedings of the Canadian Dam Association Meeting.* 3–8 October, Whistler. pp. 245–258. Canadian Dam Safety Association, Toronto.

Moss A, Diachenko S, Townsend P (2006) Interaction between the block cave and the pit slopes at Palabora mine. *Journal of the Southern African Institute of Mining and Metallurgy* **106**, 479–484.

Mossop D (2014) *Integrated Slope Monitoring*. Anglo American, Johannesburg.

Mphathiwa N (2012) Design principles for a survey slope monitoring system. Research Project submitted to the Faculty of Engineering and the Built Environment. University of the Witwatersrand, South Africa.

Mphathiwa N, Cawood FT (2014) Design principles for optimizing an established survey slope monitoring system. *Journal of the Southern African Institute of Mining and Metallurgy* **114**, 463–470.

Mufundirwa A, Fujii Y, Kodama J (2010) A new practical method for prediction of geomechanical failure-time. *International Journal of Rock Mechanics and Mining Sciences* **47**, 1079–1090. doi:10.1016/j.ijrmms.2010.07.001

Mura JC, Gama FF, Paradella WR, Negrão P, Carneiro S, de Oliveira CG, Brandão WS (2018) Monitoring the vulnerability of the dam and dikes in Germano Iron Mining Area after the collapse of the tailings dam of Fundão (Mariana-MG, Brazil) using DInSAR techniques with TerraSAR-X data. *Remote Sensing* **10**(10), 1507. doi:10.3390/rs10101507

Mysiorek J, Onsel E, Stead D, Rosser N (2019a) Engineering geological characterization of the 2014 Jure Nepal landslide: an integrated field, remote-sensing-virtual/mixed reality approach. In *Proceedings of the 53rd US Rock Mechanics/ Geomechanics Symposium*. 23–26 June, New York. ARMA-2019-38.

Mysiorek J, Onsel E, Stead D, Rosser N (2019b) Engineering geological characterization of the 2014 Jure Landslide, Nepal: an interactive mixed-reality approach to slope characterization, hazard identification and mitigation. In *Proceedings of GeoStJohns: Canadian Geotechnical Society Annual Conference*. 29 September–2 October, St Johns, Newfoundland.

National Coal Board – Mining (1975) *Subsidence Engineers' Handbook*. National Coal Board, London.

Neuzil C (1993) Low fluid pressure within the Pierre Shale: a transient response to erosion. *Water Resources Research* **29**(7), 2007–2020. doi:10.1029/93WR00406

Newcomen W, Dick G (2015) An update to strain-based pit wall failure prediction method and a justification for slope monitoring. In *Proceedings, Slope Stability 2015*, pp. 139–150. Cape Town.

Newcomen WH, Murray C, Shwydiuk L (2003) Monitoring pit wall deformations in real time at Highland Valley Copper. In *Proceedings of the Fourth International Conference on Computer Applications in the Minerals Industry*, Calgary.

Ngidi S, Boshoff P (2007) Cave management and secondary breaking practices at Palabora Mining Company. *Journal of the Southern African Institute of Mining and Metallurgy* **107**, 783–789.

Nicoll S, Bar N, du Plessis P (2017) Managing risks on an active haul road adjacent to a propagating subsidence zone at Telfer Gold Mine. *Procedia Engineering* **191**, 172–178. doi:10.1016/j.proeng.2017.05.169

Noon D (2003) Slope stability radar for monitoring mine walls. *Proceedings of the Mining Risk Management Conference*. 9–12 September, Sydney. pp. 267–276. Australasian Institute of Mining and Metallurgy, Melbourne.

Noon D, Harries N (2007) Slope stability radar for managing rock fall risks in open cut mines. In *Proceedings of the Large Open Pit Mining Conference*. 10–11 September, Perth. Australasian Institute of Mining and Metallurgy, Melbourne.

Noon D, Harries N, Rowley K (2006) Case studies of slope stability radar used in open cut mines. In *Proceedings of the International Symposium on Stability of Rock Slopes in Open Pit Mining and Civil Engineering*. 3–6 April, Cape Town. pp. 335–342. South African Institute of Mining and Metallurgy, Johannesburg.

Noon D, Bellet P, Campbell L (2011) *A New and Easy-to-Use Slope Radar Tool for Mine Production Crews*. GroundProbe, Brisbane.

O'Connor KM, Dowding CH (1999) *GeoMeasurements by Pulsing TDR Cables and Probes*. CRC Press, Boca Raton, FL.

Onsel E, Donati D, Stead D, Chang O (2018) Applications of virtual and mixed reality in rock engineering. In *Proceedings of the 52nd US Rock Mechanics/Geomechanics Symposium*. 17–20 June, Seattle. ARMA-2018-798.

Onsel E, Chang O, Mysiorek J, Donati D, Stead D, Barnett W, Zorzi L (2019) Applications of mixed and virtual-reality techniques in site characterization. In *Proceedings of the 26th Symposium of the Vancouver Geotechnical Society*. 31 May, Vancouver. Vancouver Geotechnical Society, Vancouver.

Peck RB (1969) Ninth Rankine Lecture: advantages and limitations of the observational method in applied soil mechanics. *Geotechnique* **19**(2), 171–187. doi:10.1680/geot.1969.19.2.171

Peck RB (1984) Observation and instrumentation: some elementary considerations – 1983 postscript. In *Judgement in Geotechnical Engineering: The Professional Legacy of Ralph B. Peck*, pp. 128–130. Wiley, New York.

Petley DN, Bulmer MH, Murphy W (2002) Patterns of movement in rotational and translational landslides. *Geology* **30**(8), 719–722. doi:10.1130/0091-7613(2002)030<0719:POMIRA>2.0.CO;2

Pienaar A, Joubert A (2011) Guidelines for deriving alarm settings based on pre-determined criteria using the movement and surveying radar (MSR). *Reutech Mining Case Studies and Papers*, 1–10.

Pieraccini M (2013) Real beam vs synthetic aperture radar for slope monitoring. In *Proceedings of the Progress in Electromagnetics Research Symposium*. 12–15 August, Stockholm. pp. 1627–1632.

Preene M (2012) Groundwater control. In *ICE Manual of Geotechnical Engineering Vol. 2: Geotechnical Design, Construction and Verification*. (Eds J Burland, T Chapman, H Skinner, M Brown) pp. 1173–1190. ICE Publishing, London.

Rabus B, Eberhardt E, Stead D, Ghuman P, Nadeau C, Woo K, Severin J, Styles T, Gao F (2009) Application of InSAR to constrain 3-D numerical modelling of complex discontinuous pit slope deformations. In *Proceedings of the International Symposium on Rock Slope Stability in Open Pit and Civil Engineering*. 9–11 November, Santiago. (Ed. J Read) University de los Andes, Santiago.

Ratliff TA (2003) *The Laboratory Quality Assurance System: A Manual of Quality Procedures and Forms*. John Wiley & Sons, New York.

Read J, Stacey P (2009) *Guidelines for Open Pit Slope Design*. CRC Press/AA Balkema, Boca Raton, FL.

Read J, Beale G, Ruest M, Robotham M (2013) Introduction. In *Guidelines for Evaluating Water in Pit Slope Stability*. (Eds G Beale, J Read) pp. 1–18. CSIRO Publishing, Melbourne.

Rocscience (2019) *HoloLens: Rocscience in Augmented Reality – Turn your Slide3 and RS3 Visualizations into Experiences*. RocScience News.

Rödelsperger S, Läufer G, Gerstenecker C, Becker M (2010) Monitoring of displacements with ground-based microwave interferometry: IBIS-S and IBIS-L. *Journal of Applied Geodesy* **4**(1), 41–54. doi:10.1515/jag.2010.005

Rose ND, Hungr O (2007) Forecasting potential rock slope failure in open pit mines using the inverse-velocity method. *International Journal of Rock Mechanics and Mining Sciences* **44**(2), 308–320. doi:10.1016/j.ijrmms.2006.07.014

Rose N, Scholz M (2009) Analysis of complex deformation behaviour in large open pit mine slopes using the Universal Distinct Element Code (UDEC). In *Proceedings of the International Symposium on Rock Slope Stability in Open Pit and Civil Engineering*. 9–11 November, Santiago. (Ed. J Read) University de los Andes, Santiago.

Rose ND, Scholz M, Burden J, King M, Maggs C, Havaej M (2018) Quantifying transitional rock mass disturbance in open pit slopes related to mining excavation. In *Proceedings of the XIV International Congress on Energy and Mineral Resources*. 11–13 April, Seville. pp. 1273–1288. Asociación Nacional de Ingenieros de Minas, Seville.

Rutledge D (2010) Accuracy versus precision: a primer on GPS truth. *GPS World* **21**(5), 42–49.

Salvoni M (2017) Rock damage assessment in a large unstable slope from microseismic monitoring: MMG Century Mine case study. PhD thesis. University of Western Australia, Australia.

Sato S, McParland MA, Henschel M (2012) *Cut 8 Infrastructure Monitoring: InSAR Monitoring at Jwaneng Mine, Debswana, Milestone 1*. Technical Report. MDA Geospatial Services, Ottawa.

Senfaute G, Duperret A, Lawrence JA (2009) Micro-seismic precursory cracks prior to rock-fall on coastal chalk cliffs: a case study at Mesnil-Val, Normandie, NW France. *Natural Hazards and Earth System Sciences* **9**, 1625–1641. doi:10.5194/nhess-9-1625-2009

Severin J, Eberhardt E, Leoni L, Fortin S (2014) Development and application of a pseudo-3D pit slope displacement map derived from ground-based radar. *Engineering Geology* **181**, 202–211. doi:10.1016/j.enggeo.2014.07.016

Singer J, Thuro K (2009) Development of a continuous 3D-monitoring system for unstable slopes using time domain reflectometry. In *Engineering Geology for Tomorrow's Cities*. (Eds MG Culshaw, HJ Reeves, I Jefferson, TW Spink) Geological Society, London.

Sjöberg J (1996a) Analysis of large-scale rock slopes. PhD thesis. Luleå University of Technology, Sweden.

Sjöberg J (1996b) *Large-Scale Slope Stability in Open Pit Mining: A Review*. Technical Report 1996:10T. Division of Rock Mechanics, Luleå University of Technology, Sweden.

Slope Indicator (2006) *Digitilt Inclinometer Probe 50302599 Manual*. Slope Indicator, Washington, DC.

Sobolev GA (2011) Seismicity dynamics and earthquake predictability. *Natural Hazards and Earth System Sciences* **11**, 445–458. doi:10.5194/nhess-11-445-2011

Soldata (2015) *GEOSCOPE Software: Real-Time Data Management System for Geotechnical, Structural and Environmental Monitoring*. Soldata, Melbourne.

Stacey TR, Xianbin Y, Armstrong R, Keyter GJ (2003) New slope stability considerations for deep open pit mines. *Journal of the South African Institute of Mining and Metallurgy* **103**(6), 373–389.

Stark TD, Choi H (2008) Slope instrumentation for landslides. *Landslides* **5**(3), 339–350. doi:10.1007/s10346-008-0126-3

Stead D, Eberhardt E, Coggan JS (2006) Developments in the characterization of complex rock slope deformation and failure using numerical modelling techniques. *Engineering Geology* **83**(1–3), 217–235. doi:10.1016/j.enggeo.2005.06.033

Stead D, Donati D, Wolter A, Sturzenegger M (2019) Application of remote sensing to the investigation of rock slopes: experience gained and lessons learned. *ISPRS International Journal of Geo-Information* **8**(7), 296. doi:10.3390/ijgi8070296

Strouth A, Eberhardt E (2006) The use of LiDAR to overcome rock slope hazard data collection challenges at Afternoon Creek, Washington. In *Proceedings of the 41st US Symposium on Rock Mechanics: 50 Years of Rock Mechanics – Landmarks and Future Challenges*. 17–21 June, Golden, CO. American Rock Mechanics Association. e-Paper 06-993.

Sturzenegger M, Stead D (2009) Close-range terrestrial digital photogrammetry and terrestrial laser scanning for discontinuity characterization on rock cuts. *Engineering Geology* **106**, 163–182. doi:10.1016/j.enggeo.2009.03.004

Su MB, Liao CH (2006) Application of TDR cables for landslide monitoring. In *Proceedings of the 3rd International Symposium and Workshop on Time Domain Reflectometry for Innovative Soils Applications*. 17–20 September, West Lafayette, IN. pp. 1–12. Purdue University, West Lafayette.

Sullivan TD (1993) Understanding pit slope movements. In *Proceedings of the Australian Conference on Geotechnical Instrumentation and Monitoring in Open Pit and Underground Mining*. 21–23 June, Kalgoorlie, Western Australia. pp. 435–446. AA Balkema, Rotterdam.

Tatone BS, Grasselli G (2010) A new 2D discontinuity roughness parameter and its correlation with JRC. *International Journal of Rock Mechanics and Mining Sciences* **47**(8), 1391–1400. doi:10.1016/j.ijrmms.2010.06.006

Terzaghi K (1950) Mechanism of landslides. In *Application of Geology to Engineering Practices*. Berkey Volume. pp. 83–123. Geological Society of America, Boulder, CO.

Terzaghi K, Peck RB (1948) *Soil Mechanics in Engineering Practice*. Wiley, New York.

Thomas HW (2011) Slope stability monitoring: a guide for practicing mine surveyors. Research Project submitted to the Faculty of Engineering and the Built Environment. University of the Witwatersrand, South Africa.

Thompson C (2005) *Importance of Good-Quality Surveying Control Network in an Automated Slope Stability Monitoring Programme*. Circular. Geosystems Africa.

Torlatch J (1996) New legislative directions. In *Proceedings of Minesafe*. 9–13 September, Perth. pp. 117–122. Chamber of Minerals and Energy of Western Australia, Perth.

Trifu CI, Shumila V, Leslie I (2008) Application of joint seismic event location techniques at Chuquicamata open pit mine, Chile. In *Proceedings of the 5th International Conference and Exhibition on Mass Mining*. 9–11 June, Luleå. (Eds H Schunnesson, E Nordlund) pp. 943–952. Luleå University of Technology, Sweden.

Tuckey ZS (2012) An integrated field mapping-numerical modelling approach to characterising discontinuity persistence and intact rock bridges in large open pit slopes. MSc thesis. Simon Fraser University, Canada.

Tuckey Z, Stead D (2016) Improvements to field and remote sensing methods for mapping discontinuity persistence and intact rock bridges in rock slopes. *Engineering Geology* **208**, 136–153. doi:10.1016/j.enggeo.2016.05.001

Vaziri A, Moore L, Ali H (2010) Monitoring systems for warning of impending failures in slopes and open pit mines. *Natural Hazards* **55**, 501–512. doi:10.1007/s11069-010-9542-5

Watt IB (1996) *Monitoring Surveys at Letlhakane and Orapa Open Pit Mines*. Consulting Report.

Wei WS (2019) *Time Series Analysis: Univariate and Multivariate Methods*. Pearson Education, Boston.

Wesseloo J, Sweby GJ (2008) Microseismic monitoring of hard rock slopes. In *Proceedings of the Southern Hemisphere International Rock Mechanics Symposium*. 16–19 September, Perth. (Eds Y Potvin, J Carter, A Dyskin, R Jeffrey) pp. 433–450. Australian Centre for Geomechanics, Perth.

Wesseloo J, Dight PM, Potvin Y (2009) *High-Resolution Seismic Monitoring in Open Pit Mine*. Report No. 279. MERIWA, Australian Centre for Geomechanics, Perth.

Whiteman DS (2010) The Smart Marker system: a new tool for measuring underground ore body flow in block and sub-level mines. *Mining Technology* **119**(3), 182–192. doi:10.1179/174328610X12820409992499

Widzyk-Capehart E, Hölck C, Fredes O, Pedemonte I (2015) Implementation of novel subsurface deformation sensing device for open pit slope stability monitoring: the Networked Smart Markers System. In *Proceedings of the 23rd International Symposium on Mine Planning and Equipment Selection*. 11–13 November, Johannesburg. pp. 203–212. South African Institute of Mining and Metallurgy, Johannesburg.

Wild CJ, Seber GAF (2000) *Chance Encounters: A First Course in Data Analysis*. John Wiley & Sons, New York.

Wilkins R, Bastin G, Chrzanowski A (2003) ALERT: a fully automated real-time monitoring system. In *Proceedings of the 11th FIG Symposium on Deformation Measurements*. 25–28 May, Santorini. (Eds SC Stiros, S Pytharouli) Patras University, Greece.

Willenberg H, Evans KF, Eberhardt E, Spillmann T, Loew S (2008) Internal structure and deformation of an unstable crystalline rock mass above Randa (Switzerland): Part II – three-dimensional deformation patterns. *Engineering Geology* **101**(1-2), 15–32. doi:10.1016/j.enggeo.2008.01.016

Williams P, Floyd J, Chitombo G, Maton T (2009) Design implementation. In *Guidelines for Open Pit Slope Design*. (Eds J Read, P Stacey) pp. 265–326. CSIRO Publishing, Melbourne.

Woo KS, Eberhardt E, Rabus B, Stead D, Vyazmensky A (2012) Integration of field characterization, mine production and InSAR monitoring data to constrain and calibrate 3-D numerical modelling of block caving-induced subsidence. *International Journal of Rock Mechanics and Mining Sciences* **53**, 166–178. doi:10.1016/j.ijrmms.2012.05.008

World Hydrological Cycle Observation System (2004) *Levelling and Surveying*. Whycos. <http://www.whycos.org/fck_editor/upload/File/Pacific-HYCOS/Surface_Waters/Levelling_and_surveying.pdf>.

Wyllie DC (2017) *Rock Slope Engineering: Civil Applications*. 5th edn. CRC Press, Boca Raton, FL.

Yang DY, Mercer RA, Brouwer KJ, Tomlinson C (2011) Managing pit slope stability at the Kemess South Mine: changes over time. In *Proceedings of the International Symposium on Rock Slope Stability in Open Pit Mining and Civil Engineering*. 18–21 September, Vancouver. Paper 271. Canadian Rock Mechanics Association.

Zavodni ZM (2001) Time-dependent movements of open pit slopes. In *Proceedings of Slope Stability in Surface Mining*. 26–28 February, Denver. (Eds WA Hustrulid, MK McCarter, DJA van Zyl) pp. 81–87. Society for Mining, Metallurgy and Exploration, CO.

Zavodni ZM, Broadbent CD (1978) Slope failure kinematics. In *Proceedings of the 19th US Symposium on Rock Mechanics*. 1–3 May, Reno, NV. ARMA-1978-0102.

Zavodni ZM, Broadbent CD (1980) Slope failure kinematics. *CIM Bulletin* **73**(816), 69–74.

Zhang L, Einstein HH (2000) Estimating the intensity of rock discontinuities. *International Journal of Rock Mechanics and Mining Sciences* **37**(5), 819–837. doi:10.1016/S1365-1609(00)00022-8

Index